INTERNATIONAL AGENCY FOR RESEARCH ON CANCER (WHO)
COMMISSION OF THE EUROPEAN COMMUNITIES

# NON-OCCUPATIONAL EXPOSURE TO MINERAL FIBRES

EDITORS: J. BIGNON, J. PETO & R. SARACCI

IARC SCIENTIFIC PUBLICATIONS
N° 90
LYON 1989

# NON-OCCUPATIONAL EXPOSURE

# TO MINERAL FIBRES

## INTERNATIONAL AGENCY FOR RESEARCH ON CANCER

The International Agency for Research on Cancer (IARC) was established in 1965 by the World Health Assembly, as an independently financed organization within the framework of the World Health Organization. The headquarters of the Agency are at Lyon, France.

The Agency conducts a programme of research concentrating particularly on the epidemiology of cancer and the study of potential carcinogens in the human environment. Its field studies are supplemented by biological and chemical research carried out in the Agency's laboratories in Lyon and, through collaborative research agreements, in national research institutions in many countries. The Agency also conducts a programme for the education and training of personnel for cancer research.

The publications of the Agency are intended to contribute to the dissemination of authoritative information on different aspects of cancer research. A complete list is printed at the back of the book.

This volume contains the proceedings of a symposium held on 8-10 September 1987 in Lyon, and organized jointly by:

> International Agency for Research on Cancer
>
> WHO Regional Office for Europe
>
> International Programme on Chemical Safety
>
> Commission of European Communities
>
> French Ministry of the Environment (Service de la Recherche des Etudes et de Traitement de l'Information sur l'Environnement)

*Cover illustration:* Scanning electron micrographs of crocidolite, glass wool, basaltic glass fibre and sepiolite samples after various treatments (see pp. 134-139). Reproducted by permisson of G. Larsen.

WORLD HEALTH ORGANIZATION

INTERNATIONAL AGENCY FOR RESEARCH ON CANCER

# Non-occupational Exposure to Mineral Fibres

*Edited by*
J. Bignon, J. Peto and R. Saracci

**IARC Scientific Publications No. 90**

International Agency for Research on Cancer, Lyon, France

1989

Published by the International Agency for Research on Cancer,
150 cours Albert Thomas, 69372 Lyon Cedex 08, France

Publication No. EUR 12068 of the
Commission of the European Communities,
Scientific and Technical Communication Unit,
Directorate-General Telecommunications,
Information Industries and Innovation, Luxembourg

©International Agency for Research on Cancer, 1989
©ECSC-EEC-EAEC, Brussels-Luxembourg, 1989

Distributed by Oxford University Press, Walton Street, Oxford OX2 6DP, UK

Distributed in the USA by Oxford University Press, New York

All rights reserved. No part of this publication may be reproduced,
stored in a retrieval system, or transmitted, in any form or by any means,
electronic, mechanical, photocopying, recording, or otherwise,
without the prior permission of the copyright holder.

The authors alone are responsible for the views expressed
in the signed articles in this publication.

Neither the Commission of the European Communities nor any
person acting on behalf of the Commission is responsible for the use
which might be made of the following information.

ISBN 92 832 1190 1

ISSN 0300-5085

Printed in the United Kingdom

# CONTENTS

Foreword .................................................. xi

## I. INTRODUCTION ........................................ 1

Mineral fibres in the non-occupational environment
    *J. Bignon* ............................................. 3

## II. EXPERIMENTAL DATA ................................. 31

Mineral fibre carcinogenesis: experimental data relating to the importance of fibre type, size, deposition, dissolution and migration
    *J.M.G. Davis* ......................................... 33

Recent results of carcinogenicity bioassays of fibres and other particulate materials
    *C. Maltoni & F. Minardi* .............................. 46

Particulate-state carcinogenesis: a survey of recent studies on the mechanisms of action of fibres
    *M.C. Jaurand* ........................................ 54

Modification of fibrous Oregon erionite and its effects on *in vitro* activity
    *R.C. Brown, R. Davies & A.P. Rood* ................... 74

Mechanisms of fibre-induced superoxide release from alveolar macrophages and induction of superoxide dismutase in the lungs of rats inhaling crocidolite
    *B.T. Mossman, K. Hansen, J.P. Marsh, M.E. Brew, S. Hill, M. Bergeron & J. Petruska* ........................................ 81

Brief inhalation of chrysotile asbestos induces rapid proliferation of bronchiolar-alveolar epithelial and interstitial cells
    *A.R. Brody, P.D. McGavran & L.H. Overby* ............. 93

Production of oxygen radicals by the reduction of oxygen arising from the surface activity of mineral fibres
    *H. Pezerat, R. Zalma, J. Guignard & M.C. Jaurand* ..... 100

Experimental studies on ingested fibres
  I. Chouroulinkov .................................................... 112
Effect of long-term ingestion of asbestos fibres in rats
  R. Truhaut & I. Chouroulinkov ..................................... 127
Experimental data on *in vitro* fibre solubility
  G. Larsen .......................................................... 134
Adsorption of polycyclic aromatic hydrocarbons on to asbestos and man-made mineral fibres in the gas phase
  P. Gerde & P. Scholander .......................................... 140
Comparative study of the effect of chrysotile, quartz and rutile on the release of lymphocyte activating factor (interleukin 1) by murine peritoneal macrophages *in vitro*
  D. Godelaine & H. Beaufay ......................................... 149
Biological effects of asbestos fibres on rat lung maintained *in vitro*
  F.M. Michiels, G. Moëns, J.J. Montagne & I. Chouroulinkov .......... 156
Translocation of subcutaneously injected chrysotile fibres: potential cocarcinogenic effect on lung cancer induced in rats by inhalation of radon and its daughters
  G. Monchaux, J. Chameaud, J.P. Morlier, X. Janson, M. Morin, & J. Bignon ........................................................ 161
Transplantation and chromosomal analysis of cell lines derived from mesotheliomas induced in rats with erionite and UICC chrysotile asbestos
  L.D. Palekar, J.F. Eyre & D.L. Coffin .............................. 167
Carcinogenicity studies on natural and man-made fibres with the intraperitoneal test in rats
  F. Pott, M. Roller, U. Ziem, F.-J. Reiffer, B. Bellmann, M. Rosenbruch & F. Huth .......................................... 173
Toxicity of an attapulgite sample studied in vivo and in vitro
  A. Renier, J. Fleury, G. Monchaux, M. Nebut, J. Bignon & M.C. Jaurand ...................................................... 180
Effect of choline and mineral fibres (chrysotile asbestos) on guinea-pigs
  A.P. Sahu .......................................................... 185
Cytotoxicity and carcinogenicity of chrysotile fibres from asbestos-cement products
  F. Tilkes & E.G. Beck .............................................. 190
Qualitative and quantitative evaluation of chrysotile and crocidolite fibres with infrared spectrophotometry: application to asbestos cement products
  F. Valerio & D. Balducci ........................................... 197

## III. FIBRE LEVEL MEASUREMENTS .............................. 205

Fibre levels in lung and correlation with air samples
  B.W. Case & P. Sébastien ........................................ 207

Non-occupational malignant mesotheliomas
  A.R. Gibbs, J.S.P. Jones, F.D. Pooley, D.M. Griffiths & J.C. Wagner .. 219

Incidence of ferruginous bodies in the lungs during a 45-year period and mineralogical analysis of the core fibres and uncoated fibres
  S. Shishido, K. Iwai & K. Tukagoshi ............................ 229

Airborne mineral fibre levels in the non-occupational environment
  W.J. Nicholson ................................................. 239

Airborne asbestos levels in non-occupational environments in Japan
  N. Kohyama .................................................... 262

Airborne asbestos fibre levels in buildings: a summary of UK measurements
  G.J. Burdett, S.A.M.T. Jaffrey & A.P. Rood ...................... 277

Levels of atmospheric pollution by man-made mineral fibres in buildings
  A. Gaudichet, G. Petit, M.A. Billon-Galland & G. Dufour ........ 291

Exposure to ceramic man-made mineral fibres
  J.J. Friar & A.M. Phillips ........................................ 299

Comparative studies of airborne asbestos in occupational and non-occupational environments using optical and electron microscope techniques
  J. Cherrie, J. Addison & J. Dodgson ............................. 304

Alveolar and lung fibre levels in non-occupationally exposed subjects
  G. Chiappino, K.H. Friedrichs, A. Forni, G. Rivolta & A. Todaro ...... 310

Fibre content of lung in amphibole- and chrysotile induced mesothelioma: implications for environmental exposure
  A. Churg & J.L. Wright ........................................ 314

Levels of airborne MMMF in dwellings in the UK: results of a preliminary survey
  S.A.M.T. Jaffrey, A.P. Rood, J.W. Llewellyn & A.J. Wilson .......... 319

Smoking and the pulmonary mineral particle burden
  P.-L. Kalliomäki, O. Taikina-Aho, P. Pääkkö, S. Anttila,
  T. Kerola, S.J. Sivonen, J. Tienari & S. Sutinen .................. 323

Fibre type and burden in parenchymal tissues of workers occupationally exposed to asbestos in the United States
  A.M. Langer & R.P. Nolan ..................................... 330

Airborne mineral fibre concentrations in an urban area near an asbestos-cement plant
    A. Marconi, G. Cecchetti & M. Barbieri .......................... 336

Non-asbestos fibre content of selected consumer products
    J.C. Méranger & A.B.C. Davey ................................... 347

Mineral fibres and dusts in lungs of subjects living in an urban environment
    L. Paoletti, L. Eibenschutz, A.M. Cassano, M. Falchi, D. Batisti, C. Ciallella & G. Donelli ............................................ 354

Measurement of inorganic fibrous particulates in ambient air and indoors with the scanning electron microscope
    K. Rödelsperger, U. Teichert, H. Marfels, K. Spurny, R. Arhelger & H.-J. Woitowitz ................................................. 361

Asbestos fibre release by corroded and weathered asbestos cement products
    K.R. Spurny ....................................................... 367

## IV. EPIDEMIOLOGICAL DATA ................................... 373

Effects on health of non-occupational exposure to airborne mineral fibres
    M.J. Gardner & R. Saracci ........................................ 375

Relation of environmental exposure to erionite fibres to risk of respiratory cancer
    L. Simonato, R. Baris, R. Saracci, J. Skidmore & R. Winkelmann ...... 398

Bilateral pleural plaques in Corsica: a marker of non-occupational asbestos exposure
    C. Boutin, J.R. Viallat, J. Steinbauer, G. Dufour & A. Gaudichet ....... 406

Mesothelioma in Cyprus
    K. McConnochie, L. Simonato, P. Mavrides, P. Christofides, R. Mitha, D.M. Griffiths & J.C. Wagner ............................ 411

Epidemiological observations on mesothelioma and their implications for non-occupational exposure
    J.C. McDonald, P. Sébastien, A.D. McDonald & B. Case ............. 420

Epidemiological studies on ingested mineral fibres: gastric and other cancers
    M.S. Kanarek ..................................................... 428

Asbestos fibre content of lungs with mesotheliomas in Osaka, Japan: a preliminary report
    K. Morinaga, N. Kohyama, K. Yokoyama, Y. Yasui, I. Hara, M. Sasaki, Y. Suzuki & Y. Sera ..................................... 438

Correlation between lung fibre content and disease in East London asbestos factory workers
  *J.C. Wagner, M.L. Newhouse, B. Corrin, C.E. Rossiter & D.M. Griffiths* .................................................. 444
Effect on health of man-made mineral fibres in kindergarten ceilings
  *A. Rindel, C. Hugod, E. Bach & N.O. Breum* ....................... 449

## V. RISK EVALUATION .............................................. 455
Fibre carcinogenesis and environmental hazards
  *J. Peto* ............................................................ 457
Development and use of asbestos risk estimates
  *J.M. Hughes & H. Weill* ........................................... 471
Estimations of risk from environmental asbestos in perspective
  *B.T. Commins* ..................................................... 476
Mesotheliomas — asbestos exposure and lung burden
  *G. Berry, A.J. Rogers & F.D. Pooley* ............................. 486
Public perception of risk and its consequences: the case of a natural fibrous mineral deposit
  *G. Major & G.F. Vardy* ............................................ 497

## VI. CONCLUDING SESSION ......................................... 509
Mineral fibres in the non-occupational environment: concluding remarks
  *R. Doll* ........................................................... 511

Index ................................................................ 519

# FOREWORD

The International Agency for Research on Cancer has had a long-standing interest and involvement in the elucidation of the cancer risk arising from exposure to mineral fibres, both natural and artificial. This is reflected in a number of Agency publications, ranging from the early volumes on the Biological Effects of Asbestos (1972) through the volumes on the Biological Effects of Mineral Fibres (1981), and several volumes (1, 14, 42) in the Monographs series which have addressed and updated the evidence concerning carcinogenic carcinogenic risk for agents such as asbestos, silica, erionite fibres and man-made mineral fibres. A number of these publications include the direct contributions that scientists in the Agency have been making, particularly over the last 10 years, to epidemiological studies on the role of man-made mineral fibres (MMMF), and on erionite in relation to mesothelioma. Recent decades have also witnessed an improvement in environmental conditions, particularly in a number of occupational settings, bringing about an effective reduction in the levels of exposure to airborne asbestos fibres in many countries. Because of this improvement in understanding of how cancers relate to exposure to mineral fibres and the lowering of the exposure levels, the focus of scientific interest has been shifting in recent years to the risks associated with long-term low-level exposures. This is a particularly thorny problem to investigate, requiring both direct observation of population groups exposed to low concentration of fibres and extrapolation from groups exposed at high levels. The combination of these approaches, as well as the solved and unsolved problems that they give rise to, are well illustrated by the papers in the present volume. Covering the ground from environmental measurements to risk assessment, they offer a comprehensive and up-to-date view that should contribute to a better appreciation and control of health risks from exposure to mineral fibres in the general environment.

Lorenzo Tomatis, MD
Director, IARC

# I. INTRODUCTION

# MINERAL FIBRES IN THE NON-OCCUPATIONAL ENVIRONMENT

## J. Bignon

*INSERM U 139, Hôpital Henri Mondor, Créteil, France*

That long-term non-occupational exposure to mineral fibres does have health effects can be deduced from the fact that a high frequency of pleural calcifications and mesothelioma has been observed in people living in certain rural areas of southern Europe where the soil contains fibrous rocks. These people (of both sexes) have probably been exposed since childhood to airborne dusts derived from the soil. While this type of environmental exposure seems unavoidable, the extent to which the general population of industrialized countries is non-occupationally exposed to airborne and waterborne mineral fibres remains a matter of controversy. The uncertainties are even greater in respect of health effects related to low dose levels in environmental settings. We are therefore facing a difficult challenge and it is our task to investigate this problem as thoroughly and accurately as possible, in order to find solutions that are both reasonable and acceptable. In this context, the decision of the US Environmental Protection Agency in January 1986 to ban progressively all asbestos-containing products in the USA seems unnecessary and unjustifiable scientifically, particularly when it is claimed that this measure will avoid 1900 deaths due to asbestos-related cancers during the next 15 years. Moreover, this decision does not take into account the uncertainties concerning the potential carcinogenic risks of non-asbestos natural or synthetic mineral fibres, which are already largely used as substitutes for asbestos, even though their toxicity has not been adequately assessed.

I do not propose to review here the world literature on the health effects of asbestos and non-asbestos mineral fibres, since several extensive reviews have been published during the last five years, covering all types of mineral fibres and all types of exposure, including non-occupational (Becklake, 1976; Doll & Peto, 1985; National Research Council, 1984; World Health Organization, 1986a,b; Ontario Royal Commission, 1984; Environmental Protection Agency, 1986).

In order to assess the potential human health hazard in relation to past, present and future exposure to mineral fibres from non-occupational sources, the following points must be considered: (1) what types of fibrous materials are we going to deal with? (2) what are the major industrial and natural sources of non-occupational exposure to fibrous dusts? (3) how great are such exposures? (4) what kind of people are exposed: the entire population or specific groups? (5) what diseases have been found to be associated with environmental exposure to mineral fibres? (6) are we dealing with a real public health problem or has it been exaggerated by scientists and environmentalists?

## Types of materials considered

According to geologists, the term fibre covers all fibrous inorganic and organic materials having a length:diameter ratio greater than 3:1. However, as we now know that the fibrous mineral particles 'critical' for health are those which are thin and long, regulatory agencies tend to include under this term all those particles, whether fibrous, prismatic or acicular, that have aspect ratios of about 10:1 and even greater (Flanigan, 1977; Zoltai, 1979).

**Natural fibres**

The main natural fibres are shown in Figure 1.

**Fig. 1. Classification of natural fibres**

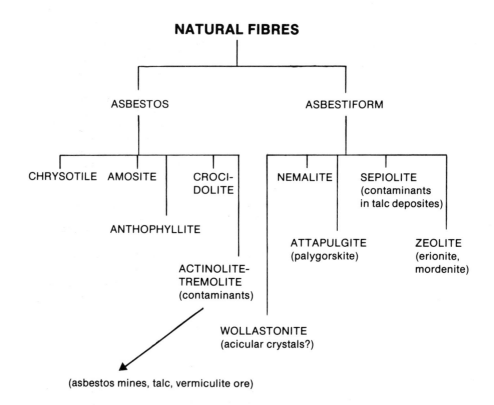

*Asbestos*

Asbestos is the name given to a group of fibrous silicates which are used commercially because of their high tensile strength and resistance to heat and chemicals. There are 3 commercial varieties, namely chrysotile, which belongs to the serpentine group of minerals, and crocidolite and amosite, which belong to the amphibole family. Three other amphibole types — anthophyllite, actinolite, tremolite — are not commercially exploited but can contaminate other commercial mineral deposits, e.g., in chrysotile asbestos mines, particularly in Quebec, and sheet silicate (talc and vermiculite) mines in the USA (Lockey et al., 1984). These well-known asbestos minerals have different fibrous properties and chemical formulae (National Research Council, 1984; World Health Organization, 1986b) (Table 1).

**Table 1. Commercial asbestos**

| Mineral | Mineral group | Chemical formula | Remarks |
|---|---|---|---|
| Chrysotile | Serpentine | $(MgFe)_6(OH)_8Si_4O_{10}$ | White asbestos — the most commonly used variety |
| Riebeckite (crocidolite)[a] | Amphibole | $Na_2Fe_2^{3+}Fe_3^{2+}(OH)_2Si_8O_{22}$ | Blue asbestos. Its use is becoming increasingly subject to restrictions |
| Anthophyllite | Amphibole | $(MgFe)_7(OH)_2Si_8O_{22}$ | Usually a low-quality asbestos. No longer commercially used. Present as a contaminant in amosite and in some chrysotile and talc deposits |
| Cummingtonite-grunerite (amosite) | Amphibole | $Mg_7(OH)_2Si_8O_{22}$<br>$Fe_7(OH)_2Si_8O_{22}$ | Brown asbestos. Still extensively used, particularly in the USA |
| Actinolite-tremolite | Amphibole | $Ca_2Fe_5(OH)_2Si_8O_{22}$<br>$Ca_2Mg_5(OH)_2Si_8O_{22}$ | Not commercially used. Common contaminant of amosite. May contaminate chrysotile, talc and vermiculite deposits |

[a]Commercial name

*Asbestiform fibres*

The term 'asbestiform fibres' covers all natural crystals that have the same fibrosity as commercial asbestos. Five varieties are listed in Table 2, of which only three are in commercial use: (1) *attapulgite* (palygorskite), a fibrous clay mineral with a high magnesium content produced in the USA, France, Spain, Senegal and Turkey in amounts exceeding one million tons per year (Bignon et al., 1980; National Research Council, 1984; IARC, 1987). This fibrous mineral, which is highly absorbent, is used in many products: oil and grease absorbents, pet litters, drilling muds, paints, drugs, cosmetics, pesticides. Fibres from the US, French and Senegalese deposits are short and thin (0.1–2.5 $\mu$m in length), whereas those from other deposits (Spain) are longer (30% of fibres longer than 5 $\mu$m). The importance of this size difference in relation to carcinogenesis will be discussed later; (2) *sepiolite* — this mineral is another variety of fibrous clay which has more or less the same commercial uses as attapulgite; (3) *wollastonite*, a calcium metasilicate which has a fibrous acicular structure. It is used as a filler in paints and in plastics, as a component of ceramics and, more recently, in insulation as a substitute for asbestos.

**Table 2. Asbestiform fibres**

| Mineral | Mineral category and chemistry | Remarks |
| --- | --- | --- |
| Nemalite (fibrous brucite)[a] | Lamellar hydroxide | Relatively rare. Contaminant of chrysotile deposits |
| Palygorskite (Attapulgite)[a] | Clay mineral. Lamellar magnesium silicate with triple subchain structure | Consists principally of short fibres. Used in drilling mud, fertilizers, industrial, commercial and domestic waste absorbents, drugs, cosmetics and insecticides |
| Sepiolite (Meerschaum)[a] | Clay mineral. Lamellar magnesium silicate with triple subchain structure | Used in adsorption granules. Meerschaum is also used for making pipes |
| Erionite | Zeolite. Fibrous aluminosilicate | Occurs as a fibrous mineral constituent of tuffaceous (volcanic ash) deposits. Similar in dimensions to asbestos. Environmental contaminant in central Turkey |
| Wollastonite | Long-chain silicates ($CaSiO_3$) | Usually acicular but may also occur as asbestiform fibres. Produced in Norway. Used as a substitute for asbestos and in ceramics. |

[a]Commercial name.

To these three commercial varieties of asbestiform minerals we can add fibrous rutile (Germine, 1985) and we must pay special attention to a fibrous zeolite, *erionite*, which is the naturally occurring asbestiform mineral associated with the very high incidence of mesothelioma in certain villages in Turkey. The fibres derived from this volcanic fibrous rock are similar in dimensions to those of asbestos (Suzuki, 1982).

## Synthetic fibres

There are more than 70 varieties of man-made mineral (or vitreous) fibres (MMMF or MMVF) (Figure 2) which, although amorphous silicates, have all or many of the asbestiform properties, particularly morphology, size and flexibility. The world consumption of about 5 million tons of MMMF emphasizes their commercial importance, which is still increasing, particularly as substitutes for asbestos.

**Fig. 2. Classification of synthetic fibres**

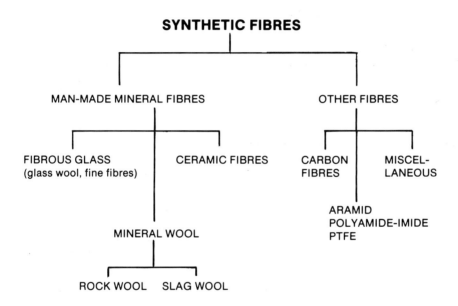

MMMF can be roughly divided into *glass fibres* (approximately 80% of MMMF production), used largely in building insulation; *mineral wools*, either rock wool or slag wool (approximately 10–15% of MMMF production), mostly used in acoustic and thermal insulation; and *ceramic fibres* (1–2% of all MMMF), which are both stronger and less soluble than glass and mineral wool fibres. This last group includes aluminosilicates, zirconium, boron and silicon carbide fibres. They are resistant to high temperatures and are used mainly in advanced technology. MMMF are usually coated with a binder, which contains an oily substance for lubrication purposes, or a resin (generally thermoplastic) for binding and/or cationic surface-active agents for adhesion. These additives must be taken into account when assessing the toxicity of these fibres. Other man-made fibres, such as carbon, graphite, alumina, potassium titanate, phosphate and organic fibres, are produced in relatively small quantities. We are therefore facing a very complicated problem and one that is difficult to assess, since more than 150 products are currently available on the world market as substitutes for asbestos, containing either MMMF or organic fibres (Peters & Peters, 1986).

The figures for the consumption and production of 3 types of mineral fibres in the USA, namely asbestos, attapulgite and MMMF, indicate clearly that the amounts of asbestos are decreasing while those of natural or artificial substitutes are increasing (National Research Council, 1984). The use of unregulated asbestiform minerals, either natural or synthetic, is increasing, but these minerals may have the same biological and toxicological properties as the commercial varieties of asbestos.

## *Conditions of exposure to fibrous dusts*

### Origin of fibres

The fibrous dusts which contaminate the environment (air, water, food) are derived either from natural or industrial sources (National Research Council, 1984).

*Natural sources*

Fibres are produced by the erosion of asbestos or asbestiform rocks widely dispersed throughout the earth's crust. Only a few of these deposits are commercially exploited; thus, while chrysotile is found in many serpentine rocks all over the world, there are only 30-40 workable commercial mines, mainly in Quebec and the USSR. Asbestiform minerals, such as anthophyllite, tremolite, erionite, attapulgite and sepiolite, may also contribute to fibre emissions as a result of natural weathering.

*Industrial sources*

Fibres are emitted during the extraction, processing and use of fibrous materials. In industrialized countries, this is the main source of mineral fibres in air and water. Since the beginning of the century, asbestos consumption has risen to about 5 million tons/year, mostly chrysotile (95%) (Becklake, 1976). For this reason, we are probably all environmentally exposed to asbestos, but at different concentrations, depending on the place of residence.

As legislative restrictions were progressively imposed on the use of asbestos and asbestos products, substitutes for asbestos in fibre-cement, brake linings, insulation and in many other applications were rapidly developed in recent years. However, this does not necessarily mean that inhalable fibrous dusts (diameter <3 $\mu$m) have disappeared. In fact, the urban outdoor fibrous aerosol is increasingly becoming a mixture of different minerals, often difficult to identify, even by analytical transmission electron microscopy (ATEM). This indicates the heterogeneous nature of environmental pollution by mineral fibres.

## Types of non-occupational exposure

There are essentially three types of non-occupational exposure, of which the first is community outdoor exposure, which in turn can be divided into rural exposure, urban ambient exposure and neighbourhood industrial exposure.

### Community outdoor exposure

*Rural exposure.* Rural exposure to mineral fibres is demonstrated by the fact that pleural plaques and mesotheliomas have been observed in rural populations possibly exposed to airborne or waterborne fibres released from natural deposits at the surface of the soil. This type of exposure can be enhanced by such human activities as agriculture, construction, quarrying and public works, and by geological and climatic factors. However, the precise origin and total amount of asbestos and asbestiform fibres released are not accurately known. Indeed, as will be discussed later, fibre measurements in the ambient air have not provided completely satisfactory data from the point of view of an understanding of dose-response relationships.

*Urban ambient exposure.* The contamination of the ambient air, particularly in large cities, can be deduced from the constant finding of asbestos fibres in the lungs of urban dwellers (Bignon *et al.*, 1970; Churg & Warnock, 1977). However, as discussed later, the background levels of such urban air pollution are extremely low, and we do not know to what extent they contribute to the low asbestos fibre pulmonary burden observed in the non-occupationally exposed general population. Non-asbestos mineral fibres are also present in the lung in the general population (Churg, 1983).

*Neighbourhood industrial exposure.* Airborne asbestos fibre concentrations were measured in various mining areas in Canada and South Africa in 1983 and 1984 and in the vicinity of asbestos processing plants in Austria (World Health Organization, 1986b). These data indicated that the concentrations of airborne asbestos fibres longer than 5 $\mu$m in the vicinity of industrial sources were within the ranges of those observed in cities (from 0.1 to 10 fibres/litre) (f/litre). Higher concentrations have been observed under certain conditions (up to 300-600 f/litre downwind of an asbestos mill in South Africa). So far, no measurements have been published on airborne fibre concentrations in the neighbourhood of plants manufacturing or processing either non-asbestos fibres or MMMF. It is likely that these emissions are quantitatively less important than those observed in the vicinity of asbestos processing plants.

### Community indoor exposure

Indoor sources of airborne mineral fibres are shown in Table 3.

The dirty clothes of asbestos workers have been found to be an important source of indoor asbestos pollution, with measured concentrations of up to 5000 ng/m$^3$ (Nicholson, this volume, pp. 239-261). This type of exposure is disappearing, as workers no longer go back home wearing their working clothes.

The daily use of domestic equipment (such as hair dryers, ovens, etc), hobbies and the use of pet litters (attapulgite, sepiolite) at home are potential sources of indoor fibre release. However, no measurements of such indoor contamination are currently available.

**Table 3. Indoor sources of airborne mineral fibres**

| Source | Type of fibre |
|---|---|
| *Clothing* | |
| Contaminated work clothes of asbestos workers | Asbestos |
| *Domestic uses* | |
| Household equipment (hair dryers, ovens) | Asbestos, MMMF, attapulgite, sepiolite |
| Hobbies | |
| Pet litters | |
| *Construction materials* | |
| Friable: sprayed or trowelled insulation | Asbestos, MMMF |
| Non-friable: | |
| Pipe and boiler insulation | |
| Insulation board, asbestos-cement board | |
| Vinyl-asbestos floor tiles | |
| Paint | |
| Paper | |
| Textiles | |

The emission of mineral fibres from fibrous materials used for building construction is the main source of indoor fibre pollution at the present time. Non-friable materials containing fibres are not really a problem because the fibres are sealed inside the compact material (cement, plastic). A problem will develop only when the building is demolished or renovated. Friable asbestos is commonly found in buildings where the walls and ceilings have been sprayed or trowelled with asbestos-containing materials. This technique was widely used for thermal or acoustic insulation or even for decorative purposes between 1950 and 1973 in the USA and more recently in Europe. In France, the spraying of asbestos-containing materials in buildings was prohibited in 1977. Many school buildings contain friable asbestos materials and these constitute a major public health concern, particularly since large numbers of children (several million in western Europe) are potentially exposed. Moreover, children may be exposed to higher levels than adults because, during their activities, they damage the surface of the walls and redisperse fibres deposited on the floor. In addition, at this age, the breathing rate is high. A precise evaluation of the magnitude of this health hazard is needed in order to find the appropriate solutions (Dewees, 1987). It is certainly possible to remove the friable materials efficiently with proper working methods and proper protection. In contrast, inappropriate removal methods increase the fibrous dust levels inside buildings. In many cases, sealing the material is all that is necessary.

As sprayed asbestos is banned in a number of countries, numerous substitutes, and particularly fibrous glass, are being increasingly used for insulation purposes. As it is probable that some of these substitute fibres may increase the risk of lung cancer (see p. 15), it will be necessary to investigate more thoroughly the public health impact of this type of exposure, based on measurement data in different occupational and non-occupational settings. This will make it possible to design future epidemiological studies for assessing accurately the potential risk of low-dose exposure. At the present time, a mixture of different types of fibrous materials, e.g., chrysotile, amosite and glass fibres, is usually found in indoor air samples from buildings (Sébastien et al., 1982). In such cases, the sources of the different types of fibre are usually easily identified.

*Consumer exposure*

Asbestos contamination of food and beverages is another way in which fibres can enter the human body. Fibres have been found in drinking-water in several cities around the world, and fibre counts of up to 100 million per litre have been reported, the fibres being mainly short fibrils, below 1 $\mu$m in length (World Health Organization, 1986b).

## Assessing non-occupational exposure

Assessment of the health risks from exposure to fibres in the general environment (air, water) necessitates, on the one hand, the quantitation of the lifetime exposure of individuals and, on the other, the estimation of the size of the population affected by such exposure. In general, high lifetime exposures affect relatively few people, while moderate or low lifetime exposures affect large numbers of people, most of whom have been environmentally exposed (Suta & Levine, 1979).

When the size of the exposed population has been estimated, a number of parameters will have to be taken into account: age (adults, children), sex, smoking habits (number of cigarettes smoked and passive exposure to tobacco smoke), other occupational and domestic exposures, place of residence, health status. It will also be necessary to define how the population was exposed: by what route (inhalation, ingestion), from what sources (natural or industrial), under what conditions (outdoors, indoors, use of manufactured products), to what types of fibres (asbestos or non-asbestos, MMMF), to fibres of what sizes (length, diameter), and possibly to what associated compounds. Was the exposure continuous or were there episodes of high exposure?

Asbestiform fibres found in the ambient environment are usually thin fibrils, shorter than 5 $\mu$m, and many of them are not detectable by phase-contrast light microscopy, which is not effective at concentrations of less than about 0.1 f/cm³. The most accurate tool is the ATEM, which is capable of detecting and identifying all fibres not detectable optically. The scanning electron microscope, which provides an intermediate magnification, makes it possible to investigate a larger sample size, as recently shown in a comparative study of different methods (Burdett & Jaffrey, 1986). In the 1970s, results were expressed in terms of the mass of asbestos fibres per unit volume of air or fluid (Nicholson et al., 1975; Sébastien et al., 1980, 1982). Although some inaccuracy is involved in determining the optical equivalent of TEM mass measurements, it is assumed that 30 μg/m³ is approximately equivalent to 1 f/m³ (fibres >5 μm in length). Table 4 shows the concentration of asbestos fibres in air expressed as the number of fibres >5 μm/litre for different airborne exposures, ranging from remote rural areas to the work-place. The indoor air concentrations in buildings with substantial amounts of friable sprayed asbestos have been found to be between 1 and 5 f/litre in 1985. The US Environmental Protection Agency (1986) concluded that 50% of all concentrations of airborne asbestos fibres in US schools were in the range 0.003–3 f/litre. Although uncertainties exist concerning the estimation of airborne asbestos concentrations in buildings, it is clear that asbestos exposure in non-occupational settings is at very low concentrations, about 3 orders of magnitude less than the 2000 f/litre corresponding to the present standard limit value for occupational exposure.

**Table 4. Concentrations of fibres in air[a]**

| Area | Asbestos | Total inorganic |
|---|---|---|
| Remote rural areas | <0.1 | 1.4 |
| Village without asbestos-cement roofing | <0.1 | 4.5 |
| Large city: | | |
|   Residential areas | 0.2–11 | 20–50 |
|   Road crossing with heavy traffic | 0.9 | 60 |
|   Expressway | 3.3 | – |
| Neighbourhood of an asbestos-cement plant: | | |
|   300 m downwind | 2.2 | 90 |
|   1000 m downwind | 0.6 | 60 |
| Indoors: | | |
|   Buildings without asbestos | <0.1 | – |
|   Buildings with friable sprayed asbestos | 1–40 | – |
|   50% of schools in the USA | 0.003–3 | – |
| Work-places | 100–100 000[b] | – |

[a]Expressed as number of fibres >5 μm per litre. Adapted from National Research Council (1984), Environmental Protection Agency (1986) and World Health Organization (1986b)

[b]The standard limit value is 2000 f/litre and is to be reduced to 1000 f/litre.

There are at present few data concerning MMMF concentrations in outdoor and indoor air. This subject is reviewed in this volume by Gaudichet (pp. 291-298). Concentrations seem to be lower than those observed for asbestos.

Thus, assessing the 'critical' concentrations to which the general population is exposed is the crucial problem. We have to consider the following questions: What is the best standardized method to be used for fibre assessment in the environment: optical or electron microscopy or other techniques? How should we express the results: mass or number per unit of volume of air or liquid? Should we limit identification and counting to the 'critical' fraction of fibrous dusts, i.e., fibres more than 5 µm in length and less than 3 µm in diameter?

## What has been learnt from occupational exposure and from laboratory studies

### Asbestos fibres

*Human studies*

During the first 60 years of this century, physicians, pathologists and epidemiologists contributed to the description of the main asbestos-associated diseases (Becklake, 1976; Selikoff & Lee, 1978): asbestosis at the beginning of the century, then, in the 1950s, lung cancer and, in the 1960s, mesothelioma (Wagner *et al.*, 1960) and pleural plaques (Kiviluoto, 1960). More recently, the question was raised as to a possible relationship between occupational exposure to asbestos and laryngeal and gastrointestinal cancers (National Research Council, 1984; World Health Organization, 1986b).

Since the 1970s, epidemiologists have described dose-response relationships in several cohorts of workers employed in industries where fibre concentrations have been measured in the work-place (Doll & Peto, 1985; National Research Council, 1984; World Health Organization, 1986b). At the 1979 IARC meeting (Wagner, 1980), epidemiologists agreed that this response was linear, although there were no data on the response to low-dose exposures and there was evidence that the steepness of the curve could vary, depending on the type of asbestos industry and/or processing concerned (Doll & Peto, 1985; Peto *et al.*, 1985; McDonald *et al.*, 1980, 1983).

In the 1980s, several epidemiological studies have demonstrated that exposure to amphibole fibre types, particularly crocidolite, was more strongly associated with mesothelioma than was exposure to chrysotile (Wagner *et al.*, 1980; McDonald & Fry, 1982). This finding may be related to the greater durability of amphibole fibres in the lung, as compared with the rapid dissolution of chrysotile (Doll & Peto, 1985). This has been demonstrated in both humans and animals (Wagner *et al.*, 1974; Sébastien *et al.*, 1986).

In the last ten years, it has also become clear that fibres are present in air and in drinking-water, but there are practically no data from which dose-response relationships can be calculated for non-malignant and malignant diseases after exposure to such low environmental doses.

*Experimental studies*

Of the numerous experimental studies carried out in order to understand the mechanisms of fibre-related fibrogenesis and carcinogenesis, the most useful are those in which an attempt was made to define the physicochemical parameters of fibres that are most closely related to biological responses and disease induction.

The historic study by Stanton *et al.* (1977) showed that, following the intrapleural inoculation of fibres of various types, the materials most highly carcinogenic to the pleura were those of length >4 µm and diameter <1.5 µm. Subsequently, Stanton *et* al. (1981) extended their study to cover a large number of fibres of different types and sizes and showed that the most highly carcinogenic were those of length >8 μm and diameter <0.25 μm. Pott et al. (1980) obtained similar results (length >4 μm). However, these dimensions do not constitute an absolute boundary between effect and no effect, and we do not know what percentage of such fibres in a given fibrous aerosol is necessary for inducing lung or pleural cancer. On the other hand, Bertrand and Pezerat (1980) have calculated from the data of Stanton that an aspect ratio (length/diameter) equal to or greater than 10 may be the most significant parameter in determining whether fibres are carcinogenic or not.

Length, diameter and/or aspect ratio are not the only fibre parameters involved in carcinogenicity, as indicated by several findings: the lower carcinogenic potential of acid-leached chrysotile cannot be explained solely by size modifications (Morgan et al., 1977; Monchaux et al., 1981). There are some cases where, as with Spanish sepiolite, long fibres did not induce mesothelioma after intraperitoneal injection (Wagner et al., 1980). Other parameters, such as surface properties (charge, free radicals), strength and durability, must play an important complementary role in toxicity and carcinogenicity.

**Asbestiform fibres**

To date, there is a paucity of data concerning occupational exposure to commercial asbestiform fibres, such as fibrous clays (attapulgite, sepiolite) and wollastonite. The published data concerning these 3 minerals and erionite have recently been reviewed (IARC, 1987). The conclusions can be formulated as follows.

*Attapulgite (palygorskite)*

At the present time, only two human studies are available, both on workers exposed in the Georgia (USA) attapulgite mine (IARC, 1987). An increased mortality from lung cancer was noted among the small group of workers with long-term, high-level exposure. However, these studies are unsatisfactory, as information on cigarette smoking was not obtained.

Several animal and *in vitro* studies are now available. Briefly, we can say that samples of attapulgite consisting of short fibres were not found to be carcinogenic in the rat after intrapleural injection, nor genotoxic to mammalian cells (Jaurand et al., 1987). By contrast, samples with long fibres and a high aspect ratio did induce mesothelioma in more than 50% of rats after intraperitoneal injection (Pott et al., 1974), or after intrapleural injection (Wagner, 1982); palygorskite was used in the first study and fibrous Spanish attapulgite in the second.

It is concluded that the evidence for the carcinogenicity of attapulgite for humans is inadequate and that there is only limited evidence of its carcinogenicity for experimental animals (IARC, 1987).

*Sepiolite*

Only one publication has presented clinical and radiological evidence of pulmonary fibrosis (small irregular opacities) in 10 of 63 workers engaged in trimming sepiolite in Eskisehir, Turkey. They had been employed from 1 to 30 years (mean, 11.9 years). However, more than half of these cases came from dusty rural regions, and radiological examination of the inhabitants of 4 villages in Eskisehir, where sepiolite has been mined and processed for more than 100 years, showed no evidence of pleural disease (Baris et al., 1980). There are currently no adequate data on the carcinogenic potential of sepiolite in either humans or experimental animals (IARC, 1987).

*Wollastonite*

There are few data on wollastonite. According to the studies of Shasby et al. (1979) and Huuskonen et al. (1983), lung fibrosis and pleural plaques have been found in workers exposed to wollastonite. Elsewhere, Stanton et al. (1981) showed that one sample of fibrous wollastonite was not carcinogenic whereas 3 samples of acicular wollastonite were carcinogenic in rats after intrapleural implantation. In the review mentioned above, it was concluded that there was inadequate evidence for the carcinogenicity of wollastonite to humans (IARC, 1987).

*Erionite*

There are many human and experimental data indicating the very high carcinogenic potential of the fibrous zeolite, erionite (IARC, 1987); these will be discussed later.

*Man-made mineral fibres*

Two large epidemiological studies, which have been extended considerably during the last 5 years, have been carried out to determine the incidence of malignant and non-malignant disease among cohorts of workers manufacturing MMMF (17 plants in the USA and 13 plants in Europe). These data and others, both epidemiological or experimental, were presented at the WHO conference in Copenhagen on MMMF (World Health Organization, 1987). Sir Richard Doll presented an overview of the conference in these terms: there was an increased mortality due to lung cancer in production and maintenance workers employed in the early days of the rock, slag and glass wool industry, but this excess was not due solely to fibre, since the majority of workers were smokers and were exposed to other carcinogens present in the occupational environment. These confounding factors could have contributed to the excess of lung cancer, particularly during the early years of production. Without a quantified dose-response relationship, it was not possible to evaluate the risk of lung cancer in association with different grades of exposure. 'However, current exposures to mean levels of 0.2 respirable fibres per ml or less seem unlikely to cause any detectable excess in lung cancer rates' (World Health Organization, 1987).

The carcinogenic potential of MMMF has been confirmed in laboratory animals (see World Health Organization, 1986a, and United Nations Environment Programme, 1986). Wagner et al. (1984) reported the results of inhalation and intrapleural inoculation of rats with various types of MMMF. Intrapleural inoculation may cause mesothelioma. However, after inhalation of MMMF in rodents, lung cancer and pulmonary fibrosis are rarely observed. Fibre dimensions are critical parameters for these two responses. However, it is probable that the durability of fibres in the lung also plays an important role. In this connection, experimental studies have suggested that short, thin glass wool and rock wool fibres tend to disappear after deposition in the lung, while longer and thicker ones seemed to become shorter and thinner with time (Davis et al., 1984).

In all these animal studies designed to assess the carcinogenic potential of MMMF, positive control animals exposed to chrysotile have been used. From the crude data, particularly those expressed per unit mass of dust measured in the air of the inhalation chambers (Wagner et al., 1984), the number of tumours produced by chrysotile would seem to be much greater than that produced by MMMF. However, the results of this study would have been different if the number of tumours had been expressed in terms of the number of fibres present in the inhalation chamber (Peto, this volume, pp. 457-470).

## Human diseases associated with non-occupational exposure to mineral fibres

In contrast to the vast amount of highly relevant epidemiological studies on workers occupationally exposed to mineral fibres (in several of which dose-response relationships for different types of asbestos exposure were firmly established), there are few or no data on non-occupational exposures. Two types of studies have been designed in order to assess the risks associated with non-occupational exposure: (*a*) case-control and cohort studies, where individuals with well identified and sometimes quantified exposures were compared with controls without such exposure; and (*b*) ecological-epidemiological studies, which essentially compared the cancer incidence in different areas where the exposure conditions for the general population had been assessed. The latter are less powerful than case-control studies because of the large number of confounding variables, which are difficult to eliminate. Moreover, the true excess cancer risk is probably underestimated in such studies, because of population movements over a latent period of several decades (Polissar, 1980; Botha et al., 1986).

Current levels of non-occupational exposure to mineral fibres are so low that there is practically no opportunity to observe diseases related to high-dose exposure, such as asbestosis. Lung cancer may be caused by many etiological factors, the most important being cigarette smoking. Doll and Peto (1985) have shown the high specificity of this carcinogen for lung cancer. Exposure to airborne asbestos fibres in non-smokers has been found to be associated with only a slight excess of lung cancer. In contrast, in workers who were cigarette smokers, exposure to asbestos dramatically increased the risk of lung cancer (Hammond et al., 1979a; McDonald, 1980). Moreover, the problem of lung cancer risk associated with asbestos exposure is complicated by the controversy as to the need for pulmonary fibrosis to be present before lung cancer can develop in subjects exposed to asbestos (Browne, 1986). All these uncertainties mean that lung cancer is probably a non-specific disease in the context of non-occupational exposure to low concentrations of airborne mineral fibres. It is noticeable that Neuberger et al. (1984) did not find any evidence of an excess of lung cancer in an Austrian town where there was asbestos contamination of air and water from natural tremolite deposits and where pleural plaques were endemic. These authors also failed to find any excess risk of stomach cancer in this town, while Botha et al. (1986) did find an excess mortality from stomach cancer in people non-occupationally exposed in the crocidolite mining districts of South Africa.

By contrast, pleural and peritoneal malignant mesothelioma represent highly specific diseases, found in association with direct or indirect occupational or non-occupational asbestos exposure in about 80-85% of cases. A dose-response relationship exists between asbestos exposure and the incidence of this tumour, which has been found even in association with low doses or short-duration exposure to asbestos, even in occupational settings (Hobbs et al., 1980). Case-control studies and cohort studies have demonstrated the strong association of mesothelioma with several types of fibres, particularly those of the amphibole group (Doll & Peto, 1985). In descending order, the carcinogenic mesothelioma potential in humans is as follows: first erionite, then crocidolite, tremolite and amosite, and lastly chrysotile. There is no known cofactor for the induction of mesothelioma, the incidence of this malignancy being completely independent of tobacco smoking and related only to years since first exposure and to age at first exposure (Peto et al., 1982; Peto, 1984).

Hyalin or calcified pleural plaques are benign lesions which seem to be good markers of past occupational and non-occupational exposure to mineral fibres. The diagnosis is easily made on postero-anterior, lateral or oblique chest X-rays. The sensitivity and specificity of the X-ray diagnosis of pleural plaques has been discussed recently (Järvholm et al., 1986). Asbestos exposure seems to be the main cause of pleural plaques (Kiviluoto, 1960; Järvholm et al., 1986). This appears clearly when exposed groups are compared with control groups, but the dose-response relationships with cumulative doses of asbestos have not been clearly demonstrated. It seems that pleural plaques may be associated with low-dose and short-duration asbestos exposure or with low asbestos fibre pulmonary burden. They seem to be more frequently encountered in association with certain categories of fibres: erionite, tremolite, anthophyllite (Hillerdal, 1981). This is in conflict with the fact that fibrils of chrysotile have been predominantly found in the parietal pleura of humans (Sébastien et al., 1979). Pleural plaques are time-related, their prevalence, number and density increasing with years since first exposure to mineral fibres, or with age in the case of environmental exposure. However, the specificity of pleural plaques is not absolute since, as with mesothelioma, about 20% of cases are not associated with asbestos exposure but are probably related to other unknown factors, which need further investigation. Tobacco smoke by itself does not cause pleural plaques but increases their prevalence by a factor of 2 in asbestos-exposed workers who are smokers or ex-smokers, whereas there is no increase in non-smokers (Järvholm, 1986). There is some controversy as to whether pleural plaques are indicative of an increased risk of asbestos-related cancer. A causal relationship between pleural plaques and mesothelioma does not seem to have been established. By contrast, the high frequency (20-30%) of pleural plaques found in association with lung cancer in surgical series is very striking. In these cases, low concentrations of mineral fibres are usually found in lung parenchyma (unpublished data). The significance of these findings is unknown.

## Exposure to airborne mineral fibres

### Outdoor community exposures

*Rural population exposure*

A high prevalence of calcified pleural plaques has been observed in populations living in rural areas where deposits of asbestos or asbestiform fibres exist. Table 5 lists the countries where such pleural calcifications have been observed and the types of mineral fibres concerned. In some of these areas an abnormal incidence of mesothelioma has also been observed. The most critical situation is exemplified by two villages in Cappadocia, Karain and Ürgüp (Turkey), where the death rate from mesothelioma was found to be much higher than in the worst cohorts of workers exposed to crocidolite (Baris *et al.*, 1978, 1981). A few cases of mesothelioma have also been observed in Corsica, and may be associated with community rural exposure (Boutin *et al.*, this volume, pp. 406-410).

**Table 5. Countries where pleural plaques have been observed and type of mineral fibre concerned**

| Country | Reference | Type of pleural plaques | Meso-thelioma | Type of mineral fibre |
|---|---|---|---|---|
| Finland | Kiviluoto, 1960 | Calcifications | 0 | Anthophylllite |
| South Africa | Wagner *et al.*, 1960 | Thickenings | ++ | Crocidolite |
| Bulgaria | Burilkov & Michailova, 1970, 1972 | Calcifications | 0 | Anthophyllite + tremolite + sepiolite |
| Czechoslovakia | Navratil & Trippe, 1972 | Calcifications | 0 | Chrysotile |
| Turkey | Yazicioglu *et al.*, 1976<br>Yazicioglu *et al.*, 1980<br>Baris *et al.*, 1978<br>Rohl *et al.*, 1982 | Calcifications<br>Calcifications<br>Calcifications<br>Calcifications | 0<br>+<br>+<br>0 | Tremolite + chrysotile |
| Turkey | Artvinli & Baris, 1979<br>Baris *et al.*, 1981 | Calcifications | +++ | Erionite |
| Greece (Metsovo) | Bazas *et al.*, 1981, 1985 | Calcifications | 0 | Tremolite |
| Greece (Metsovo) | Constantopoulos *et al.*, 1985 | Calcifications | ++ | Tremolite |
| France | Boutin *et al.*, 1986 | Calcifications | + | Tremolite |

Paradoxically, in air samples taken in two of these areas (Cappadocia, Corsica), fibres were found only at very low concentrations, 2 or 3 orders of magnitude less than in work-places (Baris *et al.*, 1981; Billon-Galand *et al.*, 1988). The low concentrations reported do not seem to be totally realistic, since the counts of ferruginous bodies in the sputum of residents in the two Turkish villages with a high incidence of mesothelioma indicated a relatively high pulmonary burden, increasing with the number of years spent in those villages (Sébastien *et al.*, 1984). It is possible that people were also exposed domestically, e.g., when they used white stucco prepared from crushed volcanic rocks. Yazicioglu *et al.* (1980) have reported pleural calcifications, pleural mesotheliomas and lung cancers apparently caused by tremolite dusts used as stucco.

*Neighbourhood exposure*

Calcified pleural plaques have also been observed in people living in the neighbourhood of industrial sources. The first cases were found in the vicinity of anthophyllite mines in Finland (Kiviluoto, 1960) and Bulgaria (Zolov *et al.*, 1967). Similar observations were made in populations living in the neighbourhood of an actinolite mine in Austria (Neuberger *et al.*, 1982) and of an asbestos factory in Czechoslovakia (Navratil & Trippe, 1972). In contrast, a cohort study conducted by Hammond *et al.* (1979b) in male residents in the neighbourhood of an amosite factory in Paterson, NJ, USA, did not show any evidence of increased risk attributable to this type of exposure.

The proportion of mesothelioma cases associated with neighbourhood asbestos exposure varied in different series, depending on the fibre type. Thus, in South Africa, mesothelioma incidence was very high in the crocidolite mining areas, very low around the amosite mines and apparently undetectable in the chrysotile areas of Zimbabwe and Swaziland (Wagner, 1963; Webster, 1977).

Data from national mesothelioma registers are not completely concordant: in two studies (Newhouse & Thompson, 1965; Bohlig & Hain, 1973), many unexposed cases were found to have lived close to asbestos factories or shipyards where mixed types of asbestos (amphiboles+chrysotile) were used. In contrast, in registers from France (Bignon *et al.*, 1979) and Canada (McDonald & McDonald, 1980), the percentage of such cases was very low (around 1%). Thus, in the mesothelioma register conducted by these last authors, only 2 out of the 254 cases of mesothelioma recorded in Quebec between 1960 and 1978 lived within 33 km of the chrysotile mines and mills. However, two ecological epidemiological studies (Pampalon *et al.*, 1982; Siemiatycki, 1983), based on the analysis of cancer incidence data from the Quebec Tumour Registry, found that the risk of mesothelioma for residents of asbestos mining communities was 1.5–8 times greater than for residents in rural areas of Quebec. The higher risks in males were attributed, in part, to occupational exposure. There was an increased risk of cancer of the pleura in both sexes, which decreased with increasing distance of residence from the asbestos mines. However, the authors emphasized the limitations of their study, due to the lack of information concerning confounding factors.

The conclusion reached in the WHO document (World Health Organization, 1986b) with regard to neighbourhood exposure was as follows: 'The risk of pleural plaques and mesothelioma may be increased in populations residing in the vicinity of asbestos mines or factories. By contrast, there is no evidence that the risk of lung cancer could be increased in similarly exposed populations'. These findings correspond to exposures at a time when factories emitted high levels of airborne fibres. Many countries now have regulations prohibiting such emissions. Levels outside factories at the present time need to be determined, since they may be useful in future epidemiological studies.

*General population exposure*

As indicated by the available data concerning fibre concentrations in the urban atmosphere and in the lungs of unexposed city dwellers, it is clear that everyone is exposed to asbestos, but at low concentrations. Moreover, these fibres are mostly short, thin fibrils, probably not of the 'critical' size for health, although they can adsorb and transport carcinogenic chemicals.

National mesothelioma registers can provide some indication of the possible health impact of pollution of the ambient air by mineral fibres. It has been found in this way that, in several countries (Canada, USA, UK, France), the mortality due to mesothelioma during the last 40–50 years showed a male/female sex ratio of about 1 for deaths before the age of 40–50. After this age, male mortality rose steeply (up to 10% per year) and the sex ratio increasing correspondingly. This trend was due to the increased number of cases associated with past occupational exposure to asbestos, which results from the increase in asbestos consumption during the last 40 years. This finding indicates that at least two factors must play a role in the genesis of mesothelioma; one that has been clearly identified is previous asbestos exposure, as found in male asbestos workers; the other, which has apparently the same weight in both sexes, is still unidentified.

The study of mesothelioma cases without specific asbestos exposure, and therefore with low asbestos lung burden, showed the same trend in sex ratio, i.e., close to 1 (Bignon, unpublished data). Moreover, these cases were 10 years younger than mesothelioma cases associated with occupational asbestos exposure, as if the disease had been induced during childhood (Hirsch *et al.*, 1982). The same trend was noted in the sex ratio of endemic pleural mesotheliomas in Karain (Turkey), where exposure started immediately after birth (Saracci *et al.*, 1982). Moreover, whereas annual mesothelioma mortality is largely determined by time since first exposure to asbestos, irrespective of age or cigarette smoking, the incidence of mesothelioma among those with asbestos exposure is proportional to age, as if 'exposure' had started with birth (Peto, 1984). As lung cancer in non-smokers is being increasingly associated with passive smoking, mesothelioma without specific exposure to mineral fibres may be related to environmental exposure to mineral fibres; this is probably the case in the two 'mesothelioma villages' of Cappadocia, where people are exposed to erionite fibres from birth. However, other factors can play a role, such as unknown carcinogens or genetic predisposition to early cancer induction in the mesothelium.

This may have been the case in the rare published familial cases of mesothelioma (Hirsch et al., 1982; Jaurand, this volume, pp. 54-73), although the role of other familial or domestic factors cannot be excluded. It is also possible that the two mechanisms (environmental exposure to mineral fibres and cofactors) may act synergistically; for instance, exposure to low concentrations of mineral fibres during childhood might sensitize genetically predisposed individuals so that they develop mesothelioma 40 years later. On the other hand, the role of age was suggested by the study of Peto et al. (1982), who found, in a cohort of asbestos workers, that the incidence of mesothelioma was higher when asbestos exposure began earlier in life. Our recent analysis of the cases in the French mesothelioma register gave the same result, since 40% of the cases associated with asbestos exposure had their first occupational exposure at between 11 and 20 years of age. These findings increase the concern about asbestos pollution in schools.

In view of all these uncertainties, the role of 'trivial' environmental exposure to mineral fibres in the causation of malignant mesothelioma requires further study. More accurate quantification of mineral fibre exposure in occupational and non-occupational settings would greatly improve estimates of mesothelioma risk at low-dose exposure. In addition, account must be taken of fibre type as well as of age at first exposure. On the other hand, animal studies must be used to explore the dose-response curve, particularly for the very low doses of fibre exposure.

## Indoor exposures

### Domestic exposures

The survey conducted by Anderson et al. (1976, 1979) on household contacts of surviving asbestos workers showed a high frequency of chest X-ray abnormalities (pleural plaques and parenchymal fibrosis) and some cases of mesothelioma. In mesothelioma registers, such household contacts have been found retrospectively in many cases (McDonald & McDonald, 1980; Bignon & Brochard, 1986; Vianna & Polan, 1978). We do not know the exact fibre concentrations to which these cases were exposed, but high concentrations were possible when handling dirty clothes.

Domestic exposure to asbestos fibres released from household equipment (hair dryers, electric heaters) or from the weathering of paints and white stucco on house walls has been suggested as the cause of some cases of idiopathic mesothelioma. However, such exposure has neither been definitely proved nor quantitatively assessed. No known human health effects are associated with exposure to attapulgite or sepiolite fibres released from carelessly handled pet litters. There is also no indication that lung fibrosis and/or mesothelioma have occurred in cats directly exposed to this dusty litter.

### Exposure inside insulated buildings

With regard to health effects of indoor asbestos pollution, there is only one published epidemiological survey to date; this relates to a cohort of workers (blue and white collar) permanently employed at the University of Paris in buildings where the sprayed asbestos materials were friable and had deteriorated and where concentrations of 1000–2000 ng/m³ have been measured. The chest X-rays, read according to the ILO classification (International Labour Office, 1980), showed no significant parenchymal or pleural changes after 10 years' work (8 hours a day, 5 days a week, 46 weeks a year) in offices or laboratories containing sprayed asbestos. No case of mesothelioma has been observed. However, sufficient time has not yet elapsed for such a study to give reliable results, exposure having begun only in 1965 (Cordier et al., 1987).

Indoor pollution by MMMF has only recently been identified. We are far from being able to evaluate any possible health hazard, especially since dose-response relationships for lung cancer are inadequately known at the present time.

## Ingestion of mineral fibres

Ten years ago, the discovery of significant contamination of drinking-water, beverages and food by asbestos fibres raised the question of a possible increased risk of gastrointestinal cancers in the general population. This is a complex question and one that is difficult to investigate because of numerous confounding factors. Even in cohorts occupationally exposed to asbestos, this increased risk of gastrointestinal cancer, which varied from cohort to cohort, has not been fully confirmed (Acheson & Gardner, 1983). Moreover, animal experiments have failed so far to demonstrate any obvious carcinogenic effects after the ingestion of different types of asbestos (Chouroulinkov, this volume, pp. 112-126).

Some 15 epidemiological studies have been conducted in various parts of North America during the last 10 years. The results of most of these studies have been reviewed by Toft et al. (1984) and are presented in the WHO document (World Health Organization, 1986b). Most failed to show any consistent evidence of an association between cancer incidence and ingestion of asbestos-contaminated drinking-water. Only the ecological-epidemiological study conducted in the San Francisco Bay area indicated a significant association between asbestos in drinking-water and the incidence of gastrointestinal cancers (Kanarek et al., 1980; Conforti et al., 1981). However, this study did not assess individual exposures and may have been biased by other confounding factors. Thus the fact that exposure was recent, mostly intermittent and at variable concentrations was not adequately taken into account. In the more powerful case-control study conducted in the Puget Sound area by Polissar et al. (1982, 1984), there was no consistent evidence of any cancer risk from the ingestion of asbestos in drinking-water. Thus the vast majority of the studies conducted to date failed to indicate a consistent excess of gastrointestinal and other cancers among residents of areas where the drinking-water supplies were contaminated with asbestos or asbestiform fibres.

## Risk assessment

The following three steps are generally necessary in assessing health risks from a specific environmental exposure: (*a*) identification of the toxic substance and its adverse effects in humans; (b) determination of the dose-response relationships, particularly for subjects occupationally exposed to high or moderate doses; and (c) then, if the relationship is linear, extrapolation to low-dose environmental exposures.

As far as asbestos is concerned, measurements of current indoor exposures will provide data from which the existing risk and the projected future risk can be established. However, the long latency period of asbestos-related diseases makes it necessary to wait at least 20 years or more before the future risk from present exposure levels can be confirmed. In addition, the types and characteristics of fibres, conditions of exposure, age at first exposure, duration of exposure and individual susceptibility to disease are different in non-occupational as compared with occupational settings. All these differences make risk assessment at the present time extremely uncertain.

Nevertheless, such quantitative and qualitative risk assessments should be useful in making policy judgements concerning the level of risks related to asbestos in non-occupational settings and in comparing this risk with other risks commonly encountered. Three studies have been published, one by the National Research Council (1984), corrected by Aroesty and Wolf in 1986, one by Doll and Peto (1985) and one by Hughes and Weill (1986). As there is some epidemiological evidence that chrysotile is a less potent carcinogen than amphiboles, at least for mesothelioma, Hughes and Weill (1986) gave different risk assessments for chrysotile and amphiboles. Doll and Peto (1985) estimated the lifetime risk of cancer at 10 per million for children exposed for 8 hours per day, 5 days per week, for the 10 years from age 8 to 18 in a school where asbestos fibre concentrations of 0.5 f/litre were present. Hughes and Weill (1986), accepting a 6-fold higher exposure (3 f/litre) during 6 years spent in school, made the following assessment for the various fibre types: for an exposure to mixed fibres (chrysotile+amphibole), a total of 15 cancers per million (0.9–3.7 lung cancer and 6.6–26.4 mesothelioma); for an exposure to chrysotile alone, a total of 4.5 cancers per million (0.9–3.7 lung cancer and 1.3–5.3 mesothelioma). These findings are one order of magnitude less than those given by Doll and Peto (1985) and those of the National Research Council (1984). In addition, Hughes and Weill (1986) compared the risk of death from asbestos-related cancer with other death risks in relation to daily activities, obtaining the following figures for the annual risk of death (per million): studying in a school containing sprayed asbestos, 0.25; cycling to school from 10 to 14 years of age, 15; inhalation or ingestion of foreign bodies, 15; playing football at school, 10; chronic smoking, 1200; passive smoking for 2 months, 1. The risk associated with exposure to low concentrations of fibres in schools must therefore be seen in its proper perspective. Nevertheless, it must neither be underestimated nor denied, and the authorities should be alert to the existence of this problem.

## Conclusions

Ideally, all fibres would be eliminated from the environment, or at least their length would be reduced as much as possible so that they became almost globular. However, all the problems would not thereby be solved, since it has recently been reported that quartz has carcinogenic potential (IARC, 1987). Mankind must therefore learn to live with certain risks, to some of which it has been exposed for thousands of years. The more that is known about them, the easier it will be to control and avoid them. As far as mineral fibres in the non-occupational environment are concerned, the risk that gives rise to the greatest concern is that associated with pollution inside buildings, since it affects large numbers of children and is therefore of major importance both to the media and politically. In spite of the optimistic forecasts of some epidemiologists, it is likely that some mesothelioma cases will be observed in association with such exposures. They will be rare, but we must be medically, scientifically, humanly and politically ready for them.

## References

Achard, S., Perderiset, M. & Jaurand, M.C. (1987) Sister chromatid exchanges in rat pleural mesothelial cells treated with crocidolite, attapulgite or benzo[3,4]pyrene. *Br. J. Ind. Med.*, *44*, 281-283

Acheson, E.D. & Gardner, M.J. (1983) *Asbestos: The Control Limit for Asbestos*, London, Her Majesty's Stationery Office

Anderson, H.A., Lilis, R., Daum, S.M., Fischbein, A.S. & Selikoff, I.J. (1976) Household contact asbestos neoplastic risk. *Ann. N.Y. Acad. Sci.*, *271*, 311-323

Anderson, H.A., Lilis, R., Daum, S.M. & Selikoff, I.J. (1979) Asbestosis among household contacts of asbestos factory workers. *Ann. N.Y. Acad. Sci.*, *330*, 387-399

Aroesty, J. & Wolf, K. (1986) Risk from exposure to asbestos. *Science*, *234*, 923

Artvinli, M. & Baris, Y.I. (1979) Malignant mesotheliomas in a small village in the Anatolian region of Turkey. An epidemiologic study. *J. Natl Cancer Inst.*, *63*, 17-23

Baris, Y.I., Sahin, A.A., Özesmi, M., Kerse, I., Özen, E., Kolaçan, B., Altinörs, M. & Göktepeli, A. (1978) An outbreak of pleural mesothelioma and chronic fibrosing pleurisy in the village of Karain/Ürgüp in Anatolia. *Thorax*, *33*, 181-192

Baris, Y.I., Sahin, A.A. & Erkan, M.L. (1980) Clinical and radiological study in sepiolite workers. *Arch. Environ. Health*, *35*, 343-346

Baris, Y.I., Saracci, R., Simonato, L., Skidmore, J.W. & Artvinli, M. (1981) Malignant mesothelioma and radiological chest abnormalities in two villages in central Turkey. *Lancet*, *ii*, 984-987

Bazas, T., Bazas, B., Kitas, D., Gilson, J.C. & McDonald, J.C. (1981) Pleural calcification in Northwest Greece. *Lancet*, *ii*, 254

Bazas, T., Oakes, D., Gilson, J.C., Bazas, B. & McDonald, J.C. (1985) Pleural calcification in Northwest Greece. *Environ. Res.*, *38*, 239-247

Becklake, M.R. (1976) State of the art: asbestos-related diseases of the lung and other organs: their epidemiology and implications for clinical practice. *Am. Rev. Respir. Dis.*, *114*, 187-227

Bertrand, R. & Pezerat, H. (1980) Fibrous glass: Carcinogenicity and dimensional characteristics. In: Wagner, J.C., ed., *Biological Effects of Mineral Fibres*, Vol. 2 (*IARC Scientific Publications No. 30*), Lyon, International Agency for Research on Cancer, pp. 901-911

Bignon, J. & Brochard, P. (1986) Rapport sur le registre national des mésotheliomes. Paris, Ministry of Health (unpublished)

Bignon, J., Goni, J., Bonnaud, G., Jaurand, M.C. & Pinchon, M.C. (1970) Incidence of pulmonary ferruginous bodies in France. *Environ. Res.*, *3*, 430-442

Bignon, J., Sébastien, P., Di Menza, L., Nebut, M. & Payan, H. (1979) French register of mesotheliomas [in French]. *Rev. Fr. Mal. Resp.*, 7, 223-241

Bignon, J., Sébastien, P., Gaudichet, A. & Jaurand, M.C. (1980) Biological effects of attapulgite. In: Wagner, J.C., ed., *Biological Effects of Mineral Fibres*, Vol. 1 (*IARC Scientific Publications No. 30*), Lyon, International Agency for Research on Cancer, pp. 163-181

Billon-Galland, M.A., Dufour, G., Gaudichet, A., Boutin, C. & Viallat, J.R. (1988) Environmental airborne asbestos pollution and pleural plaques in Corsica. In: *Inhaled Particles VI* (in press)

Bohlig, H. & Hain, E. (1973) Cancer in relation to environmental exposure, type of fibre, dose, occupation, and duration of exposure. In: Bogovski, P., Gilson, J.C., Timbrell, V. & Wagner, J.C., eds, *Biological Effects of Asbestos* (*IARC Scientific Publications No. 8*), Lyon, International Agency for Research on Cancer, pp. 217-221

Botha, J.L., Irwig, L.M. & Strebel, P.M. (1986) Excess mortality from stomach cancer, lung cancer, asbestosis and/or mesothelioma in crocidolite mining districts in South Africa. *Am. J. Epidemiol.*, 123, 30-40

Boutin, C., Viallat, J.R., Steinbauer, J., Massey, D.G., Charpin, D. & Mouries, J.C. (1986) Bilateral pleural plaques in Corsica: a non-occupational asbestos exposure marker. *Eur. J. Resp. Dis.*, 69, 4-9

Browne, K. (1986) Is asbestos or asbestosis the cause of the increased risk of lung cancer in asbestos workers? *Br. J. Ind. Med.*, 43, 145-149

Burdett, G.J. & Jaffrey, S.A.M.T. (1986) Airborne asbestos concentrations in buildings. *Ann. Occup. Hyg.*, 30, 185-199

Burilkov, T. & Michailova, L. (1970) Asbestos content in the soil and endemic pleural asbestosis. *Environ. Res.*, 3, 443-451

Burilkov, T. & Michailova, L. (1972) On the soil sepiolite content in regions with endemic pleural calcification [in German]. *Int. Arch. Arbeitsmed.*, 29, 95-101

Churg, A. (1983) Nonasbestos pulmonary mineral fibres in the general population. *Environ. Res.*, 31, 189-200

Churg, A. & Warnock, M.L. (1977) Analysis of the cores of ferruginous (asbestos) bodies from the general population. I. Patients with and without lung cancer. *Lab. Invest.*, 37, 280-286

Conforti, P.M., Kanarek, M.S., Takson, L.A., Cooper, R.C. & Murcho, J.C. (1981) Asbestos in drinking-water and cancer in the San Francisco Bay Area: 1969-1974 incidence. *J. Chron. Dis.*, 34, 211-224

Constantopoulos, S.H., Goudevenos, J.A., Saratzis, N., Langer, A.M., Selikoff, I.J. & Moutsopoulos, H.M. (1985) Metsovo lung: pleural calcification and restrictive lung function in northwestern Greece. Environmental exposure to mineral fiber as etiology. *Environ. Res.*, 38, 319-331

Cordier, S., Ameille, J., Brochard, P., Bignon, J., Proteau, J. & Lazar, P. (1987) Epidemiologic investigation of respiratory effects related to environmental exposure to asbestos inside insulated buildings. *Arch. Environ. Health*, 42, 303-309

Davis, J.M.G., Addison, J., Bolton, R.E., Donaldson, K., Jones, A.D. & Wright, A. (1984) The pathogenic effects of fibrous ceramic aluminium silicate glass administered to rats by inhalation or peritoneal injection. In: *Biological Effects of Man-made Mineral Fibres*, Vol. 2, Copenhagen, World Health Organization Regional Office for Europe, pp. 303-322

Dewees, D.N. (1987) Does the danger from asbestos in buildings warrant the cost of taking it out? *Am. Sci.*, 75, 285

Doll, R. & Peto, J. (1985) *Effects on Health of Exposure to Asbestos*, London, Her Majesty's Stationery Office

Environmental Protection Agency (1986) *Airborne Asbestos Health Assessment Update*, Washington DC (EPA/600,8-84/003F)

Flanigan, E.M. (1977) Crystal structure and chemistry of natural zeolites. In: Boles, J.R. & Mumpton, F.A.M., eds, *Mineralogy and Geology of Natural Zeolites*, Washington DC, Mineralogical Society of America, pp. 19-52

Germine, M. (1985) Fibrous rutile from Franklin, New Jersey. *Miner. Rec.*, *16*, 483-484

Hammond, E.C., Selikoff, I.J. & Seidman, H. (1979a) Asbestos exposure, cigarette smoking and death rates. *Ann. N.Y. Acad. Sci.*, *330*, 473-490

Hammond, E.C., Garfinkel, L., Selikoff, I.J. & Nicholson, W.J. (1979b) Mortality experience of residents in the neighborhood of an asbestos factory. *Ann. N.Y. Acad. Sci.*, *330*, 417-422

Hillerdal, G. (1981) Non-malignant asbestos pleural disease. *Thorax*, *36*, 669-675

Hirsch, A., Brochard, P., de Cremoux, H., Erkan, L., Sébastien, P., Di Menza, L. & Bignon, J. (1982) Features of asbestos-exposed and unexposed mesothelioma. *Am. J. Ind. Med.*, *3*, 413-422

Hobbs, M.S.T., Woodward, S.D., Murphy, B., Musk, A.W. & Elder, J.E. (1980) The incidence of pneumoconiosis, mesothelioma and other respiratory cancer in men engaged in mining and milling crocidolite in Western Australia. In: Wagner, J.C., ed., *Biological Effects of Mineral Fibres*, Vol. 2 (*IARC Scientific Publications No. 30*), Lyon, International Agency for Research on Cancer, pp. 615-625

Hughes, J.M. & Weill, H. (1986) Asbestos exposure — quantitative assessment of risk. *Am. Rev. Respir. Dis.*, *133*, 5-13

Huuskonen, M.S., Tossavainen, A., Kosinen, H., Zitting, A., Korhonen, O., Nickels, J., Korhonen, K. & Vaaranen, V. (1983) Wollastonite exposure and lung fibrosis. *Environ. Res.*, *30*, 291-304

IARC (1987) *IARC Monographs on the Evaluation of the Carcinogenic Risk of Chemicals to Humans*, Vol. 42, *Silica and some Silicates*, Lyon, International Agency for Research on Cancer

International Labour Office (1980) *Classification of Radiographs for Pneumoconiosis*, Geneva

Järvholm, B., Arvidsson, H., Bake, B., Hillerdal, G. & Westrin, C.G. (1986) Pleural plaques —asbestos — ill-health. *Europ. J. Respir. Dis.*, *68*, 1-59

Jaurand, M.C., Fleury, J., Monchaux, G., Nebut, M. & Bignon, J. (1987) Pleural carcinogenic potency of mineral fibers (asbestos, attapulgite) and their cytotoxicity on cultured cells. *J. Natl Cancer Inst.*, *79*, 797-804

Kanarek, M.S., Conforti, P.M., Jackson, L.A., Cooper, R.C. & Murchio, J.C. (1980) Asbestos in drinking water and cancer incidence in the San Francisco Bay Area. *Am. J. Epidemiol.*, *112*, 54-72

Kiviluoto, R. (1960) Pleural calcification as a roentgenologic sign of non-occupational endemic anthophyllite-asbestosis. *Acta Radiol.*, *194*, 1-67

Lockey, J.E., Brooks, S.M., Jarabek, A.M., Khoury, P.R., McKay, R.T., Carson, A., Morrison, J.A., Wiot, J.F. & Spitz, H.B. (1984) Pulmonary changes after exposure to vermiculite contaminated with fibrous tremolite. *Am. Rev. Respir. Dis.*, *129*, 952-958

McDonald, A.D. & McDonald, J.C. (1980) Malignant mesothelioma in North America. *Cancer*, *46*, 1650-1656

McDonald, A.D. & Fry, J.S. (1982) Mesothelomia and fiber type in three American asbestos factories: preliminary report. *Scand. J. Work Environ. Health*, *8*, Suppl., 53-58

McDonald, A.D., Fry, J.S., Woolley, A.J. & McDonald, J.C. (1983) Dust exposure and mortality in an American chrysotile textile plant. *Br. J. Ind. Med.*, *40*, 361-367

McDonald, J.C. (1980) Asbestos and lung cancer: has the case been proven? *Chest*, *78*, 374S-376S

McDonald, J.C., Liddell, F.D.K., Gibbs, G.W., Eyssen, G.E. & McDonald, A.D. (1980) Dust exposure and mortality in chrysotile mining, 1910-75. *Br. J. Ind. Med.*, *37*, 11-24

Monchaux, G., Bignon, J., Jaurand, M.C., Lafuma, J., Sébastien, P., Masse, R., Hirsch, A. & Goni, J. (1981) Mesotheliomas in rats following inoculation with acid-leached chrysotile asbestos and other mineral fibres. *Carcinogenesis*, 2, 229-236

Morgan, A., Davies, P., Wagner, J.C., Berry, G. & Holmes, A. (1977) The biological effects of magnesium-leached chrysotile asbestos. *Br. J. Exp. Pathol.*, 58, 465-473

National Research Council (1984) *Asbestiform Fibres. Non Occupational Health Risks*, Washington DC, National Academy Press

Navratil, M. & Trippe, F. (1972) Prevalence of pleural calcification in persons exposed to asbestos dust and in the general population in the same district. *Environ. Res.*, 5, 210-216

Neuberger, M., Raeber, A. & Friedl, H.P. (1982) Epidemiology of asbestos-linked lung diseases in Austria. In: *Proceedings, XVI Congress of the Austrian Society for Lung Diseases*, Vienna, Hoffman, pp. 127-130

Neuberger, M., Kundi, M. & Friedl, H.P. (1984) Environmental asbestos exposure and cancer mortality. *Arch. Environ. Health*, 39, 261-265

Newhouse, M.L. & Thompson, H. (1965) Mesothelioma of pleura and peritoneum following exposure to asbestos in the London area. *Br. J. Ind. Med.*, 22, 261-269

Nicholson, W.J., Rohl, A.N. & Weisman, I. (1975) *Asbestos Contamination of the Air in Public Buildings*, Research Triangle Park, NC, Environmental Protection Agency (EPA 450/3-76-004)

Ontario Royal Commission (1984) *Report of the Royal Commission on Matters of Health and Safety Arising from the Use of Asbestos in Ontario*, Toronto, Ontario Government Bookshop

Pampalon, R., Siemiatycki, J. & Blanchet, M. (1982) Environmental asbestos pollution and public health in Quebec. *Union Med. Can.*, 111, 475-489

Peters, B.J. & Peters, G.A. (1986) *Sourcebook on Asbestos Diseases*, Vol. 1, *Medical, Legal and Engineering Aspects*. Chapter 7. Asbestos Substitutes, New York, Garland, pp. 175-194

Peto, J. (1984) Dose and time relationships for lung cancer and mesothelioma in relation to smoking and asbestos exposure. In: Fischer, M. & Meyer, E., eds, *Zur Beurteilung der Krebsgefahr durch Asbest*, Vol. 2, Berlin, München, Springer Verlag, pp. 126-132

Peto, J., Seidman, H. & Selikoff, I.J. (1982) Mesothelioma mortality in asbestos workers: implications for models of carcinogenesis and risk assessment. *Br. J. Cancer*, 45, 124-135

Peto, J., Doll, R., Hermon, C., Binns, W., Clayton, R. & Goffe, T. (1985) Relationship of mortality to measures of environmental asbestos pollution in an asbestos textile factory. *Ann. Occup. Hyg.*, 29, 305-355

Polissar, L. (1980) The effect of migration on comparison of disease rates in geographic studies in the United States. *Am. J. Epidemiol.*, 111, 175-182

Polissar, L., Severson, R.K., Boatman, E.S. & Thomas, D.B. (1982) Cancer incidence in relation to asbestos in drinking-water in the Puget Sound region. *Am. J. Epidemiol.*, 116, 314-328

Polissar, L., Severson, R.K. & Boatman, E.S. (1984) A case-control study of asbestos in drinking water and cancer risk. *Am. J. Epidemiol.*, 119, 456-471

Pott, F., Huth, F. & Friedrichs, K.H. (1974) Tumorigenic effect of fibrous dusts in experimental animals. *Environ. Health Perspect.*, 9, 313-315

Pott, F., Huth, F. & Spurny, K. (1980) Tumour induction after intraperitoneal injection of fibrous dusts. In: Wagner, J.C., ed., *Biological Effects of Mineral Fibres*, Vol. 1 (*IARC Scientific Publications No. 30*), Lyon, International Agency for Research on Cancer, pp. 337-347

Rohl, A.N., Langer, A.M., Moncure, G., Selikoff, I.J. & Fischbein, A. (1982) Endemic pleural disease associated with exposure to mixed fibrous dust in Turkey. *Science*, 216, 518-520

Saracci, R., Simonato, L., Baris, Y., Artvinli, M. & Skidmore, J. (1982) The age-mortality curve of endemic pleural mesothelioma in Karain, central Turkey. *Br. J. Cancer*, *45*, 147-149

Sébastien, P., Janson, X., Bonnaud, G., Riba, G., Masse, R. & Bignon, J. (1979) Translocation of asbestos fibres through respiratory tract and gastrointestinal tract according to fibre and size. In: Lemen, R. & Dement, J., eds, *Dusts and Disease*, Park Forest South, IL, Pathotox Publishers, pp. 65-85

Sébastien, P., Billon-Galland, M.A., Dufous, G. & Bignon, J. (1980) *Measurement of Asbestos Air Pollution inside Buildings sprayed with Asbestos*, Washington DC, Environmental Protection Agency (EPA-560/13-80-026)

Sébastien, P., Martin, M. & Bignon, J. (1982) Indoor airborne asbestos pollution: from the ceiling and the floor. *Science*, *216*, 1410-1413

Sébastien, P., Bignon, J., Baris, Y.I., Awad, L. & Petit, G. (1984) Ferruginous bodies in sputum as an indication of exposure to airborne mineral fibers in the mesothelioma villages of Cappadocia. *Arch. Environ. Health*, *39*, 18-23

Sébastien, P., Begin, R., Case, B.W. & McDonald, J.C. (1986) Inhalation of chrysotile dusts. In: Wagner, J.C., ed., *Accomplishments in Oncology*, Philadelphia, Lippincott, pp. 19-29

Selikoff, I.J. & Lee, D.H.K. (1978) *Asbestos and Disease*, New York, Academic Press

Shasby, D.M., Petersen, M., Hodous, T., Boehlcke, B. & Merchant, J. (1979) Respiratory morbidity of workers exposed to wollastonite through mining and milling. In: Lemen, R. & Dement, J.M., eds, *Dusts and Disease*, Park Forest South, IL, Pathotox Publishers, pp. 251-256

Siemiatycki, J. (1983) Health effects on the general population: mortality in the general population in asbestos mining areas. In: *Proceedings of the World Symposium on Asbestos*, Montreal, Canadian Asbestos Information Center, pp. 337-348

Stanton, M.F., Layard, M., Tegeris, A., Miller, E., May, M. & Kent, E. (1977) Carcinogenicity of fibrous glass: pleural response in the rat in relation to fiber dimension. *J. Natl Cancer Inst.*, *58*, 587-603

Stanton, M.F., Layard, M., Tegeris, A., Miller, E., May, M., Morgan, E. & Smith, A. (1981) Relation of particle dimension to carcinogenicity in amphibole asbestoses and other fibrous minerals. *J. Natl Cancer Inst.*, *67*, 965-975

Suta, B. & Levine, R.J. (1979) Nonoccupational asbestos emissions and exposures. In: Michaels, L. & Chissi, S.S., eds, *Asbestos: Properties, Applications and Hazards*, Vol. 1, New York, Wiley and Sons, pp. 171-205

Suzuki, Y. (1982) Carcinogenic and fibrogenic effects of zeolites: preliminary observations. *Environ. Res.*, *27*, 433-445

Toft, P., Meek, M.E., Wigle, D. & Meranger, J.C. (1984) Asbestos in drinking-water. *CRC Crit. Rev. Environ. Control*, *14*, 151-197

United Nations Environment Programme (1986) International Programme on Chemical Safety (IPCS). *Environmental Health Criteria for Man-made Mineral Fibres* (UNEP/ILO/WHO).

Vianna, J.C. & Polan, A.K. (1978) Non-occupational exposure to asbestos in malignant mesothelioma in females. *Lancet*, *i*, 1061-1063

Wagner, J.C. (1963) *Asbestos dust and malignancy*. In: *Proceedings, XIV International Congress of Occupational Health*, pp. 1066-1067

Wagner, J.C., ed. (1980) *Biological Effects of Mineral Fibres (IARC Scientific Publications No. 30)*, Lyon, International Agency for Research on Cancer

Wagner, J.C. (1982) Health hazards of substitutes. In: *Proceedings of the World Symposium on Asbestos*, Montreal, Canadian Asbestos Information Center, pp. 244-266

Wagner, J.C., Sleggs, C.A. & Marchand, P. (1960) Diffuse pleural mesothelioma and asbestos exposure in the north-western Cape Province. *Br. J. Ind. Med., 17*, 260-271

Wagner, J.C., Berry, G., Skidmore, J.W. & Timbrell, V. (1974) The effects of the inhalation of asbestos in rats. *Br. J. Cancer, 29*, 252-269

Wagner, J.C., Berry, G. & Pooley, F.D. (1980) Carcinogenesis and mineral fibres. *Br. Med. Bull., 36*, 53-56

Wagner, J.C., Berry, G.B., Hill, R.J., Munday, D.E. & Skidmore, J.W. (1984) Animal experiments with MMM(V)F — Effects of inhalation and intrapleural inoculation in rats. In: *Biological Effects of Man-made Mineral Fibres*, Copenhagen, World Health Organization Regional Office for Europe, pp. 209-233

Webster, J. (1977) Methods by which mesothelioma may be diagnosed. In: Glen, H.W., ed., *Proceedings of Asbestos Symposium, Johannesburg 1977*, Randburg, National Institute for Metallurgy, pp. 3-8

WHO (1986a) Les fibres minérales artificielles sur les lieux de travail. *Symposium international, Copenhague, 21–29 octobre 1986, Rapport sommaire.* ICP/OCH 125(5) 68420. Copenhagen, World Health Organization Regional Office for Europe

WHO (1986b) *Asbestos and Other Natural Mineral Fibres* (Environmental Health Criteria No. 53), Geneva, World Health Organization

WHO (1987) Proceedings WHO/IARC Conference on Man-made Mineral Fibres in the Working Environment. *Ann. Occup. Hyg., 31*, 517-822

Yazicioglu, S. (1976) Pleural calcification associated with exposure to chrysotile asbestos in south east Turkey. *Chest, 70*, 43-47

Yazicioglu, S., Ilcayto, R., Balci, K., Sayli, B.S. & Yorulmaz, B. (1980) Pleural calcification, pleural mesotheliomas and bronchial cancers caused by tremolite dust. *Thorax, 35*, 564-569

Zolov, C., Bourilkov, T. & Babadjov, L. (1967) Pleural asbestosis in agricultural workers. *Environ. Res., 1*, 287-292

Zoltai, T. (1979) Asbestiform and acicular mineral fragments. *Ann. N.Y. Acad. Sci., 330*, 621-643

# II. EXPERIMENTAL DATA

# MINERAL FIBRE CARCINOGENESIS: EXPERIMENTAL DATA RELATING TO THE IMPORTANCE OF FIBRE TYPE, SIZE, DEPOSITION, DISSOLUTION AND MIGRATION

### J.M.G. Davis

*Institute of Occupational Medicine, Edinburgh, UK*

*Summary.* Fibre type, fibre size, deposition, dissolution and migration are all factors of importance in mineral fibre carcinogenesis. These factors are, however, so interrelated that only fibre size can be considered on its own to any extent. When dusts are injected into the pleural or peritoneal cavities, the most carcinogenic samples, producing the most mesotheliomas, are those containing the most long, thin fibres. When very short fibre samples of both amosite and chrysotile recently became available for comparison with long fibre preparations of the same materials, short fibres were found to be much less fibrogenic and carcinogenic than long fibres. The same studies provided important information on fibre deposition and dissolution. Short fibre samples of both asbestos varieties penetrated to the pulmonary parenchyma more easily than long ones but, after deposition, short fibres were cleared more quickly. Very much less chrysotile was present in lung tissue at the end of one year's dusting and clearance during the following 6 months was very much faster. The long fibre chrysotile, which would be expected to be resistant to mechanical clearance, was removed from the lungs much more quickly than short fibre amosite, which was easily phagocytosed by macrophages. This indicates that rapid chrysotile removal from lung tissue is due at least in part to fibre dissolution. The phenomenon of chrysotile dissolution probably explains why this asbestos type has been shown to be extremely carcinogenic in rats but seems less carcinogenic than the amphiboles in humans. Fibres may remain in lung tissue for the 1–2 years necessary to cause tumours in rats but this is too short a time for the much longer lived humans. Only very few fibres penetrate the walls of the gut following massive asbestos ingestion, although a few of these can subsequently be found disseminated to other organs. Fibres are disseminated to other organs much more effectively after inhalation. One area where fibre dissemination has been suggested as being very important is that of transport from the lung tissue to the pleural cavity, but in rats, direct fibre penetration to the pleura occurs very rarely and the exact mechanism by which inhaled fibres reach the sites where they can produce mesotheliomas remains one of the most important subjects for future research.

Experimental studies have shown that tumour production following dust inhalation can be related to the type and size of fibres as well as to the ease of fibre deposition in the lung and to subsequent dissolution and migration of fibres. These factors are frequently interrelated and difficult to separate from one another experimentally.

Many studies have concentrated on the effects of fibre size, and the work of Pott (1978, 1983), Pott and Friedrichs (1972), Stanton et al. (1977), Stanton and Layard (1978) and Stanton and Wrench (1972) has shown that, following intrapleural or intraperitoneal injection of dust, mesothelioma production is related to the number of long, thin fibres. It has been suggested that the most dangerous fibres are those $>8$ $\mu$m in length and $<0.25$ $\mu$m in diameter but that fibres of all types in this size range, including man-made mineral fibres (MMMF) as well as asbestos, are equally carcinogenic.

Apart from the production of mesotheliomas following intrapleural implantation, fibre length is important in short-term *in vitro* toxicity to macrophages (Forget et al., 1986; Tilkes & Beck, 1983) and, following intratracheal injection, long fibres produce pulmonary fibrosis while short fibres do not (Wright & Kuschner, 1977).

Until recently, there has been no evidence that fibre size was equally important in pulmonary carcinogenesis following dust inhalation, since suitably sized dusts had not been available in the large quantities needed for inhalation work. It is now possible, however, to obtain samples of amosite and chrysotile asbestos produced by sedimentation techniques so that the majority of fibres are extremely short. These have been used for full lifespan inhalation studies in rats, and their pathogenic effects compared with those of dust clouds generated from the same batches of asbestos in such a way as to contain as many long fibres as possible (Davis et al., 1986a, and unpublished data).

The long-fibre amosite dust cloud, at a respirable dose level of 10 mg/m$^3$, contained 1110 fibres per ml (f/ml) $>10$ $\mu$m in length as seen by the phase-contrast optical microscope (PCOM) while the short-fibre amosite cloud contained only 12 fibres per ml as long as this. The long-fibre chrysotile cloud contained 1930 f/ml $>10$ $\mu$m in length while the short-fibre cloud contained 330. This short-fibre chrysotile cloud had been prepared in the laboratories of the Institute for Research and Development on Asbestos in Canada by the technique reported by Jolicoeur et al. (1981). Originally, when used for producing milligram quantities, this process had proved capable of separating a chrysotile fraction with no fibres $>8$ $\mu$m in length. Unfortunately, when used to produce more than 1 kg of dust, it was less successful but the resulting chrysotile preparation is still the shortest that has been used in long-term inhalation studies.

The levels of pulmonary interstitial fibrosis produced by these amosite and chrysotile dust clouds and the numbers of pulmonary tumours found in these studies are shown in Tables 1 and 2. In each case, the long-fibre dust cloud has been significantly more fibrogenic and more carcinogenic, the short amosite dust, in fact, producing neither tumours nor detectable fibrosis. These findings indicate that fibres $<5$ $\mu$m in length may be innocuous in lung tissue, since the level of pulmonary fibrosis and the tumours produced by the 'short-fibre' chrysotile are explicable by the number of long fibres present.

These findings are of considerable importance in relation to the likely hazard from asbestos contamination of the non-occupational environment. Where asbestos fibres are found in the normal urban atmosphere, and especially in buildings containing

**Table 1. Percentage of lung tissue occupied by interstitial fibrosis in rats of advanced age treated with long- and short-fibre samples of amosite and chrysotile asbestos**[a]

| Type of fibre | Percentage of lung occupied by interstitial fibrosis[b] | |
|---|---|---|
| | Amosite | Chrysotile |
| Long | 11.0 (0.4–34.6) | 12.6 (6.4–23.9) |
| Short | 0.15 (0–2.8) | 2.4 (0.3–6.7) |

[a]All rats which survived to within 2 months of the end of their respective studies were included in the estimates of interstitial fibrosis. They were therefore aged between 30 and 33 months and group numbers varied between 10 and 23.

[a]Figures in parentheses are ranges

**Table 2. Numbers of pulmonary tumours and pleural mesotheliomas produced in rats by the inhalation of long- and short-fibre samples of amosite and chrysotile**[a]

| Type of fibre | Pulmonary tumours and pleural mesotheliomas | |
|---|---|---|
| | Amosite | Chrysotile |
| Long | 13 | 22 |
| Short | 0 | 7 |

[a]The number of animals in each group available for examination varied between 40 and 42.

asbestos-based insulation, almost all detectable fibres are <5 μm in length (Ontario Royal Commission, 1984). This means that the common practice of expressing levels of asbestos contamination as ng/m³ of air gives little indication of hazard. What is needed is information on the number of long fibres present, and fibre counts similar to those undertaken in routine factory monitoring should be undertaken.

Fibre type and levels of fibre deposition and dissolution in lung tissue are so interrelated that they must be considered together. The importance of fibre deposition and dissolution is, in fact, closely related to a major problem in evaluating the importance of fibre type in disease production. While human epidemiology suggests that amphibole fibres are the most dangerous, animal experiments have indicated that chrysotile can be at least as hazardous; in some studies, chrysotile has even been found to be more fibrogenic and carcinogenic than either amosite or crocidolite (Bolton et al., 1982a; Davis et al., 1978; Wagner et al., 1973, 1974).

Reports of asbestos-related pathology from animal inhalation studies were particularly surprising when it was found that, following periods of exposure to equal masses of respirable dust, much more amphibole than chrysotile was always found in lung tissue (Middleton et al., 1977; Wagner & Skidmore, 1965). Frequently, after a one-year inhalation period, 10 times more amphibole than chrysotile was present in rat lungs. Timbrell (1973) suggested that these findings were due to the curly nature of long chrysotile fibres, which made it more difficult for them to penetrate the smallest bronchial tubes than for the straight amphibole fibres. This suggestion has been difficult to confirm, but fibre length is certainly an important factor in both pulmonary deposition and clearance of both chrysotile and the amphiboles. Table 3 shows the amounts of both long- and short-fibre samples of chrysotile and amosite present in rat lungs at the end of a one-year inhalation period at a dose of 10 mg/m$^3$ of respirable dust and 6 months after the end of dusting.

**Table 3. Lung burdens of asbestos in rats treated for 12 months with a respirable mass dose of 10 mg/m$^3$ of amosite and chrysotile as long- and short-fibre preparations**[a]

| Time after end of dusting (months) | Lung asbestos burden ($\mu$g) | | | |
|---|---|---|---|---|
| | Long amosite | Short amosite | Long chrysotile | Short chrysotile |
| 0 | 3570 | 5640 | 350 | 1020 |
| 6 | 3080 | 4470 | 161 | 109 |

[a]The figures are the means for groups of 4 animals

For both asbestos types, more short-fibre dust is present in lung tissue at the end of dusting, indicating that long fibres are less likely to penetrate into the lung parenchyma. During the 6-month period following the end of dusting, however, short fibres of both chrysotile and amosite are removed from lung tissue faster than long fibres. While both asbestos types behave in a similar manner in respect to fibre length in pulmonary deposition and clearance, there are nevertheless very important quantitative differences between chrysotile and amosite.

As in previously reported studies, much more amosite than chrysotile is present at the end of dusting with both long- and short-fibre preparations. In these studies, roughly five times more short-fibre amosite than short-fibre chrysotile was present, the ratio increasing to 10 for the long-fibre material. Large differences were also found in clearance rates. While 14% of long-fibre amosite and 20% of short-fibre amosite was cleared during the six months following the end of dusting, the comparable figures for long- and short-fibre chrysotile were 55% and 90%. From these findings it appears that differences between the lung burdens for similarly sized preparations of chrysotile and amosite are due mainly to the rapid removal of chrysotile. This process will be taking place during any prolonged period of dust exposure, most short fibres deposited during the first six months being cleared during the last six months of a one-year inhalation period.

There is no evidence at present to prove that this very rapid removal of chrysotile from lung tissues is not due to enhanced mechanical clearance, but it is difficult to explain why macrophages should preferentially clear chrysotile rather than similarly sized particles of amphibole, which they can certainly phagocytose with equal ease. The most likely explanation is that chrysotile undergoes chemical dissolution in lung tissues. It has been known for many years that chrysotile rapidly loses its magnesium in water or physiological saline, while the amphiboles show little chemical reactivity at all (Hodgson, 1986). A factor which may aid this process is the fact that chrysotile fibres rapidly separate into their individual fibrils in tissues and, after the loss of magnesium, these extremely fine filaments will be very fragile. In old rats, nearly 2 years after the end of dusting, no chrysotile could be found in macrophages or in contact with any other type of pulmonary cell (Davis et al., 1986b). The very small amount present was in areas of acellular fibrous tissue or in basement membranes which were often greatly thickened. Both these sites may be backwaters of chemical activity where fibre dissolution is retarded.

These experimental findings are in agreement with those of many studies on the lung dust content of asbestos workers at autopsy. Almost all publications in this field record more amphibole than chrysotile, even though in some cases it is known with certainty that chrysotile exposure during life was far greater than exposure to amphibole (Gylseth et al., 1983; Pooley, 1976).

The experimental studies may also offer an explanation for one apparent anomaly in the human lung dust studies. Le Bouffant et al. (1976) reported that, where amphibole was found to be the predominant fibre in the lung parenchyma, relatively more chrysotile was present in the pleura. If the pleural tissues examined in this study were fibrosed, as was frequently the case in asbestos workers, then chrysotile may have been retained in this chemically inactive environment while it was removed from the lung parenchyma. Le Bouffant's finding may have little relevance to pleural disease, however, since almost all recorded fibres were very short.

MMMF, such as glass fibre, slag wool and rock wool, which are often used as asbestos substitutes, have been shown to undergo dissolution in chemical environments similar in character to lung tissue (Förster, 1984; Kleinholz & Steinkopf, 1984; Leineweber, 1984), and this probably explains why there is little evidence of pulmonary disease in workers producing or handling these materials in industry (Saracci, 1986). Some types of MMMF appear to be very durable in lung tissue, however, and heavy exposure to these may well represent a hazard to health. This is a matter which should be examined carefully during the development phase of new fibre materials.

The last aspect to be considered is that of the migration of fibres throughout the body, an area where relatively little information is available. Following asbestos inhalation, fibres are certainly transported to the pulmonary-associated lymph-nodes in large amounts. As would be expected, short fibres are transported most readily (Bolton, 1979) but, in rats treated with amphibole asbestos, many fibres >20 µm and up to 50 µm in length can be seen in the mediastinal and hilar nodes by light microscopy.

Following the inhalation of asbestos, fibres can be found in small but significant amounts in many of the organs of the body, which they presumably reach by lymphatic transport. Table 4 shows the results obtained from the ashed tissues of a group of 4 rats treated with crocidolite asbestos in our laboratory. Following a one-year inhalation period at a dose level of 10 mg/m³, the animals were allowed to survive for their full lifespan, although the study was terminated at 950 days. Tissues were ashed in nascent oxygen at low temperature and residues examined by scanning electron microscopy. Small but notable numbers of fibres were found in most of the organs examined, although the mediastinal lymph-nodes were an exception since they were heavily laden with fibres. The relatively large numbers of fibres found in the gut tissues does not indicate fibre penetration following ingestion. Fibres cleared from the lung via the bronchial tubes are swallowed, so that after 12 months of dusting all the rats would have had a low level of asbestos in their gut lumina at all times. While gut contents were carefully washed away before the tissues were ashed, there was little hope of eliminating all fibres trapped among the convoluted gut surfaces.

**Table 4. Numbers of crocidolite fibres per sample per rat in ashed tissues after crocidolite inhalation**[a]

| Tissue | Conversion factor[b] | Animal age (days) | | | |
| --- | --- | --- | --- | --- | --- |
| | | 448 | 950 | 950 | 950 |
| Liver | 5450 | 10 | ND | 7 | 15 |
| Spleen | 120 | 0 | 6 | 3 | ND |
| Kidney | 170 | 19 | 8 | 9 | 11 |
| Selected gut | 1400 | 36 | 17 | 24 | 12 |
| Mesenteric lymph-nodes | 140 | 12 | 16 | 9 | 22 |
| Cervical lymph-nodes | 150 | 2 | 9 | 0 | 4 |
| Mediastinal lymph-nodes | 50 | ++ | ++ | ++ | ++ |

[a] A group of 4 rats were treated with crocidolite by inhalation for a one-year period at a dose level of 10 mg/m³. The animals were allowed to survive for their full lifespan, but the study was terminated at 950 days.

[b] This allows for the fact that only a fraction of ashed residue of the whole organ was examined by scanning electron microscopy at a magnification of 10 000, e.g., a conversion factor of 5450 means that only 1/5450 of the ash was so examined.

ND, not done.

Similar studies were undertaken with rats that had ingested asbestos in their diets for long periods of time; the findings from 13 rats treated with crocidolite are given in Table 5. The number of fibres found in the tissues examined was far lower than following dust inhalation. In this study, gut tissues were not examined until at least 2 months after the animals had ceased receiving asbestos in their diet; this accounts for the lack of figures for gut tissues prior to 924 days. The results shown in Tables 4 and 5 were previously reported by Bolton (1979) and Bolton et al. (1982b) and the studies covered amosite and chrysotile asbestos as well as crocidolite. The results for amosite in both studies were very similar to those for crocidolite, although fewer chrysotile fibres were detected in tissues following both inhalation and ingestion.

**Table 5. Numbers of crocidolite fibres per sample per rat in ashed tissues after long-term ingestion of crocidolite**

| Tissue | Conversion factor[a] | Animal age (days) | | | | | | | | | | | |
|---|---|---|---|---|---|---|---|---|---|---|---|---|---|
| | | 604 | 870 | 871 | 871 | 877 | 924 | 927 | 927 | 933 | 933 | 940 | 940 | 940 |
| Lungs | 351 | 1 | ND[b] | 1 | 0 | 0 | 0 | 0 | 0 | 0 | 2 | 0 | 0 | 0 |
| Liver | 5200 | 0 | 1 | 0 | 0 | 0 | 0 | 0 | 0 | 0 | 0 | 1 | ND | 1 |
| Spleen | 120 | 0 | 0 | 0 | 0 | 0 | 0 | 0 | 0 | 0 | 0 | 0 | ND | 0 |
| Kidney | 176 | 3 | 0 | 0 | 0 | 0 | 0 | 3 | 1 | 0 | 0 | ND | 5 | 0 |
| Gut | 1164 | ND | ND | ND | ND | ND | ND | 0 | 4 | 2 | 0 | 0 | 0 | 2 |
| Mesentery | 141 | 0 | 1 | 0 | 0 | ND | 0 | 0 | 0 | 0 | 0 | 1 | 0 | 0 |
| Omentum | 30 | 0 | 0 | 0 | 0 | 0 | 0 | 0 | 1 | 0 | 1 | 5 | 0 | 0 |
| Thoracic body wall | 12 640 | 0 | 0 | ND | 0 | ND | 0 | ND | 0 | 0 | 0 | 0 | 0 | ND |
| Peritoneal body wall | 6000 | 0 | 0 | ND | 0 | 0 | 0 | ND | 0 | 0 | 0 | 0 | 0 | ND |

[a]This allows for the fact that only a fraction of ashed residue of the whole organ was examined by scanning electron microscopy at a magnification of 10 000, e.g., a conversion factor of 351 means that only 1/351 of the ash was examined

[b]ND, not done.

Whether or not the small amounts of asbestos found in tissue residues following dust ingestion are present as a direct result of penetration of the gut wall or of technical limitations in the ashing or microscope techniques is still a matter of debate. While some workers have reported fibre penetration of the gut wall (Cunningham & Pontefract, 1973; Hallenbeck & Patel-Mandlik, 1979; Storeygard & Brown, 1977), others have failed to find evidence of this process (Gross et al., 1974). It seems obvious that, even if penetration occurs, it is a relatively rare event and the tissue content of 'ingested' fibres will always be extremely low for all organs.

The area where fibre migration may be the most important is in mesothelioma production. It is generally assumed that pleural mesotheliomas at least are caused by fibres which migrate directly through the visceral pleural surface, and small amounts of fibre have been reported in both mesotheliomas (Le Bouffant et al., 1976) and pleural plaques (Le Bouffant et al., 1973). In experimental inhalation studies with asbestos, fibres are plentiful in the most peripheral alveoli bordering on the pleura, so that the distance to be covered in migration to the pleural space is no more than a few hundred μm. In spite of this, however, fibre penetration of the external elastic lamina of the lung appears to be a rare event. Rats treated with both chrysotile and amphibole asbestos frequently develop, in advanced age, areas of pleural fibrosis and vesicular hyperplasia where the spaces in the collagenous fibrous tissues are lined with flattened mesothelial cells (Figure 1). From their histological pattern, these lesions look as though they should be mesothelioma precursors, but are present in almost all old rats treated with asbestos and, in these inhalation studies, mesothelioma production is a relatively rare event.

**Fig. 1. Area of vesicular hyperplasia on the pleural surface of a rat treated with tremolite asbestos. While the lung tissue contains large amounts of dust, none is visible in the pleural lesions. H.-E. × 150.**

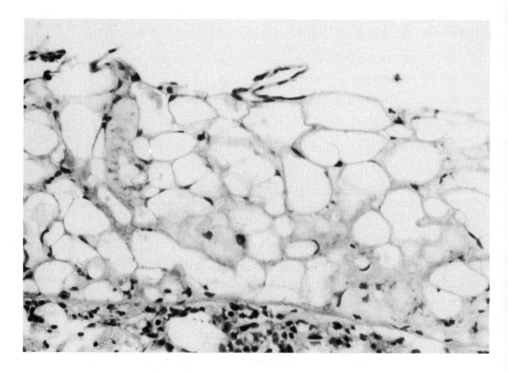

Chrysotile asbestos is difficult to see in normal light microscope preparations, but a limited amount of examination by transmission electron microscopy of these areas of vesicular hyperplasia has failed to demonstrate the presence of any chrysotile fibres at all. The amphibole asbestos types can be seen by light microscopy, and in our studies deposits of tremolite and amosite were often seen close to and apparently in contact with the external elastic lamina of the lung just below areas of vesicular pleural hyperplasia (Figure 2). Once again, however, almost no fibre penetration of the pleura could be demonstrated. It appears likely, therefore, that these pleural lesions may be caused by inflammatory or growth-stimulating products which diffuse from the damaged lung parenchyma (Pitt, 1985) rather than by direct contact with asbestos fibres. This appears particularly likely since, in recent inhalation studies in which rats were treated with coal and quartz mixtures, similar areas of pleural vesicular hyperplasia were found and, once again, these lesions contained no dust. Both these particulate dusts did, however, manage to penetrate the pleural surface in quite large amounts in some areas, where they formed fibrosing granulomas similar to the multiple 'silicotic' nodules found throughout the lung parenchyma (Figure 3). This pleural penetration appears to be a characteristic of quartz rather than of all particulate dusts, since inhalation of titanium dioxide by rats does not result in pleural

**Fig. 2.** Area from the surface of rat lungs treated with tremolite dust as seen by light microscopy with an oil immersion lens. Outside the internal elastic lamina of the lung, which was delineated by Weigert's elastin stain, is an area of vesicular pleural hyperplasia. Closely apposed to the inside of the lamina are several deposits of tremolite fibres. The position of the lamina is indicated by arrows labelled L and the tremolite deposits by arrows labelled T. × 480.

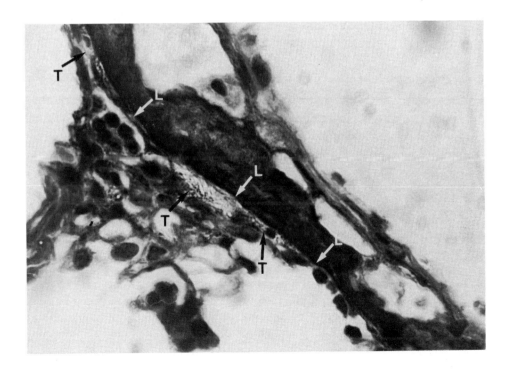

**Fig. 3.** Section of a fibrosing granuloma on the visceral pleura of a rat lung treated with coalmine dust containing 20% quartz. Numerous dust particles are visible within the pleural granulation tissue (A) as well as in a subpleural parenchymal nodule (B). The position of the internal elastic lamina of the lung is indicated by arrows labelled L. H.-E. × 240.

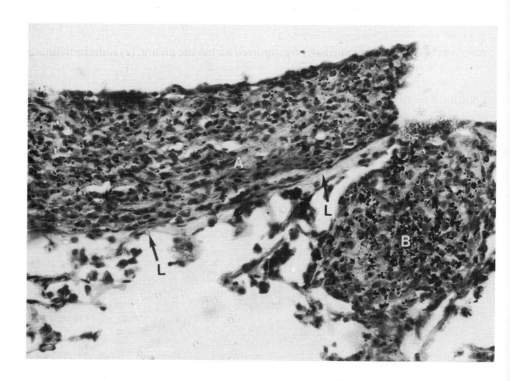

granulomas and quartz is known to be preferentially cleared to lymph-nodes (Vincent et al., 1987). We have not studied the ability of pure crushed coal without quartz to penetrate the pleural surface, but coal is not a very reactive dust and would be expected to behave like titanium dioxide. Consequently, the presence of coal in the rat pleural granulomas may indicate that it is carried through the pleura by a process initiated and maintained by the reactive quartz particles. If this is so, then it is possible that quartz could facilitate the passage of fibrous dusts across the pleural membrane as well.

Mesotheliomas related to crocidolite or amosite seldom occur in more than 10% of even heavily exposed working populations, and this may be due largely to the difficulties of pleural penetration. However, one fibrous dust, erionite, appears to cause mesotheliomas with great facility, either in humans subjected to low environmental doses (Baris et al., 1979) or in rats treated by inhalation, where 100% of animals developed these tumours (Wagner et al., 1985). We have examined a sample of the

Oregon erionite used by Wagner in these studies and found that its general appearance and fibre size distribution is extremely close to that of UICC crocidolite, which causes only rare mesotheliomas when inhaled by rats. Erionite, therefore, represents a fibre where carcinogenicity is not solely related to fibre morphology, and it may be that the surface chemistry of erionite fibres is such as to stimulate their transport across the pleural membrane in the manner of quartz. Erionite is an anomaly, but for the industrially used fibres of crocidolite and amosite there appear to be the following possibilities: (1) the mesotheliomas that they induce are caused by the extremely small amounts of dust that are normally transported across the pleura; (2) some individuals, perhaps because of concomitant non-asbestos-related pulmonary disease, transport more fibres across the pleura than others; or (3) these individuals have inhaled, in addition to asbestos, dusts with quartz-like properties of pleural penetration which aid fibre transport. Certainly, the transport of fibres across the pleural surface remains one of the areas of fibre-related pathology where future research should be concentrated.

## References

Baris, Y.I., Artvinli, M. & Sahin, A.A. (1979) Environmental mesotheliomata in Turkey. *Ann. N.Y. Acad. Sci.*, *330*, 423-432

Bolton, R.E. (1979) *Toxicity of asbestos in laboratory animals with special reference to the carcinogenicity of ingested and disseminated fibre*, Ph.D. thesis, Edinburgh University

Bolton, R.E., Davis, J.M.G., Donaldson, K. & Wright, A. (1982a) Variations in the carcinogenicity of mineral fibres. *Ann. Occup. Hyg.*, *26*, 569-582

Bolton, R.E., Davis, J.M.G. & Lamb, D. (1982b) The pathological effects of prolonged asbestos ingestion in rats. *Environ. Res.*, *29*, 134-150

Cunningham, H.M. & Pontefract, R.D. (1973) Asbestos fibres in beverages, drinking water and tissues. *J. Assoc. Off. Anal. Chem.*, *56*, 976-981

Davis, J.M.G., Beckett, S.T., Bolton, R.E., Collings, P. & Middleton, A.P. (1978) Mass and number of fibres in the pathogenesis of asbestos-related lung disease in rats. *Br. J. Cancer*, *37*, 673-688

Davis, J.M.G., Addison, J., Bolton, R.E., Donaldson, K., Jones, A.D. & Smith, T. (1986a) The pathogenicity of long versus short fibre samples of amosite asbestos administered to rats by inhalation and intraperitoneal injection. *Br. J. Exp. Pathol.*, *67*, 415-430

Davis, J.M.G., Bolton, R.E., Brown, D. & Tully, H.E. (1986b) Experimental lesions in rats corresponding to advanced human asbestosis. *Exp. Mol. Pathol.*, *44*, 207-221

Forget, G., Lacroix, M.J., Brown, R.C., Evans, P.H. & Sirois, P. (1986) Response of perfused alveolar macrophages to glass fibres. Effect of exposure duration and fibre length. *Environ. Res.*, *39*, 124-135

Förster, H. (1984) The behaviour of mineral fibres in physiological solutions. In: *Biological Effects of Man-made Mineral Fibres*, Vol. 2, Copenhagen, World Health Organization, Regional Office for Europe, pp. 27-60

Gross, P., Harley, R.A., Swinburne, L.M., Davis, J.M.G. & Greene, W.B. (1974) Ingested mineral fibres. Do they penetrate and cause cancer? *Arch. Environ. Health*, *29*, 341-347

Gylseth, B., Mowe, G. & Wannag, A. (1983) Fibre type and concentration in the lungs of workers in an asbestos cement factory. *Br. J. Ind. Med.*, *40*, 375-379

Hallenbeck, W.H. & Patel-Mandlik, K.J. (1979) Presence of fibres in the urine of a baboon gavaged with chrysotile asbestos. *Environ. Res.*, *20*, 335-340

Hodgson, A.A. (1986) *Scientific Advances in Asbestos*, Crowthorne, Anjalena Publications

Jolicoeur, C., Roberge, P. & Fortier, J. (1981) Separation of short fibres from bulk chrysotile asbestos fibre materials: analysis and physico-chemical characterization. *Can. J. Chem.*, 59, 1140-1148

Klingholz, R. & Steinkopf, B. (1984) The reactions of MMMF in a physiological model fluid and in water. In: *Biological Effects of Man-made Mineral Fibres*, Vol. 2, Copenhagen, World Health Organization, Regional Office for Europe, pp. 60-87

Le Bouffant, L., Martin, J.C., Durif, S. & Daniel, H. (1973) Structure and composition of pleural plaques. In: Bogovski, P., Gilson, J.C., Timbrell, V. & Wagner, J.C., eds, *Biological Effects of Asbestos (IARC Scientific Publications No. 8)*, Lyon, International Agency for Research on Cancer, pp. 249-258

Le Bouffant, L., Bruyère, S., Martin, J.C., Tichoux, G. & Normand, C. (1976) Some observations on asbestos fibres and other mineral particles found in the lungs of asbestosis patients [in French]. *Rev. Fr. Mal. Respir.*, 4 (Suppl. 2), 121-140

Leineweber, J.P. (1984) Solubility of fibres *in vitro* and *in vivo*. In: *Biological Effects of Man-made Mineral Fibres*, Vol. 2, Copenhagen, World Health Organization, Regional Office for Europe, pp. 87-102

Middleton, A.P., Beckett, S.T. & Davis, J.M.G. (1977) A study of the short-term retention and clearance of inhaled asbestos using UICC standard reference samples. In: Walton, W.H., ed., *Inhaled Particles, IV*, Oxford, Pergamon Press, pp. 247-257

Ontario Royal Commission (1984) *Report of the Royal Commission on Matters of Health and Safety Arising from the Use of Asbestos in Ontario*, Toronto, Ontario Government Bookshop, pp. 565-566

Pitt, M.L.M. (1985) *Alterations in pleural cavity cell populations in mice exposed to asbestos by inhalation*, Ph.D. thesis, University of Witwatersrand

Pooley, F.D. (1976) An examination of the fibrous mineral content of asbestos lung tissue from the Canadian chrysotile mining industry. *Environ. Res.*, 12, 281-298

Pott, F. (1978) Some aspects on the dosimetry of the carcinogenic potency of asbestos and other fibrous dusts. *Staub-Reinhalt. Luft*, 38, 486-490

Pott, F. (1983) Animal experiments with mineral fibres. In: *Short and Thin Mineral Fibres. Proceedings of a Symposium, Stockholm 1982*, Arbete och Hälsa, pp. 133-161

Pott, F. & Friedrichs, K.H. (1972) Tumours in rats after intraperitoneal injection of asbestos dusts. *Naturwissenchaften*, 59, 318-332

Saracci, R. (1986) Contributions to the IARC study on mortality and cancer incidence among man-made mineral fibre production workers. *Scand. J. Work Environ. Health*, 12 (Suppl. 1), 5-85

Stanton, M.F. & Layard, M. (1978) The carcinogenicity of fibrous materials. In: Gravatt, C.C., La Fleur, P.D. & Heinrich, K.F.J., eds, *Workshop on Asbestos: Definition and Measurement Methods*, Washington DC, National Measurement Laboratory (*National Bureau of Standards Special Publication 506*), pp. 143-151

Stanton, M.F. & Wrench, C. (1972) Mechanisms of mesothelioma induction with asbestos and fibrous glass. *J. Natl Cancer Inst.*, 48, 797-821

Stanton, M.F., Layard, M., Tegeris, A., Miller, M., May, M. & Kent, E. (1977) Carcinogenicity of fibrous glass: pleural response in the rat in relation to fibre dimension. *J. Natl Cancer Inst.*, 58, 587-603

Storeygard, A.R. & Brown, A.L. (1977) Penetration of the small intestinal mucosa by asbestos fibres. *Mayo Clin. Proc.*, 52, 809-812

Tilkes, F. & Beck, E.G. (1983) Macrophage functions after exposure to mineral fibres. *Environ. Health Perspect.*, 51, 67-72

Timbrell, V. (1973) Physical factors as etiological mechanisms. In: Bogovski, P., Gilson, J.C., Timbrell, V. & Wagner, J.C., eds, *Biological Effects of Asbestos (IARC Scientific Publications No. 8)*, Lyon, International Agency for Research on Cancer, pp. 295-304

Vincent, J.H., Jones, A.D., Johnston, A.M., McMillan, C., Bolton, R.E. & Cowie, H. (1987) Accumulation of inhaled mineral dust in the lung and associated lymph nodes: implications for exposure and dose in occupational lung disease. *Ann. Occup. Hyg., 31,* 375-393

Wagner, J.C. & Skidmore, J.W. (1965) Asbestos dust deposition and retention in rats. *Ann. N.Y. Acad. Sci., 132,* 77-86

Wagner, J.C., Berry, G. & Timbrell, V. (1973) Mesothelioma in rats after inoculation with asbestos and other materials. *Br. J. Cancer, 28,* 173-185

Wagner, J.C., Berry, G., Skidmore, J.W. & Timbrell, V. (1974) The effects of the inhalation of asbestos in rats. *Br. J. Cancer, 29,* 252-269

Wagner, J.C., Skidmore, J.W., Hill, R.T. & Griffiths, D.M. (1985) Erionite exposure and mesothelioma in rats. *Br. J. Cancer, 51,* 727-730

Wright, G.W. & Kuschner, M. (1977) The influence of varying lengths of glass and asbestos fibres on tissue response in guinea pigs. In: Walton, W.H., ed., *Inhaled Particles, IV,* Oxford, Pergamon Press, pp. 455-472

# RECENT RESULTS OF CARCINOGENICITY BIOASSAYS OF FIBRES AND OTHER PARTICULATE MATERIALS

## C. Maltoni & F. Minardi

*Institute of Oncology, Bologna, Italy*

*Summary.* Different types of natural, commercial and modified asbestos fibres were tested in a highly standardized manner by peritoneal injection into rats and mice in order to assess their carcinogenicity.

Differences in mesotheliomatogenic effect were found between the various materials tested. Of particular interest is the finding that treatment of the fibres with phosphorus oxychloride and heating to 300°C markedly reduces the carcinogenicity of chrysotile fibres.

## *Introduction*

A systematic and integrated study involving long-term experimental bioassays on particulate materials was started in January 1981 at the Bologna Institute of Oncology and is still in progress. The study covers a variety of fibrous and non-fibrous, natural and man-made materials, present in the occupational and/or general environment. Some of the materials studied are of major industrial importance.

The study is aimed at:

— identifying new potentially carcinogenic materials;
— assessing, in quantitative terms, the level of carcinogenic risk of a given material, and comparing the risks represented by different materials (assessment of the relative carcinogenic risk);
— helping to predict the target organs;
— defining the role in carcinogenesis of the physical and chemical characteristics of the test compounds.
— determining the role of different biological and experimental factors affecting the neoplastic response and, consequently, shedding some light on the pathogenesis of the possible oncogenic effects;
— helping to reconstruct the natural history of the tumours which may be induced by the test compounds.

Information on test materials and animals and the experimental procedures is given in Table 1. In view of the aims of the study, the experimental conditions are strictly standardized.

## Table 1. Test materials, animals, method of treatment and experimental procedure

### Test materials

Different types of natural, modified natural and man-made materials including: crocidolite (UICC), chrysotile (Canada, UICC), chrysotile (Rhodesia, UICC), chrysotile (California), amosite (UICC), anthophyllite (UICC), commercial chrysotiles (3 samples), modified chrysotiles (4 types), asbestos-cement, crystalline silica, amorphous silica, alumina, wollastonite, talc (2 samples), kaolin, bentonite, natural zeolites (13 minerals), man-made zeolites (20 types), rock wool (4 samples), carbon fibres, synthetic fibres (2 types).

### Test animals

Groups of 40-100 (20-50 males and 20-50 females), 6-8-week-old Sprague-Dawley rats. For erionite and modified chrysotiles, also groups of 40 (20 males and 20 females) 8-week-old Swiss mice. A total of 10 760 rats and 900 mice have been tested.

### Method of treatment

Intraperitoneal (i.pe.) (for all materials apart from carbon fibres), intrapleural (i.pl.) and subcutaneous (s.c.) (for most of the materials) injection, and ingestion (gavage) and intratracheal instillation (in the case of detergent 4A zeolite).

### Experimental procedure

The animals are kept under observation (body weight and clinical controls) until they die naturally or until 2 years from the beginning of the experiments. All animals undergo complete autopsy and systematic histopathological examination.

---

A summary of the experimental work so far completed or in progress is given in Table 2.

In this paper the most recent results are presented of the bioassays of several types of natural, commercial and modified asbestos, administered as a single intraperitoneal (i.pe.) injection, to rats and, in a few instances, to mice.

The modified asbestos consisted of Canadian chrysotile of different fibre lengths and from different sources, treated with phosphorus oxychloride and heated at 300°C. This type of treatment was used in order to bring about physicochemical changes which might reduce the carcinogenicity of the natural material.

In rats, the peritoneum proved to be more responsive than the pleura to the mesotheliomatogenic effect of asbestos; this is the opposite of what is observed with erionite (Table 3).

The complete final results of the basic experiments on the effects of the asbestos materials studied in rats at a single dose of 25 mg, are available. Data on the effects of some forms of modified asbestos, as single administered doses of 10, 5 or 1 mg, to rats and mice, and of 25 mg to mice, are preliminary and based solely on gross examination, since the experiments were started only 76 weeks ago.

**Table 2. Summary of experimental work so far completed or in progress**

| Materials | Injection | | | Ingestion and intratracheal instillation |
|---|---|---|---|---|
| | i.pe. | i.pl. | s.c. | |
| Crocidolite (UICC) | + | + | + | |
| Chrysotile (Canada, UICC) | + | + | + | |
| Chrysotile (Rhodesia, UICC) | + | | | |
| Chrysotile (California) | + | | | |
| Amosite (UICC) | + | | | |
| Anthophyllite (UICC) | + | | | |
| Commercial chrysotiles (3 types)[a] | + | + | | |
| Modified chrysotiles (4 types)[a] | + | + | | |
| Asbestos-cement | + | + | + | |
| Crystalline silica | + | | + | |
| Amorphous silica | + | | + | |
| Alumina | + | + | + | |
| Wollastonite | + | + | + | |
| Talc (2 samples) | + | + | + | |
| Kaolin | + | | + | |
| Bentonite | + | + | | |
| Natural zeolites (13 minerals) | + | + | + | |
| Man-made A zeolites (2 types) | + | + | + | + |
| Man-made X zeolite | + | + | + | |
| Man-made Z zeolite (17 types) | + | + | + | |
| Rock wool (4 samples) | + | + | | |
| Carbon fibres[b] | | | + | |
| Synthetic fibres (2 types) | + | + | + | |
| Water (controls) | + | + | + | |

[a]Experiments still in progress
[b]Implantation.

**Table 3. Final results of tests on crocidolite, chrysotile (Canada) and sedimentary erionite[a]**

| Material | Peritoneal mesotheliomas | | | Pleural mesotheliomas | | |
|---|---|---|---|---|---|---|
| | Tumour-bearing animals | | Average latency time (weeks) | Tumour-bearing animals | | Average latency time (weeks) |
| | No. | % | | No. | % | |
| Crocidolite | 39 | 97.5 | 59.5 | 18 | 45.0 | 104.8 |
| Chrysotile (Canada) | 32 | 80.0 | 92.2 | 26 | 65.0 | 111.1 |
| Sedimentary erionite | 20 | 50.0 | 106.1 | 35 | 87.5 | 64.2 |
| Water (controls) | 0 | – | – | 0 | – | – |

[a]Sprague-Dawley rats (20 males and 20 females) were given a single intraperitoneal and intrapleural injection of the material (25 mg in 1 ml of water) and kept under observation for their full lifespan.

## Materials and methods

The asbestos materials, obtained from Mount Sinai School of Medicine (New York, USA) (natural asbestos), Associazione Cemento-Amianto (Balangero, Italy) (asbestos-cement) and CERAM-SNA (Canada) (commercial and modified asbestos), were injected into the peritoneal cavity of 8-week-old Sprague-Dawley rats and Swiss mice. Groups of 40 animals (20 males and 20 females) were used for each species, material and dose level.

The animals were examined 3 times daily for general behaviour and were weighed and examined for gross changes every 2 weeks.

m In experiment BT 2101 (Table 4), the animals were kept under observation for their full lifespan. In experiment BT 2106 (Table 5), the animals were kept for up to 104 weeks, at which time the survivors were sacrificed. Experiments BT 2111 (Table 6) and BT 2112 (Table 7) will also be terminated at 104 weeks.

**Table 4. Final results of tests on various types of natural asbestos and asbestos-cement (experiment BT 2101)**[a]

| Material | Peritoneal mesotheliomas | | |
|---|---|---|---|
| | Tumour-bearing animals | | Average latency time (weeks) |
| | No. | % | |
| Crocidolite (UICC) | 39 | 97.5 | 59.5 |
| Chrysotile (Canada, UICC) | 32 | 80.0 | 92.2 |
| Chrysotile (Rhodesia, UICC) | 33 | 82.5 | 89.7 |
| Chrysotile (California) | 29 | 72.5 | 85.3 |
| Amosite (UICC) | 36 | 90.0 | 66.7 |
| Anthophyllite (UICC) | 35 | 82.5 | 73.3 |
| Asbestos-cement | 21 | 52.5 | 99.7 |
| Water (controls) | 0 | – | – |

[a]Sprague-Dawley rats (20 males and 20 females) were given a single intraperitoneal injection of the material (25 mg in 1 mg of water) and kept under observation for their full lifespan.

A complete autopsy is performed on all animals, whether dying naturally or sacrificed. A histopathological examination is carried out on the peritoneum (site of injection), brain, thymus, lungs, liver, spleen, kidneys, adrenals, stomach, uterus, gonads, mesenteric, mediastinal and subcutaneous lymph-nodes, and all pathological organs and tissues.

The strictly standardized experimental conditions and procedures allow quantitative comparisons between different groups in the same experiment and between different experiments.

**Table 5. Final results of tests on various types of chrysotile and other non-asbestos fibres (experiment BT 2106)**[a]

| Material | | Treatment | Peritoneal mesotheliomas | | |
|---|---|---|---|---|---|
| Code | Type | | Tumour-bearing animals | | Average latency time (weeks) |
| | | | No. | % | |
| Chr 1 | Paperbestos 5 – 100 M + 200 M[b] | – | 33 | 82.5 | 78.3 |
| Chr 2 | Paperbestos 5 – 100 M + 200 M[b] | POCl$_3$ + 300°C | 17 | 42.5 | 88.6 |
| Chr 3 | Paperbestos 5 – 100 M + 200 M from asbestos-latex paper[b] | – | 17 | 42.5 | 87.5 |
| Chr 4 | Paperbestos 5 – 100 M + 200 M from asbestos-latex paper[b] | POCl$_3$ + 300°C | 15 | 37.5 | 94.7 |
| Chr 5 | Short chrysotile 7 (by water fractionation of Paperbestos 5)[b] | – | 29 | 72.5 | 76.4 |
| Chr 6 | Short chrysotile 7 (by water fractionation of Paperbestos 5)[b] | POCl$_3$ + 300°C | 3 | 7.5 | 97.6 |
| Chr 7 | Chrysotile (Canada, UICC) | – | 30 | 75.0 | 76.1 |
| Chr 8 | Chrysotile (Canada, UICC) | POCl$_3$ + 300°C | 20 | 50.0 | 90.0 |
| 9 | Rock-wool fibres | – | 3 | 7.5 | 79.8 |
| 10 | Kevlar fibres | – | 0 | – | – |
| 11 | Water (controls) | – | 0 | – | – |

[a]Sprague-Dawley rats (20 males and 20 females) were given a single intraperitoneal injection of one material (25 mg in 1 ml of water) and kept under observation for 104 weeks.
[b]Supplied by IRDA.

**Table 6. Gross findings after 76 weeks in tests on various types of chrysotile and other non-asbestos fibres (experiment BT 2111)**[a]

| Material | | Treatment | Dose (mg) | Peritoneal mesotheliomas | | |
|---|---|---|---|---|---|---|
| Code | Type | | | Tumour-bearing animals | | Average latency time (weeks) |
| | | | | No. | % | |
| Chr 5 | Short chrysotile 7 (by water fractionation of Paperbestos 5)[b] | - | 10 | 11 | 27.5 | 57.0 |
| | | | 5 | 4 | 10.0 | 59.7 |
| | | | 1 | 0 | - | - |
| Chr 6 | Short chrysotile 7 (by water fractionation of Paperbestos 5)[b] | $POCl_3$ + 300°C | 10 | 0 | - | - |
| | | | 5 | 0 | - | - |
| | | | 1 | 0 | - | - |
| Chr 7 | Chrysotile (Canada, UICC) | - | 10 | 14 | 35.0 | 61.6 |
| | | | 5 | 11 | 27.5 | 60.3 |
| | | | 1 | 0 | - | - |
| Chr 8 | Chrysotile (Canada, UICC) | $POCl_3$ + 300°C | 10 | 3 | 7.5 | 61.0 |
| | | | 5 | 1 | 2.5 | 67.0 |
| | | | 1 | 1 | 2.5 | 64.0 |
| 9 | Fiberfrax | - | 10 | 5 | 12.5 | 65.2 |
| | | | 5 | 0 | - | - |
| | | | 1 | 0 | - | - |
| 10 | Kevlar fibres | - | 10 | 0 | - | - |
| | | | 5 | 0 | - | - |
| | | | 1 | 0 | - | - |
| 11 | Water (controls) | - | 0 | 0 | - | - |

[a]Sprague-Dawley rats (20 males and 20 females) were given a single intraperitoneal injection of the material (in 1 ml of water) and kept under observation for 104 weeks.
[b]Supplied by IRDA.

The bioassays of experiment BT 2106 (the first on modified asbestos) were performed in such a way that the nature of the materials tested was disclosed only when the biological data had already been obtained.

**Table 7. Gross findings after 76 weeks in tests on various types of chrysotile (experiment BT 2112)**[a]

| Material | | Treatment | Dose (mg) | Peritoneal mesotheliomas | | |
|---|---|---|---|---|---|---|
| Code | Type | | | Tumour-bearing animals | | Average latency time (weeks) |
| | | | | No. | % | |
| Chr 5 | Short chrysotile 7 (by water fractionation of Paperbestos 5)[b] | - | 25 | 6 | 15.0 | 65.7 |
| | | | 10 | 3 | 7.5 | 64.7 |
| | | | 5 | 1 | 2.5 | 71.0 |
| | | | 1 | 0 | - | - |
| Chr 6 | Short chrysotile 7 (by water fractionation of Paperbestos 5)[b] | $POCl_3$ + 300° C | 25 | 0 | - | - |
| | | | 10 | 0 | - | - |
| | | | 5 | 0 | - | - |
| | | | 1 | 0 | - | - |
| Chr 7 | Chrysotile (Canada, UICC) | - | 25 | 6 | 15.0 | 58.5 |
| | | | 10 | 5 | 12.5 | 61.0 |
| | | | 5 | 0 | - | - |
| | | | 1 | 0 | - | - |
| Chr 8 | Chrysotile (Canada, UICC) | $POCl_3$ + 300° C | 25 | 3 | 7.5 | 70.0 |
| | | | 10 | 1 | 2.5 | 56.0 |
| | | | 5 | 0 | - | - |
| | | | 1 | 0 | - | - |
| 11 | Water (controls) | - | 0 | 0 | - | - |

[a] Swiss mice (20 males and 20 females) were given a single intraperitoneal injection of the material (in 1 ml of water) and kept under observation for 104 weeks.

[b] Supplied by IRDA.

## Results

The results of the experiments are presented in Tables 4–7.

The bioassays of different natural asbestos fibres and of asbestos-cement, at a dose of 25 mg, show that the mesotheliomatogenic effect (evaluated on the basis of the incidence and latency time of tumours) of the materials tested decreases (Table 4) in the order:

(1) crocidolite;
(2) amosite;
(3) anthophyllite;
(4) chrysotiles;
(5) asbestos-cement.

The chemical and heat treatment of asbestos fibres, tested at a dose of 25 mg, markedly reduces the carcinogenicity of 'short chrysotile 7' and to a lesser extent 'Paperbestos 5', while that of 'UICC Canada chrysotile' is reduced still less. The carcinogenic potency of fibres from 'asbestos-latex-paper' is lower than that for 'Paperbestos 5' and seems little affected by the chemical and heat treatment. These results may be due to the insulating effect of latex (Table 5).

The reduction in carcinogenic potency of 'short chrysotile 7' and also of 'UICC Canada chrysotile' seems also to be confirmed at the present stage of the study by the preliminary incomplete results of experiments BT 2111 and BT 2112, in which the materials are being tested at various doses on rats and mice (Tables 6 and 7).

## Conclusions

The data reported here seem to indicate that experimental long-term bioassays may be used as a tool for quantifying the carcinogenic potency of particulate materials and, in the present case, specifically of different types of asbestos.

The findings presented may provide the basis for further analytical studies aimed at clarifying the possible mechanisms of asbestos carcinogenesis. They are also of technological interest; however, at present, it is difficult to predict their implications for the asbestos industry and, in general, for the manufacture and use of other fibres.

## Acknowledgements

The study was supported in part by the EEC (Contracts Nos ENV/347/I-(S) and ENV/755/I) and in part by the Institut de Recherche et de Développement sur l'Amiante, Canada

# PARTICULATE-STATE CARCINOGENESIS: A SURVEY OF RECENT STUDIES ON THE MECHANISMS OF ACTION OF FIBRES

## M.C. Jaurand

*INSERM U.139, CHU Henri Mondor, Créteil, France*

*Summary.* Animal experiments using, in addition to asbestos, erionite and man-made mineral fibres, have confirmed that fibre dimension is an important factor in the carcinogenic potency of fibres. However, it seems to be established that it is not the only parameter of importance and that fibre type and physicochemistry play a role. *In vitro* experiments have provided new information on the relationship between fibre size and effect: long, thin fibres are more effective in inducing the transformation of certain cell types and in producing injury arising from oxygen species.

Different *in vitro* experiments have tested chromosomal and genetic damage induced by fibres, especially asbestos. Depending on the cell type, chromosomal damage and aneuploidy have been observed and interactions with the genomic material have been demonstrated. Some conflicting data in this area might be explained by differences in cell type used or in methods used to prepare the asbestos samples.

At present, a working hypothesis for the mechanisms of carcinogenesis induced by fibres can be suggested. Mineral or synthetic fibres deposited in the lung are first processed by macrophages, which eliminate short fibres. The remaining fibres may be ingested by cells that are potentially transformable, the longest fibres possibly being phagocytosed preferentially. Ingested fibres may produce DNA damage in both resting and dividing cells, either by the direct production of oxygen radicals, by the formation of clastogenic factor or by direct chromosome interaction in cells in mitosis.

A missegregation of chromosomes can result from interactions between fibres and the mitotic spindle. Even if these processes take place slowly, they can occur over a long period of time due to the persistence of the fibres in the tissue. The carcinogenic potency of a fibre, therefore, will be dependent on its durability, its dimensions and its physicochemical properties.

Asbestos fibres appear to be pluripotent in inducing chromosome abnormalities as well as some responses shared with compounds such as promoters. Thus, these fibres can be considered as potent complete carcinogens. The potency of a given fibre in a specific target tissue will be the sum of both its initiating and its promoting effects.

It seems that new concepts must be formulated to account for the mechanisms of action of fibres. 'Particulate state' carcinogenesis includes the possibility of multiple hits occurring during the time that the particles remain within the target tissue and will therefore be time-dependent.

## Introduction

Numerous studies have been carried out to investigate the mechanisms of action of either natural or man-made fibres. The data can be found in the literature related to toxicology, lung pathology, cancer, biophysics, biochemistry and physical chemistry, showing the wide range of research needed to study the toxicology of these particles.

The historical development of experimental studies can be divided into three parts. In the mid-1960s, several authors studied the carcinogenic and fibrogenic potency of asbestos fibres in the lung in animal experiments (Holt et al., 1965; Wagner & Berry, 1969). In vitro, fibre cytotoxicity was determined mainly using several types of macrophages and red blood cells (Koshi et al., 1968; Pernis et al., 1966; Schnitzer & Pundsack, 1970; Selikoff & Churg, 1965). Later, the biological effects of ingested asbestos in animals were studied and in vitro experiments were performed with other cell types, such as fibroblasts, pleural mesothelial cells and epithelial cells from the lung or from the gastrointestinal tract (Brown et al., 1980; Lee, 1974). From the beginning, the various authors were concerned with the intrinsic fibre parameters capable of explaining the cellular reactions (Beck et al., 1971; Hayashi, 1974; Schnitzer, 1974; see Langer, 1985, for review) or with the possible interactions between asbestos and other factors, such as oils or benzo[a]pyrene (BP) (Harington & Roe, 1965; Miller et al., 1965).

During the past decade, studies of cellular responses of cultured cells to fibres have been carried out (Beck & Bignon, 1985; Second International Workshop on the *In Vitro* Effects of Mineral Dusts, 1983; Summary Workshop on Ingested Asbestos, 1983). These studies were carried out to elucidate the mechanisms of cytotoxicity, fibrogenicity and carcinogenicity of fibres. They have emphasized the role of the fibre parameters involved in cellular responses (see Langer, 1985, for review) and have shown that the multiple biological effects of fibres depend on their physical structure and physicochemical properties.

Recently, significant new studies have been reported dealing with the events required for the neoplastic conversion of normal cells treated with fibres, as well as the cellular events important in fibrogenesis (e.g., release of inflammatory factors and of growth factors, cell interactions). This review will focus on new information from these recent experiments dealing with the mechanisms of action of fibres in inducing cytotoxicity and carcinogenesis.

In order to understand the mechanism of action of fibres, the following three questions must be answered. What intrinsic properties of fibres are responsible for toxicity? Why do these properties act as they do? How do they exert their effects?

In general, the direct effects of fibres on cells have been studied. However, it must be noted that experiments have shown that intratracheal instillation of amosite into hamsters resulted in early morphological changes in pleural mesothelial cells in the absence of direct fibre contact, suggesting a distance effect (Dodson & Ford, 1985). The significance of these indirect effects remains to be determined.

## Fibre parameters involved in the mechanism of action of fibres

### Physical state

From results reported previously, it is generally accepted that fibre shape and size are both important parameters in fibre-related carcinogenesis. The effect of size is extensively discussed in this volume by J.M.G. Davis (p. 33-45), but some data of interest relating to the presumed mechanisms of carcinogenesis will be reported here.

Animal experiments have recently shown a high incidence of mesotheliomas in rats (about 100%) following intrapleural inoculation of zeolite (Wagner et al., 1985), thus confirming previous data obtained after intraperitoneal injection (Suzuki & Kohyama, 1984). It is noteworthy that the percentage of thin (diameter <0.25 $\mu$m) and long (length >8 $\mu$m) fibres, suggested by Stanton et al. (1981) as being the most carcinogenic, was very low in these two animal experiments. According to Stanton et al. (1987), a high correlation between carcinogenic potency and size was found with fibres of length greater than 4 $\mu$m and diameter up to 1.5 $\mu$m. Even when these parameters were taken into consideration, the percentage of fibres in this size range remained low with one type of zeolite (about 10%). In addition, non-fibrous zeolite induced mesothelioma following intrapleural inoculation (5%) or inhalation (~4%) (Wagner et al., 1985).

Extensive studies have also been performed using man-made mineral fibres. A carcinogenic effect was found with glass microfibres containing a low percentage of long, thin fibres (Wagner et al., 1984). It is interesting to note that the number of particles in this sample was one-quarter of that in the chrysotile sample used as a positive control. In addition, the number of fibres was much lower, since only 56% of the particles were fibres, as compared with 96% in the chrysotile sample. These findings make the data difficult to interpret for purposes of comparison of the carcinogenic potency of the two samples. If account is taken of the effects of a given amount of fibres of a given size range, glass fibres might therefore be more active on a per fibre basis than chrysotile fibres by this route of exposure.

On the other hand, Pott et al. (1984) reported that slag wool which contained significant numbers of long thin fibres (90% of the fibres were of diameter <0.28 $\mu$m and 10% of the fibres of length >10 $\mu$m) did not produce a significant incidence of mesotheliomas following intrapleural injection in rats.

Overall, these results seem to indicate that fibre dimension is not the only parameter playing a role in carcinogenesis or that only a very small number of fibres satisfying certain criteria are required to induce mesothelioma.

Other recent data have confirmed Stanton's hypothesis as to the higher carcinogenic potency of long fibres as compared with shorter fibres (Stanton et al., 1977, 1981). Pott et al. (1984) found that JM 100 glass fibres, which are shorter and thicker than JM 104, induced fewer peritoneal mesotheliomas after intraperitoneal injection in rats. Davis et al. (1984) showed that a fibrous ceramic aluminium silicate glass containing relatively few long fibres produced a low rate of mesothelioma. It has also been reported that an attapulgite sample with a mean fibre length of less than 1 $\mu$m did not induce mesothelioma (Jaurand et al., 1987a). However, while there does seem to be a relationship between fibre dimension and carcinogenicity for a given type of fibre, it appears that, for different fibres, the same number within the same size range can have a different carcinogenic potential. This indicates a role not only for fibre size but also for other fibre properties. In a recent animal experiment, it was demonstrated in our laboratory that 3 chrysotile samples of different size did not have the same carcinogenic potency following intrapleural inoculation in rats (Jaurand et al., 1987a). A short sample (mean length 1.25 $\mu$m) induced 50% fewer mesotheliomas than another sample with longer fibres (mean length 3.21 $\mu$m). When the results obtained from this experiment and from other experiments previously reported (Monchaux et al., 1981) were analysed, the data for chrysotile did not give a good fit with Stanton's curve of mesothelioma incidence, although, for amphiboles, the data were more or less consistent with this curve.

The mechanisms whereby 'long' fibres are more potent than 'short' ones are not clear, but some hypotheses can be proposed. Stanton et al. (1977) suggested that the lower carcinogenic potency of short fibres might be due to their preferential engulfment by macrophages. Recently, Morgan and Holmes (1984), studying the clearance of glass fibres in rats, confirmed that fibres of length 5 $\mu$m are cleared quite efficiently. In these experiments, the critical length above which clearance was impaired was >10 $\mu$m and <30 $\mu$m. Hesterberg et al. (1986) reported that Syrian hamster embryo (SHE) cells in vitro, which can be transformed after fibre treatment, concentrated the longest fibres. One possible mechanism is therefore the clearing of short, less active, fibres by macrophages and the ingestion of long, more active, fibres by other cells (possibly the replicative target cells). This could explain the observation of a progressive increase in the mean length of fibres retained in the lung tissue, with both chrysotile (Roggli & Brody, 1984) and crocidolite (Roggli et al., 1987), as well as the finding of McFadden et al. (1986) that cigarette smoking increases the retention of short fibres, presumably as a result of damage to macrophage-mediated asbestos clearance.

The higher potency of long as compared with short fibres may be related to the physical or physicochemical properties of the fibres. Hesterberg et al. (1986) demonstrated that long glass fibres induced more transformants in SHE cell cultures than short ones, possibly due to the physical impairment of chromosome segregation, as will be discussed later. Mossman et al. (1986) observed that long fibres produced oxygen species in hamster epithelial cells, presumably by means of a cell-derived mechanism or one involving the physicochemical state of the fibres (Zalma et al., 1987), while short fibres did not.

These new experiments confirm previous results indicating that, while the carcinogenic potency of fibres is size-dependent, other parameters such as their chemistry and physical chemistry, may also be of importance. The important new data will be discussed below.

New experiments have also been designed to study the in vitro effects of fibres of differing size on cultured cells. Brown et al. (1986) have compared the activities of 13 carcinogenic fibrous samples, using A549 cells and V79/4 cells to assess cytotoxicity. They found a relationship between fibre length or diameter and cytotoxicity. Hesterberg and Barrett (1984) have demonstrated that long fibres were more cytotoxic and more transforming for SHE cells than short fibres. In the same way, thin fibres were more potent than coarse fibres, on a per weight basis, when the same end-points were studied. The results of these *in vitro* experiments are, therefore, in agreement with those obtained *in vivo*.

**Physical chemistry**

Several epidemiological studies have suggested that amphiboles are more carcinogenic in humans than chrysotile (Acheson *et al.*, 1982). One important parameter which can partly explain the differences in the human carcinogenic potency of chrysotile versus amphiboles is their durability in lung tissue. Using intratracheal instillations in hamsters, Kimizuka *et al.* (1987) confirmed previous data on the higher stability of amphiboles as compared with chrysotile. This does not necessarily imply that chrysotile has a lower carcinogenic risk than amphiboles, since the fibres may exert their effect in the early stages of carcinogenesis. In experiments with several types of man-made mineral fibres, it was confirmed that the durability of the fibres differed substantially, depending on the nature of the fibre. In several experiments, the effects of glass fibres, slag wool, and rock wool were compared (WHO/IARC, 1984). Slag-wool fibres were much more sensitive than the other fibres to chemical degradation in the presence of biological fluids. Pott *et al.* (1984) reported that chemically treated glass fibres gave a reduced incidence of mesothelioma. In these experiments, the authors studied the durability of different fibre types for over 4 years after intraperitoneal injection in rabbits. Glass fibres were relatively stable, while slag wool showed visible corrosion; it was therefore concluded that slag wool collapsed fairly quickly. This fact may be important in explaining the failure of these fibres to induce mesothelioma.

Morgan and Holmes (1984) studied the durability of glass fibres in rat lungs. They reported that the dissolution rate depended on the fibre size. Within months, long glass fibres (60 $\mu$m) exhibited a reduction in their diameter, while short fibres (5 $\mu$m and 10 $\mu$m) were only slightly solubilized. Not all the fibre types had the same solubility; thus rock wool was not solubilized. Fibre size modification was also reported by Wagner *et al.* (1984), who found that the diameter of a glass wool sample was reduced in rat lungs. It is assumed that, for a given fibre type, the fibre dimensions play an important role in carcinogenesis, but it should be noted that this parameter can change with time, depending on the physicochemical properties determining the stability of that type.

If it is accepted that the size, chemistry, and surface state of a fibre affect its carcinogenicity, the question arises as to how these parameters are involved. Data related to the cellular effects of fibres relevant to carcinogenesis provide some information on this matter.

## *Genotoxicity of fibres* in vitro

Until recently, it was believed that, as shown by bacterial assays, asbestos fibres were not mutagenic (Chamberlain & Tarmy, 1977). Chromosome damage was reported with some cell types but not with others, and the genotoxicity of asbestos fibres was a subject of controversy. Further studies have now been carried out using *in vitro* cell cultures and human biological samples. Recent data on the chromosome analysis of human mesothelial cells from tumours have indicated that some chromosome changes may be non-random (Gibas *et al.*, 1986). Popescu *et al.* (1987) reported changes involving chromosome 3 (non-random deletion) in 7 out of 9 cases of human mesothelioma, some of them being asbestos-related. The genotoxicity of mineral fibres is therefore of particular interest.

## Numerical chromosome changes induced by fibres

Microtubule polymerization and condensation are responsible for the orderly segregation of chromosomes between daughter cells. Interference with cytoskeletal components injures the dividing cell and may produce aneuploidy. The normal equal distribution of chromosomes is the result of several events, including microtubule elongation from two centrosomes (Mazia, 1984), chromosome attachment to microtubules (Nicklas *et al.*, 1982), and orderly arrangement of the kinetochore microtubules. The presence of mineral or synthetic fibres within a mitotic cell might impair the normal operation of cell division, resulting in aneuploidy.

Recently, Hesterberg and Barrett (1985) reported that asbestos fibres induced anaphase abnormalities in crocidolite-treated Syrian hamster embryo cells. Oshimura *et al.* (1984, 1986) also observed that asbestos induced chromosomal changes, including aneuploidy and polyploidy, in Syrian hamster embryo cells. A non-random numerical change was observed in 6 out of 8 asbestos-induced immortal cell lines at the earliest passage examined. This chromosome change may be of importance in asbestos-related cell transformation (Oshimura *et al.*, 1986).

The induction of aneuploidy and polyploidy was reported several years ago (Lavappa *et al.*, 1975; Sincock *et al.*, 1982; Sincock & Seabright, 1975) with Chinese hamster cells. Similar results have been recently observed with rat pleural mesothelial cells (Jaurand *et al.*, 1983a, 1986a), human mesothelial cells (Lerchner *et al.*, 1985) and rat tracheal epithelial cells (Hesterberg *et al.*, 1987) treated with asbestos, as well as in CHO cells treated with crocidolite or erionite (Kelsey *et al.*, 1986) and Chinese hamster cells exposed to erionite or to asbestos (Palekar *et al.*, 1987). In contrast, with bronchial human epithelial cells, there was no significant induction of numerical chromosome changes by asbestos (Kodama *et al.*, 1987).

The mechanism whereby asbestos fibres induce these chromosome changes may be related to the particle shape, size and to some physicochemical properties of asbestos, such as its adsorption potency. The presence of a fibrous solid structure in the dividing cell may interfere with the movement of the chromosomes and induce aneuploidy as well as polyploidy. In addition, asbestos fibres, especially chrysotile, are capable of adsorbing a number of proteins (Jaurand *et al.*, 1983a; Morgan, 1974), as well as DNA (Touray *et al.*, 1987), and this fact may help to explain the interactions with microtubules or chromosomes. The ability of asbestos fibres to adsorb membranes may also be of importance since the mitotic apparatus contains a mass of membranes (Mazia, 1984). Direct interaction between chromosomes from rat pleural mesothelial cells and asbestos fibres has recently been reported (Wang *et al.*, 1987); chromosome damage was more frequently observed with chrysotile than with crocidolite, in agreement with the respective adsorption capacity of these fibres.

It should be noted that, in experiments on the induction of numerical chromosome changes in which chrysotile was compared with amphiboles, the effects of chrysotile were always greater than those of amphiboles on a per weight basis. This may be related to the different numbers of fibres in the samples, but could also be related to the different numbers of fibres ingested by the cells, as recently reported by Hesterberg *et al.* (1987) with rat tracheal epithelial cells; more intracellular chrysotile fibres were found than crocidolite fibres.

Together with shape and physicochemical properties, fibre size may also be important from the point of view of effects on chromosomes. It is conceivable that a long fibre will have a higher probability of interacting with the chromosomes or with components of the spindle apparatus than a short one. It is therefore of interest that Hesterberg *et al.* (1986) have shown that the mean length of glass fibres ingested by SHE cells was higher than that of the original sample, indicating a preferential phagocytosis of the longest fibres. It should also be noted that the aneuploidy-inducing capacity was substantially higher for a sample of long fibres than for shorter fibres when equal numbers of intracellular fibres per cell were present.

**Clastogenic effects of fibres**

The clastogenic effect of asbestos fibres has previously been reported using different cell types, including CHO (Babu *et al.*, 1980; Sincock *et al.*, 1982; Sincock & Seabright, 1975), human lymphocyte cells (Valerio *et al.*, 1983) and Syrian hamster embryo cells, where the incidence of chromosomal aberrations was low (Lavappa *et al.*, 1975). The structural changes consisted of chromosome as well as chromatid damage, including breaks and fragments.

Recent studies have confirmed the clastogenicity of asbestos in CHO cells (Kelsey *et al.*, 1986) and lung hamster cells (Palekar *et al.*, 1987); in addition, erionite was also found to be clastogenic. Chrysotile fibres were also clastogenic for rat pleural mesothelial cells (Jaurand *et al.*, 1986a). Weak clastogenicity was reported in SHE cells treated with chrysotile (Oshimura *et al.*, 1984) and in human bronchial epithelial cells, but neither crocidolite nor fibreglass significantly enhanced the incidence of chromosome aberrations in SHE cells. No clastogenic effect was observed by Hei *et al.* (1985) in C3H 10T½ cells treated with amphiboles.

The differences observed may be due to the cell type as well as to the fibre sample used in these experiments. Sincock *et al.* (1982) found that primary human fibroblasts or human lymphoblastoid lines, in contrast to Chinese hamster ovary cells, failed to show chromosome aberrations and polyploidy following treatment with asbestos or glass fibres. These authors thought that this might be due both to a greater proficiency of excision repair following DNA damage of human cells, and to a difference in the relative uptake of fibres by the cells. It is well known that, at least in experimental animals, different samples of chrysotile do not have the same carcinogenic potency (Jaurand *et al.*, 1987a; Wagner *et al.*, 1973). In addition, unpublished data have shown that different batches of a given sample of chrysotile may have different surface properties (H. Pezerat, personal communication).

The mechanism whereby fibres induce chromosome damage is not clear. Several categories of chemicals, for example, alkylating or cross-linking agents, are known to interact directly with DNA. Radiation can also damage DNA by producing oxygen-derived radicals. *In vitro*, asbestos fibres induce the formation of hydroxyl radicals (Weitzman & Graceffa, 1984; Zalma *et al.*, 1987). Such radicals might arise from oxygen reduction at the electron donor site on the surface of asbestos (Bonneau *et al.*, 1986). Zalma *et al.* (1987) reported that the number of hydroxyl radicals formed by amphiboles is strongly dependent on the treatment undergone by the fibres, and fresh but moderate grinding of amphiboles strongly enhanced OH˙ production, possibly due to the uncovering of new reactive sites. The level of OH˙ production by Canadian chrysotile was estimated to be $10^4$ per fibre for an average size of 5 $\mu$m $\times$ 0.1 $\mu$m.

Using another *in vitro* system, Kasai and Nishimura (1984) reported DNA damage by asbestos in the presence of hydrogen peroxide. The authors observed a hydroxylation of the C-8 position of guanine residues in calf thymus DNA. It is very likely that this occurs via a reaction involving the OH˙ radical.

The significance of these *in vitro* cell-free experiments or their relationship to the effects observed in cultured cells or *in vivo* is difficult to establish. DNA may possibly be damaged via oxy-radical formation or by direct interaction between DNA and asbestos. However, it is not clear how OH˙ produced at the fibre surface reaches the DNA molecule; in addition, culture media and biological fluids contain proteins and molecules which can be adsorbed and cover the fibre surface, thus 'poisoning' the fibres' reactive site for radical formation as well as for DNA adsorption. Nevertheless, it has been found that the 0.22 $\mu$m filtered culture medium of chrysotile-treated rat pleural mesothelial cells contains clastogenic factor(s) for human lymphocytes (Jaurand *et al.*, 1987b). A possible effect of any short fibres remaining was excluded by using controls. Since the formation of clastogenic factor(s) has been related to the production of oxy-radicals in certain diseases (Emerit *et al.*, 1980), such radicals may be involved in the formation of clastogenic factors by mesothelial cells. This is the first report on asbestos-induced chromosome damage possibly mediated by radical derivatives. Recent studies have reported the production of oxy-radicals by asbestos-treated cultured cells, namely macrophages (Goodglick & Kane, 1986; Hansen & Mossman, 1987) and hamster tracheal epithelial cells (Mossman *et al.*, 1986). In these experiments, the use of scavengers of oxygen metabolites prevented cytotoxicity, suggesting a relationship between oxy-radical formation and cell injury.

## Induction of gene mutation

In Chinese hamster cells treated with crocidolite, Huang *et al.* (1978) observed a statistically significant increase in the frequency of mutation at the hypoxanthine-guanine phosphoribosyl transferase locus (HGPRT) with median doses but not with low or higher ones. In later studies, Huang (1979) reported a weak mutagenic effect; the significance of the induced mutation depended on the statistical test used to evaluate the data.

Recently, no mutation was reported at the same locus in CHO cells by Kenne et al. (1986), who concluded that crocidolite was not mutagenic in these cells (the mutagenic frequency was nevertheless 400% of the control rate with an intermediate dose of 500 µg per dish). Using SHE cells, Oshimura et al. (1984) reported no mutagenic enhancement at either the HGPRT or the ouabain locus with chrysotile or crocidolite at doses which induce cell transformation. Erionite and crocidolite were not mutagenic in a lymphoblastoid cell line (Kelsey et al., 1986).

The mutagenesis of *Escherichia coli* has recently been studied again, using an alkali-rich analogue of tremolite (Cleveland, 1984). It was found that a significantly higher frequency of mutant colonies appeared following treatment with fibres. This enhancement was much higher in the presence of added S-9 rat liver homogenate; the mutagenic potency was tentatively attributed to impurities in the asbestos sample. Since the sample used in this experiment was ground by hand, one possible explanation could be the opening up of fresh sites for the reduction of oxygen, as discussed above.

Rita and Reddy (1986) recently showed that chrysotile did not induce abnormalities in germ cells after oral administration in mice. In this specific case it is possible that the distance of the fibres from the target DNA exceeded the critical distance.

**Induction of sister chromatid exchanges**

Mutagenic compounds have also been reported to induce sister chromatid exchange (SCE). Perry and Evans (1975), in an extensive study, found that 12 out of 14 mutagens tested increased SCE levels in CHO cells. Using the same cell type, Garrano et al. (1978) reported a relationship between mutation at the HGPRT locus and SCE. This relationship was not constant and some compounds were highly mutagenic but poor inducers of SCE, while the opposite was observed with other compounds. Other studies have shown that many mutagens, clastogens, or carcinogens can induce SCE (Hollstein et al., 1979). Recent data on SCE induction by fibres concern rat pleural mesothelial cells treated with crocidolite or with attapulgite (Achard et al., 1987). Crocidolite induced a significant enhancement of SCE but attapulgite failed to induce any. However, the level of SCE increase was much lower than that observed with BP. Kelsey et al. (1986) reported that erionite at multiple doses induced SCE in CHO cells; crocidolite produced a significant increase in only one experiment at one experimental dose. Conflicting data have been previously reported with asbestos fibres, since Casey (1983) and Price-Jones et al. (1980) did not find SCE enhancement with CHO-K cells, human fibroblasts, human lymphoblastoid cells or Chinese hamster lung cells (V79-4). In contrast, Livingston et al. (1980) reported an increase in SCE following treatment of CHO cells with crocidolite.

**Induction of DNA damage**

DNA damage can be detected by using methods for assessing DNA repair in which DNA double- or single-strand breaks are measured. Previous reports of studies on several cell types (epithelial and mesothelial cells and fibroblasts) failed to show any strand breaks after treatment with asbestos (Harris et al., 1985; Mossman et al., 1983).

Another method of measuring DNA repair is by means of unscheduled DNA synthesis (UDS). Poole *et al.* (1983) reported that erionite fibres enhanced UDS in C3H 10T½ and A549 cells. Using primary cultures of hepatocytes, Denizeau *et al.* (1985a,b) reported that Canadian chrysotile and xonotlite (a fibrous calcium silicate used as an asbestos substitute) failed to elicit UDS. In addition, chrysotile did not interfere with the genotoxic effect of dimethylnitrosamine or 2-acetylaminofluorene. Using rat pleural mesothelial cells, our group reported that UDS was enhanced in $G_0/G_1$ arrested cells (Jaurand *et al.*, 1986b and unpublished data). DNA repair is a valuable end-point for the evaluation of chemical genotoxicity but has not been validated with solid compounds. Further investigations are necessary to explain these findings. Nevertheless, it should be noted that the mechanism of UDS stimulation starts with the enzyme recognition of DNA damage which might result in a conformational change. Specific endonucleases have been found in *E. coli* but have not yet been characterized in eukaryotic cells (Friedberg, 1985). Strand breaks by OH˙ or clastogenic factor, but also the affinity of the fibre for DNA or for certain proteins, may play a role in relation to DNA alteration.

The data on UDS seem conflicting; the differences do not seem related to questions of techniques since two methods of assaying UDS, namely, liquid scintillation and autoradiography, have been used in these experiments. Both cell types (hepatocytes and mesothelial cells) are able to ingest chrysotile fibres, as reported elsewhere (Fleury *et al.*, 1983; Jaurand *et al.*, 1979). However, the quantity of fibres ingested is not known, the events involved in phagocytosis may be different in the different cell types, and it remains to be determined whether DNA damage by mineral fibres results from early, intermediate or late phagocytosis or is caused by other unrelated events. If, for example, plasma-membrane-fibre interactions are involved (e.g., via lipid peroxidation) it is conceivable that different cell types have different responses. In addition, as discussed previously in connection with the clastogenicity of fibres, different cell types differ in their ability to repair DNA. Hepatocytes are good indicators of chemically induced DNA repair, but it is not known whether fibre-induced repair is similar.

## Concluding remarks on the genotoxicity of fibres

Overall, the results reported on the genotoxicity of fibres, especially asbestos, are conflicting and make it difficult to reach a clear conclusion as to the ability of asbestos to induce genetic damage. The reasons for these differences are unknown. However, it should be noted that the experiments reported have been carried out with cells originating from different species and tissues; the authors used a number of protocols, several different types of particles, and different methods to prepare the samples. In addition, it is known that mutations at the HGPRT or ouabain loci are induced only by point mutations or chromosome deletions. Other types of genetic events involved in carcinogenesis (e.g., aneuploidy and chromosome rearrangements) may not be detected by this method.

It is nevertheless clear that asbestos has a significant clastogenic effect on some cultured cells but that this is small when compared with that of other mutagenic or carcinogenic chemicals. Another effect of asbestos is the induction of aneuploidy in certain cell types; this may be an important step in the early stages of carcinogenesis. It should also be noted that in several studies the dose-effect relationship was found to be biphasic. No enhancement was found at low concentrations; it became significant only when the fibre concentration was increased. At higher concentrations, no effect was observed, possibly because of cytotoxicity. This fact makes it difficult to reach any definite conclusion. The mechanism whereby asbestos is cytotoxic to cultured cells is not necessarily related to genetic damage. A recent study by Hesterberg et al. (1987) showed that nuclear damage and mitotic arrest were not related to cell lysis in rat tracheal epithelial cells. Other phenomena must be of greater importance, e.g., interaction with cell components or oxy-radicals in some cell types.

## *Transformation of cultured cells by fibres* in vitro

The transformation of SHE cells by either asbestos fibres or glass fibres has recently been reported (Hesterberg et al., 1986; Hesterberg & Barrett, 1984). In these papers, the importance of fibre dimension was demonstrated, the highest carcinogenic potency being associated with the longest fibres. The problem of comparisons between the effects of different fibres was discussed by the authors. Such comparisons may be made either on a per weight or a per fibre basis, but may give different results because the number of fibres per unit weight differs from one sample to another. It is nevertheless true that long fibres were more effective than the short fibres (milled long fibres) used in these experiments in inducing transformation. In addition, this difference was not related to differences in the number of fibres ingested, since this was slightly higher in the case of the short fibres (Hesterberg et al., 1986). Fibre size affects interpretation of the data in a different way when diameter is taken into consideration, as extensively discussed by Hesterberg et al. (1986). These authors showed that the transformation induced by thin fibres was greater than that induced by thick fibres on a per weight basis. However, on a per fibre basis, the opposite was true: thick fibres were more effective than thin fibres. These results are in agreement with the previous findings on the cytotoxicity of thick fibres versus thin fibres reported by Brown et al. (1979).

The transformation potency of fibres depends on the cell type used, since Brown et al. did not observe transformation of C3H 10T½ cells following asbestos application while Patérour et al. (1985) reported the transformation of rat pleural mesothelial cells and Hesterberg and Barrett (1984) the transformation of SHE cells. These results indicate that some fibres can act as a complete carcinogen in some cell types; however, it is not known whether some secondary factors involved in the progression of transformation are necessary to allow the expression of the transformed phenotype or if it is due solely to the effect of the particles.

Interestingly, Hei et al. (1984), in a study on C3H 10T½ transformation, reported that amphiboles, which were not by themselves inducers of cell transformation, significantly enhanced transformation when the cells were treated first with asbestos and subsequently with X-rays. In contrast, when the cells were exposed to X-rays before treatment with asbestos, there was no enhancement of the transformation as observed with 12-*O*-tetradecanoylphorbol 13-acetate (TPA) in other systems.

## Molecular events reported after asbestos treatment

Few data have been reported concerning the molecular effects of fibres. To my knowledge, the first report concerned the oncogene expression of rat pleural mesothelial cells treated with BP and chrysotile in a two-stage model (Paterour *et al.*, 1985; Tobaly *et al.*, 1985). DNA analysis showed similar *Eco*RI patterns of the treated cells, but significant variations in the relative intensities of the bands were found for the c-*myc* oncogene. One interesting point was a selective decrease by a factor of 10 in the c-*myc* RNA of the cells treated with BP and chrysotile when compared with that found with each component in isolation.

Dubes and Mack (1988) have tested the ability of asbestos to mediate transfection by viral RNA; they found that several samples of asbestos did so and that transfection was more effective when the asbestos was mixed with viral RNA rather than added separately. These findings may be associated with the sorptive properties of asbestos with regard to DNA reported elsewhere (Touray *et al.*, 1987). It is likely that RNA can also be adsorbed on asbestos, thereby facilitating RNA penetration within the cell. However, this assumption remains to be confirmed.

## Conclusions

The multistage nature of chemical carcinogenesis has been demonstrated and it must be now determined in which steps particulate carcinogens are involved. Asbestos fibres cannot be regarded as a pure promoter according to the definition of Blumberg *et al.* (1983) recently reviewed by Shubik (1986). From the recent data obtained *in vitro* as well as in experimental animals, it seems that fibres such as asbestos act as complete carcinogens. The concept of tumour promotion has been developed from experimental studies with mouse skin using chemicals as initiators and phorbol esters as promoters (Berenblum & Shubik, 1947). The promoters used were not highly carcinogenic by themselves and the initiators were carcinogens used at non-carcinogenic doses. This two-stage model, later converted into a multistage one (Slaga, 1983), has been applied *in vitro* using TPA as promoter (Lasne *et al.*, 1974; Mondal *et al.*, 1976). Except in one study (Paterour *et al.*, 1985), asbestos has not been used *in vitro* as a promoter and no promoting activity has been found. Moreover, when asbestos was applied after radiation no transformation of C3H cells was observed, in contrast to what occurred when it was applied before. However, as demonstrated by Mossman (1983) and Mossman *et al.* (1985), asbestos fibres do have some effects similar to those of classical promoters.

Several studies have shown a synergism between asbestos and a chemical carcinogen (Brown *et al.*, 1983; DiPaolo *et al.*, 1983). However, Paterour *et al.* (1985) did not find any synergism between BP and chrysotile; there was even inhibition of growth in soft agar and tumorigenesis when BP-treated cells were further treated with chrysotile. The ingestion by cells of BP-coated fibres results in an enhancement of BP uptake and metabolism by tracheal explants from hamsters (Mossman et al., 1985). The differences between the results of the various experiments may be due to differences in their design. When BP is coated on to asbestos, the latter may increase the local concentration of BP by its carriage of BP molecules.

The studies reported above were performed because of earlier reports on the enhancement of the risk of lung cancer in insulation workers (Hammond et al., 1979), where the combined risk (smoking + asbestos exposure) was multiplicative when compared to that of unexposed non-smokers. However, this does not seem true for workers exposed to lower doses, where the risk was increased in non-smokers to a greater extent than in smokers (Berry et al., 1985; IARC, 1986). According to these authors, the lung cancer risk was higher for those who had never smoked, but this has not been confirmed. These studies also confirmed that no association exists between mesothelioma formation and smoking; this may be because carcinogenic compounds from tobacco smoke do not reach the pleura and does not imply a specific effect at the pleural level.

With regard to the use of BP to assess the synergistic effect between chemical carcinogens and asbestos, it should be noted that, after intratracheal injection of chrysotile plus BP in hamsters, Miller et al. (1965) found an increase in the number of tumours induced as compared with those induced by any compound alone but there was no such increase with amosite. BP is not the only component of tobacco smoke; other compounds in such smoke are highly carcinogenic and many promoters are present (Feron et al., 1985). The number of mutagenic events produced by tobacco smoke condensate in the Ames or SCE tests is much greater than that predicted on the basis of the amount of BP contained in the sample (Evans, 1981). Compounds such as catechols and nicotine-derived nitrosamines are powerful carcinogens (Hoffman et al., 1985). Interestingly, these latter authors report that amosite or chrysotile sprayed with nicotine and subsequently exposed to either nitrogen dioxide or cigarette smoke accelerated nitrosamine formation but that the amounts formed were greater with chrysotile than with amosite. This emphasizes once again the role of the physical chemistry of the fibres. However, in one experiment, Denizeau et al. (1985b) did not find any enhancement by chrysotile of nitrosamine genotoxicity in hepatocytes.

It seems that new concepts will have to be developed to account for the carcinogenicity of mineral dusts and other particulate matter. Until now, the models and mechanisms proposed were based on concepts arising from the study of chemical carcinogens or radiation. When compared to other carcinogenic substances, one remarkable feature of fibres is their persistence in the tissues for a long time after inhalation. Even if the damage caused by a chemical is long lasting, the molecule itself will disappear as a result of metabolism or of other processes. In contrast, fibres may be considered as a pluripotent carcinogen, due both to their pleiotropic effects and to their ability to exert effects for a long period of time. The effect of time must therefore be taken into consideration for such resident carcinogens, and specifically for 'particulate state' carcinogens. Even a substance of low carcinogenic potency may be able to induce detectable rates of cancer after long exposure. In addition, unstable particles such as chrysotile or certain glass fibres may have a reduced *in vivo* effect when compared with stable particles, depending on their characteristics and rate of dissolution.

From the early studies beginning with animal experiments and investigations of cell cytotoxicity, research on the carcinogenicity of fibres has moved towards the determination of the cellular events involved in fibre-induced cell transformation. It now appears that studies should focus on the molecular level. These approaches are of great interest in understanding the mechanisms of carcinogenesis related to fibres and will allow comparison with chemically induced or spontaneous neoplastic transformation.

Further experiments are necessary to establish the mechanisms of carcinogenesis. At present, it may be suggested that fibres are carcinogenic by virtue of their size and physicochemical state, inducing genetic damage by their physical presence, by the adsorption of other molecules or by radical production. The persistence of the fibres may allow multiple hits with time, resulting in neoplastic progression.

While it is evident that cellular studies are of fundamental importance in explaining these mechanisms, it must be noted that investigations on the fibre properties involved in these processes are needed. Only both types of studies will allow us to understand the relevant properties of new fibres or particulates in industrial or commercial use.

*In vitro* experiments may make it possible to determine the relative potential of fibres as well as the absolute potential of a given fibre type with regard to toxicity and to understand its mechanism of action. However, that may be different from the 'effective' potential *in vivo*, which depends on both the intrinsic fibre properties and the host environment (clearance, retention, biological attack, etc.).

## Acknowledgements

I would like to thank J. Carl Barrett for his critical reading of the manuscript and Helena Bonner for typing it.

## References

Achard, S., Perderiset, M. & Jaurand, M.C. (1987) Sister chromatid exchanges in rat pleural mesothelial cells treated with crocidolite, attapulgite, or benzo 3-4 pyrene. *Br. J. Ind. Med.*, 44, 281-283

Acheson, E.D., Garner, M.J., Pippard, E.C. & Grime, L.P. (1982) Mortality of two groups of women who manufactured gas masks from chrysotile and crocidolite. A 40 years follow-up. *Br. J. Ind. Med.*, 39, 344-348

Babu, K.A., Lakkad, B.C., Nigam, S.K., Bhatt, D.K., Kainik, A.B., Thakore, N.K., Kashyap, N.S. & Chatterjee, S.K. (1980) *In vitro* cytological and cytogenic effects of an Indian variety of chrysotile asbestos. *Environ. Res.*, 21, 416-422

Beck, E.G. & Bignon, J., eds (1985) In vitro *Effects of Mineral Dusts (NATO ASI Series, Series G: Ecological Sciences)*, Vol. 3, Berlin, Springer-Verlag, pp. 1-548

Beck, E.G., Holt, P.F. & Nasrallah, E.T. (1971) Effects of chrysotile and acid-treated chrysotile on macrophage cultures. *Br. J. Ind. Med.*, 28, 179-185

Berry, G., Newhouse, M.L. & Antonis, P. (1985) Combined effect of asbestos and smoking on mortality from lung cancer and mesothelioma in factory workers. *Br. J. Ind. Med.*, *42*, 12-18

Blumberg, P.M., Delclos, K.B., Dunn, J.A., Jaken, S., Leach, K.L. & Yek, E. (1983) Phorbol ester receptors and the *in vitro* effects of tumour promoters. *Ann. N.Y. Acad. Sci.*, *407*, 303-313

Bonneau, L., Suquet, H., Malard, C. & Pezerat, M. (1986) Studies on surface properties of asbestos. I. Active sites on surface of chrysotile and amphiboles. *Environ. Res.*, *41*, 251-267

Brown, R.C., Chamberlain, M., Davies, R., Gaffen, J. & Skidmore, J.W. (1979) *In vitro* biological effects of glass fibres. *J. Environ. Pathol. Toxicol.*, *2*, 1369-1383

Brown, R.C., Gormley, J.P., Chamberlain, M. & Davis, R., eds (1980) *The in vitro Effects of Mineral Dusts*, New York, Academic Press

Brown, R.C., Poole, A. & Flemming, G.T.A. (1983) The influence of asbestos dust on the oncogenic transformation of C3H 10T½ cells. *Cancer Lett.*, *18*, 221-227

Brown, G.M., Cowie, H., Davis, J.M.G. & Donaldson, K. (1986) *In vitro* assays for detecting carcinogenic mineral fibres: a comparison of two assays and the role of fibre size. *Carcinogenesis*, *7*, 1971-1974

Casey, G. (1983) Sister chromatid exchange in cell kinetics in CHO-K1 cells, human fibroblasts and lymphoblastoid cells exposed *in vitro* to asbestos and glass fibers. *Mutat. Res.*, *116*, 367-377

Chamberlain, M. & Tarmy, E.M. (1977) Asbestos and glass fibers in bacterial mutation tests. *Mutat. Res.*, *43*, 159-164

Cleveland, N.G. (1984) Mutagenesis of *Escherichia coli* (CSH50) by asbestos (41954). *Proc. Soc. Exp. Biol. Med.*, *177*, 343-346

Davis, J.M.G., Addison, J., Bolton, R.E., Donaldson, K., Jones, A.D. & Wright, A. (1984) The pathogenic effects of fibrous ceramic aluminium silicate glass administered to rats by inhalation or peritoneal injection. In: *Biological Effects of Man-made Mineral Fibres,* Vol. 2, Copenhagen, World Health Organization, Regional Office for Europe, pp. 303-322

Denizeau, F., Marion, M., Chevalier, G. & Cote, M. (1985a) Inability of chrysotile asbestos fibers to modulate the 2-acetylaminofluorene-induced UDS in primary cultures of rat hepatocytes. *Mutat. Res.*, *155*, 83-90

Denizeau, F., Marion, M., Chevalier, G. & Cote, M. (1985b) Genotoxicity of dimethylnitrosamine in the presence of chrysotile asbestos UICC B and xonotlite. *Carcinogenesis*, *6*, 1815-1817

DiPaolo, J.A., De Marinis, A.J. & Doniger, J. (1983) Asbestos and benzo[*a*]pyrene synergism in the transformation of Syrian hamster embryo cells. *Pharmacology*, *27*, 65-73

Dodson, R.F. & Ford, J.F. (1985) Early response of the visceral pleura following asbestos exposure: an ultrastructural study. *J. Toxicol. Environ. Health*, *15*, 673-686

Dubes, G.R. & Mack, L.R. (1988) Asbestos-mediated transfection of mammalian cell cultures. *In Vitro* (in press)

Emerit, I., Michelson, A., Levy, A., Cames, J. & Emerit, J. (1980) Chromosome breaking agent of low molecular weight in human systemic lupus erythematosis. Protection effect of superoxide dismutase. *Hum. Genet.*, *55*, 341-344

Evans, H.J. (1981) Cigarette smoke induced DNA damage in man. *Prog. Mutat. Res.*, *2*, 111-128

Feron, V.J., Kuper, C.F., Spit, B.J., Renzel, P.G.T. & Woutersen, R.A. (1985) Glass fibers and vapour phase components of cigarette smoke as cofactors in experimental carcinogenesis. In: Mass, M.J., Kaufman, D.G., Siegfried, J.M., Steele, V.E. & Nesnow, S., eds, *Carcinogenesis. A Comprehensive Survey*, Vol. 8, New York, Raven Press, pp. 93-118

Fleury, J., Cote, N.G., Marion, G., Chevalier, G. & Denizeau, F. (1983) Interaction of asbestos fibers with hepatocytes: An ultrastructural study. *Toxicol. Lett.*, *19*, 15-22

Friedberg, E.C. (1985) Excision repair. In: Friedberg, E.C., ed., *DNA Repair*, New York, W.H. Freeman, pp. 141-374

Garrano, A.V., Thompson, L.H., Lindl, P.A. & Minker, J.L. (1978) Sister chromatid exchanges as an indicator of mutagenesis. *Nature, 271*, 551-553

Gibas, Z., Li, F.P., Antman, K.H., Bernal, S., Staehl, R. & Sandberg, A.A. (1986) Chromosome changes in mesothelioma. *Cancer Genet. Cytogenet., 20*, 191-201

Goodglick, L.A. & Kane, A.B. (1986) Role of reactive oxygen metabolites in crocidolite asbestos toxicity to mouse macrophages. *Cancer Res., 46*, 5558-5566

Hammond, E.C., Selikoff, I.J. & Seidman, H. (1979) Asbestos exposure, smoking and cell death rates. *Ann. N.Y. Acad. Sci., 330*, 473-490

Hansen, K. & Mossman, B.T. (1987) Generation of superoxide ($O_2^-$) from alveolar macrophages exposed to asbestiform and nonfibrous particles. *Cancer Res., 47*, 1681-1686

Harington, J.S. & Roe, F.J.C. (1965) Studies of carcinogenesis of asbestos fibers and their natural oils. *Ann. N.Y. Acad. Sci., 132*, 439-450

Harris, C.C., Lechner, J.F., Yoakum, G.H., Amstad, P., Korba, B.E., Gabrielson, E., Graftstrom, R., Shamsuddin, A. & Trump, B.F (1985) *In vitro* studies of human lung carcinogenesis. In: Barrett, J.C. & Tennant, R.W., eds, *Carcinogenesis*, Vol. 9, New York, Raven Press, pp. 257-269

Hayashi, H. (1974) Cytotoxicity of heated chrysotile. *Environ. Health Perspect., 9*, 267-270

Hei, T.K., Hall, E.J. & Osmak, R.S. (1984) Asbestos, radiation and oncogenic transformation. *Br. J. Cancer, 50*, 717-720

Hei, T.K., Geard, C.R., Osmak, R.S. & Travisano, M. (1985) Correlation of *in vitro* genotoxicity and oncogenicity induced by radiation and asbestos fibres. *Br. J. Cancer, 52*, 591-597

Hesterberg, T.W. & Barrett, J.C. (1984) Dependence of asbestos- and mineral dust-induced transformation of mammalian cells in culture on fibre dimension. *Cancer Res., 44*, 2170-2180

Hesterberg, T.W. & Barrett, J.C. (1985) Induction by asbestos fibers of anaphase abnormalities: mechanism for aneuploidy induction and possibly carcinogenesis. *Carcinogenesis, 6*, 473-476

Hesterberg, T.W., Butterick, C.J., Oshimura, M., Brody, A. & Barrett, J.C. (1986) Role of phagocytosis in Syrian hamster cell transformation and cytogenetic effects induced by asbestos and short and long glass fibers. *Cancer Res., 46*, 5795-5802

Hesterberg, T.W., Ririe, D.G., Barrett, J.C. & Nettesheim, P. (1987) Mechanisms of cytotoxicity of asbestos fibres in rat tracheal epithelial cells in culture. *Toxicol. in vitro, 1*, 59-65

Hoffman, D., Melikian, A., Adams, J.D., Brunneman, K.D. & Haley, N.J. (1985) New aspects of tobacco carcinogenesis. In: Mass, M.J., Kaufman, D.G., Siegfried, J.M., Steele, V.E. & Nesnow, S., eds, *Carcinogenesis. A Comprehensive Survey*, Vol. 8, New York, Raven Press, pp. 239-256

Hollstein, M., McCann, J., Angelosanta, F.A. & Nichols, W.W. (1979) Short-term tests for carcinogens and mutagens. *Mutat. Res., 65*, 133-226

Holt, P.F., Mills, J. & Young, D.K. (1965) Experimental asbestosis with four types of fibers. *Ann. N.Y. Acad. Sci., 132*, 87-97

Huang, S.L. (1979) Amosite, chrysotile and crocidolite asbestos are mutagenic to Chinese hamster lung cells. *Mutat. Res., 68*, 265-274

Huang, S.L., Saggioro, D., Michelman, H. & Malling, H.V. (1978) Genetic effects of crocidolite asbestos in Chinese hamster lung cells. *Mutat. Res., 57*, 225-232

IARC (1986) *IARC Monographs on the Evaluation of the Carcinogenic Risk of Chemicals to Humans*, Vol. 38, *Tobacco Smoking*, Lyon, International Agency for Research on Cancer, pp. 199-375

Jaurand, M.C., Kaplan, H., Thiollet, J., Pinchon, M.C., Bernaudin, J.F. & Bignon, J. (1979) Phagocytosis of chrysotile fibers by pleural mesothelial cells in culture. *Am. J. Pathol., 94*, 529-532

Jaurand, M.C., Baillif, R., Thomassin, J.H., Magne, L. & Touray, J.C. (1983a) X-ray photoelectron spectroscopy and chemical study of the adsorption of biological molecules on chrysotile asbestos surface. *J. Colloid Interf. Sci., 95*, 1-9

Jaurand, M.C., Bastie-Sigeac, I., Bignon, J. & Stoebner, P. (1983b) Effect of chrysotile and crocidolite on the morphology and growth of rat pleural mesothelial cells. *Environ. Res., 30*, 255-269

Jaurand, M.C., Kheuang, L., Magne, L. & Bignon, J. (1986a) Chromosomal changes induced by chrysotile fibres or benzo-3,4-pyrene in rat pleural mesothelial cells. *Mutat. Res., 169*, 141-148

Jaurand, M.C., Renier, A., Achard, S., Kheuang, L., Magne, L. & Bignon, J. (1986b) *In vitro* genotoxic effects of benzo[*a*]pyrene (BP) and of mineral fibres on rat pleural mesothelial cells. *In Vitro, 22*, 17A

Jaurand, M.C., Fleury, J., Monchaux, G., Nebut, M. & Bignon, J. (1987a) Pleural carcinogenic potency of mineral fibers (asbestos, attapulgite) and their cytotoxicity on cultured cells. *J. Natl Cancer Inst., 79*, 797-804

Jaurand, M.C., Pezerat, H. & Emerit, I. (1987b) Oxy-radical formation by asbestos-treated mammalian cells and induction of a clastogenic factor. In: *Proceedings, Meeting of the Society for Free Radical Research, University of California*, 6-7 February

Kasai, M. & Nishimura, S. (1984) DNA damage induced by asbestos in the presence of hydrogen peroxide. *Gann, 75*, 841-844

Kelsey, K.T., Yano, E., Liber, H.L. & Little, J.B. (1986) The *in vitro* genetic effects of fibrous erionite and crocidolite asbestos. *Br. J. Cancer, 54*, 107-114

Kenne, K., Ljungquist, S. & Ringertz, N.R. (1986) Effects of asbestos fibers on cell division, cell survival, and formation of thioguanine-resistant mutants in Chinese hamster ovary cells. *Environ. Res., 39*, 448-464

Kimizuka, G., Wang, N.S. & Hayashi, Y. (1987) Physical and microchemical alterations of chrysotile and amosite asbestos in the hamster lung. *J. Toxicol. Environ. Health, 21*, 251-264

Kodama, Y., Hesterberg, T.W., Maness, S.C., Iglehart, J.D. & Boreiko, C.J. (1987) Asbestos-induced cytogenetic effects in cultured human bronchial epithelial cells. *Proc. Am. Soc. Cancer Res., 28*, 85

Koshi, K., Hayashi, H. & Sakabe, H. (1986) Cell toxicity and hemolytic action of asbestos dust. *Ind. Health (Kawasaki), 6*, 69-79

Langer, A. (1985) Physico-chemical properties of minerals relevant to biological activities: state of the art. In: Beck, E.G. & Bignon, J., eds, In Vitro *Effects of Mineral Dusts*, (*NATO ASI Series, Series G: Ecological Sciences*), Vol. 3, Berlin, Springer-Verlag, pp. 9-24

Lasne, C., Gentil, A. & Chouroulinkov, I. (1974) Two stage transformation of rat fibroblasts in tissue culture. *Nature, 242*, 490-491

Lavappa, K.S., Fu, N.R. & Epstein, S.S. (1975) Cytogenetic studies on chrysotile asbestos. *Environ. Res., 10*, 165-173

Lechner, J.F., Tokiwa, T., Leveck, M., Benedict, W.F., Banchs-Schlegel, S., Yeager, M., Banerjee, A. & Harris, C.C. (1985) Asbestos-associated chromosomal changes in human mesothelial cells. *Proc. Natl. Acad. Sci. USA, 82*, 3884-3888

Lee, D.H.K. (1974) Proceedings of the joint NIEHS-EPA conference on 'Biological effects of ingested asbestos'. *Environ. Health Perspect., 9*, 113-338

Livingston, G.K., Rom, W.N. & Morris, M.V. (1980) Asbestos induced sister chromatid exchanges in cultured Chinese hamster ovarian cells. *J. Environ. Pathol. Toxicol., 4*, 373-382

Mazia, D. (1984) Centrosomes and mitotic poles. *Exp. Cell Res., 153*, 1-15

McFadden, D., Wright, J.L., Wiggs, B. & Churg, A. (1986) Smoking inhibits asbestos clearance. *Am. Rev. Respir. Dis.*, *133*, 372-374

Miller, L., Smith, W.E. & Berliner, S.W. (1965) Tests for effect of asbestos on benzo[a]pyrene carcinogenesis in the respiratory tract. *Ann. N.Y. Acad. Sci.*, *132*, 489-500

Monchaux, G., Bignon, J., Jaurand, M.C., Lafuma, J., Sébastien, P., Massé, R., Hirsch, A. & Goni, J. (1981) Mesothelioma in rats following inoculation with acid leached chrysotile asbestos and other mineral fibres. *Carcinogenesis*, *2*, 229-236

Mondal, S., Brankow, D.W. & Heidelberger, C. (1976) Two stage chemical carcinogenesis in cultures of C3H 10T½ cells. *Cancer Res.*, *36*, 2254-2260

Morgan, A. (1974) Adsorption of human serum albumin on asbestiform minerals and its application to the measurement of surface areas of dispersed samples of chrysotile. *Environ. Res.*, *7*, 330-341

Morgan, A. & Holmes, A. (1984) The deposition of MMMF in the respiratory tract of the rat, their subsequent clearance, solubility *in vivo* and protein coating. In: *Biological Effects of Man-made Mineral Fibres*, Vol. 2, Copenhagen, World Health Organization, Regional Office for Europe, pp. 1-17

Mossman, B.T. (1983) *In vitro* approaches for determining mechanisms of toxicity and carcinogenicity by asbestos in the gastrointestinal and respiratory tracts. *Environ. Health Perspect.*, *53*, 155-161

Mossman, B.T., Eastman, A., Landesman, J.M. & Bresnick, E. (1983) Effects of crocidolite and chrysotile asbestos on cellular uptake and metabolism of benzo[a]pyrene in tracheal epithelial cells. *Environ. Health Perspect.*, *51*, 331-335

Mossman, B.T., Cameron, G.S. & Yotti, L.P. (1985) Cocarcinogenic and tumor promoting properties of asbestos and other minerals in tracheobronchial epithelium. In: Moss, M.J., Laufman, D.G., Siegfried, J.M., Steele, V.E. & Nesnow, S., eds, *Carcinogenesis, A Comprehensive Survey*, Vol. 8, New York, Raven Press, pp. 217-238

Mossman, B.T., Marsh, J.P. & Shatos, M.A. (1986) Alteration of superoxide dismutase activity in tracheal epithelial cells by asbestos and inhibition of cytotoxicity by antioxidants. *Lab. Invest.*, *54*, 204-212

Nicklas, R.B., Kubai, D.F. & Hays, T.S. (1982) Spindle microtubulesects with cell transformation of Syrian hamster embryo cells in culture. *Cancer Res.*, *44*, 5017-5022

Oshimura, M., Hesterberg, T.W. & Barrett, J.C. (1986) An early non-random karyotypic change in immortal Syrian hamster cell lines transformed by asbestos: trisomy of chromosome 11. *Cancer Genet. Cytogenet.*, *22*, 225-237

Palekar, L.D., Eyre, J.F., Most, B.M. & Coffin, D.L. (1987) Metaphase and anaphase analysis of V79 cells exposed to erionite, UICC chrysotile and UICC crocidolite. *Carcinogenesis*, *8*, 553-560

Patérour, M.J., Bignon, J. & Jaurand, M.C. (1985) *In vitro* transformation of rat pleural mesothelial cells by chrysotile and/or benzo[a]pyrene. *Carcinogenesis*, *6*, 523-529

Pernis, B., Vigliani, E.G., Marchisio, M.A. & Zanardi, S. (1966) Observations on the effect of asbestos on cells *in vitro*. *Med. Lavoro*, *59*, 561-575

Perry, P. & Evans, H.J. (1975) Cytological detection of mutagen-carcinogen exposure by sister chromatid exchanges. *Nature*, *258*, 121-125

Poole, A., Brown, R.C., Turner, J.C., Skidmore, J.W. & Griffiths, D.M. (1983) *In vitro* genotoxic activities of fibrous erionite. *Br. J. Cancer*, *47*, 697-705

Popescu, N.C., Amsbaugh, S.C., Chahinian, P.A. & DiPaolo, J.A. (1987) Nonrandom deletions of chromosome 3 (p 14-21) in human malignant mesothelioma. *Proc. Am. Soc. Cancer Res.*, *28*, 37

Pott, F., Schlipkoter, H.W., Ziem, V., Spurny, K. & Huth, F. (1984) New results from implantation experiments with mineral fibres. In: *Biological Effects of Man-made Mineral Fibres*, Vol. 2, Copenhagen, World Health Organization, Regional Office for Europe, pp. 286-302

Price-Jones, M.J., Gubbings, G. & Chamberlain, M. (1980) The genetic effects of crocidolite asbestos; a comparison of chromosome abnormalities and sister chromatid exchanges. *Mutat. Res.*, *79*, 331-336

Rita, P. & Reddy, P.P. (1986) Effect of chrysotile asbestos fibers on germ cells of mice. *Environ. Res.*, *41*, 139-143

Roggli, V.G. & Brody, A.R. (1984) Changes in numbers and dimensions of chrysotile asbestos fibers in lungs of rats following short-term exposure. *Exp. Lung Res.*, *7*, 134-147

Roggli, V.L., George, M.H. & Brody, A.R. (1987) Clearance and dimensional changes of crocidolite asbestos fibers isolated from lungs of rats following short-term exposure. *Environ. Res.*, *42*, 94-105

Schnitzer, R.J. (1974) Modification of biological surface activity of particles. *Environ. Health Perspect.*, *9*, 261-266

Schnitzer, R.J. & Pundsack, F.L. (1970) Asbestos hemolysis. *Environ. Res.*, *3*, 1-13

Second International Workshop on the *In Vitro* Effects of Mineral Dusts (1983) *Environ. Health Perpect.*, *51*, pp. 1-385

Selikoff, J.J. & Churg, J. (1965) Biological effects of asbestos. *Ann. N.Y. Acad. Sci.*, *132*, 1-766

Shubik, P. (1984) Progression and promotion. *J. Natl Cancer Inst.*, *73*, 1005-1011

Sincock, A.M. & Seabright, M. (1975) Induction of chromosome changes in hamster cells by exposure to asbestos fibres. *Nature*, *257*, 56-58

Sincock, A.M., Delhanty, J.D.A. & Casey, G. (1982) A comparison of the cytogenetic response to asbestos and glass fibre in Chinese hamster and human cell lines. *Mutat. Res.*, *101*, 257-268

Slaga, T.J. (1983) Multistage skin tumor promotion and specificity of inhibition. In: Slaga, T.J., ed., *Mechanisms of Tumour Promotion*, Vol. II, *Tumor Promotion and Skin Carcinogenesis*, Boca Raton, FL, CRC Press, pp. 189-196

Stanton, M.F., Layard, M., Tegeris, A., Miller, E., May, M. & Kent, E. (1977) Carcinogenicity of fibrous glass: pleural response in the rat in relation to fiber dimension. *J. Natl Cancer Inst.*, *58*, 587-603

Stanton, M.F., Layard, M., Tegeris, A., Miller, E., May, M., Morgan, E. & Smith, A. (1981) Relation of particle dimension to carcinogenicity in amphibole asbestoses and other fibrous minerals. *J. Natl Cancer Inst.*, *67*, 965-975

Summary Workshop on Ingested Asbestos (1983) *Environ. Health Perspect.*, *53*, 1-204

Suzuki, Y. & Kohyama, N. (1984) Malignant mesothelioma induced by asbestos and zeolite in the mouse peritoneal cavity. *Environ. Res.*, *35*, 277-292

Tobaly, J., Salle, M., Paterour, M.J., Jaurand, M.C., Bignon, J. & Ravicovitch-Ravier, R. (1985) Preliminary results on oncogenes in rat pleural mesothelial cells transformed *in vitro* by benzo[*a*]pyrene and/or chrysotile. In: Beck, E.G. & Bignon, J., eds, In Vitro *Effects of Mineral Dusts*, Vol. 3 (*NATO ASI Series, Series G: Ecological Sciences*), Berlin, Springer-Verlag, pp. 209-214

Touray, J.C., Baillif, P., Jaurand, M.C., Bignon, J. & Magne, L. (1987) Comparative study of adsorption of DNA on chrysotile and phosphorylated chrysotile (chrysophosphate) [in French]. *J. Can. Chim.*, *65*, 508-511

Valerio, F., DeFerrari, M., Ottagio, L., Repetto, E. & Santi, L. (1983) Chromosomal aberrations induced by chrysotile and crocidolite in human lymphocytes *in vitro*. *Mutat. Res.*, *133*, 397-402

Wagner, J.C. & Berry, G. (1969) Mesotheliomas in rats following inoculation with asbestos. *Br. J. Cancer*, *23*, 567-581

Wagner, J.C., Berry, G. & Timbrell, V. (1973) Mesotheliomata in rats after inoculation with asbestos and other materials. *Br. J. Cancer*, *28*, 173-185

Wagner, J.C., Berry, G.B., Hill, R.J., Munday, D.E. & Skidmore, J.W. (1984) Animal experiments with MMM(V)F — effects on inhalation and intraperitoneal inoculation in rats. In: *Biological Effects of Man-made Mineral Fibres*, Vol. 2, Copenhagen, World Health Organization, Regional Office for Europe, pp. 209-233

Wang, N.S., Jaurand, M.C., Magne, L., Kheuang, L., Pinchon, M.C. & Bignon, J. (1987) The interactions between asbestos fibers and metaphase chromosomes of rat pleural mesothelial cells in culture. *Am. J. Pathol.*, *126*, 343-349

Weitzman, S.A. & Graceffa, R. (1984) Asbestos catalyzes hydroxyl and superoxide radical generation from hydrogen peroxide. *Arch. Biochem. Biophys.*, *228*, 373-376

WHO/IARC (1984) *Biological Effects of Man-made Mineral Fibres,* Vol. 2, Copenhagen, World Health Organization, Regional Office for Europe

Zalma, R., Bonneau, L., Jaurand, M.C., Guignard, J. & Pezerat, H. (1987) Formations of oxy-radicals by oxygen reduction arising from the surface activity of asbestos. *Can. J. Chem.* (in press)

# MODIFICATION OF FIBROUS OREGON ERIONITE AND ITS EFFECTS ON *IN VITRO* ACTIVITY

R.C. Brown & R. Davies

*MRC Toxicology Unit, Carshalton, UK*

A.P. Rood

*HSE Occupational Medicine and Hygiene Laboratories, London, UK*

*Summary.* A sample of erionite, a fibrous zeolite, was modified by milling to reduce the number and length of the fibres and by extraction with cyclohexane. The *in vitro* activities of this mineral were found to depend on the presence of long fibres. The erionite contained fewer of these fibres than the UICC asbestos samples but, unlike these materials, erionite can cause the transformation of C3H 10T½ cells. Erionite did not increase the activities of benzo[*a*]pyrene in this cell transformation assay. The cytotoxic activities of both asbestos and erionite have a similar dependence on the number of long fibres. Extraction with cyclohexane did not affect the activity of erionite.

## *Introduction*

Endemic mesothelioma in several Turkish villages has been attributed to the inhalation of zeolite fibres (Baris *et al.*, 1981). The fibrous zeolite mineral erionite has been tested for carcinogenicity and found to cause mesotheliomata in mice (Suzuki, 1982; Suzuki *et al.*, 1980) and in rats (Maltoni *et al.*, 1982). In rats, these tumours occurred at much higher rates than those caused by any other fibrous dust yet examined (Wagner *et al.*, 1985) despite the fact that erionite contains fibres of size similar to those in the UICC sample of crocidolite.

It has been reported that the *in vitro* cytotoxicity of erionite was of the same order as that of crocidolite, that this was related to the number of long fibres, and that there were no differences in the inherent activity of these fibres (Brown *et al.*, 1980). However, erionite has a number of genotoxic activities not found in other fibrous materials (Poole *et al.*, 1983, 1986). It was therefore decided to modify fibrous erionite in a number of ways and to determine the effect that this had on these *in vitro* activities.

## *Materials and methods*

### Preparation of erionite samples

A sample of erionite from Rome, Oregon, USA, was obtained and a respirable sample prepared as described previously (Poole *et al.*, 1983).

Since, in this previous study, it was suggested that the oncogenic activity of erionite was the result of contamination of the fibre by hydrocarbons or other carcinogens, a sample of this dust was extracted in a soxhlet with cyclohexane and then dried *in vacuo*. Observation under the electron microscope showed that this treatment did not affect the size distribution or fibre content of this dust.

Two samples of the fibre were milled to obtain a non-fibrous product. One was reduced to this state in a stainless steel disk mill and the other was ground in an agate planetary ball mill.

**Tissue culture materials**

Culture media and fetal calf-serum were obtained from Flow Laboratories, Irvine, Scotland. Benzo[*a*]pyrene (BP) was obtained from Sigma Chemical Co., Poole, England.

*In vitro* **toxicity**

Toxicity studies were carried out on V79-4 cells using the methods previously described (Chamberlain & Brown, 1978). The survival of C3H 10T½ cells was estimated by measuring the total protein in cultures treated in parallel with those used in the transformation experiments; survival was estimated after 7–10 days.

**Cell transformation assay**

C3H 10T½ cells derived from mouse embryo fibroblasts (Reznikoff *et al.*, 1973a) were used between passages 10 and 12. These cells were cultured in Dulbecco's modification of Eagle's minimum essential medium (DMEM), with a concentration of bicarbonate of 3.6 g/l. To permit equilibration with a gas phase of 8% $CO_2$ in air, the medium was supplemented with heat-inactivated fetal calf serum (10% v/v), and penicillin/streptomycin (200 units/50 $\mu$g/ml) was added. Samples of cells (5 ml, 200 cells per ml) from subconfluent cultures were distributed among 25-cm² tissue-culture flasks (Falcon), which were incubated overnight at 37°C with caps screwed on lightly to allow for equilibration of the gas phase. The cultures were treated, 24 h after plating, with suspensions of erionite (autoclaved dry, suspended in DMEM and sonicated just prior to addition). As a positive control, BP was dissolved in acetone and added to the cultures (final concentration of acetone <0.5%). Where mixtures of the 2 agents were used, these were added at the same time.

The cultures were left for 48 h at 37°C, after which time a medium change was made. The medium was then changed twice weekly until the cells reached confluence; thereafter, the concentration of serum was reduced to 5% and medium changes made weekly. After 6 weeks, the cultures were fixed in buffered formalin (10%), stained in methylene blue (1%), and scored for type 2 and 3 transformed foci, using the criteria described in Reznikoff *et al.* (1973b).

## Results

**Size distribution and cytotoxicity**

Electron microscopic examination of dispersed samples of the dust showed that less than 10% of fibres in the intact sample were more than 5$\mu$m long and that this sample contained a total of $3.8 \times 10^3$ fibres per µg (f/µg) of dust. Both milled samples contained about $1 \times 10^3$ f/µg and none of those examined exceeded 5 µm in length. Transmission electron micrographs of the intact and one of the milled samples are shown in Figure 1.

The elemental composition and electron diffraction patterns of particles from all the samples were similar both to one another and to those of fibres from the Turkish village of Karain.

The intact fibres used in this study had a size distribution similar to that used previously (Poole et al., 1983) and contained fewer fibres (and particularly a smaller proportion of long fibres) than the sample used by Wagner et al. (1985). Since all these samples were prepared in the same laboratory from the same starting material, this is particularly surprising. The small number of long fibres led to this sample having only a low cytotoxicity to V79-4 cells. Milling the fibre reduced this considerably (Table 1).

### Table 1. Survival of V79-4 cells[a]

| Concentration of erionite (µg/ml) | % survival | |
|---|---|---|
| | Long fibres | Short fibres |
| 5 | 86 ± 7.6 | 102 ± 13.7 |
| 10 | 89 ± 10.3 | 108 ± 12.6 |
| 20 | 75 ± 4.5 | 96 ± 9.8 |
| 50 | 61 ± 7.4 | 94 ± 7.0 |

[a]Determined by the method of Chamberlain and Brown (1978). The 95% confidence limits are also shown. The long fibres were those in the intact sample, while the short fibres were produced by milling in a disk mill

**Transformation assay**

The results are presented in Table 2, and show that exposure to erionite caused an increase in the number of transformed foci as compared with the negative control cultures. This sample was less active than that used previously (Poole et al., 1983). The milled non-fibrous sample caused fewer of the type-3 foci to appear than the intact fibres, but was more active in producing the less-transformed (and harder to score) type-2 foci.

This experiment was repeated with similar results (Table 3); extraction of the erionite with cyclohexane did not reduce its ability to cause transformation. In this experiment, the interaction of the fibres with BP was also investigated, and all of the dusts were tested in combination with it. In no case was the activity of the erionite and BP mixtures significantly greater than would be expected from a simple additive interaction.

Fig. 1. Transmission electron micrographs of the intact fibre (A) and of the sample milled in the planetary ball mill (B). The other milled sample was similar.

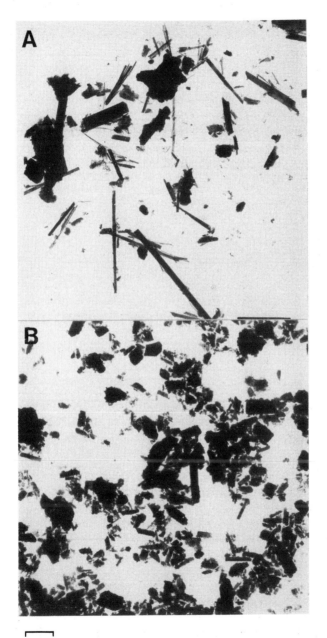

2.5 μm

## Table 2. Transformation of C3H 10T½ cells (first experiment)[a]

| Treatment | Survival (%) | No. of flasks | No. of foci | |
|---|---|---|---|---|
| | | | Type 3 | Type 2 |
| Controls | 100 | 39 | 2 | 14 |
| Benzo[a]pyrene (0.2 µg/ml) | 64 | 38 | 34 | 21 |
| Benzo[a]pyrene (1.0 µg/ml) | 31 | 35 | 38 | 27 |
| Long erionite (30 µg/ml) | 43 | 44 | 11 | 40 |
| Short erionite (30 µg/ml) | 87 | 40 | 2 | 111 |

[a]Estimated as described in the text. The number of type-2 and type-3 transformed foci was determined by the method of Reznikoff et al. (1973b)

## Table 3. Transformation of C3H 10T½ cells (second experiment)[a]

| Treatment | No. of type-3 foci | No. of type-2 foci | No. of flasks |
|---|---|---|---|
| **Controls** | 0 | 16 | 40 |
| **Benzo[a]pyrene:** | | | |
| 0.1 µg/ml | 5 | 15 | 20 |
| 0.5 µg/ml | 12 | 13 | 20 |
| **Erionite:** | | | |
| Long (20 µg/ml) | 6 | 34 | 20 |
| Long (40 µg/ml) | 4 | 49 | 19 |
| Short (20 µg/ml) | 1 | 31 | 20 |
| Short (40 µg/ml) | 0 | 44 | 20 |
| Extracted (20 µg/ml) | 4 | 30 | 20 |
| Extracted (40 µg/ml) | 6 | 30 | 20 |
| Long (20 µg/ml) + BP (0.1 µg/ml) | 6 | 32 | 20 |
| Long (40 µg/ml) + BP (0.1 µg/ml) | 11 | 20 | 20 |
| Short (20 µg/ml) + BP (0.1 µg/ml) | 6 | 49 | 20 |
| Short (40 µg/ml) + BP (0.1 µg/ml) | 3 | 33 | 20 |
| Extracted (20 µg/ml) + BP (0.1 µg/ml) | 3 | 15 | 20 |
| Extracted (40 µg/ml + BP (0.1 µg/ml) | 14 | 8 | 20 |

[a]Estimated as described in the text. The number of type-2 and type-3 transformed foci was determined by the method of Reznikoff et al. (1973b).
BP, benzo[a]pyrene.

## Discussion

The extremely high oncogenic activity of erionite raises several problems. In animal experiments, the activity of this dust is much greater than that of any asbestos sample (Wagner et al., 1985). The activity of erionite in the human populations of the Turkish villages may also be greater than that of other mineral fibres, though in that case exposure also occurs at an earlier age than is common for asbestos. In producing *in vitro* transformation, this mineral fibre seems to have a qualitatively different effect from those expressed (or rather absent) in asbestos samples.

The oncogenic activity of most fibres depends on the number of long, thin fibres inoculated or implanted; this relationship also holds for many *in vitro* effects of mineral fibres (see for example Brown et al., 1978). Milling the erionite destroys its ability to cause the transformation of C3H 10T½ cells, and thus the relationship between fibre size and activity seems also to hold for this end-point. Transformation cannot be explained simply as the result of the presence of these long fibres, since there are fewer of them in the present sample than is the case for the UICC samples of asbestos.

The relationship between fibre size and number and the reduction of the plating efficiency of V79-4 cells seems simpler. The activity of erionite is less than, for example, that of crocidolite asbestos. Thus it seems that, while both the cytotoxicity and transforming ability of erionite are dependent on the presence of long, thin fibres, the quantitative relationships for the two activities are not the same. Fewer fibres of erionite are needed for the expression of the transforming activity than for that of the cytotoxicity.

The fact that extraction with cyclohexane did not affect the activity of the erionite may mean that its carcinogenic activity is not due to an adsorbed carcinogen. However, extraction with this solvent could still leave a more strongly polar material on the fibre surface or any type of carcinogen within its porous structure, since the pores of this mineral are too small to be accessible to this solvent. At the same time, they are also too small to contain most types of carcinogen.

The absence of a synergism between erionite and the polycyclic carcinogen BP is in contrast with the 'promoter-like' effect of asbestos on both the action of this carcinogen (Brown et al., 1983) and radiation (Hei et al., 1984). Erionite might lack this promoting effect and be a true initiator in this transformation assay, whereas asbestos only acts as a promoter.

The large number of type-2 foci produced by the milled erionite is difficult to explain; this is the first time that such a response has been reported, though we have subsequently found that other non-fibrous zeolites produce the same result.

## References

Baris, Y.I., Saracci, R., Simonato, L., Skidmore, J.W. & Artvinli, M. (1981) Malignant mesothelioma and radiological abnormalities in two villages in central Turkey. *Lancet, i*, 984-986

Brown, R.C., Chamberlain, M., Griffiths, D.M. & Timbrell, V. (1978) The effect of fibre size on the *in vitro* biological activity of three types of amphibole asbestos. *Int. J. Cancer, 22*, 721-727

Brown, R.C., Chamberlain, M., Davies, R. & Sutton, G.T.A. (1980) The *in vitro* activities of pathogenic mineral dusts. *Toxicology, 17*, 143-147

Brown, R.C., Poole, A. & Fleming, G.A. (1983) The influence of asbestos dust on the oncogenic transformation of C3H 10T½ cells. *Cancer Lett., 18*, 221-227

Chamberlain, M. & Brown, R.C. (1978) The cytotoxic effects of asbestos and other mineral dust in tissue culture cell lines. *Br. J. Exp. Pathol., 59*, 183-189

Hei, T.K., Hall, E.J. & Osmak, R.S. (1984) Asbestos, radiation and oncogenic transformation. *Br. J. Cancer, 50*, 717-721

Maltoni, C., Minardi, F. & Morisi, L. (1982) The relevance of the experimental approach in the assessment of the oncogenic risks from fibrous and non-fibrous particles. The ongoing project at Bologna. *Med. Lav., 73*, 393-397

Poole, A., Brown, R.C., Turver, C.J., Skidmore, J.W. & Griffiths, D.M. (1983) *In vitro* genotoxic activities of fibrous erionite. *Br. J. Cancer, 47*, 697-705

Poole, A., Brown, R.C. & Rood, A.P. (1986) *In vitro* activities of a highly carcinogenic mineral fibre — potassium octatitanate. *Br. J. Exp. Pathol., 67*, 289-296

Reznikoff, C.A., Bertram, J.S., Brankow, D.W. & Heidelberger, C. (1973a) Quantitative studies of chemical transformation of cloned C3H mouse embryo cells sensitive to post confluence inhibition of cell division. *Cancer Res., 33*, 3239-3249

Reznikoff, C.A., Brankow, D.W. & Heidelberger, C. (1973b) Establishment and characterization of a cloned line of C3H mouse embryo cells sensitive to post confluence inhibition of division. *Cancer Res., 33*, 3231-3238

Suzuki, Y. (1982) Carcinogenic and fibrogenic effects of zeolites. Preliminary observations. *Environ. Res., 27*, 433-445

Suzuki, Y., Rohl, A.N., Langer, A.M. & Selikoff, I.J. (1980) Mesothelioma following intraperitoneal administration of zeolite. *Fed. Proc., 39*, 3

Wagner, J.C., Skidmore, J.W., Hill, R.J. & Griffiths, D.M. (1985) Erionite exposure and mesothelioma in rats. *Br. J. Cancer, 51*, 727-730

# MECHANISMS OF FIBRE-INDUCED SUPEROXIDE RELEASE FROM ALVEOLAR MACROPHAGES AND INDUCTION OF SUPEROXIDE DISMUTASE IN THE LUNGS OF RATS INHALING CROCIDOLITE

B.T. Mossman[1], K. Hansen[2], J.P. Marsh[1], M.E. Brew[1], S. Hill[1], M. Bergeron[1] & J. Petruska[1]

[1] Department of Pathology, University of Vermont
College of Medicine, Burlington, VT, USA
[2] Nordic Gentocle Ltd, Copenhagen, Denmark

*Summary.* Asbestos resembles the phorbol ester, 12-*O*-tetradecanoylphorbol 13-acetate (TPA), in its ability to elicit release of superoxide ($O_2^-$) from rodent alveolar macrophages (AM) *in vitro*. In addition, superoxide dismutase (SOD), the antioxidant enzyme scavenging $O_2^-$, is increased in cultures of tracheobronchial epithelial cells and lung fibroblasts after exposure to either crocidolite or chrysotile asbestos. Our objectives here were to determine: (1) the chemical and physical properties of asbestos important in the generation of $O_2^-$ from rat AM; and (2) the effects of $O_2^-$ in comparison with asbestos on biosyntheses of collagen and non-collagen protein in rat lung fibroblasts *in vitro*. We were also interested in whether increased production of SOD occurred in the lungs of rats after inhalation of crocidolite asbestos.

To determine whether $O_2^-$ was elicited in response to a variety of asbestiform fibres, AM lavaged from Fischer 344 rat lungs were exposed *in vitro* to equivalent non-toxic amounts of crocidolite asbestos, erionite, Code 100 fibreglass, sepiolite, and their non-fibrous analogues, riebeckite, mordenite and glass particles. In addition, sized preparations of long ($>10$ $\mu$m) and short ($<2$ $\mu$m) asbestos were introduced at identical concentrations to determine whether length of fibres is critical in $O_2^-$ release. The amount of $O_2^-$ released from AM in response to dusts was then determined by measuring SOD-inhibitable reduction of cytochrome C. All asbestiform fibres caused a significant ($p<0.05$) increase in generation of $O_2^-$ from epithelial cells, whereas non-fibrous particles were less active at comparable concentrations. Experiments with long ($>10$ $\mu$m) versus short ($<2$ $\mu$m) chrysotile showed that long fibres caused a more striking, dosage-dependent release of $O_2^-$. To determine whether $O_2^-$ plays a role in the causation of fibrotic lung disease, rat lung fibroblasts were exposed to a biochemical generation system (xanthine-xanthine oxidase) for $O_2^-$ before quantitation of cell-associated collagen and non-collagen protein at 24, 48 and 72 h thereafter. At the latter time periods, significant increases in total collagen per ng DNA were observed. In comparison with controls, the generation system for $O_2^-$ also caused an initial decrease in synthesis of non-collagen protein followed by increases in synthesis of non-collagen protein at 48 and 72 h.

Using a rapid onset inhalation model of asbestosis developed in this laboratory, we assessed the amounts of SOD (both the Cu-Zn and Mn-containing forms) in the lungs of Fischer 344 rats exposed to crocidolite asbestos (10 mg/m³ air; 6 per day) for 1, 3, 6 and 9 days. At 3, 6 and 9 days, SOD in the lungs of crocidolite-exposed rats was increased ($p < 0.05$) above levels observed in sham-exposed animals. Taken in context, results here suggest that enhanced release of $O_2^{-\cdot}$ and a compensatory increase in antioxidant enzymes occur in cells of the respiratory tract following exposure to asbestiform minerals. Active oxygen species generated upon phagocytosis of fibres by AM or by direct interaction of 'target' cells with asbestos cause alterations in normal cell differentiation, e.g., increases in collagen biosynthesis, which may contribute to lung disease.

## Introduction

Exposure to asbestos, a family of hydrated silicate fibres (where a fibre is defined as a particle of length-to-diameter ratio $>3:1$), is associated with the development of malignant (mesothelioma, bronchogenic carcinoma) and fibrotic (asbestosis) lung disease. The precise mechanisms involved in the pathogenesis of asbestos-associated diseases are unclear, but work by this laboratory (Hansen & Mossman, 1987; Mossman *et al.*, 1986a,b; Shatos *et al.*, 1987) and others (Goodglick & Kane, 1986; Weitzman & Weitberg, 1985) suggests that the generation of active oxygen species is involved in inflammation and cell damage. For example, crocidolite-induced toxicity to rodent tracheobronchial epithelial cells (Mossman *et al.*, 1986a), lung fibroblasts (Shatos *et al.*, 1987), and alveolar (Shatos *et al.*, 1987) or peritoneal (Goodglick & Kane, 1986) macrophages *in vitro* can be prevented by the administration of scavengers of oxygen metabolites such as superoxide dismutase (SOD), catalase and mannitol. Crocidolite, chrysotile and amosite asbestos all cause lipid peroxidation of cell membranes (Gulumian *et al.*, 1983; Weitzman & Weitberg, 1985) which can be ameliorated by the use of desferroxamine, an iron chelator (Weitzman & Weitberg, 1985). Moreover, lipid peroxidation is further enhanced upon addition of NADPH and asbestos, crocidolite being a more potent stimulus than chrysotile at similar concentrations (Fontecave *et al.*, 1987).

The release of oxygen free radicals from cells of the immune system, specifically polymorphonuclear leucocytes (PMN) and macrophages, occurs after a number of toxic insults and is associated with the development of an inflammatory response (Freeman & Crapo, 1982). Since alveolar macrophages (AM) accumulate in the lungs of animals exposed to asbestos (Craighead & Mossman, 1982), we were interested in the characteristics (chemistry, geometry, length) of fibres important in eliciting $O_2^{-\cdot}$ from these cell types. Moreover, we wished to determine what effects generation of $O_2^{-\cdot}$ might have on a 'target' cell of disease, such as the lung fibroblast affected in asbestosis. Specifically, we tested the hypothesis that $O_2^{-\cdot}$ would cause changes in cell synthesis of collagen, a protein accumulating in the lung during fibrotic lung disease. To this end, we measured collagen and non-collagen protein synthesis in a normal line (RL-82) of rat lung fibroblasts after exposure to either crocidolite asbestos, an agent stimulating collagen biosynthesis in these cells (Mossman *et al.*, 1986b), or xanthine and xanthine oxidase, a reaction mixture generating $O_2^{-\cdot}$.

SOD, an enzyme scavenging $O_2^-$, is increased in cells of the respiratory tract after exposure to asbestos *in vitro* (Mossman et al., 1986a; Shatos et al., 1986). Another goal of the studies described here was to determine whether production of SOD is enhanced in animals after inhalation of crocidolite asbestos. Presumably, this might represent a protective mechanism to combat the increase in $O_2^-$ generated by AM in response to asbestos (Hansen & Mossman, 1987).

## Materials and methods

### Isolation of AM and exposure to particulates

Male golden Syrian hamsters (EHS strain), 6–8 weeks old, and Fischer 344 rats were anaesthetized with phenobarbital and lavaged as described previously (Hansen & Mossman, 1987). The lavage fluid from 6–8 rodents was centrifuged for 10 min at 900 g, and the cell pellet resuspended in Medium 199 (GIBCO, Grand Island, NY) containing 10% heat-inactivated fetal bovine serum. Over 98% of the cell yield consisted of AM as determined by differential cell counts. AM were plated in 12-well dishes at $5 \times 10^5$ or $10^6$ cells per well and maintained at 37°C in a NAPCO 5300 incubator (95% air, 5% $CO_2$) for 2 h, at which time non-adherent cells were decanted before exposure to particulates. Because hamster AM require the addition of opsonized Zymosan to detect a measurable increase in $O_2^-$ in response to particulates (Hansen & Mossman, 1987), these cell types were exposed to dusts for 2 h in serum-containing medium followed by removal of medium and subsequent exposure of cells to opsonized Zymosan A, heat-killed yeast particles (Sigma Chemical Company, St Louis, MO) in the $O_2^-$ reaction mixture (see below) for 1 h. In contrast, rat cells were exposed to dusts for 1 h in the $O_2^-$ reaction mixture. Particulates were dispersed by ultrasonication and suspended in serum-containing medium prior to their addition to cells and examined at a range of concentrations (2.5–25 μg/cm² dish).

### Sources of particulates

Several fibres, including crocidolite asbestos ($Na_2O.Fe_2O_3.3FeO.8SiO_2.H_2O$, UICC reference sample), Oregon erionite (($Na_2,K_2,Ca,Mg)_{4,5}(Al_9Si_{27}O_{72}).27H_2O$, a gift from Dr Reg Davies, MRC Toxicology Unit, Carshalton, Surrey, UK, and Code 100 fibreglass ($SiO_2$, Manville Corp., Denver, CO), were added to rat and hamster AM at non-toxic concentrations, as determined by exclusion of the vital dye, nigrosin, at 24 h after exposure to particulates. Preparations of their non-fibrous, chemically similar analogues, namely riebeckite ($Na_2O.Fe_2O_3.3FeO.8SiO_2.H_2O$, Wards National Science Est., Rochester, NY), mordenite (($Ca,Na_2,K_2)_4Al_8Si_{40}O_{96}.28H_2O$, a gift from Dr Robert Emerson, Department of Pathology, University of Vermont) and glass beads ($SiO_2$, 1–4 μm diameter, Particle Information Service, Inc.) were used in comparative studies. Suspensions of sepiolite ($Mg_4(SiO_5)_3(OH).6H_2O$, Minerals Research, Clarkstown, NY) consisted of mixed fibres and particles (Hansen & Mossman, 1987). The geometry and dimensions of all preparations of dusts have been reported previously (Hansen & Mossman, 1987; Woodworth et al., 1983).

### Assay for $O_2^-$

AM alone and with addition of dusts were incubated in a reaction mixture containing Earle's balanced salt solution (GIBCO, Grand Island, NY) and 40 $\mu$M cytochrome C (Type VI, Sigma Chemical Co., St Louis, MO) for 1 h. At this time, the supernatant was decanted into microfuge tubes on ice and centrifuged at 8000 g to remove debris. All groups ($N$=3–5 per determination) were run with and without the addition of 100 $\mu$g/ml bovine erythrocyte SOD. Reduction of cytochrome C was measured spectrophotometrically at 550 nm, and results calculated as $\mu$mol reduced cytochrome C per $5 \times 10^5$ or $10^6$ cells per h, using a 0.029 mM extinction coefficient for cytochrome C. Analysis of variance by the Dunnett's procedure was used to determine statistically significant differences between groups (Duncan, 1955). Cell-free reaction mixtures to which fibres were added served as additional controls in all experiments.

### Measurement of collagen and non-collagen protein synthesis in rat lung fibroblasts (RL-82 cells) after exposure to $O_2^-$

To determine whether $O_2^-$ alters protein synthesis in fibroblasts of the lung, the normal RL-82 cell line isolated from the lung of a Fischer 344 rat was obtained from Dr Marlene Absher, Department of Medicine, University of Vermont. Cells were maintained in minimal essential medium (MEM, GIBCO, Grand Island, NY) containing 10% fetal calf serum and exposed at near confluence to xanthine (50 $\mu$M) and xanthine oxidase (0.5 and 1 units per ml, Sigma Chemical Co., St Louis, MO). Controls consisted of untreated cells and those to which xanthine (50 $\mu$M) alone had been added. At 24-h intervals, ascorbate ($10^{-4}$M) was added to dishes. Cells were then pulsed with [$^3$H]proline (10 $\mu$Ci/ml medium) for 2 h, homogenized and boiled for 10 min to inactivate proteases, and assayed in the presence and absence of purified bacterial collagenase (Advance Biofactures Corp., USA, 40 units per assay tube) to determine newly synthesized collagen and non-collagen protein. Results, expressed as cpm per ng DNA, were examined by the Student's $t$-test, adjusting for multiple comparisons between groups.

### Determination of SOD in the lungs of Fischer 344 rats after inhalation of crocidolite asbestos

To ascertain whether increased production of SOD, the endogenous scavenger enzyme of $O_2^-$, occurs in the lung after inhalation of asbestos, male Fischer 344 rats, 6–8 weeks old, were exposed to NIEHS crocidolite asbestos (10 mg/m³ air, 6 h per day) for 1, 3, 6 and 9 days. On the morning after completion of the exposure, the lungs were perfused with phosphate-buffered saline, removed from asbestos-exposed and sham control animals ($N$=4 per group), minced in potassium phosphate buffer (pH 7.8) containing 0.1 mM EDTA, and cell-free extracts prepared using a Polytron homogenizer (Brinkmann Instrument, Westbury, NY). The activity of total endogenous SOD (both Cu-Zn and Mn-containing forms) was then determined by methods described previously (Mossman et al., 1986a). The Student's $t$-test, adjusting for multiple group comparisons, was used to determine the statistical significance of the results.

## Results

### Release of $O_2^-$ from AM response to particulates

Both hamster and rat AM phagocytosed fibres and particles immediately after their addition to cultures. Because hamster AM did not generate $O_2^-$ in the absence of dusts, and production of $O_2^-$ in these cells was minimal in response to crocidolite asbestos (Hansen & Mossman, 1987), pretreatment with Zymosan was necessary to demonstrate measurable responses to particulates. Figure 1 shows significant ($p<0.05$) increases in release of $O_2^-$ after exposure to all of the fibres tested, whereas riebeckite and glass beads, non-fibrous analogues of crocidolite, and Code 100 fibreglass, were ineffective at both comparable and higher concentrations. The results obtained from experiments using rat AM were similar (Figure 2). Although increased release of $O_2^-$ occurred in response to all particulates, several-fold higher concentrations of particles (in comparison to fibres) were required for measurable amounts of $O_2^-$ to be detected. For example, addition of crocidolite at 2.5 and 5.0 $\mu g/cm^2$ per dish caused a significant ($p<0.01$) dose-dependent elevation of $O_2^-$ in comparison with unexposed AM (Figure 2A). In contrast, its non-fibrous analogue, riebeckite, caused less significant ($p<0.05$) increases in $O_2^-$ at 5-fold higher concentrations, i.e., 25 $\mu g/cm^2$ per dish. Similar trends were also observed using Code 100 fibreglass (Figure 2B) and erionite (Figure 2C) in comparison with their non-fibrous derivatives, glass beads and mordenite, respectively. Thus, the fibrous geometry of asbestiform dusts appears important in the generation of $O_2^-$ from AM.

The length of the fibre also affects release of $O_2^-$ from cells. Table 1 shows generation of $O_2^-$ from rat AM after exposure to long ($>10$ $\mu m$) or short ($<2$ $\mu m$) fibres of Mansville chrysotile asbestos. Long ($>10$ $\mu m$) fibres caused a more marked increase in production of $O_2^-$ than short ($<2$ $\mu m$) fibres at higher concentrations, although generation of $O_2^-$ in response to both fibre preparations was enhanced in comparison with control cells. As has been observed previously with the use of crocidolite (Hansen & Mossman, 1987), AM show only a small increase in response to chrysotile asbestos when compared with opsonized Zymosan.

### Collagen and non-collagen protein synthesis after exposure of RL-82 lung fibroblasts to xanthine-xanthine oxidase or crocidolite asbestos

Whether release of $O_2^-$ by AM alters the normal differentiation of other cell types in the lung is of critical importance to an understanding of the pathogenesis of asbestos-associated disease. To address this possibility, we assessed collagen and non-collagen protein synthesis in rat lung fibroblasts at 24, 48 and 72 h after addition of xanthine and xanthine oxidase, a reaction mixture generating $O_2^-$. In comparison with untreated and xanthine controls, a decrease ($p<0.01$) in non-collagen protein synthesis is observed at 24 h with use of 0.5 and 1.0 units/ml xanthine oxidase (Figure 3) at 24 h. Alternatively, significant increases in non-collagen protein synthesis are observed at 48 and 72 h. Similar trends were observed with regard to collagen sythesis (Figure 4). Both non-collagen and collagen protein synthesis declined after 24 h in control and xanthine-treated cells as these cells became confluent.

**Fig. 1. Release of $O_2^-$ from hamster alveolar macrophages (AM) exposed to fibres and particles**

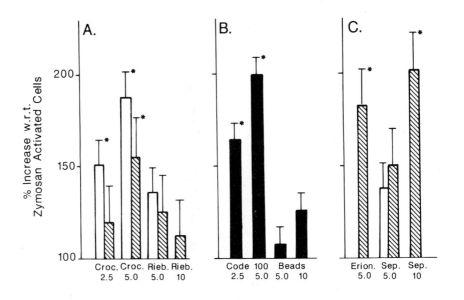

Various dusts were added for 2 h in culture medium before addition of Zymosan (0.25 mg/ml) to the reaction mixture consisting of Earle's balanced salt solution and 40 µM cytochrome C. The amount of $O_2^-$ released (mean ± SEM; $N = 4$ observations per group) is presented as a percentage increase as compared with values observed in cells exposed to Zymosan alone; □ and ▨ indicate the results of duplicate experiments; * = increased significantly ($p < 0.05$) in comparison with controls exposed to Zymosan alone. Reprinted with permission from Hansen and Mossman (1987) *Cancer Res.*, **47**, 1681-1686.

**Table 1. Release of superoxide ($O_2^-$) from rat alveolar macrophages (AM) in response to long (>10 µm) and short (<2 µm) chrysotile asbestos**

| Stimulus | $O_2^-$ Generation (µM cytochrome C per h per $10^6$ cells)[a] |
|---|---|
| None (control) | 1.59 ± 0.25 |
| Zymosan (opsonized) | 10.44 ± 1.65[b] |
| Long chrysotile: | |
| 0.5 µg/cm² | 2.84 ± 0.35[b] |
| 1.0 µg/cm² | 5.44 ± 0.66[b] |
| 2.0 µg/cm² | 6.87 ± 0.69[b] |
| Short chrysotile: | |
| 0.5 µg/cm² | 2.99 ± 0.85 |
| 1.0 µg/cm² | 4.19 ± 1.89[b] |
| 2.0 µg/cm² | 4.14 ± 2.3[b] |

[a]Mean ± SEM of 2-3 observations per group in duplicate or triplicate experiments.
[b]Increased significantly ($p < 0.05$) in comparison with untreated controls.

**Fig. 2. Release of $O_2^-$ from rat alveolar macrophages (AM) exposed to fibres and particles**

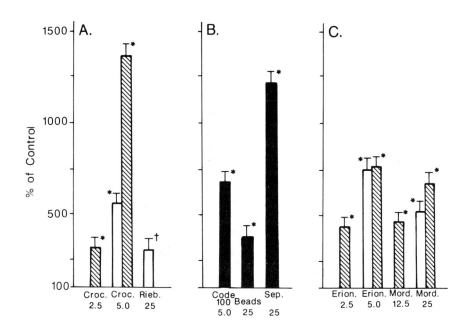

The amount of $O_2^-$ released (mean ± SEM; $N$ = 3–5 observations per group) is presented as a percentage increase as compared with values observed in unexposed controls; * = increased significantly ($p < 0.01$) in comparison with controls; † = increased significantly ($p < 0.05$) in comparison with controls. Modified from Hansen and Mossman (1987) *Cancer Res., 47*, 1681-1686

### Levels of SOD in the lungs of Fischer 344 rats exposed to asbestos

If release of $O_2^-$ by AM or other cell types occurred in the lung after exposure to asbestos, one would expect a compensatory increase in endogenous SOD, the enzyme scavenging $O_2^-$. In confirmation of this hypothesis, the lungs of rats inhaling crocidolite asbestos showed significant elevation of SOD as early as 3 days after initial exposure (Figure 5). These increases became more striking with time.

## *Discussion*

The data presented support the concept that oxygen free radicals are produced in the initial inflammatory response to asbestos and other fibres in the lung. Several mechanisms of generation of $O_2^-$ and other reactive species appear possible. For example, Weitzman and Graceffa (1984) have shown, using electron spin resonance spectroscopy, that incubation of amosite, chrysotile or crocidolite asbestos with hydrogen peroxide leads to production of $O_2^-$ and the hydroxyl radical (OH'). Since generation of $O_2^-$ and OH' from hydrogen peroxide does not occur when fibres treated with desferroxamine, an iron chelator, are used, these investigators suggest

**Fig. 3. Effects of xanthine and xanthine oxidase on non-collagen protein synthesis in RL-82 lung fibroblasts.**

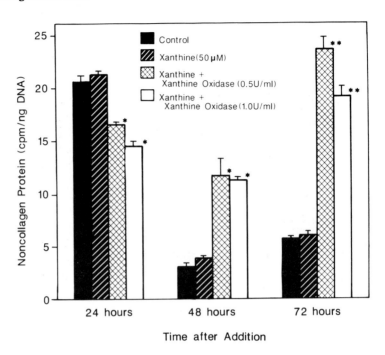

Values represent the mean ± SEM of a typical experiment. All experiments were performed in triplicate; * = significantly different ($p < 0.01$) from xanthine (50 μM) group; ** = significantly different ($p < 0.001$) from xanthine (50 μM) group.

that iron on the surface of fibres serves to catalyse the reaction. Pezerat and colleagues (Pezerat et al. (pp. 100-111); Zalma et al., 1987) have also demonstrated the formation of OH· from asbestos and other iron-containing minerals in a cell-free system by reduction of oxygen arising from the surface of the fibres. Thus, direct generation of active oxygen species by fibres appears to be an iron-dependent process occurring in the absence of cellular contact, e.g., from unphagocytosed asbestos fibres residing in the lung. This observation provokes intriguing speculations concerning the nature of ferruginous or asbestos bodies in the lungs of man. These coated fibres are apparently created after deposition of haemosiderin on fibres by macrophages, and are a hallmark of exposure to asbestos (Craighead & Mossman, 1982). Although the biological significance of asbestos body formation is unclear, most scientists view the process as a protective response whereby the active surfaces of fibres are contained. In contrast, the results of recent experimental studies (Goodglick & Kane, 1986; Shatos et al., 1987; Weitzman & Graceffa, 1984; Weitzman & Weitberg, 1985; Zalma et al., 1987) suggest that iron drives reactions favouring the production of active oxygen species and subsequent cytotoxicity to cells.

**Fig. 4. Effects of xanthine and xanthine oxidase on collagen protein synthesis in RL-82 lung fibroblasts**

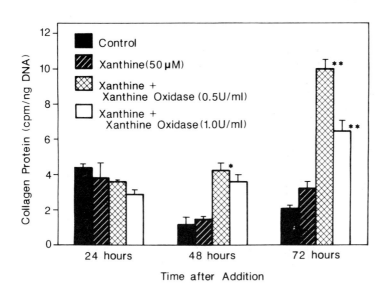

Values represent the mean ± SEM of a typical experiment. All experiments were performed in triplicate; * = significantly different ($p < 0.01$) from xanthine (50 μM) group; ** = significantly different ($p < 0.001$) from xanthine (50 μM) group.

We show here another mechanism of generation of oxygen free radicals involving incomplete phagocytosis of fibres by AM. Clearly, longer fibres are more apt to generate $O_2^-$ in comparison with shorter fibres or particulates which can be effectively incorporated into phagosomes and lysosomes (Mossman et al., 1977). In initial experiments, we observed a significant decrease (in comparison with controls) in $O_2^-$ release into medium by AM exposed to short (<2 μm) chrysotile fibres. Presumably, $O_2^-$ release under these circumstances occurred into the lysosome, since we were unable to measure it extracellularly. In contrast to generation of $O_2^-$ by fibres directly, production of active oxygen species by 'frustrated' phagocytosis of longer fibres is not an iron-dependent process as it occurs in response to a number of non-iron-containing fibres of similar dimensions to asbestos.

Oxygen free radicals induce a plethora of biochemical changes in cells including peroxidation of lipids and damage to DNA (Freeman & Crapo, 1982). These phenomena are linked causally to cytotoxicity. To determine whether non-toxic concentrations of $O_2^-$ could produce changes in differentiation of cells that might contribute to disease, we exposed normal lung fibroblasts (RL-82 cells) to low

**Fig. 5. Superoxide dismutase (SOD) in the lungs of sham- and crocidolite-asbestos-exposed Fischer 344 rats**

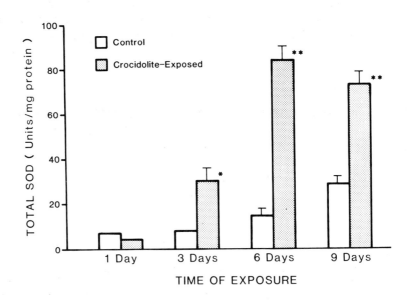

Values represent the mean ± SEM of 3-4 rats per group. Experiments were performed in duplicate; * = significantly increased ($p<0.05$) in comparison with sham controls; ** = significantly increased ($p<0.01$) in comparison with sham controls

concentrations, i.e., those not causing cytotoxic alterations in cells over a 24-h period, of xanthine and xanthine oxidase. Based on our prior results with crocidolite (Mossman et al., 1986b) showing an increased ratio of total cell-associated collagen to non-collagen protein in these cell types, we speculated that the generation of $O_2^-$ by asbestos fibres might be responsible for these biological effects. We therefore measured collagen and non-collagen protein synthesis in cultures exposed to an $O_2^-$-generating system in an attempt to establish a cause-and-effect relationship. The results indicate that $O_2^-$ alone causes an increase in collagen and non-collagen synthesis in RL-82 cells at 48 and 72 h. This time frame of response corresponds to the times when increased synthesis of proteins is observed after exposure to crocidolite asbestos (Mossman et al., 1986b). We were unable to demonstrate any changes in the incorporation of [$^3$H]thymidine in asbestos (Mossman et al., 1986b) or xanthine-xanthine oxidase-exposed RL-82 cells (data not shown), indicating that increased proliferation of these cell types does not occur under these circumstances.

As described in the Introduction to this paper, a number of *in vitro* studies support the concept that asbestos-induced cytotoxicity is linked to the formation of oxygen free radicals. Using a rodent inhalation model of disease (Mossman et al., 1985), we report here a rapid increase in the amount of SOD, an enzyme scavenging $O_2^-$, in the lungs of crocidolite-exposed rats. These results provide indirect evidence that increased production of $O_2^-$ occurs in the lungs of animals and humans exposed to asbestos. Presumably, the increase in total SOD represents a protective mechanism to combat the production of active oxygen species by asbestos. We are currently studying the regulation of antioxidant enzyme genes in cells of the lung in an attempt to answer this question.

## Acknowledgements

This research was supported by grant No. PHS RO133501 from the National Cancer Institute, grant No. PHS RO103878 from the National Institute of Environmental Health Sciences and a SCOR grant from the National Heart, Blood and Lung Institute. We thank Virginia Kelleher for typing this manuscript and Judith Kessler and David Hardwick for preparing illustrative materials.

## References

Craighead, J.E. & Mossman, B.T. (1982) The pathogenesis of asbestos-associated diseases. *New Engl. J. Med., 306*, 1446-1455

Duncan, D.B. (1955) Multiple range and F tests. *Biometrics, 11*, 1-23

Fontecave, M., Mansuy, D., Jaouen, M. & Pezerat, H. (1987) The stimulatory effects of asbestos on NADPH-dependent lipid peroxidation in rat liver microsomes. *Biochem. J., 241*, 561-565

Freeman, B.A. & Crapo, J.P. (1982) Biology of disease: free radicals and tissue injury. *Lab. Invest., 47*, 412-426

Goodglick, L.A. & Kane, A.B. (1986) Role of reactive oxygen metabolites in crocidolite asbestos toxicity to mouse macrophages. *Cancer Res., 46*, 5558-5566

Gulumian, M., Sardanios, F., Kilroe-Smith, T. & Ockerse, G. (1983) Lipid peroxidation in microsomes induced by crocidolite fibers. *Chem. Biol. Interact., 44*, 111-118

Hansen, K. & Mossman, B.T. (1987) Generation of superoxide from alveolar macrophages exposed to asbestiform and nonfibrous particles. *Cancer Res., 47*, 1681-1686

Mossman, B.T., Ley, B.W., Kessler, J. & Craighead, J.E. (1977) Interaction of crocidolite asbestos with hamster respiratory mucosa in organ culture. *Lab. Invest., 36*, 131-137

Mossman, B.T., Shatos, M.A., Marsh, J.P., Doherty, J.M., Ashikaga, T., Hacker, M.P., Absher, M. & Hemenway, D. (1985) Contrasting patterns of early lung injury in asbestosis and silicosis. *Am. Rev. Respir. Dis., 131*, A187

Mossman, B.T., Marsh, J.P. & Shatos, M.A. (1986a) Alteration of superoxide dismutase activity in tracheal epithelial cells by asbestos and inhibition of cytotoxicity by antioxidants. *Lab. Invest., 54*, 204-212

Mossman, B.T., Gilbert, R., Doherty, J., Shatos, M., Marsh, J. & Cutroneo, K. (1986b) Cellular and molecular mechanisms of asbestosis. *Chest, 89*, 160-161S

Mossman, B.T., Marsh, J.P., Hardwick, D., Gilbert, R., Hill, S., Sesko, A., Shatos, M., Doherty, J., Weller, A. & Bergeron, M. (1987) Approaches to prevention of asbestos-induced lung disease using polyethylene glycol (PEG)-conjugated catalase. *J. Free Radical Biol. Med., 2*, 335-338

Shatos, M.A., Orfeo, T., Doherty, J. & Mossman, B.T. (1986) Manganese (Mn) form of superoxide dismutase (SOD) is increased in rat lung fibroblasts exposed to asbestos. *In Vitro, 22*, 48A

Shatos, M.A., Doherty, J.M., Marsh, J.P. & Mossman, B.T. (1987) Prevention of asbestos-induced cell death in rat lung fibroblasts and alveolar macrophages by scavengers of active oxygen species. *Environ. Res.*, *44*, 103-116

Weitzman, S.A. & Graceffa, P. (1984) Asbestos catalyzes hydroxyl and superoxide radical generation from hydrogen peroxide. *Arch. Biochem. Biophys.*, *228*, 373-376

Weitzman, S.A. & Weitberg, A.B. (1985) Asbestos-catalyzed lipid peroxidation and its inhibition by desferroxamine. *Biochem. J.*, *225*, 259-262

Woodworth, C., Mossman, B.T. & Craighead, J.E. (1983) Induction of squamous metaplasia in organ cultures of hamster trachea by naturally occurring and synthetic fibres. *Cancer Res.*, *43*, 4906-4911

Zalma, R., Bonneau, L., Jaurand, M.C., Guignard, J. & Pezerat, H. (1987) Formation of oxy-radicals by oxygen reduction arising from the surface activity of asbestos. *Can. J. Chem.* (in press)

# BRIEF INHALATION OF CHRYSOTILE ASBESTOS INDUCES RAPID PROLIFERATION OF BRONCHIOLAR-ALVEOLAR EPITHELIAL AND INTERSTITIAL CELLS

A.R. Brody, P.D. McGavran & L.H. Overby

*Laboratory of Pulmonary Pathobiology,
National Institute of Environmental Health Sciences,
Research Triangle Park, NC, USA*

*Summary.* Inhalation of asbestos fibres causes a progressive interstitial pulmonary fibrosis. To understand the basic cellular mechanisms which lead to this disease, we have studied the earliest proliferative events at the bronchiolar-alveolar regions of rats and mice exposed to chrysotile asbestos for 5 h. Animals were injected with tritiated thymidine 4 h prior to sacrifice at varying times ranging from immediately after cessation of exposure to one month post exposure. Light microscopic autoradiography showed that air-exposed control animals never had more than 1% of cells labelled. Rats and mice studied immediately after exposure also had normal numbers of labelled cells. However, between 12 and 48 h post exposure, asbestos-exposed animals exhibited up to 4-fold increases in the percentages of labelled epithelial and interstitial cells. Normal labelling returned by 8 days after exposure and was maintained through the one-month period studied. We conclude that inhalation of chrysotile asbestos induces rapid and highly significant increases in proliferation of epithelial and interstitial cells of the bronchiolar-alveolar regions where asbestos fibres were initially deposited.

## *Introduction*

It has become clear that inhalation of asbestos fibres causes interstitial pulmonary fibrosis (Selikoff & Lee, 1978). Consequently, in recent years, occupational exposures to high concentrations of asbestos dust have become rare. However, there are increasing efforts to remove asbestos products already in place, while mining and manufacturing continue at a limited pace. All of these activities must release some amounts of aerosol dust into the environment. The occurrence of brief high-concentration exposures during demolition of old buildings and removal of friable insulation may also be possible, and both workers and local inhabitants could inhale large numbers of fibres. The pathobiological sequelae of such exposures are not at all clear. By studying brief exposures of these types in an animal model, we have learned that the fibres which reach the alveolar level provoke rapid cellular responses, such as epithelial uptake of fibres (Brody *et al.*, 1981), complement activation (Warheit *et al.*, 1985, 1986) and consequent macrophage migration (Warheit *et al.*, 1984, 1985, 1986). In this paper we review our recent findings (Brody & Overby, 1988) on the earliest proliferative responses of epithelial and interstitial cells at the sites where the initial lesions of asbestosis are manifested (Chang et al., 1988).

## Materials and methods

Male (CD White) rats or mice (normal B10.D2 or complement-deficient B10.D2/oSn), 8 weeks old, were exposed (5 animals per group) to room air (sham) or to an aerosol of chrysotile asbestos (10 mg/m$^3$ respirable mass) for 5 h in open cages as previously described (Warheit et al., 1984, 1985). Following recovery periods of 0, 19, 24, 33, 48 h and 1, 2 and 4 weeks after inhalation, tritiated thymidine ($^3$HTdR) (2$\mu$Ci/g) was administered intraperitoneally (i.p.) to the asbestos- and sham-exposed animals. They were sacrificed 4 h after administration of the $^3$HTdR by injection of 1 ml sodium pentabarbitol (50 mg/ml) i.p. Animals were then perfused through the vasculature (by the pulmonary artery) with fixative (1% glutaraldehyde, 1% paraformaldehyde) at 23 cm H$_2$O pressure for 5 min. The whole lungs were removed from the chest and immersed in fixative overnight.

After the tissues were fixed, slices (2 × 5 × 10 mm) were taken at right angles to each mainstem bronchus from both right and left lungs and post-fixed with 0.5% osmium tetroxide in veronal acetate buffer. The slices were embedded in soft Epox 812 and polymerized at 60°C for 15 h. These plastic blocks were softened on a warming tray at 40°C, and slices 0.5 mm thick were cut parallel to the large tissue face with a double-edged razor blade. A terminal bronchiole with its attached alveolar ducts, including the bifurcation between the ducts, was selected under a dissecting microscope and cut out of the warmed plastic block. Two terminal bronchiole-alveolar duct regions per animal were selected for a total of 10 anatomical units per treatment group. The selected tissue was glued on to a BEEM blank with epoxy glue. For light microscope autoradiography, 0.6-$\mu$m sections were cut with a diamond knife on a MT5000 microtome, and from the first two blocks sectioned at each time point, a thin section for electron microscopy was cut adjacent to the thick section taken for autoradiography. The thin sections were mounted on slotted grids (0.2 × 1 mm), stained with uranyl acetate and lead citrate, and photographed with a JEOL 100CX electron microscope. The autoradiography thick sections were placed on glass slides, coated with Ilford L4 emulsion, exposed for 3 weeks, developed with D-19 (Kodak, Rochester, NY) and stained with toluidine blue. For each treatment group, labelled nuclei (6 or more grains over a nucleus) were counted by light microscopy at 100× magnification in the epithelium and interstitium of the following 4 anatomical units of the terminal bronchiole-alveolar duct regions: (1) terminal bronchiole; (2) alveolar duct walls between the terminal bronchiole and first alveolar duct bifurcation; (3) first alveolar duct bifurcation; and (4) ducts distal to the bifurcation (Figure 1).

Data were analysed by Wilcoxon's rank sum test for non-parametric values, and statistical significance was reached at a level of $p<0.05$.

## Results and discussion

As previously described for normal animals (Shami et al., 1986; Tryka et al., 1986), sham-exposed rats and mice showed no increase in tritiated thymidine ($^3$HTdR)

**Fig. 1. Light micrograph of the lung of a mouse 48 h after exposure to chrysotile asbestos**

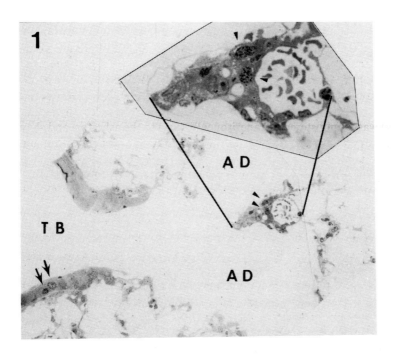

Bronchiolar epithelial cells (arrows) are clearly labelled with $^3$HTdR. The terminal bronchiole (TB) leads into alveolar ducts (AD), and the first alveolar duct bifurcation exhibits labelled epithelial and interstitial cells (arrowheads). These are best observed in the enlargement of the bifurcation.

incorporation at any time post exposure. All anatomical regions had a labelling level of about 1% (Figure 2).

A 5-hour exposure to chrysotile asbestos induced clear incorporation of $^3$HTdR by about 20 h post exposure (Figures 1 and 2). Immediately after exposure, percentages of labelled cells remained at normal levels (Figure 2). These percentages rose significantly through a 12–20-h period and reached a peak by 48 h post exposure (Figure 2). Percentages of labelled cells declined to normal levels by one week after exposure and remained normal through the one-month period studied (Figure 2).

The cell types labelled included cuboidal epithelial cells and attenuated interstitial cells of the terminal bronchioles (Figure 1) as well as Type II epithelial and undefined interstitial cells at the alveolar level (Figure 1). Here, cells of the proximal alveolar ducts and first alveolar duct bifurcations exhibited significantly increased levels of $^3$HTdR incorporation. No increases were observed in any cells distal to the first duct bifurcations. These findings are consistent with ultrastructural morphometric studies on rats showing that epithelial and interstitial cells of the first alveolar duct

**Fig. 2. Incorporation of tritiated thymidine ($^3$HTdR) after exposure to air (sham) and an aerosol of chrysotile asbestos (10 mg/m$^3$ respirable mass)**

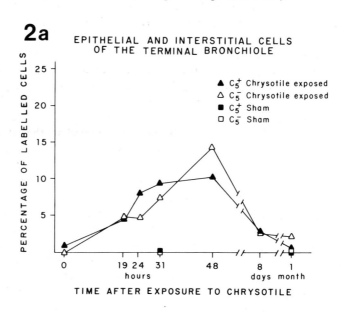

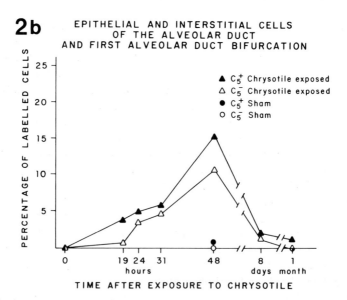

Note the clear time-dependent increase in HTdR-labelled cells in both the bronchioles and alveolar ducts of asbestos-exposed mice. Sham-exposed animals never exhibited more than 1% labelling at any time point. $C_5+$ = complement normal; $C_5-$ = complement deficient.

bifurcations are increased in number and volume only 48 h after a one-hour exposure to chrysotile asbestos (Chang et al., 1988). Interestingly, these morphometric studies show that an interstitial lesion at the bifurcations progresses through the one-month period studied, but tissues distal to the first alveolar duct bifurcations apparently remain unaffected.

The autoradiographic findings presented here and the morphometric studies support our hypothesis that asbestos fibres originally deposited on the surfaces of alveolar duct bifurcations (Brody & Roe, 1983) induce rapid cellular responses and consequent development of an interstitial lesion (Chang et al., 1988). The basic biological mechanisms which mediate the progression of the lesion are uncertain. It is clear that increased numbers of macrophages are attracted to the sites of fibre deposition (Warheit et al., 1984, 1985), and some of the cells which have incorporated $^3$HTdR are apparently interstitial macrophages, several of which contain asbestos (Brody & Overby, 1988). Since the complement-deficient mice exhibit a reduced macrophage response to asbestos inhalation (Warheit et al., 1985), we hoped that autoradiographic studies would shed some light on the role macrophages might play in the early cell responses. At this point in our studies, no apparent differences have been discerned between the normal and complement-deficient mice at the early time points (Figure 2). However, it was most interesting to learn that, by one month post exposure, the complement-deficient mice exhibited a significantly reduced interstitial lesion compared with the normal controls (McGavran et al., 1987). Further autoradiographic and ultrastructural comparisons are being carried out on this interesting model system.

Thus, at least two alternative, although not mutually exclusive, hypotheses exist to explain mechanisms controlling the pathogenesis of the asbestos-induced cellular alterations. First, it is clear that inhaled asbestos fibres are deposited at all levels of the respiratory tract (Morgan et al., 1975). It is conceivable that the fibres on epithelial membranes which are taken into the cytoplasm of Type I alveolar epithelial cells (Brody et al., 1981) could directly provoke a mitogenic response by the Type II epithelium. Even though Type II cells very rarely take up fibres (Pinkerton et al., 1984), it is the reaction of Type I cells which can dictate a proliferative response in Type II cells. This has been shown clearly in studies where injury to Type I cells by oxygen or $NO_2$ (Crapo et al., 1984) results in a mitogenic response by Type II cells. The fact that the Type I cells rapidly become thicker (Chang et al., 1988), together with the development of a small but significant leak of serum proteins into the lung (Warheit et al., 1986), support the notion that the inhaled asbestos could induce a mitogenic effect through direct membrane interactions with epithelial cells. On the other hand, direct effects on the bronchiolar epithelium are more difficult to envision. These cells have not been shown to phagocytose asbestos fibres *in vivo*, and they are covered by a layer of secreted proteins, lipids and carbohydrates.

A second possibility to be considered as a source of the mitogenic stimulus is the population of pulmonary macrophages which responds to the inhaled asbestos (Warheit et al., 1984, 1985). An alveolar macrophage-derived factor has been shown to induce tritiated thymidine incorporation by Type II cells *in vitro* (Leslie et al., 1986), and macrophage-derived growth factors for fibroblasts have been clearly demonstrated (Ross et al., 1986). Recently, work in our laboratory has shown that pulmonary macrophages secrete a platelet-derived growth factor homologue which binds specifically to receptors on rat lung fibroblasts (Kumar et al., 1988). Since increased numbers of pulmonary macrophages accumulate in the bronchiolar-alveolar regions during the 48-h post-exposure period (Warheit et al., 1985, 1986), it is conceivable that these cells synthesize and secrete mitogenic factors which induce the epithelial and interstitial cells to incorporate tritiated thymidine during proliferation. This hypothesis remains to be proven; however, the anatomical distribution of alveolar and interstitial macrophages appears to be appropriate. It is clear that asbestos fibres cause early significant incorporation of $^3$HTdR by a variety of pulmonary cells.

## Conclusions

We have shown that a brief exposure to chrysotile asbestos fibres induces a proliferative response by epithelial and interstitial cells associated with sites of initial dust deposition in rats and mice. These findings correlate with earlier morphometric studies showing progression of an interstitial lesion. This shows that brief exposures, such as those that might occur during demolition or removal of asbestos-containing structures, induce cellular responses and subsequent bronchiolar alveolar lesions. Whether or not similar lesions will occur in humans and progress, with pathological sequelae, has not yet been determined.

## Acknowledgements

We are indebted to the highly co-operative individuals from Northrop Services, Inc., who operate the NIEHS Inhalation Exposure Facility under contract from the National Toxicology Program.

## References

Brody, A.R. & Overby, L.H. (1988) Incorporation of tritiated thymidine by epithelial and interstitial cells in bronchiolar-alveolar regions of asbestos-exposed rats. *Am. J. Pathol.* (in press)

Brody, A.R. & Roe, M.W. (1983) Deposition pattern of inorganic particles at the alveolar level in the lungs of rats and mice. *Am. Rev. Respir. Dis.*, *128*, 724-729

Brody, A.R., Hill, L.H., Adkins, B. & O'Connor, R. (1981) Chrysotile asbestos inhalation in rats: deposition pattern and reaction of alveolar epithelium and pulmonary macrophages. *Am. Rev. Respir. Dis.*, *123*, 670-679

Chang, L.Y., Hill, L.H., Brody, A.R. & Crapo, J.D. (1988) Progressive lung cell reactions and extracellular matrix production after a brief exposure to asbestos. *Am. J. Pathol.* (in press)

Crapo, J.D., Barry, B.E., Chang, L. & Mercer, R.R. (1984) Alterations in lung structure caused by inhalation of oxidants. *J. Toxicol. Environ. Health*, *13*, 301-321

Kumar, R.K., Bennett, R.A. & Brody, A.R. (1988) A homologue of platelet-derived growth factor produced by rat alveolar macrophages. *FASEB J.*, *2*, 2272-2277

Leslie, C.L., McCormick-Shannon, K., Cook, J.L. & Mason, R.J. (1986) Macrophages stimulate DNA synthesis in rat alveolar Type II cells. *Am. Rev. Respir. Dis.*, *132*, 1246-1252

McGavran, P.D., Butterick, C.J., Overby, L.H. & Brody, A.R. (1987) Asbestos-induced incorporation of $^3$H-thymidine in bronchiolar-alveolar cells and development of an interstitial lesion. *Am. Rev. Respir. Dis.*, *135*, A165

Morgan, A., Evans, J.C., Evans, R.J., Hounam, R.F., Holmes, A. & Doyle, S. (1975) Studies on the deposition of inhaled fibrous material in the respiratory tract of the rat and its subsequent clearance using radioactive tracer techniques. II. Deposition of UICC standard reference samples of asbestos. *Environ. Res.*, *10*, 196-207

Pinkerton, K.E., Pratt, P.C., Brody, A.R. & Crapo, J.D. (1984) Fiber localization and its relationship to lung reactions in rats after chronic exposure to chrysotile asbestos. *Am. J. Pathol.*, *117*, 484-493

Ross, R., Raines, E.W. & Bowen-Pope, D.F. (1986) The biology of platelet-derived growth factor. *Cell*, *46*, 155-169

Selikoff, I.J. & Lee, D.H. (1978) *Asbestos and Disease*, New York, Academic Press

Shami, S.G., Evans, M.J. & Martinez, L.A. (1986) Type II cell proliferation related to migration of inflammatory cells into the lung. *Exp. Mol. Pathol.*, *44*, 344-352

Tryka, A.F., Witschi, H., Gosslee, D.G., McArthur, A.H. & Clapp, N.K. (1986) Patterns of cell proliferation during recovery from oxygen injury. *Am. Rev. Respir. Dis.*, *133*, 1055-1059

Warheit, D.B., Chang, L.Y., Hill, L.H., Hook, G.E., Crapo, J.D. & Brody, A.R. (1984) Pulmonary macrophage accumulation and asbestos-induced lesions at sites of fiber deposition. *Am. Rev. Repir. Dis.*, *129*, 301-310

Warheit, D.B., George, G., Hill, L.H., Snyderman, R. & Brody, A.R. (1985) Inhaled asbestos activates a complement-dependent chemoattractant for macrophages. *Lab. Invest.*, *52*, 505-514

Warheit, D.B., Hill, L.H., George, G. & Brody, A.R. (1986) Time course of chemotactic factor generation and the corresponding macrophage response to asbestos inhalation. *Am. Rev. Respir. Dis.*, *134*, 128-133

# PRODUCTION OF OXYGEN RADICALS BY THE REDUCTION OF OXYGEN ARISING FROM THE SURFACE ACTIVITY OF MINERAL FIBRES

### H. Pezerat, R. Zalma & J. Guignard

*Laboratoire de Réactivité de Surface et Structure,
Université Pierre et Marie Curie, Paris, France*

### M.C. Jaurand

*INSERM U 139, CHU Henri Mondor, Créteil, France*

*Summary.* According to certain hypotheses, the production of oxygen radicals within the biological medium (the phenomenon of oxidative stress) may play an important role in fibrosis and in certain steps of carcinogenesis. The mineral fibres of various materials are capable of participating in this phenomenon, owing to the reducing nature of their surface activity, so that OH˙ radicals can be produced from oxygen in 3 steps. The surface activity of inorganic materials which are insoluble or only very slightly soluble is due to the presence of electron donor active sites, generally linked to $Fe^{2+}$ ions found in the neighbourhood of the surface. In biological systems, these sites may emerge on the surface as a result of the partial dissolution of the particle, the action of a biological reducing agent, the phenomenon of deposition on the surfaces or cation exchange. We have explored the reducing properties of the surfaces of a certain number of mineral fibres, in aqueous buffer medium, by electron paramagnetic resonance (EPR) measurement of the adduct with the radical-trapping agent 5,5′-dimethyl-1-pyrrolidine-$N$-oxide (DMPO), produced from the radicals initially formed (OH˙ or R˙). We have found certain fibres to be highly effective in producing radicals from dissolved oxygen (Canadian chrysotile, nemalite, freshly ground amphiboles) while others have little effect. The reducing activity of certain fibres may be markedly increased by prior treatment in the presence of a ferrous salt (as in the case of erionite) or by the addition of glutathione to the reaction medium (as in the case of UICC crocidolite). It is suggested that the carcinogenic activity of certain inorganic materials at the pulmonary level is the result of their surface reducing properties. These reducing properties may either be present at the time of inhalation or acquired in the biological medium. This hypothesis is not in conflict with the observation of the role of the dimensional characteristics of fibres in mesothelioma.

## Introduction

Studies on chemical carcinogenesis have shown that the genotoxicity of an organic compound is in general the result of the oxidation or alkylation of the macromolecules of the genome by an electrophilic entity produced by the metabolism of the compound.

For many inorganic compounds, including the different varieties of asbestos, which are slightly or very slightly soluble in biological media, the intracellular activity will be essentially the surface activity of a phagocytosed particle. The surface activity of a solid is, in turn, determined by the nature of the surface active sites. These sites may be electron donors or acceptors, thus possessing reducing or oxidizing properties; they may also exhibit acidic or basic properties (being electron pair acceptors or donors).

As far as chemical carcinogenesis is concerned, we can therefore propose a model for the genotoxic activity, shown schematically in Figure 1 for a mineral particle which has been phagocytosed.

**Fig. 1. Model of genotoxic activity of phagocytosed mineral particle. For explanation of symbols, see text.**

In this model $S_M$ represents the surface of a particle, capable of reacting with a target molecule Z present in the cell so as to yield a genotoxic entity $Z^*$ which is electrophilic and capable of leading to the oxidation or alkylation of the macromolecules of the genome. $S'_M$ symbolizes the passivated surface after reaction with the molecule Z.

Of course, the real course of events will be more complex and will include some intermediate steps between Z and $Z^*$. It may also include complementary steps involving reactive cellular species, allowing the regeneration of $S_M$ from $S'_M$.

In the case of the various asbestos materials, previous work (Bonneau & Pezerat, 1983; Bonneau et al., 1986a) has shown that the surface active sites are essentially electron donor sites (basic and reducing in character) with a relatively high density of about $10^{17}$ to $10^{18}$ per m². We have shown elsewhere (Zalma et al., 1986) that the surface basic sites enable us to convert a molecule (Z) which is slightly acidic, such as fluorene, into fluorenone, passing through a carbanion. But carbanions, being non-electrophilic entities, do not, in general, fit the required genotoxic model except, perhaps, when the basic sites on the solid are very strong; we shall come back to this point with respect to erionite.

We have therefore explored the capacity of the reducing surface sites, choosing as the target (Z) oxygen dissolved in an aqueous medium. Oxygen is present in constant concentration in all cells. This being the case, the toxic entity ($Z^*$) will be the hydroxyl radical OH˙, formed at the end of the 3-step sequence shown in equations (1) – (3).

The OH˙ radical, a powerful oxidizing agent, can react in two different ways, of which the first is with a reducing site on the surface, $S_M$, according to the deactivation reaction (4).

$$O_2 \xrightarrow{+e^-} O_2^- \quad (1)$$

$$O_2^- \xrightarrow{+e^- + 2H^+} H_2O_2 \quad (2)$$

$$H_2O_2 \xrightarrow{+e^-} OH^\cdot + OH^- \quad (3)$$

$$OH^\cdot \xrightarrow{+e^- + H^+} H_2O \quad (4)$$

The second reaction is with a molecule, RH, present in the medium, as follows:

$$OH^\cdot + RH \longrightarrow H_2O + R^\cdot \quad (5)$$

Reactions (4) and (5) are competitive. One or the other may become preponderant, depending on the concentrations of the reducing sites and RH, and the individual rate constants of these reactions.

The combination of $OH^\cdot$ and $R^\cdot$ constitutes a satisfactory model of the electrophilic entity $Z^*$. It has also the advantage of being in accordance with various observations by biologists on the peroxidation of membrane lipids by asbestos (Fontecave et al., 1987; Gulumian et al., 1983; Weitzman & Weitberg, 1985) and the production of oxy-radicals following the treatment of cell cultures with asbestos (Mossman et al., 1986; Turver et al., 1985).

In assessing the capacity of a mineral particle to reduce oxygen dissolved in an aqueous medium, account must be taken of the phenomena of passivation and activation of the surface active sites.

The origin of the reducing surface activity of mineral fibres is essentially due to the presence of $Fe^{2+}$ and the reducing sites may be the surface OH groups or water molecules, bound to an underlying octahedral $Fe^{2+}$. These $Fe^{2+}$ ions are oxidized in air to $Fe^{3+}$, this process taking place still more rapidly in an aqueous medium, leading to a structural reorganization to a limited depth. Passivation may, however, be caused in other ways, e.g., by the adsorption of certain biological macromolecules or by the deposition of a mineral coating on the fibres impeding any access of oxygen to the active sites.

Activation can be the result of any of the following processes: grinding, which creates fresh, unoxidized surfaces; dissolution of the superficial passivated sheet in the model or in the biological medium; or, e.g., the action of reducing agents present in the medium, which convert the $Fe^{3+}$ near the surface into $Fe^{2+}$.

For a particle, therefore, the extra- and intracellular surface reducing activity will be a function of:

— the surface area accessible to the reactive agents, which can often be assumed to be the surface area measured using the nitrogen adsorption isotherm (BET) method;
— the density of the reducing surface sites, which is related to the composition and structure of the mineral, and also to the competition between activation and passivation;
— the strength of these reducing sites, which is linked to their configuration.

## Materials and methods

### Asbestos

The following were used:

— UICC and commercial samples;
— six Canadian chrysotile samples representative of various Quebec mines, supplied by Dr C. Jolicoeur;
— two samples of short-fibre chrysotile (Johns Manville Mine, Asbestos, Quebec), the former obtained by sedimentation and the latter by the treatment with phosphorus oxychloride (chrysophosphate) of the former. These samples were also supplied by Dr C. Jolicoeur.

### Other materials

These included the following:

— Commercial samples: attapulgite (Senegal), fine fibreglass (borosilicate, Johns-Manville code 104, provided by Dr F. Pott) and potassium titanate fibres (Tismos D., Otsuka Chemicals);
— mineralogical samples: wollastonite (asbestos mine, Quebec), magnetite (Norway, laboratory code 329, containing only a few per cent of chlorite, a usual contaminant of magnetite) and erionite (Oregon);
— synthetic samples: magnetite-maghemite ($Fe_2O_3\gamma$, containing only 2% of $Fe^{2+}$).

### Methods

The experimental methods have been described in detail in other papers from our group (Zalma et al., 1987). The reactivity of the material (45 mg) in the reduction of oxygen dissolved in aqueous medium (2 ml) is studied at 37°C, in the absence of light, at pH 7.4 (potassium phosphate) by using the following reactions:

$$OH^{\cdot} + HCO_2^- \longrightarrow CO_2^- + H_2O \qquad (6)$$

$$CO_2^- + DMPO \longrightarrow (DMPO, CO_2^-)^o \qquad (7)$$

The formate anion, $HCO_2^-$ (1 ml of M solution of sodium formate), selected for its very high rate constant in reaction (6), can be replaced by other reactants RH, such as ethanol, acetone, dimethylsulfoxide, etc.

DMPO (5,5'-dimethyl-1-pyrroline-$N$-oxide, 1 ml of a 100 mM solution) is a radical trapping agent, which yields a radical adduct (DMPO, $CO_2$)$O_2^-$ in reaction (7). The half-life of this adduct is of the order of one hour under our experimental conditions, permitting a convenient quantitative measurement by electron paramagnetic resonance. The time of addition of DMPO is taken as time origin. Aliquots are withdrawn after 25 and 60 min, filtered (0.65 $\mu$m) and immediately examined by EPR. All the results given in Tables 1, 2 and 3, with intensity of the signal (DMPO, $CO_2$)$^{\cdot -}$ are based on the same arbitrary scale, which can be considered as an index of the activity of the solid material in the reduction of oxygen to $OH^{\cdot}$. Standardization of the number of $OH^{\cdot}$ radicals formed and consumed in reactions (6) and (7) is effected with a solution of diphenyl-$\alpha\alpha$-picryl-$\beta$-hydrazyl (DPPH) in benzene. A signal with an intensity of 1000 on our scale corresponds to the formation of $2 \times 10^{14}$ radicals (DMPO, $CO_2$)$^{\cdot -}$ per mg of the material.

### Table 1. Relative activity of materials in the reduction of $O_2$ to $OH^{\cdot a}$

| Material | (DMPO, $CO_2$)$^{\cdot -}$ signal intensity | |
|---|---|---|
| | After 25 min | After 60 min |
| **Inactive** | | |
| Attapulgite (Senegal) | 0 | 7 |
| Wollastonite (Asbestos, Quebec) | 0 | 27 |
| Magnetite-maghemite (MAG 2) | 0 | 0 |
| Magnetite (mineralogical sample) | 35 | 37 |
| Chrysotile UICC (A, Rhodesian) | 25 | 30 |
| Crocidolite UICC | 25 | 40 |
| Amosite UICC | 30 | 50 |
| Fine glass fibres (J.M. 104) | 33 | 32 |
| Potassium titanate fibres (Tismos D) | 0 | 14 |
| Erionite (Oregon) | 45 | 20 |
| **Active** | | |
| Chrysotile (commercial) | 1800 | 1500 |
| Chrysotile UICC (B, Canada) | 1200 | 600 |
| Nemalite (Asbestos) | 200 | 3000 |
| 6 chrysotile samples from different Canadian mines | 900-2700 | 1000-2200 |
| **Controls** | 10 | 15 |

[a]The samples of wollastonite, magnetite (mineralogical sample), nemalite, erionite and commercial chrysotile were slightly ground several weeks before use. Prior washings with hot benzene to remove adsorbed organic impurities do not significantly change the results. All intensities are given to ± 20%

### Table 2. Reactivity of materials after activation[a]

| Materials and treatment | (DMPO, $CO_2$)$^{\overline{\cdot}}$ signal intensity | |
|---|---|---|
| | After 25 min | After 60 min |
| **(A) Activation by grinding** | | |
| Crocidolite (commercial) | 700 | 1000 |
| Crocidolite UICC | 640 | 600 |
| Amosite UICC | 850 | 1000 |
| **(B) Activation by $Fe^{2+}$ exchange** | | |
| Erionite | 700 | 900 |
| **(C) Activation by GSH treatment** | | |
| Chrysotile UICC (A, Rhodesian) | 110 | 130 |
| Magnetite-maghemite (MAG 2) | 80 | 130 |
| Crocidolite UICC (not ground) | 50 | 240 |

[a]The controls are the same as in Table 1. Amphiboles were hand ground in an agate mortar for 1 min. Cation exchange for erionite was effected by treatment in aqueous medium (pH = 3.2) with a 0.1M solution of $FeCl_2$, stirring for 3 h, and then washing with distilled water. Activation by glutathione (GSH), a weak reducing agent, was effected by addition of this compound directly to the reactor ($0.78 \times 10^{-3}$ mol l$^{-1}$) together with buffer, mineral and formate; DMPO was added a few minutes afterwards. All intensities are given to ±30%.

[b]The material was used 5 min after grinding.

### Table 3. Reactivity of materials after passivation[a]

| Material and treatment | (DMPO, $CO_2$)$^{\overline{\cdot}}$ signal intensity | |
|---|---|---|
| | After 25 min | After 60 min |
| Chrysotile UICC (A, Canada) after 16 h in buffer medium | 20 | 15 |
| Crocidolite UICC, ground, then 2 h in buffer and 1 h ultrasonic treatment | 15 | 15 |
| Crocidolite UICC ground, then exposed to air: | | |
| for 1 day | 220 | 180 |
| for 20 days | 230 | 170 |
| Chrysotile, Quebec, short fibres | 490 | 225 |
| 'Chrysophosphate' obtained from above short fibres | 135 | 75 |

[a]The controls are the same as in Table 1 and intensities are given to ± 20%.

## Results

The results obtained with mineral fibres, without treatment, after activation treatment and after passivation treatment, are shown in Tables 1-3.

Magnetite (a non-fibrous material) was studied because it is a frequent contaminant of many of the chrysotile samples. Nevertheless, it should be noted that the very weak activity of the mineralogical sample is probably due to the presence of a very small amount of another mineral, namely chlorite. The studies of Zalma (1988), and Costa et al. (unpublished data), in fact, show that the activity of magnetite increases very strongly as a function of the chlorite content of the mineralogical samples.

In tests conducted with very active solids under the usual conditions, but with the radical trap added to the filtrate (i.e., after incubation of the solid in the medium and subsequent filtration), radical formation was not detected, showing that the activity, at least in essentials, is localized on the solid-liquid interface.

In experiments conducted with the addition of hydrogen peroxide, according to the technique of Weitzman and Graceffa (1984), but with formate also added to the medium, we obtained (DMPO, $CO_2$)⁻ signals whose intensities, in comparison with those obtained in the reduction of oxygen, were either of the same order or even lower for the very active materials in Table 1, such as UICC chrysotile (B), or very high for the inactive materials in Table 1 containing only $Fe^{3+}$ such as magnetite-maghemite (MAG 2).

We should also like to point out that other studies by our group (Costa et al., unpublished data) have shown that, whereas iron oxides are only slightly active in the reduction of oxygen to $OH^{\cdot}$, certain materials containing $Fe^{2+}$, in particular, phyllosilicates and carbonates, have an activity of the same order as that found for the very active asbestos varieties in Tables 1 and 2.

The addition of an iron chelator, deferroxamine, to the reaction medium leads to the passivation of the surface activity, the signal intensity of (DMPO, $CO_2$)⁻ becoming negligible (see Zalma, 1988).

If we convert our arbitrary intensity scale into the number of radicals produced, we obtain, for an amphibole (crocidolite or amosite) with a surface area of 5 m²/g and giving an intensity of 1000, about $10^5$ radicals for a fibre 5 μm in length and 0.4 μm in diameter.

## Discussion

Table 1 contains two categories of materials, namely inactive or slightly active and highly active.

The inactive or slightly active materials in our model system, i.e., those giving a signal intensity equal to or less than 20-40 are:

— either materials which do not contain $Fe^{2+}$, such as attapulgite, erionite, fine glass fibres, potassium titanate fibres and wollastonite from Asbestos (Quebec);

— or materials containing $Fe^{2+}$, but coated with a passivation sheet, resisting dissolution or deaggregation in our model medium. This is the case with magnetite, chrysotile A (where the $Fe^{2+}$ is essentially in the accompanying magnetite) and also UICC amphiboles ground several years previously.

The highly active materials in Table 1 are those in which the dissolution of the passivation sheet in our model system allows the appearance of a high density of reducing sites linked to $Fe^{2+}$. All the samples of Canadian chrysotile and nemalite belong to this category. Here, we may consider that the immersion of the mineral in the solution of potassium phosphate is equivalent to an activation treatment by partial lixiviation of the oxidized surface sheets rich in $Mg^{2+}$. This lixiviation process is accelerated by the corresponding precipitation of a double phosphate of $Mg^{2+}$ and $K^+$, as we confirmed in the case of nemalite.

We may therefore consider all the mineral samples exposed to humid air for long periods as having been passivated. Their activation will always require a certain modification of the surface sheet. By analogy with organic compounds, we can say that the reducing activity of these minerals in a biological medium will be a function of the metabolism of their surface.

The activation of amosite and crocidolite, which are richer in $Fe^{2+}$ than chrysotile, does not take place spontaneously in our model medium (Table 1). The passivation sheet, probably formed of $Fe^{3+}$ oxyhydroxycarbonates, would not be soluble in this medium. Nevertheless, the surface activation of these amphiboles will be a relatively easy process, and can be brought about by:

— grinding (cf. Table 2);

— splitting parallel to the fibre axis, a process which has been demonstrated in biological media, particularly by Cook *et al.* (1982); or

— dissolution of the passivation sheet in the biological medium by various reactants having reducing and complexing properties.

The activation of an inactive chrysotile (A, Rhodesia) in our reaction medium, is, no doubt, a more difficult process. However, the surface properties of mineral samples, apparently similar on a rapid examination by X-ray diffraction, may vary. Another sample of Rhodesian chrysotile (from other deposits or different sections of the same deposit) would probably show marked reducing activity. In the same way, while all the Canadian samples are active, those from different mines do show distinct differences in activity, linked to their accessible surface area, the nemalite content, etc. The reducing surface properties of inorganic materials are therefore very sensitive to parameters not generally taken into account during their prior characterization.

Independently of the activation modes mentioned above for the amphiboles and chrysotile, other activation phenomena may also be anticipated in a biological environment, such as:

— Enrichment of the $Fe^{2+}$ on the surfaces due to an exchange of cations. In particular, this phenomenon can be expected for materials possessing strong cation exchange capacity (CEC), such as erionite, but it remains to be determined whether this phenomenon can play a role in certain sites of the biological environment.

— Cell enrichment in iron-rich proteins (Stroink et al., 1987), with adsorption on mineral particles and corresponding changes in conformation and redox properties. This phenomenon has yet to be studied. It may be considered as the precursor to the formation of ferruginous bodies, something that takes place with all the mineral particles except silica.

— Superficial reduction of $Fe^{3+}$ in the particle by biological reducing systems (such as NADPH P450, ascorbate, etc.). Our experiments (limited to the action of glutathione) have not yet advanced sufficiently to enable us to judge whether this activation process takes place.

Neither can we exclude a combination of these three activation modes, e.g., the adsorption of iron-rich proteins, followed by a partial reduction to $Fe^{2+}$, leading either to the deposition of ferrous compounds or to an accumulation of $Fe^{2+}$ as exchangeable cations when the CEC is high. These three activation modes are mentioned here only as a hypothesis, with reference to earlier animal experiments (Pott et al., 1976; Stanton et al., 1977), which showed that materials free of $Fe^{2+}$, and even of $Fe^{3+}$ (fine glass fibres, alumina, dawsonite) are capable of inducing mesothelioma.

The number of $OH^{\cdot}$ radicals produced can also be very high for all the iron-containing materials, e.g., asbestos (Eberhardt et al., 1985; Weitzman & Graceffa, 1984) when the model system or the cellular compartment contains hydrogen peroxide (Goodglick & Kane, 1986). The bioavailability of $H_2O_2$ being low and the reactivity of asbestos in the production of $OH^{\cdot}$ radicals from $H_2O_2$ being less than that of the very common pollutants, such as haematite ($Fe_2O_3\alpha$), goethite ($FeOOH\alpha$) and magnetite ($Fe_3O_4$), it would seem difficult for this mechanism to play the principal role in cancer induction.

The passivation of surface active sites is brought about more or less rapidly by the action of atmospheric oxygen. Passivation is accelerated by treatment in an aqueous medium (as with short fibres obtained by sedimentation, for example) and even more by ultrasonic treatment. Such passivation can perhaps explain the negative results obtained in some *in vitro* experiments. In this case, a process of prior activation of the particle surface would be necessary for a better evaluation of the biological activity of the samples. We may even ask whether the evidence produced by Cleveland (1984) for the mutagenic properties of a freshly ground mineralogical sample of an amphibole was not linked to these phenomena.

Greater passivation, effected, e.g., by coating the fibres with polyphosphate (chrysophosphate), can more efficiently prevent any reactivation in a biological medium, insofar as this coating is quite insoluble and contains very little iron. Passivation of magnetite should also be emphasized, the external sheet being more probably $Fe_2O_3\gamma$ (maghemite), whose oxygen lattice is the same as that of magnetite in bulk. The perfect continuity of the two mineral species makes the inner $Fe^{2+}$ almost inaccessible.

The properties demonstrated in our studies are, in essentials, linked to the $Fe^{2+}$-$Fe^{3+}$ couple, but we cannot exclude the possibility that other redox couples, such as $Cu^+$-$Cu^{2+}$ or $Ni^+$-$Ni^{2+}$, can also play a role in other inorganic compounds, particularly if the weak valency is stabilized by ligands possessing a strongly electron-donating character. These reducing properties are not limited to fibrous particles and may perhaps explain the excess of cancers sometimes observed in miners (Axelson, 1986).

Nevertheless, fibres do have some properties not found with other types of particles. In comparison with an isometric particle, e.g., a sphere, a fibrous particle of the same material and same weight has a larger external surface. Hence, for the same density of reducing sites, a fibrous particle can produce a much greater quantity of free radicals intracellularly, more easily distributed in the cell. Based on the experimental data of Stanton, we have previously shown (Bertrand et al., 1980; Bonneau et al., 1986b) that, for samples of equal weight and having the same crystalline and chemical character, introduced into the pleural medium, the probability of obtaining pleural tumours increases with increasing average length and decreasing average diameter of the fibre samples. For any given diameter, when the length of the phagocytosed fibres is increased, the surface area also increases. A decrease in the average diameter leads to an increase in the surface area of the injected sample as a whole.

A second parameter, related to the retention time of the particles in the target biological tissue, no doubt plays a fundamental role in greatly enhancing the toxicity of fibrous particles. Wagner et al. (1984) have shown that particles having a pronounced fibrous texture are more slowly cleared from the pleural medium. An appreciable retention time may favour the activation of surfaces.

Finally, as regards inhalation, since the probability that a particle will reach the lung is a function of its aerodynamic diameter, for any given value of this diameter, the surface area of a fibre is much greater than that of an isometric particle.

Among the fibrous minerals, those which appear capable of causing oxidative stress in the lungs are those having an intense reducing activity, so that free radicals are produced in such numbers that the intracellular defence mechanisms are swamped. In a mesothelial medium, the dimensional characteristics of fibres are such as to increase the retention time, so that complementary activation processes may occur, as compared with the simple dissolution of the passivated surface.

The highly carcinogenic nature of erionite in the mesothelial environment may be due to three special characteristics of this mineral:

(i) an internal surface (accessible to reactants of small dimensions) which is large as compared with the external surface, due to the presence of channels distributed throughout the material, permitting the rapid diffusion of the reactants;

(ii) a high cation exchange capacity, which enables strongly reducing cations, such as $Fe^{2+}$, to remain in the channels, thus conferring a strong reducing activity towards oxygen;

(iii) the presence of highly basic sites, which may perhaps allow electron exchange without transition metal cations being involved (Poole et al., 1983, 1986). Such a mechanism is currently being investigated.

## Conclusions

The phenomenon of oxidative stress, induced by the production of OH· radicals in a biological environment during the reduction of oxygen by the surface active sites of certain inorganic minerals, both fibrous and non-fibrous, can play an important role in some of the steps leading to fibrogenesis and carcinogenesis.

In the absence of direct contact between the fibre and the genome, it is very unlikely that the OH· radicals, which have a very short lifetime, will be able to attack the macromolecules of the genome. On the other hand, it is much more probable that there is a significant production of R· radicals from relay molecules (RH) present in the medium, with chain reactions and a considerable increase in the half-life of the free radicals. The RH can either be present in the cell or be produced by cellular degradation. These molecules can also be xenobiotics, in which case they will play a synergistic role with mineral particles (in cigarette smoke, for example).

A physicochemical mechanism has been proposed here to account for the genotoxic activity of certain inorganic materials, including the mineral fibres. Our conclusions now need to be confirmed (or invalidated) by establishing correlations with the results of *in vitro* and *in vivo* experiments, taking into account the phenomena of passivation and activation of the surfaces. While our proposal has the advantage of being in agreement with what is already known about the surface properties of minerals, further studies are still needed before it can be accepted.

## Acknowledgements

The authors thank C. Jolicoeur for providing a number of samples, B. Morin for technical assistance and L. Bonneau for skilfully developing the radical trapping technique. Valuable discussions with Dr Mansuy and A. Picot are acknowledged. Thanks are also due to CNRS (Action spécifique programmée, Programme interdisciplinaire de recherche sur les technologies, le travail, l'emploi, les modes de vie) for financial support.

## References

Axelson, O. (1986) Epidemiology of occupational cancer: mining and ore processing. In: *Prevention of Occupational Cancer*, Geneva, International Labour Office, pp. 135-149

Bertrand, R. & Pezerat, H. (1980) Fibrous glass: carcinogenicity and dimensional characteristics. In: Wagner, J.C., ed., *Biological Effects of Mineral Fibres*, Vol. 2, (*IARC Scientific Publications No. 30*), Lyon, International Agency for Research on Cancer, pp. 901-911

Bonneau, L. & Pezerat, H. (1983) Study of electron donor and acceptor sites on the surface of asbestos fibres. [in French]. *J. Chim. Phys.*, 80, 275-280

Bonneau, L., Suquet, H., Malard, C. & Pezerat, H. (1986a) Active sites on surface of chrysotile and amphiboles. *Environ. Res.*, 41, 251-267

Bonneau, L., Malard, C. & Pezerat, H., (1986b) Role of dimensional characteristics and surface properties of mineral fibers in the induction of pleural tumors. *Environ. Res.*, 41, 268-275

Cleveland, M.G. (1984) Mutagenesis of *Escherichia coli* (CSH50) by asbestos. *Proc. Soc. Exp. Biol. Med.*, 177, 343-346

Cook, P.M., Paletar, L.D. & Coffin, D.L. (1982) Interpretation of the carcinogenicity of amosite asbestos and ferroactinolite on the basis of retained fiber dose and characteristics in vivo. *Toxicol. Lett.*, *13*, 151-158

Eberhardt, M.K., Roman-Franco, A.A. & Quiles, M.R. (1985) Asbestos-induced decomposition of hydrogen peroxide. *Environ. Res.*, *37*, 287-292

Fontecave, M., Mansuy, D., Jaouen, M. & Pezerat, H. (1987) The stimulatory effects of asbestos NADPH-dependent lipid peroxidation in rat liver microsomes. *Biochem. J. 241*, 561-565

Goodglick, L.A. & Kane, A.B. (1986) Role of reactive oxygen metabolites in crocidolite asbestos toxicity to mouse macrophages. *Cancer Res.*, *46*, 5558-5566

Gulumian, M., Sardianos, F., Kilroe-Smith, T. & Ockerse, G. (1983) Lipid peroxidation in microsomes induced by crocidolite fibers. *Chem. Biol. Interact.*, *44*, 111-118

Mossman, B.T., Marsh, J.P. & Shatos, M.A. (1986) Alteration of superoxide dismutase activity in tracheal epithelial cells by asbestos and inhibition of cytotoxicity by antioxidants. *Lab. Invest.*, *54*, 204-212

Poole, A., Brown, R.C., Turver, C.J., Skidmore, J.W. & Griffiths, D. (1983) *In vitro* genotoxic activities of fibrous erionite. *Br. J. Cancer*, *47*, 697-705

Poole, A., Brown, R.C. & Rood, A.P. (1986) The *in vitro* activities of a highly carcinogenic mineral fibre — potassium octatitanate. *Br. J. Exp. Pathol.*, *67*, 289-296

Pott, F., Friedrichs, K.H. & Hulth, F. (1976) Results of animal experiments concerning the carcinogenicity of fibrous dusts and their interpretation with regard to carcinogenesis in humans. *Zentralbl. Bakt. I. Abt. Orig. Reihe B*, *162*, 467-505

Stanton, M.F., Layard, M., Tegeris, A., Miller, E., May, H. & Kent, E. (1977) Carcinogenicity of fibrous glass: Pleural response in the rat in relation to fiber dimension. *J. Natl Cancer Inst.*, *58*, 587-603

Stroink, G., Lim, D. & Dunlap, R.A. (1987) A Mossbauer study of autopsied lung tissue of asbestos workers. *Phys. Med. Biol.*, *32*, 203-211

Turver, C.J., Poole, A. & Brown, R.C. (1985) Lipid peroxidation and the generation of malondialdehyde in crocidolite-treated cell cultures. In: Beck, E.G. & Bignon, J., eds, In vitro *Effects of Mineral Dusts*, Vol. 3 (*NATO ASI Series, Series G: Ecological Sciences*), pp. 267-274

Wagner, J.C., Griffiths, D.M. & Hill R.J. (1984) The effect of fibre size on the *in vivo* activity of UICC crocidolite. *Br. J. Cancer*, *49*, 453-458

Weitzman, S.A. & Graceffa, P. (1984) Asbestos catalyzes hydroxyl and superoxide radical generation from hydrogen peroxide. *Arch. Biochem. Biophys.*, *228*, 373-376

Weitzman, S.A. & Weitberg, A.B. (1985) Asbestos-catalyzeds lipid peroxidation and its inhibition by desferoxamine. *Biochem. J.*, *225*, 259-262

Zalma, R. (1988) *Contribution à l'étude de la réactivité de surface des fibres minerales. Relations possible avec leurs propriétés cancérogènes*, Thesis, Universite P. et M. Curie, Paris

Zalma, R., Bonneau, L. & Pezerat, H. (1986) Catalytic role of asbestos in fluorene oxidation. *Environ. Res.*, *41*, 296-301

Zalma, R., Bonneau, L., Jaurand, M.C., Guignard, J. & Pezerat, H. (1987) Formation of oxy-radicals by oxygen reduction arising from the surface activity of asbestos. *Can. J. Chem.* (in press)

# EXPERIMENTAL STUDIES ON INGESTED FIBRES

### I. Chouroulinkov

*Institut de Recherches Scientifiques sur le Cancer (CNRS),
Villejuif, France*

## Introduction

Occupational exposure to asbestos fibres via the respiratory route has been clearly associated with the development of pleural mesotheliomas (IARC, 1977; Peto *et al.*, 1982) and to a lesser extent with the development of lung cancer, particularly in cigarette-smoking workers (Hammond *et al.*, 1979; IARC, 1977). Moreover, certain epidemiological investigations have found evidence of mortality excess from gastrointestinal (GI) cancer in asbestos insulation workers (Doll & Peto, 1985; Selikoff *et al.*, 1979). This is not surprising, since 28% of the inhaled dust, including asbestos fibres, is transported via the mucociliary clearance mechanism to the pharynx and subsequently swallowed (Gross *et al.*, 1974). Consequently, the GI tract is indirectly the major recipient of inhaled air-borne mineral fibres. In addition, a large proportion of the population ingest asbestos fibres through drinking-water, beverages and food (IARC, 1977; Rowe, 1983). The majority (65%) of the water samples from 352 cities in the United States showed detectable amounts of asbestos fibres, including chrysotile (Millette *et al.*, 1983). The asbestos concentrations in the water supplies of 41 cities exceeded 10 million fibres per litre (National Toxicology Program, 1985). In Canadian tap-water, 2-173 million fibres per litre were found (Cunningham & Pontefract, 1971). Asbestos fibres were found in rivers and lakes, the highest concentrations being observed near places where the mining of asbestos (Province of Quebec) or taconite iron ore (Lake Superior) was being carried out (IARC, 1977). Asbestos fibres have also been found in samples of spirits (13-24 million per litre), sherries, ports, vermouth, soft drinks (1.7-12.2 million per litre), beer from various countries (1-6.6 million per litre) (IARC, 1977) and in wines (2-64 million fibres per litre (Gaudichet *et al.*, 1978).

The foregoing strongly suggests that ingestion of asbestos fibres may lead to an increase in GI and possibly systemic cancer. The problem is how to prove that a relationship exists between ingestion of asbestos fibres and such an increase and how to evaluate the risk. The solution lies with the results of epidemiological and experimental studies.

Experimental investigations of two types have been carried out with ingested asbestos fibres, the first being aimed at studying the gut clearance capacity, fibre penetration in mucosal cells, transmigration, damage to the mucosa and changes in DNA synthesis in the GI tract, liver or pancreas. The second type is concerned with the chronic and/or carcinogenic effects after long-term ingestion of asbestos fibres. Studies of this type provide the basis for evaluating the carcinogenic potential of chemicals. This paper will therefore focus on studies of long-term ingestion of asbestos fibres from the point of view of toxicity and particularly of the effects on GI carcinogenicity.

A total of 13 papers on long-term investigations of the effects of ingested mineral fibres were reviewed. Four important studies, on chrysotile in hamsters, and amosite, crocidolite and tremolite in rats, from the National Toxicology Program were unfortunately not yet available.

## Toxicity of long-term ingestion of mineral fibres

The studies on the chronic toxicity of ingested mineral fibres were all undertaken recently, so that combined carcinogenicity-chronic toxicity protocols were used.

Six long-term studies on ingested mineral fibres are available, which provide fairly complete information on chronic toxicity, 2 in the Syrian hamster and 4 in rats.

### Hamsters

Smith et al. (1980) gave amosite or taconite tailings at doses of 0.5, 5.0 and 50.0 mg/litre of water to male and female hamsters for 650 days. The median survival time at 95% confidence interval for each treated group overlapped that of the controls and the mineral fibres did not significantly alter the median survival time. The body weights of the treated animals were also not significantly different from those of controls. Finally, the frequency of histopathological lesions did not differ as between the treated and untreated groups.

The second study, on hamsters (National Toxicology Program, 1983) fed with a diet containing 1% of amosite, reported a median survival time of 55 and 84 weeks for female and male controls respectively, as compared with 60 and 80 weeks for treated females and males. Surprisingly, the survival rate is significantly higher ($p>0.01$) in hamsters fed with the diet containing amosite.

It can be concluded from these 2 studies that ingested amosite or taconite tailings are not toxic for the Syrian hamster.

### Rats

Four complete clinical and histopathological studies have been carried out on rats given mineral fibres orally.

Bolton et al. (1982) gave Wistar Han rats 250 mg per week of chrysotile, amosite or crocidolite in a margarine-complemented diet for 750 days. McConnell et al. (1983) studied the lifetime feeding of F344 rats with a diet containing 1% of amosite or tremolite. In another study (National Toxicology Program, 1985), F344 rats were fed during their lifetime with a diet containing 1% chrysotile. Finally, Truhaut and Chouroulinkov (this volume, pp. 127-133) studied the effects of chrysotile and of a mixture of chrysotile and crocidolite (75%/25%) fibres in palm oil in Wistar Han rats at 10, 60 and 360 mg per day for 24 months. In all these studies the survival time was comparable for treated and control groups. The body weight gain, apart from small variations, was not significantly affected by the ingestion of fibres. In contrast, consumption of the fatty vehicles, margarine (Bolton et al., 1982) and palm oil (Truhaut & Chouroulinkov, this volume pp. 127-133) was followed by a significant body weight increase as compared with that of rats fed a normal diet. However, there was no difference between rats treated with fibres and control animals.

Thus the results of all these studies are in agreement: the long-term ingestion of asbestos fibres at high doses, as in hamsters, had no toxic or adverse health effects for the treated rats.

## Carcinogenicity of long-term ingestion of mineral fibres

Of the 13 papers on the long-term effects of ingested mineral fibres, 7 involved chrysotile fibres (Table 1), 6 amosite, 2 crocidolite and 1 tremolite (Table 2). Another group of investigations (Table 3) was concerned with various mineral materials, including talc (2 studies), taconite tailings (2 studies), beach rock powder, diatomaceous earth, water sediments or water containing large amounts of amphiboles. Finally, there is the group of studies on the association between different types of fibres and between asbestos fibres and chemicals with carcinogenic potential for the GI tract (Table 4).

### Chrysotile carcinogenicity in long-term ingestion studies

The 7 studies identified, all in rats, involving ingestion of chrysotile fibres alone are summarized in chronological order in Table 1.

Gross et al. (1974) did not observe GI tumours in rats treated with chrysotile fibres (5% in water *ad libitum*), or in butter (10 mg per week, 16 weeks). Wagner et al. (1977) reported one gastric leiomyosarcoma in 32 rats receiving 100 mg per day of chrysotile in malted milk (no GI tumours in 16 controls). In the same year Cunningham et al. (1977) reported 2 studies (720 and 900 days) in which Wistar male rats received 1% chrysotile in the diet. Only in the second study were 1 ileal sarcoma and 1 colon carcinoma in 36 treated rats observed (no GI tumours in 30 controls). It appeared to the authors that the evidence for the carcinogenicity of ingested chrysotile was inconclusive. Donham et al. (1980) studied the effects of chrysotile on the colon (10% in diet); 3 colon carcinomas in 95 treated males, and 1 adenomatous polyp in 94 females were found, while 3 colon carcinomas in 155 male and 2 in 157 female controls were also observed. Although no significant increase in tumours in treated animals was observed, the authors considered that there was an increased probability of asbestos-fed animals developing lesions in the colon.

Bolton et al. (1982), in a combined chronic toxicity-carcinogenicity study with chrysotile in margarine (5 mg/g) added to the diet found no GI tumours in either the chrysotile-treated or control (margarine and normal diet) groups. However, 16 other tumours were observed in 22 treated rats and 9 in 47 control rats. The authors emphasize the significant excess of benign tumours in the chrysotile-treated group (4 mesenteric haemangiomas). The National Toxicology Program (1985) study on ingested chrysotile reached the following conclusions:

## Table 1. Summary of long-term ingestion studies with chrysotile asbestos fibres

| Reference | Dosage | Observation time (days)[a] | No. of rats treated/ controls[a] | Remarks | | |
|---|---|---|---|---|---|---|
| Gross et al., 1974 | 5% in water ad lib. | Up to 630 | 10/5 | 'No tumour production in GI[b] tract or mesothelium during the lifetime' | | |
| | 10 mg/week in butter, 16 weeks | Up to 533 | 31/24 | | | |
| Wagner et al., 1977 | 100 mg/day in malted milk, 5 days/week, 20 weeks | 619 (MS)[b] | 32/16 | 1 gastric leiomyosarcoma; no GI tumours | | |
| Cunningham et al., 1977 | 1% in diet ad lib. | Up to 720 | 7/8 | 1 peritoneal sarcoma in treated rats and 1 in controls | | |
| | 1% in diet | Up to 900 | 36/38 | 1 sarcoma (ileum), 1 carcinoma (colon); no GI tumours; other tumours: 9/11 | | |
| Donham et al., 1980 | 10% in diet | Up to 960 | M 95/155 | 3 colon carcinomas, 1 peritoneal mesothelioma; 3 colon carcinomas in controls | | |
| | | | F 94/157 | 1 adenomatous polyp; 2 colon carcinomas in controls | | |
| Bolton et al., 1980 | 250 mg/week in diet (margarine/chrysotile) (5 mg/g) | 750 | M 22/47 | No GI tumours; 4 mesenteric haemangiomas and 12 other tumours (5 malignant, 7 benign); 1 peritoneal sarcoma and 8 other tumours (5 malignant, 3 benign) in controls | | |

| | | | | Tumours of the alimentary tract | | |
|---|---|---|---|---|---|---|
| | | | | | Malignant | Benign |
| National Toxicology Program, 1985 | 1% in diet (short-range fibres) | M up to 960 F up to 1015 | M 250/88 F 250/88 | | 8/4 9/2 | 7/1 13/2 |
| | 1% in diet (IR[b] fibres) | M up to 1000 | M 250/88 | | 10/0 | 24/2 |
| | 1% in diet (IR fibres) PW[b] chrysotile-gavaged animals (0.47 mg/g bw/day, 21 days) | F up to 1020 M up to 1000 F up to 1020 | M 250/88 M 100/88 F 100/88 | | 4/1 5/0 3/0 | 6/3 7/0 3/0 |
| Truhaut & Chouroulinkov[c] | 10, 60 and 360 mg/day in palm oil (0.2, 1.2, 7.2%), 24 months | 900 | M 70, F 70 per dose | M 205 F 209 | 7 5 | 4 4 |
| | Controls: palm oil/ diet | 900 | M 70, F 70 per group | M 137 F 139 | 4 1 | 3 2 |

[a]M, male; F, female.

[b]GI, gastrointestinal tract; MS, mean survival time; PW, preweaning; IR, intermediate range.

[c]See pp. 127-133

**Table 2. Summary of long-term ingestion studies with amosite, crocidiolite and tremolite fibres**

| Reference | Dosage | Observation time (days) | Animal species | No. treated/controls[a] | Remarks |
|---|---|---|---|---|---|
| **Amosite** Smith et al., 1980 | 0.5, 5.0 and 50 mg/litre of water ad lib. | Up to 650 | Hamster | M 30, F 30 per dose | GI[b] tract: 2 (1 M, 1 F) stomach carcinomas; 4 stomach (3 M, 1 F); 2 adenomatous polyps (M); 1 peritoneal mesothelioma (M) |
|  | Controls: 2 filtered waters | Up to 650 | Hamster | M 30, F 30 per group | 6 squamous-cell papillomas (3 M, 3 F) 1 adenomatous polyp (M) |
| Ward et al., 1980 | 10 mg in 1 ml saline 3 times/week intra-gastrically for 10 weeks + saline subcutaneously | Up to 665 | Rat | M 48/0 | 5 intestinal carcinomas; 11 polyps |
| Hilding et al., 1981 | 300 mg/day in cottage cheese (50 mg/g) | 750 | Rat | 20 | 1 leiomyoma (ileum); no cottage cheese controls |
| Bolton et al., 1981 | 250 mg/week in diet + margarine/amosite | 750 | Rat | 24/24 | 1 gastric leiomyosarcoma, 1 skin fibroma; no GI tumours in controls (others: 3 carcinomas, 2 adenomas) |
| McConnell et al., 1983 | 1% in pelleted diet | Up to 1020 | Rat | M 250 F 250 | Alimentary tract: 7 carcinomas (5 M, 2 F); 4 adenomatous polyps (2 M, 2 F) |
|  | 1% in diet (PW[b] chrysotile-gavaged animals, 0.47 mg/g bw/day, 21 days) | Up to 1020 | Rat | M 100/117 F 100/117 | 4 carcinomas (2 M, 2 F); 2 adenomatous polyps (1 M, 1 F) Controls: 4 carcinomas (3 M, 1 F) 2 adenomatous polyps (1 M, 1 F) |
| National Toxicology Program, 1983 | 1% in pelleted diet | Up to 520 | Hamster | M 248/122 F 237/119 | GI tumours: 8 papillomas (4 M, 4 F); 1 (M) adenomatous polyp; controls: 1 (M) papilloma |
| **Crocidolite** Gross et al., 1974 | 5 and 10 mg/week in butter for 16 weeks | Up to 616 | Rat | 33 and 34 respectively | No GI tract tumours; 1 lymphoma |
|  | Butter (as above) | Up to 644 | Rat | 24 | No GI tract tumours; 5 other malignant tumours |
| Bolton et al., 1982 | 250 mg/week in diet + margarine/crocidolite (5 mg/g) | 750 | Rat | M 22 | No GI tract tumours; 6 other tumours |
|  | Margarine/normal diet | 750 | Rat | M 24/24 | No GI tract tumours; other tumours: 5/3 |
| **Tremolite** McConnell et al. 1983 | 1% in diet | Up to 1020 | Rat | M250 F250 | Alimentary tract 6 carcinomas, 2 papillomas, 3 polyps 6 carcinomas, 2 papillomas |
|  | Controls | Up to 1020 |  | M 118 F 118 | 4 carcinomas, 2 papillomas, 1 polyp 2 carcinomas, 1 polyp |

[a]M, male, F, female; [b]GI, gastrointestinal; PW, preweaning

Table 3. Summary of long-term ingestion studies with various mineral fibres and materials

| Reference | Test material | Dosage | Observation time (days) | Animal species | No.[a] | Remarks |
|---|---|---|---|---|---|---|
| Gibel et al., 1976 | Talc | 50 mg/kg bw/day in diet | 649 (MS)[b] | Rat | 45 | NO GI tumours; 3 liver carcinomas, 4 mammary fibroadenomas. In 49 controls: 2 liver carcinomas and 5 mammary fibroadenomas |
| Wagner et al., 1977 | Talc | 100 mg/day in malted milk, 5 days/week, 20 weeks | 614 (MS)[b] | Rat | 32 | 1 gastric leiomyosarcoma. No tumours in 16 control rats |
| Smith et al., 1980 | Taconite tailings (cummingtonite/ grunerite and quartz) Reserve Mining Corporation, Silver Bay | 0.5, 5.0 and 50.0 mg/litre of water (0.45 μm filtered ad lib. | Up to 650 | Hamster | M 30, F 30 per group | 6 forestomach papillomas (2 M, 4 F) 2 adenomatous polyps (M). Other tumours: 1 uterine leiomyosarcoma, 6 benign (5M, 1F) |
|  | Pulverized beach rock | 5.0 and 50.0 mg/litre of water (0.45 μm filtered) ad lib. | Up to 650 | Hamster | M 30, F 30 per group | 5 forestomach papillomas (3 M, 2 F), 1 colon leiomyosarcoma (M). Other tumours: 3 malignant (2 M, 1 F) 5 benign (4 M, 1 F) |
|  | Control | 0.45 μm, and 0.1 μm filtered waters (Lake Superior) | Up to 650 | Hamster | M 30, F 30 per group | 5 forestomach papillomas (3 M, 2 F), 1 (M) adenomatous polyp, 1 larynx papilloma (M). Other tumours: 1 (M) lymphoma, 6 benign (3 M, 3 F) |
| Hilding et al., 1981 | Diatomaceous earth | 20 mg/day in cottage cheese | Up to 840 | Rat | 30 | 1 peritoneal mesothelioma, 1 salivary gland carcinoma, 1 skin cancer, 2 uterine sarcomas, 13 benign (9 breast, 1 adenoma, 3 pancreas) |
|  | Water sediment (Lake Superior) | $5 \times 10^9$ amphibole fibres/ litre of water | Up to 840 | rat | 22 | 1 lung cancer, 1 skin (ear) cancer, 1 uterine sarcoma 7 benign (5 breast, 2 pancreas) |
|  | Taconite tailings (Res. Min. Co., Silver Bay) | $10 \times 10^{10}$ amphibole fibres/ litre of water | Up to 840 | Rat | 30 | 1 neck sarcoma, 1 chest wall sarcoma, 1 lymphoma, 11 benign (9 breast, 1 adrenal, 1 pancreas) |
|  | Unfiltered Duluth city tap water | $10 \times 10^7$ amphibole fibres per litre | Up to 960 | Rat | 28 | 1 salivary gland carcinoma, 1 skin cancer, 1 uterine sarcoma, 1 lymphoma, 9 benign (breast) |
|  | Filtered water | $10^6$ amphibole fibres per litre | Up to 960 | Rat | 27 | 1 forestomach carcinoma, 1 lung cancer, 1 ovary carcinoma, 5 benign (breast) |

[a] M, male; F, female.
[b] GI, gastrointestinal tract; MS = mean survival time.

## Table 4. Summary of long-term ingestion studies with asbestos fibres in combination with other materials

| Reference | Combination | Dosage | Observation time (days) | No. of rats[a] | Remarks |
|---|---|---|---|---|---|
| Ward et al., 1980 | Amosite + saline (sc) | 10 mg/ml saline intragastrically 3/week 10 weeks | Up to 665 | 48 | 5 intestinal carcinomas, 11 polypoid tumours |
| | Amosite + azoxymethane | As above 7.4 mg/kg, once a week, 10 weeks (sc) | Up to 665 | 48 | 18 intestinal carcinomas, 24 polypoid tumours |
| | Saline + azoxymethane | 1 ml intragastrically, 3/week, 10 weeks; 7.4 mg/kg, once a week 10 weeks | Up to 665 | 49 | 12 intestinal carcinomas, 37 polypoid tumours |
| McConnell et al., 1983 | Amosite + DMH[b] | 1% in diet for lifetime 7.5 mg/kg (M) 15.0 mg/kg (F) gavage once a week, 5 doses | Up to 1020 | M 175 F 175 | Animals with GI[b] tract (primary) neoplasms: M 118 (68%), F 114 (65%) |
| | DMH-control[c] | As above | Up to 1020 | M 125 F 124 | Animals with GI[b] tract (primary) neoplasms: M 92 (74%), F 77 (62%) |
| National Toxicology Program, 1985 | Chrysotile (intermediate-range) + DMH[b] (gavage) | 1% in diet 7.5 mg/kg (M) 15.0 mg/kg (F), gavage once a week, 5 doses | Up to 1020 | M 175 F 175 | Alimentary tract tumours: M: malignant 58, benign 93; F: malignant 66, benign 96 |
| | DMH controls[d] | As above | Up to 1020 | M 125 F 125 | Alimentary tract tumours: M: malignant 20, benign 63; F: malignant 45, benign 71 |
| Gibel et al., 1976 | Powdered filter material (52.6% chrysotile, 47.4% nature unknown) | 50 mg/kg bw/day in diet | 441 (MS)[e] | 42 | 2 forestomach papillomas. Other tumours: 12 malignant, 5 benign |
| | Controls | | 702 (MS)[e] | 49 | No GI[b] tumours. Other tumours: 2 malignant, 5 benign |
| Hilding et al., 1981 | Chrysotile for 210 days followed by amosite | 20 mg/day in cottage cheese 20 mg/day in cottage cheese | Up to 870 | 30 | No GI[b] tumours. Other tumours: 4 malignant (1 mesothelioma, chest wall), 17 benign (15 breast fibromas). In 28 control rats: 3 carcinomas (1 forestomach) and 5 fibromas |

| | | | | | Tumours of the alimentary tract: | | |
|---|---|---|---|---|---|---|---|
| | | | | | | Malignant | Benign |
| Truhaut & Chouroulinkov[f] | Mixture of chrysotile/ crocidolite (75%/25%) | 10, 60 and 360 mg/day in palm oil (0.2, 1.2 and 7.2%) for 720 days | 900 | M 70 F 70 per group | M 207 F 208 | 2 3 | 7 1 |
| | Controls | Palm oil and normal diet | 900 | M 70 F 70 per group | M 137 F 139 | 4 1 | 3 2 |

[a]M, male; F, female; [b]DMH, dimethylhydrazine dihydrochloride; GI, gastrointestinal tract; [c]For amosite controls, see Table 2; [e]For chrysotile controls, see Table 2; [e]MS = mean survival time; [f]See p. 127-133.

1. with short-range (SR) chrysotile fibres (1% in the diet for lifetime) there was no evidence of local (GI) or general carcinogenicity in either male or female rats;
2. with intermediate-range (IR) chrysotile fibres, there was again no evidence of local (GI) or general carcinogenicity in female rats. However, there was some evidence of carcinogenicity in male rats, as indicated by an increased incidence of adenomatous polyps (AP) in the large intestine (20 AP in 250 male rats and 0 in 88 controls). It should be noted, however, that, with the same IR fibres in 100 preweaning (PW) chrysotile-gavaged male rats, only 4 AP were observed in 100 PW female rats. This difference in response lacks any rational explanation.

Finally, in a study with chrysotile fibres in palm oil (10, 60 and 360 mg per day, 24 months), no excess of local (GI) tumours or any general tumour increase was found in treated animals (Truhaut & Chouroulinkov, this volume, pp. 127-133).

In conclusion, the results of the above studies indicate that only the intermediate-range chrysotile fibres, 1% in the diet, significantly increase the incidence of AP in the large intestine of male rats.

## Amosite carcinogenicity in long-term ingestion studies

The 6 studies identified on the long-term ingestion of amosite are summarized in chronological order in Table 2. Except for one study in which Syrian hamsters were used, rats were used in these studies.

Amosite fibres, 0.5, 5.0 and 50.0 mg/litre of water *ad libitum*, were given to Syrian hamsters (30 males and 30 females per group) for 650 days. In the treated (180) hamsters, 8 GI tumours were observed as compared with 7 in 120 controls (Smith *et al.*, 1980). In the same year, Ward *et al.* (1980) reported 5 intestinal carcinomas and 11 polypoid tumours in 48 male F344 rats after intragastric administration of amosite (10 mg/ml saline). This GI tumour frequency (32.6%) is very high. However, probably because of the absence of any real control groups, the authors concluded that the experimental evidence suggested but did not prove that oral asbestos exposure may have increased the incidence of intestinal tumours. One year later, Hilding *et al.* (1981) reported 1 ileal leiomyoma in 28 rats treated with 300 mg per day of amosite in cottage cheese (no GI tumours were found in 28 control rats). Bolton *et al.* (1982) found 1 gastric leiomyosarcoma in 24 Wistar Han male rats receiving amosite in margarine (5 mg/g) incorporated in the diet (no GI tumours in 24 margarine controls). McConnell *et al.* (1983) administered amosite (1% in diet for the lifetime) to normal and PW chrysotile-gavaged male and female rats. In all normal (500) and all PW (200) rats, 11 and 6 GI tumours, respectively, were reported; in 234 control rats, 6 GI tumours were found. Thus, no increase in GI tumour frequency related to the amosite treatment was observed. In a similar study, amosite (1% in diet) was given to male and female Syrian hamsters for their lifetime; 9 benign GI tumours (8 papillomas, 1 AP) in 485 treated hamsters were reported (only 1 papilloma was found in 241 controls) (National Toxicology Program, 1983).

In conclusion, the administration of amosite via the oral route to male and female Syrian hamsters and rats did not affect the incidence of GI tumours.

## Crocidolite and tremolite carcinogenicity in long-term ingestion studies

Two studies with crocidolite were identified. No GI tumours were observed either in Wistar rats fed with crocidolite in butter (Gross et al., 1974) or in Wistar Han rats receiving crocidolite in margarine (5 mg/g) incorporated in the diet (Bolton et al., 1982).

In the only study on tremolite, 1% in diet was given for lifetime to F344 rats and 11 and 8 GI tumours were observed in 250 male and 250 female rats respectively. In the controls, 7 and 3 GI tumours were observed in male and female rats, respectively (118 of each sex). The GI tumour frequency did not differ, therefore, as between the treated and control groups (McConnell et al., 1983).

## Long-term ingestion study of carcinogenicity of various mineral fibres and materials

The results of the studies with various mineral fibres and materials are summarized in Table 3. Most of these studies are concerned with asbestos water pollution, i.e., with conditions closer to those of real human exposure. The materials used are representative of the water pollution in areas where taconite iron mining is carried on (Lake Superior) (IARC, 1977).

Two studies deal with taconite tailings. In the first, in which taconite tailings (0.5, 5.0 and 50.0 mg/litre of filtered water) were given for lifetime to Syrian hamsters, 6 forestomach papillomas and 2 AP were observed in all 180 treated hamsters. This incidence of GI tumours was the same as that in the 2 control groups (Smith et al., 1980). In the second study with taconite tailings ($19 \times 10^{10}$ amphibole fibres per litre of water) given to Sprague Dawley rats, GI tumours were again not observed. In the same study, water sediment (Lake Superior) ($5 \times 10^9$ amphibole fibres per litre of water) and unfiltered Duluth city tap-water ($10 \times 10^7$ amphibole fibres per litre) were given to 22 and 28 rats respectively. In neither groups were any GI tumours observed, while 1 forestomach carcinoma was found in 27 rats receiving filtered water ($10^6$ amphibole fibres per litre) (Hilding et al., 1981). Smith et al. (1980) studied pulverized beach rock at 5.0 and 50.0 mg/litre of 0.45 $\mu$m filtered water in Syrian hamsters. After 650 days 9 GI tumours in all (120) hamsters were observed, so that the incidence was the same as in the control groups (7 GI tumours in 120 animals).

In addition, talc was studied in rats (50 mg/kg body weight per day in the diet) (Gross et al., 1974) and at 100 mg per day in malted milk (Wagner et al., 1977). No GI tumours were reported in either the treated rats or the controls in the first study and 1 gastric leiomyosarcoma in 32 rats (none in 16 controls) was observed in the second.

Finally, Hilding et al. (1981) administered diatomaceous earth (20 mg/day in cottage cheese) for lifetime to rats. No GI tumours were found.

## Conclusions

The various asbestos fibres, as well as possible and known water pollutants, such as taconite tailings, water sediment, unfiltered water containing large numbers of amphibole fibres, etc., did not modify GI tumour incidence in rats or hamsters after long-term ingestion.

## Long-term studies with asbestos fibres in combination with other materials

In one study (Gibel et al., 1976), powdered filter material in a diet containing 52.6% chrysotile was given to 42 rats. Two forestomach papillomas and 17 other tumours (12 malignant, 5 benign, were observed in treated animals and no GI tumours but 7 other tumours (2 malignant, 5 benign) in 49 control rats. The large number of tumours in the treated animals cannot be attributed to the chrysotile since the chemical nature of the 47.4% of filter powder was unknown.

Two studies concerned combinations of different types of asbestos fibres. In the first study, chrysotile (20 mg/day in cottage cheese) was given to 30 rats for 210 days, after which it was replaced by amosite until the end of the experiment (870 days). No GI tumours were observed. In parallel groups, 1 leiomyoma (ileum) in 20 amosite-treated rats, and 1 forestomach carcinoma in 28 control rats were observed (Hilding et al., 1981). In the second study, a mixture of chrysotile/crocidolite (75%/25%) in palm oil was given at 10, 60 and 360 mg per day to Wistar Han rats. No difference in the incidence of GI tumours was found as compared with the controls (palm oil, normal diet) or the chrysotile group (Truhaut & Chouroulinkov, this volume, pp. 127-133).

In 3 other investigations, asbestos fibres were associated with chemicals with carcinogenic potential to the GI tract, such as azoxymethane (AOM) or dimethyl-hydrazine dihydrochloride (DMH).

Ward et al. (1980) treated 48 rats with amosite (intragastric) and AOM (subcutaneous), 49 rats with AOM alone, and 48 rats with amosite, and observed 42, 49 and 16 GI tumours respectively. In view of these results, it is not easy to evaluate the effect of the combination, since there is no control group and the amosite group exhibited a large number of GI tumours.

In 2 other studies, amosite or chrysotile (1% in the diet) were combined with DMH (7.5 mg/kg for male and 15.0 mg/kg for female rats; gavage). In the first study, amosite did not increase GI tumour frequency. In the DMH-treated group, 74% and 62% of male and female rats respectively developed GI tumours. In the amosite + DMH treated group, the frequency was 68% and 65% repectively (McConnell et al., 1983). In the second study, 151 and 162 GI tumours were observed in 175 male and 175 female rats respectively, treated with chrysotile (intermediate range) + DMH. In the DMH control group there were 83 tumours in 125 male rats and 116 in 125 female rats. The conclusion reached is that 'IR chrysotile asbestos did not appear to influence the rate of neoplasia induced by DMH in the intestine' (National Toxicology Program, 1985).

Thus it can be concluded that chronic ingestion of amosite and chrysotile (intermediate-range) fibres did not influence the rate of neoplasia induced in rats either by AOM or DMH, and there is therefore no evidence of any cocarcinogenic effect of the asbestos fibres studied.

## Discussion

Analysis of all the long-term mineral fibre ingestion studies which are adequate in terms of quantity and quality shows no evidence of toxicity and very little evidence of carcinogenic potential. Only IR chrysotile fibres (1% in the diet) slightly increased the formation of AP in the large intestine of male F344 rats (National Toxicology Program, 1985). It should be noted, however, that some authors felt that there might be some relationship between colon tumour development (Donham et al., 1980) or between peritoneal haemangiomas and chrysotile ingestion (Bolton et al., 1982). At the same time, AP were not observed in another study with chrysotile or with a mixture of chrysotile and crocidolite at high doses (Truhaut & Chouroulinkov, this volume, pp. 127-133), in which an attempt was made to analyse the factors which may interfere with such tumour development. It was found that the vehicle probably plays an important role. From the analysis of the data presented in Table 5, it appears that, when fatty vehicles are used for asbestos fibre ingestion, no AP develop. When normal pelleted diet is used, AP are observed in both treated and control animals. If the chrysotile (IR, 1%) group, which gave positive results in rats (Table 5) is excluded, there is no statistically significant increase in AP frequency in treated animals. When this group is included in the total diet data, the AP frequency is significantly increased.

Table 5. Adenomatous polyps (AP) observed in long-term ingestion studies with asbestos fibres in rats[a]

| Reference | Type of asbestos | In treated rats | | In controls | |
|---|---|---|---|---|---|
| | | $M^b$ | $F^b$ | $M^b$ | $F^b$ |
| **Diet** | | | | | |
| Donham et al., 1980 | Chrysotile (10%) | 0/95 | 0/94 | 0/155 | 0/157 |
| McConnell et al., 1983 | Amosite (1%) | 3/350 | 3/350 | 1/117 | 1/117 |
| | Tremolite | 3/250 | 0/250 | 0/118 | 0/118 |
| National Toxicology Program, 1985 | Chrysotile (SR[c]) (1%) | 4/248 | 6/244 | 0/88 | 2/88 |
| | Chrysotile (IR[d]) (1%) | 20/250 | 3/250 | 0/88 | 0/88 |
| | Chrysotile (IR[d]) (1%) | 4/100[e] | 0/100[e] | | |
| **Fat** | | | | | |
| Bolton et al., 1982 | Chrysotile, amosite, crocidolite (margarine/diet) | 0/68 | - | 0/48 | - |
| Truhaut & Chouroulinkov[f] | Chrysotile, crocidolite (in palm oil) | 0/412 | 0/417 | 0/138 | 0/137 |

[a]No. of AP/No. of rats.
[b]M, male; F, female.
[c]SR, short-range fibres.
[d]IR, intermadiate-range fibres.
[e]In preweaning chrysotile-gavaged animals.
[f]See p. 127-133

The same type of analysis was applied to all GI tumours in all experiments in relation to the vehicle used (Table 6). When the data are considered in this way, it appears that, in general, asbestos fibres incorporated in the diet significantly increase the GI tumour frequency ($p<0.001$). When fat or water is used as vehicle, no evidence of any increase in tumour frequency related to asbestos fibres is found. Thus the diet used as a vehicle in the ingestion of asbestos fibres by laboratory animals plays a part in the increase in GI tumours in animals treated with asbestos fibres. The mechanism of this interaction between diet and asbestos fibres is currently unknown.

Table 6. Gastrointestinal tumours (data from all experiments) in relation to the vehicle used

| Vehicle | Treated animals | | Controls | | $p$ |
|---|---|---|---|---|---|
| | No. of tumours | No. of animals | No. of tumours | No. of animals | |
| Diet | 149 | 3204 | 28 | 1470 | 0.001 |
| Fat | 21 | 995 | 10 | 471 | NS[b] |
| Water | 25 | 570 | 9 | 175 | NS[b] |

[a]Butter, margarine or palm oil.
[b]NS, not significant.

The next stage was to determine which asbestos fibres are involved in the increase in GI tumours. The data from all the studies with regard to the type of fibre used (Table 7) give a clear response. From these data, only chrysotile fibres in long-term ingestion studies significantly increase the GI tumour frequency ($p<0.001$). The other asbestos fibres — amosite, crocidolite, etc. — did not show any such effect.

Table 7. Alimentary tract tumours observed in long-term ingestion studies with asbestos and other mineral fibres[a]

| Type of fibre | Treated animals | | Controls | | $p$ |
|---|---|---|---|---|---|
| | No. of tumours | No. of animals | No. of tumours | No. of animals | |
| Chrysotile | 125 | 1941 | 30 | 1078 | 0.001 |
| Amosite | 24 | 924 | 13 | 406 | NS[b] |
| Crocidolite | 0 | 89 | 0 | 72 | NS[b] |
| Tremolite | 17 | 500 | 10 | 236 | NS[b] |
| Other mineral fibres | 18 | 477 | 8 | 212 | NS[b] |

[a]All the experimental data (species, sexes, vehicles) have been combined.
[b]NS, not significant.

The conclusions as to the role of the diet (the vehicle) and of chrysotile fibres in GI tumour increase in long-term ingestion studies are based on the analysis of the combined results of a large number of experiments, which individually may be negative, equivocal, inconclusive or only slightly significant. The validity of such an analysis and the statistical significance of the results will depend on the factors taken (or not taken) into account, and which reduce, eliminate or increase this significance (Doll & Peto, 1985). In studies on the long-term ingestion of asbestos fibres many factors must be considered. Apart from the types of asbestos fibre used and the vehicle (diet), they include factors such as the dose, the treatment and observation periods, the species and strains of the experimental animals used, sex, age at beginning of treatment, the number of animals, laboratory conditions and possibly others. Such a multifactorial analysis was impossible to carry out at the present time. The author therefore strongly suggests that such an exercise should be conducted when the data from the National Toxicology Program on chrysotile in hamsters, and amosite, crocidolite and tremolite in rats, become available.

In the meantime, it should be borne in mind that chrysotile would appear to be a potent carcinogen for the GI tract of animals in long-term ingestion studies. This is in agreement with the finding of increased lung cancer deaths in workers exposed to chrysotile as compared with those exposed to crocidolite and with the close correlation between the standardized mortality ratios for lung and GI cancer ($r=0.916$, $p<0.001$) (Doll & Peto, 1985), and also with the results obtained in *in vitro* cell systems. Chrysotile was more cytotoxic than amosite and crocidolite *in vitro* to human embryonic intestinal cells (Reiss *et al.*, 1980). Chrysotile also induced aneuploidy in human lymphocytes *in vitro* (Valerio *et al.*, 1983) and in Syrian hamster embryo (SHE) cells (Oshimura *et al.*, 1984), induced the formation of binucleated cells (Jaurand *et al.*, 1984), as well as morphological transformation in SHE cells (Hesterberg & Barrett, 1984) and promoted the cloning efficiency of rat lung epithelial cells (Michiels *et al.*, this volume, pp. 000-000). In spite of all these convergent results, however, it would be premature to try and evaluate the risk of increased human GI cancer due to ingested chrysotile fibres. The forthcoming data from long-term ingestion studies in the USA (National Toxicology Program, 1985) and a more careful analysis of the data for both GI and other tumours may make it possible to reach more reliable conclusions as to the carcinogenicity of asbestos fibres for experimental animals and eventually for evaluation of the risk to humans.

In summary, the data on the long-term ingestion of asbestos fibres, analysed globally, indicate that chrysotile significantly increases the number of GI tumours, particularly when incorporated in the diet. Crocidolite, amosite, tremolite and other amphiboles did not exhibit such an effect. This finding is in agreement with the excess of lung cancer deaths in workers exposed to chrysotile and with the results obtained in *in vitro* cell systems. However, a re-evaluation of the long-term ingestion studies when further results are available is recommended.

## References

Bolton, R.E., Davis, J.M.G. & Lamb, D. (1982) The pathological effects of prolonged asbestos ingestion in rats. *Environ. Res.*, *29*, 134-150

Cunningham, H.M. & Pontefract, R.D. (1971) Asbestos fibres in beverages and drinking water. *Nature*, *232*, 332-333

Cunningham, H.M., Moodie, C.A., Lawrence, G.A. & Pontefract, R.D. (1977) Chronic effects of ingested asbestos in rats. *Arch. Environ. Contam. Toxicol.*, *6*, 507-513

Doll, R. & Peto, J. (1985) *Effects on Health Exposure to Asbestos. Report to the Health and Safety Commission*, London, Her Majesty's Stationery Office

Donham, K.J., Berg, J.W., Will, L.A. & Leininger, J.R. (1980) The effects of long-term ingestion of asbestos on the colon of F344 rats. *Cancer*, *45*, 1073-1084

Gaudichet, A., Bientz, M., Sébastien, P., Dufour, G., Bignon, J., Bonnaud, G. & Puisais, J. (1978) Asbestos fibres in Nimes: Relation to filtration process. *J. Toxicol. Environ. Health*, *4*, 853-860

Gibel, W., Lohs, K., Horn, K.H., Wildner, G.P. & Hoffmann, F. (1976) Animal experimental study of a carcinogenic effect of asbestos filter material after oral administration [in German], *Arch. Geschwulstforsch*, *46*, 437-442

Gross, P., Harley, R.S., Swinburne, L.M., Davis, J.M.G. & Green, W.G. (1974) Ingested mineral fibers. Do they penetrate tissue or cause cancer? *Arch. Environ. Health*, *29*, 341-347

Hammond, E.C., Selikoff, I.J. & Seidman, H. (1979) Asbestos exposure, cigarette smoking and death rates. *Ann. N.Y. Acad. Sci.*, *330*, 473-490

Hesterberg, T.W & Barrett, J.C. (1984) Dependence of asbestos and mineral dust induced transformation of mammalian cells in culture on fiber dimension. *Cancer Res.*, *44*, 2170-2180

Hilding, A.C., Hilding, D.A., Larson, D.L. & Aufderheide, A.C. (1981) Biological effects of ingested amosite asbestos, taconite tailings, diatomaceous earth and Lake Superior water in rats. *Arch. Environ. Health*, *36*, 298-303

IARC (1977) *IARC Monographs on the Evaluation of the Carcinogenic Risk of Chemicals to Man*, Vol 14, *Asbestos*, Lyon, International Agency for Research on Cancer

Jaurand, M., Bastie-Sigeac, I., Renier, A. & Bignon, J. (1983) Comparative toxicities of different forms of asbestos on rat pleural mesothelial cells. *Environ. Health Perspect.*, *51*, 153-159

McConnell, E.E., Rutter, H.A., Uliand, B.M. & Moor, J.A. (1983) Chronic effects of dietary exposure to amosite asbestos and tremolite in F344 rats. *Environ. Health Perspect.*, *53*, 27-44

Millette, J., Clark, P., Stober, J. & Rosenthal, M. (1983) Asbestos in water supplies of the United States. *Environ. Health Perspect.*, *53*, 45-48

National Toxicology Program (1983) *Lifetime carcinogenesis studies of amosite asbestos in Syrian golden hamster (feed studies)*, Washington, DC, US Department of Health and Human Services, Public Health Service (NIH Publication No. 84-2505)

National Toxicology Program (1985) *Toxicology and carcinogenesis studies of chrysotile asbestos in F344 rats (feed studies), (NIH Publication No. 86-2551)*, Washington, DC, US Department of Health and Human Services, Public Health Service

Oshimura, M., Hesterberg, T., Tsutsui, T. & Barrette, J. (1984) Correlation of asbestos-induced cytogenetic effects with cell transformation of Syrian hamster embryo cells in culture. *Cancer Res.*, *44*, 5017-5022

Peto, J., Seidman, H. & Selikoff, I.J. (1982) Mesothelioma mortality in asbestos workers: implications for models of carcinogenesis and risk assessment. *Br. J. Cancer*, *45*, 124-135

Reiss, B., Solomon, S., Weisburger, J.H. & Williams, G.M. (1980) Comparative toxicities of different forms of asbestos in a cell culture assay. *Environ. Res.*, *22*, 109-129

Rowe, J. (1983) Relative source contributions of diet and air to ingested asbestos exposure. *Environ. Health Perspect.*, *53*, 115-120

Selikoff, I.J., Hammond, E.C. & Seidman, H. (1979) Mortality experience of insulation workers in the United States and Canada, 1943-1976. *Ann. N.Y. Acad. Sci.*, *330*, 91-116

Smith, W.E., Hubert, D.D., Sobel, H.J., Peters, E.T. & Doerfler, T.E. (1980) Health of experimental animals drinking water with and without amosite asbestos and other mineral particles. *J. Environ. Pathol. Toxicol.*, *3*, 277-300

Valerio, F., de Ferrari, M., Ottagio, L., Repetto, E. & Santi, L. (1983) Chromosomal aberrations induced by chrysotile and crocidolite in human lymphocytes *in vitro*. *Mutat. Res.*, *122*, 397-402

Wagner, J.C., Berry, C., Cook, T.J., Hill, R.J., Pooley, F.D. & Skidmore, J.W. (1977) Animal experiments with talc. In: Walton, W.C., ed., *Inhaled Particles*, Vol. IV, part 2, New York, Pergamon Press, pp. 647-654

Ward, J.M., Frank, A.L., Wenk, M., Devor, D. & Tarone, R.E. (1980) Ingested asbestos and intestinal carcinogenesis in F344 rats. *J. Environ. Pathol. Toxicol.*, *3*, 301-312

# EFFECT OF LONG-TERM INGESTION OF ASBESTOS FIBRES IN RATS

### R. Truhaut

*Laboratoire de Toxicologie et Hygiène Industrielle,*
*Faculté des Sciences Pharmaceutiques, Paris, France*

### I. Chouroulinkov

*Institut de Recherches Scientifiques sur le Cancer (CNRS), Villejuif, France*

*Summary.* The effects of ingested asbestos fibres were studied in Wistar Han rats. Chrysotile and a mixture of chrysotile/crocidolite (75%/25%) in palm oil were given for 24 months to 70 males and 70 females per group (daily doses 10, 60 and 360 mg); one control group was fed with normal diet, a second with normal diet plus palm oil. The animals were observed for a further 6 months after the end of the treatment. The results indicate that ingestion of asbestos fibres at high doses had no toxic effects and did not affect animal survival; in addition, there was no evidence of carcinogenic effects.

## *Introduction*

Asbestos fibres have been found in lakes, rivers and drinking-waters, in beverages, beer, soft drinks, in food (IARC, 1977) and in wines (Gaudichet et al., 1978). This widespread human exposure via the oral route and/or via the respiratory route, the pulmonary clearance mechanism leading to swallowing of the fibres after inhalation (Gross et al., 1974), raises the question of local alimentary tract and/or systemic cancer risk. The likelihood that such a risk exists is strengthened by the relation between respiratory exposure to asbestos fibres and the development of pleural mesotheliomas (Doll & Peto, 1985; IARC, 1977). Moreover, in some epidemiological studies, an association was found between heavy asbestos exposure and the development of peritoneal tumours (Newhouse et al., 1972; Selikoff et al., 1964). However, this correlation was not confirmed by Doll and Peto (1985).

Studies on the chronic oral administration of various asbestos fibres reported before 1981 failed to find any relation between tumour incidence and treatment both in rats (Cunningham et al., 1977; Donham et al., 1980; Gross et al., 1974; Hilding et al., 1981; Wagner et al., 1977; Ward et al., 1980) and in Syrian hamsters (Smith et al., 1980). Only Gibel et al. (1976) reported an increase in malignant tumours in rats receiving lifetime treatment with filter material (50 mg/kg body weight per day) containing 52.6% of chrysotile fibres and 47.4% of unknown compounds.

In view of the importance of the problem and the widespread and unavoidable human exposure to asbestos fibres via the oral route, it was decided to study the effects in rats of the ingestion for 24 months of chrysotile and a mixture of chrysotile and crocidolite fibres at different doses, incorporated in palm oil. The experimental conditions and the results are summarized below.

## Materials and methods

The experimental protocol was drawn up and corrected after taking the advice of 12 experts, in accordance with OECD carcinogenicity study guideline No. 451 (OECD, 1981).

### Animals

Wistar Han SPF rats (Evic-Ceba, Bordeaux, France), males and females, 4–5 weeks old when received, were used. They were housed 5 per cage (polypropylene, Evic-Ceba, model BC4-52, 46.5 × 31.0 × 19.0 cm). After one week of acclimatization, the animals were distributed between the experimental groups.

### Asbestos fibres

Chrysotile and crocidolite asbestos fibres, UICC granulometry (electron microscope controlled) were provided by Eternit Industries (Vernouillet, France).

### Administration

Chrysotile alone, and a mixture of chrysotile/crocidolite (75%/25% respectively) incorporated in palm oil (Emeraude Grade, Astra-Calve, France), were given each morning to the animals, which had been starved during the night. At noon, the palm oil, with or without fibres, was removed and the normal diet (A 04 C, UAR, Villemoisson-sur-Orge, France) provided for the afternoon. A 2-month preliminary study indicated that rats regularly eat palm oil in measurable quantities. Ingestion of fibres was interrupted after 24 months and the surviving animals kept under observation for 6 further months.

### Doses and experimental groups

Eight groups of 140 (70 M, 70 F) rats were constituted as follows: group 0: controls (no treatment); group 1: controls (palm oil only); groups 2, 3 and 4: chrysotile at 10, 60 and 360 mg per day, respectively; groups 5, 6 and 7: chrysotile/crocidolite (75%/25%) also 10, 60 and 360 mg per day (Table 1). The corresponding concentrations of asbestos fibres in the palm oil were 0.2, 1.2 and 7.2% respectively. The mixtures were prepared twice a month, kept in a cold room and checked for the fibre content. Animal maintenance, the controls for toxicity, individual body weight measurement (weekly), post mortem and histopathology examinations were conducted in accordance with the OECD Good Laboratory Practice Guidelines. All macroscopic lesions and all organs and tissues from each animal were fixed for histopathological examination.

## Results

The clinical and physiological controls did not reveal any changes in animal behaviour and in the vegetative reactions, respiratory and cardiac rhythms and retinal sensitivity to light.

Survival after 18 months of treatment was 94% (minimum 82% for group 5, males, and maximum 100% for group 0, males). After 24 months, minimum survival was 60%

**Table 1. Chronic asbestos ingestion in rats: groups, doses, survival and body weights**

| Group | | Dose (mg/day) | | Survival[b] (%) | | Body weight[c] (g) | | |
|---|---|---|---|---|---|---|---|---|
| No. | Treatment | Planned | Ingested[a] | 24 months | 30 months | 12 months | 24 months | 30 months |
| *Male* | | | | | | | | |
| 0 | Controls | - | - | 71.4 | 34.3 | 550.4 | 574.1 | 523.1 |
| 1 | Palm-oil | - | - | 75.7 | 32.9 | 583.0 | 662.8 | 558.1 |
| 2 | Chrysotile | 10 | 10.4±0.1 | 74.3 | 28.6 | 588.9 | 666.1 | 544.3 |
| 3 | Chrysotile | 60 | 65.4±5.8 | 77.1 | 25.7 | 575.4 | 683.1 | 554.9 |
| 4 | Chrysotile | 360 | 397.0±32.3 | 68.6 | 30.0 | 594.3 | 705.5 | 549.2 |
| 5 | Mixture[d] | 10 | 10.4±1.1 | 64.3 | 25.7 | 597.2 | 673.5 | 548.7 |
| 6 | Mixture[d] | 60 | 65.5±7.0 | 77.1 | 24.3 | 601.9 | 679.2 | 567.3 |
| 7 | Mixture[d] | 360 | 379.0±12.4 | 77.1 | 27.1 | 595.0 | 688.9 | 563.8 |
| *Female* | | | | | | | | |
| 0 | Controls | - | - | 60.0 | 18.6 | 290.6 | 346.7 | 331.8 |
| 1 | Palm-oil | - | - | 72.9 | 22.9 | 310.1 | 414.6 | 351.1 |
| 2 | Chrysotile | 10 | 8.8±1.2 | 64.0 | 17.1 | 317.8 | 415.7 | 344.0 |
| 3 | Chrysotile | 60 | 59.8±6.8 | 68.6 | 37.1 | 312.0 | 431.4 | 367.5 |
| 4 | Chrysotile | 360 | 358.1±29.9 | 67.1 | 27.1 | 330.2 | 438.7 | 369.1 |
| 5 | Mixture[d] | 10 | 8.6±1.3 | 62.9 | 20.0 | 310.8 | 416.7 | 365.6 |
| 6 | Mixture[d] | 60 | 56.7±6.6 | 77.1 | 38.6 | 326.9 | 437.7 | 360.3 |
| 7 | Mixture[d] | 360 | 359.2±36.9 | 75.7 | 28.6 | 304.0 | 419.8 | 367.0 |

[a] Mean of monthly means ± S.D.

[b] Survival at 12 and 18 months was 100 and approximately 92%, respectively.

[c] Means of the body weight of all surviving animals in the group. In the first month the mean body weight was 117.3 and 91.4 g for males and females respectively.

[d] Chrysotile (75%) and crocidolite (25%)

(group O, F) and the maximum 77.1% for many groups (Table 1). Mortality increased markedly during the post-treatment 6-month observation period. However, at the end of the study, 30 months after the beginning of treatment, survival was still satisfactory, varying between 18 and 38% (Table 1). The main causes of death were senescence and neoplasia development.

Regularity of feeding was measured by the palm-oil and normal diet intakes. This made it possible to follow the quantity of asbestos fibres ingested and the changes in body weight. The measured daily doses, when compared with those in the study plan, are slightly higher for male rats and slightly lower for female rats (Table 1). However, if the doses are expressed in mg/kg body weight, the ingested doses are higher for female rats. If the mean annual body weight is taken as the basis for the evaluation, the ingested doses of chrysotile are approximately 20, 130 and 770 mg/kg body weight per day for males and 30, 220 and 1240 mg/kg body weight per day for females. The doses of the mixture are approximately the same, since the variation in body weight between the different groups is insignificant. It should be noted that body weights in all groups receiving palm oil, with or without fibres, when compared with those of the control group (group 0), are significantly higher ($p<0.01$) from 12 to 24 months of treatment. This difference disappears, however, after palm-oil ingestion is ended. A general increase in body fat, also affecting the liver, which regressed when the normal diet was restored, was found on histopathological examination.

**Tumour pathology**

The data on tumour pathology are summarized in Table 2. It should be noted that the proportion of animals bearing primary tumours is very high, but in agreement with previously published data for this strain of rats (von Deerberg et al., 1982). This proportion is almost identical for all groups, both controls and treated. Analysis of the number of tumours per animal for the different treatments gives the following results: for male groups: $\chi^2=20.050$, df=20, $p=0.863$ (not significant); for female groups: $\chi^2=31.728$, df=35, $p=0.627$ (not significant).

**Table 2. Chronic asbestos ingestion: numbers of tumours observed[a]**

| Group | No. of rats with tumours/ No. examined | | Tumours | | | | | | | | | | Total |
|---|---|---|---|---|---|---|---|---|---|---|---|---|---|
| | | | Alimentary tract | | Peritoneal cavity | | Respiratory tract | | Others | | All types | | |
| | | | b | m | b | m | b | m | b | m | b | m | |
| 0 Controls | M | 67/70 | 1 | 2 | 0 | 8 | 6 | 3 | 90 | 36 | 97 | 49 | 146 |
| | F | 66/68 | 0 | 1 | 0 | 1 | 2 | 0 | 51 | 85 | 53 | 87 | 140 |
| 1 Palm oil | M | 62/68 | 2 | 2 | 2 | 6 | 4 | 1 | 70 | 51 | 78 | 60 | 138 |
| | F | 66/69 | 2 | 0 | 1 | 2 | 2 | 0 | 66 | 87 | 71 | 89 | 160 |
| 2 Chrysotile (10 mg/day) | M | 63/69 | 3 | 2 | 2 | 4 | 1 | 1 | 67 | 42 | 73 | 49 | 122 |
| | F | 68/70 | 1 | 3 | 0 | 3 | 1 | 1 | 53 | 81 | 55 | 88 | 143 |
| 3 Chrysotile (60 mg/day) | M | 64/68 | 1 | 3 | 1 | 8 | 2 | 1 | 72 | 45 | 76 | 57 | 133 |
| | F | 68/70 | 2 | 2 | 1 | 3 | 1 | 2 | 74 | 69 | 78 | 76 | 154 |
| 4 Chrysotile (360 mg/day) | M | 64/68 | 0 | 2 | 2 | 4 | 4 | 0 | 64 | 41 | 70 | 47 | 117 |
| | F | 68/69 | 1 | 0 | 2 | 4 | 1 | 0 | 77 | 73 | 81 | 77 | 158 |
| 5 Mixture[b] (10 mg/day) | M | 64/69 | 2 | 0 | 0 | 7 | 1 | 1 | 77 | 46 | 80 | 54 | 134 |
| | F | 68/70 | 1 | 1 | 2 | 5 | 1 | 0 | 81 | 56 | 85 | 62 | 147 |
| 6 Mixture[b] (60 mg/day) | M | 63/68 | 1 | 0 | 0 | 5 | 3 | 1 | 74 | 48 | 78 | 54 | 132 |
| | F | 68/70 | 0 | 0 | 1 | 6 | 0 | 1 | 94 | 74 | 94 | 81 | 175 |
| 7 Mixture[b] (360 mg/day) | M | 63/70 | 4 | 2 | 2 | 8 | 1 | 0 | 90 | 29 | 97 | 39 | 136 |
| | F | 67/68 | 0 | 2 | 1 | 3 | 0 | 0 | 82 | 65 | 83 | 70 | 153 |

[a] b, benign; m, malignant.
[b] Chrysotile (75%) and crocidolite (25%).

The types of tumour observed included 7 mesotheliomas, of which 1 was peritoneal (palm oil, group 2, males); 3 in the thoracic cavity (groups 0, 3 and 6, males); 1 in the area of the salivary glands (group 4, females); and 2 in the testicles (groups 5 and 6). The way that they are distributed between the different groups (2 mesotheliomas in the control groups) and the localization indicate that these tumours are not related to the ingestion of asbestos fibres.

Differences in the incidence of tumours in the alimentary and respiratory tracts between treated and control groups are not statistically significant. The number of tumours in the peritoneal cavity, predominantly lymph-node angiomas or angiosarcomas, appears to be increased, but the increase is not statistically significant and no dose-response relationship exists.

Multifactorial analysis failed to find any difference in tumour frequency with respect to localization, type of fibre, dose and sex.

## Discussion

The clinicopathological results of this study showed that ingestion of chrysotile or a mixture of chrysotile/crocidolite (75%/25%) at high doses for 24 months did not adversely affect animal health in general, body weight gains, survival and tumour incidence.

Body weight and survival compare favourably with those found in other similar studies (Bolton *et al.*, 1982; Donham *et al.*, 1980; Hilding *et al.*, 1980; McConnell *et al.*, 1983; National Toxicology Program, 1985). Like Bolton *et al.* (1982) with margarine, we observed a significant increase in body weight related to the consumption of palm oil. After treatment was concluded, the excessive fat deposits regressed (Table 1) without any pathological effects.

Analysis of the results with respect to tumour incidence failed to show any statistically significant increase in tumour development in general, or at any specific site, such as the alimentary tract or peritoneal cavity, or in specific type of tumours, such as mesotheliomas. These results are in agreement with the data already published from long-term asbestos ingestion studies. It should, however, be mentioned that Gibel *et al.* (1976) and Ward *et al.* (1980) reported equivocal results. Gibel *et al.* (1976) reported an increase in the overall number of tumours in rats fed with chrysotile (52.6%) and filter material (47.4%) in diet. The nature of the latter material was not defined. Ward *et al.* (1980) reported 16 intestinal tumours in 49 rats gavaged with amosite in saline (10 mg/ml) 3 times weekly for 10 weeks. Unfortunately, the authors did not include a control group in the study.

A significant number of adenomatous polyps of the large intestine in male rats fed with a diet containing 1% of chrysotile, intermediate-range (IR) fibres only, has been reported (National Toxicology Program, 1985). In the groups receiving short-range chrysotile fibres, and in the preweaning chrysotile-gavaged group receiving IR fibres, the increase in the numbers of this lesion was not significant. Such adenomatous polyps were reported in other studies using diet as vehicle (Donham *et al.*, 1980; McConnell *et al.*, 1983). In the present study, no adenomatous polyps were seen in any group. It therefore seems that the palm oil used as vehicle played a protective role. Bolton et al. (1982), using a diet to which margarine had been added, failed to observe any such polyps.

In conclusion, the ingestion of chrysotile or of a mixture of chrysotile/crocidolite (75%/25%) at various doses, and even at high ones, did not adversely affect the health of rats and there was no evidence of any increase in tumours of the alimentary tract or of any general increase in tumour frequency.

## Acknowledgements

This study was carried out in the Evic-Ceba Laboratory, 33290 Blanquefort, France (Director P. Masson) and was supported by the French asbestos industry, the French Ministry for Foreign Affairs and the Government of Quebec (Canada).

## References

Bolton, R.E., Davis, J.M.G. & Lamb, D. (1982) The pathological effects of prolonged asbestos ingestion in rats. *Environ. Res., 29*, 134-150

Cunningham, H.M., Moodie, C.A., Lawrence, G.A. & Pontefract, R.D. (1977) Chronic effects of ingested asbestos in rats. *Arch. Environ. Contam. Toxicol., 6*, 507-513

von Deerberg, F., Rapp, K.G., Pitterman, W. & Rehm, S. (1980) On the tumour spectrum of Wistar Han rats [in German]. *Versuchstierkd., 22*, 267-280

Doll, R. & Peto, J. (1985) *Effects on Health of Exposure to Asbestos. Report to the Health and Safety Commission*, London, Her Majesty's Stationery Office

Donham, K.J., Berg, J.W., Will, L.A. & Leiminger, J.R. (1980) The effects of long-term ingestion of asbestos on the colon of F344 rats. *Cancer, 45*, 1073-1084

Gaudichet, A., Bientz, M., Sébastien, P., Dufour, G., Bignon, J., Bonnaud, G. & Puisais, J. (1987) Asbestos fibres in Nîmes: Relation to filtration process. *J. Toxicol. Environ. Health, 4*, 853-860

Gibel, W., Lohs, K., Horn, H.K., Wildner, G.P. & Hoffman, F. (1976) Animal experimental study of a carcinogenic effect of asbestos filter material after oral administration [in German]. *Arch. Geschwulstforsch, 46*, 437-442

Gross, P., Harley, R.S., Swinburne, L.M., Davis, J.M.G. & Green, W.G. (1974) Ingested mineral fibers. Do they penetrate tissue or cause cancer? *Arch. Environ. Health, 29*, 341-347

Hilding, A.C., Hilding, D.A., Larson, D.L. & Aufderheide, A.C. (1981) Biological effects of ingested amosite asbestos, taconite tailings, diatomaceous earth and Lake Superior water in rats. *Arch. Environ. Health, 36*, 298-303

IARC (1977) *IARC Monographs on the Evaluation of the Carcinogenic Risk of Chemicals to Man*, Vol. 14, *Asbestos*, Lyon, International Agency for Research on Cancer

McConnell, E.E., Rutter, H.A., Uliand, B.M. & Moor, J.A. (1983) Chronic effects of dietary exposure to amosite asbestos and tremolite in F344 rats. *Environ. Health Perspect., 53*, 27-44

National Toxicology Program (1985) *Toxicology and carcinogenesis studies of chrysotile asbestos in F344 rats (feed studies)*, (*NIH Publication No. 86-2551*), Washington DC, Department of Health and Human Services, Public Health Service

Newhouse, M.L., Berry, G., Wagner, J.C. & Turok, M.E. (1972) A study of the mortality of female asbestos workers. *Br. J. Ind. Med., 29*, 134-141

OECD (1981) *OECD Guidelines for Testing of Chemicals. Carcinogenicity Studies*, No. 451, Paris, Organization for Economic Co-operation and Development

Selikoff, I.J., Churg, J. & Hammond, E.C. (1964) Asbestos exposure and neoplasia. *J. Am. Med. Assoc., 188*, 22-26

Smith, W.E., Hubert, D.D., Sobel, H.J., Peters, E.T. & Doerfler, T.E. (1980) Health of experimental animals drinking water with and without amosite asbestos and other mineral particles. *J. Environ. Pathol. Toxicol., 3*, 277-300

Wagner, J.C., Berry, C., Cook, T.J., Hill, R.J., Pooley, F.D. & Skidmore, J.W. (1977) Animal experiments with talc. In: Walton, W.C., ed., *Inhaled Particles*, Vol. IV, part 2, New York, Pergamon Press, pp. 647-654

Ward, J.M., Frank, A.L., Wenk, M., Devor, D. & Tarone, R.E. (1980) Ingested asbestos and intestinal carcinogenesis in F344 rats. *J. Environ. Pathol. Toxicol., 3*, 301-312

# EXPERIMENTAL DATA ON IN VITRO FIBRE SOLUBILITY

## G. Larsen

*Institute for Mineralogy and Economic Geology, Aachen College of Technology, Aachen, Federal Republic of Germany*

*Summary.* Different types of natural and synthetic fibres have been subjected to systematic solubility tests *in vitro* in a physiological solution at 37°C. Both closed-system and open-system experiments were carried out. Atomic absorption spectrometry of the filtered fluids showed characteristic differences of solubility. Plastic fibres are practically insoluble. In contrast to glass fibres, the solubility of asbestos fibres is low. Sepiolite and wollastonite are of moderate solubility. The results were confirmed by scanning electron microscopy. Kinetic studies and extensive solubility tests led to a new exponential expression which describes the dissolution process in a closed system better than the square-root time laws often used. Moreover, this exponential model provides a new method of distinguishing between different materials by means of their initial rates of dissolution.

## Introduction

The solubilities of fibrous materials are of interest not only in their application but also from the point of view of occupational medicine. In this paper, the results of a study of the solubility of a number of different types of fibres *in vitro* are reported.

## Materials and methods

A method of carrying out solubility tests in both closed and open systems has been developed. These tests can easily be carried out and give reproducible results. The fibrous materials investigated were the natural fibres chrysotile, crocidolite, sepiolite and wollastonite, different glass fibres (glass wool, basalt wool), ceramic fibres, and two synthetic plastic fibres. The tests were carried out in Gamble's solution, which is similar in composition to lung fluid (without the organic components). Table 1 shows its composition.

In most cases the test temperature was 37°C, but some fibres were tested at higher temperatures (90°C) and higher pressures (1000 bar=14 500 psi) in autoclaves.

The duration of the experiments ranged from 1 hour to 20 weeks for closed-system and 1 hour to 2 weeks for open-system conditions.

The filtered fluids were analysed by atomic absorption spectrometry (AAS; Perkin Elmer 4000, Perkin Elmer 5000). The fibre samples were examined by

## Table 1. Chemical composition of Gamble's solution[a]

| Compound | Content of solution [g/l] |
| --- | --- |
| $MgCl_2.6H_2O$ | 0.212 |
| NaCl | 6.171 |
| KCl | 0.311 |
| $Na_2HPO_4$ | 0.148 |
| $Na_2SO_4$ | 0.079 |
| $CaCl_2.2H_2O$ | 0.255 |
| $NaCH_3COO.3H_2O$ | 1.065 |
| $NaHCO_3$ | 2.571 |

[a]In the treatment of plastic fibres, the sodium acetate and sodium sulfate were omitted.

scanning electron microscopy (SEM; JSM 35 Jeol (Japan Electron Optics Lab)) with energy dispersive spectrometry (EDS) facilities (EDS system Ortec 6230).

## Results

### Atomic absorption spectrometry results

As silicon is the most important constituent of most of the fibres under investigation, the silicon content of the fluids was predominantly determined. AAS analyses of the filtered fluids showed surface-related differences of solubility of several orders of magnitude. Depending on the type of fibre, the values ranged from a few ng silicon dissolved per $cm^2$ (chrysotile and crocidolite) to several thousands of $ng/cm^2$ silicon dissolved (glass wools). Aramide and carbon fibres (both plastic fibres) proved to be practically insoluble.

The solubilities of chrysotile and crocidolite are very low. Solubility experiments with chrysotile show that only a very small amount of silicon can be dissolved but that a larger amount of magnesium is generally released. For example, the values obtained after a 6-week shaking-table experiment (closed system) are 6 $ng/cm^2$ silicon and 160 $ng/cm^2$ magnesium dissolved. Sepiolite and wollastonite fibres were found to be of moderate solubility.

Glass fibres (glass wool, basalt wool) are of higher solubility. Depending on their chemical composition, they can be divided into two groups: glass wools with a silica content of more than 60% are more soluble than fibres of basaltic composition. In open-system experiments (continuous-flow equipment), none of the glass components reaches a saturation concentration since, under these conditions, a stable surface layer cannot be built up. Ceramic fibres release a small amount of silicon when treated with Gamble's solution. Fibres not containing $Cr_2O_3$ seem to be more resistant than those containing around 2.5% $Cr_2O_3$.

## Scanning electron microscopy results

Figures 1 and 2 show different types of treated fibres as seen by SEM. SEM examination confirmed the results obtained by AAS. As seen in Figure 1(a), aramide fibres were not corroded by treatment in physiological solution. Carbon fibres showed the same behaviour. Chrysotile and crocidolite showed no signs of surface corrosion, but split into a large number of thin fibrils (Figures 1(b) and 1(c)), with a consequent enormous increase in the sample surface. Sepiolite and wollastonite exhibited gel-like coatings (Figure 1d), which can be explained as precipitation effects after super-saturation of the test fluid.

Corrosion effects of the fibre surfaces are characteristic for all glass fibres. For example, etch pits and gel layers, which tended to become detached and break away, could be observed (Figures 2(a) and 2(b)). Some fibres showed precipitation of salts (Figure 2(c)).

The ceramic fibres differed in their solution behaviour. Thus fibres containing $Cr_2O_3$ showed corrosion effects similar to those seen with glass fibres. Their surface layers expanded in Gamble's solution and became detached as a result of the increase in volume (Figure 2(d)).

## Results of kinetic studies

The dissolution of glass and minerals under closed-system conditions has repeatedly been expressed in terms of square-root time laws. Mathematical expressions of this type suffer from the disadvantage that it is impossible to describe the way in which the concentration tends to approach a limiting value. Square-root models are therefore inadequate.

Since the dissolution process is surface-controlled between pH 7 and pH 9 (Lasaga, 1981; Stumm & Morgan, 1981), the following exponential expression can be derived to describe the dissolution process in a closed system (Feck, 1984; Forster & Feck, 1985):

$$c = c_{sat}(1 - e^{-kt})$$

where:

$c$ = concentration [ng/cm²]

$c_{sat}$ = saturation concentration [ng/cm²]

$t$ = time [d]

$k$ = exponential coefficient [1/d]

In addition, this exponential model offers a new means of distinguishing between different materials, using the initial dissolution rate expressed as the first differential coefficient at $t=0$:

$$\left. \frac{dc(t)}{dt} \right|_{t=0} = c_{sat}k$$

Fig. 1. Scanning electron micrographs of treated fibres: (*a*) aramide fibre after a 1-week stationary experiment; (*b*) crocidolite after a 6-week stationary experiment; (*c*) crocidolite after a 10-week stationary experiment; (*d*) sepiolite after a 2-week stationary experiment.

**Fig. 2.** Scanning electron micrographs of treated fibres: (*a*) marl wool after a 6-week stationary experiment; (*b*) basaltic glass fibre after a 3-week stationary experiment at 100°C and 1000 bar (autoclave); (*c*) glass wool after a 6-week stationary experiment; (*d*) ceramic fibre after a 2-week stationary experiment at 90°C.

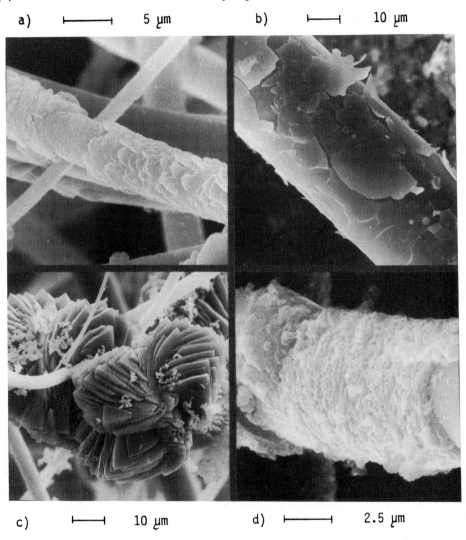

which is a useful (surface-dependent) constant for a given material. Some examples are given in Tables 2 and 3, showing the values of this constant for fibrous materials having approximately similar surfaces.

**Table 2. Initial dissolution rates of different amorphous fibrous materials**

| Fibrous material | $c_{sat}{}^k$ (Si dissolved [ng/cm² d]) |
|---|---|
| Marl wool | 25 |
| Rock wool | 33 |
| Glass wool I[a] | 144 |
| Glass wool II[a] | 860 |

[a]Glass wool I and glass wool II show differences in their chemical composition: glass wool I contains 2.46% (weight) BaO, whereas the corresponding value of glass wool II is 0.07% (weight) BaO.

**Table 3. Initial dissolution rate of two natural fibrous materials**

| Fibrous material | $c_{sat}{}^k$ [Si dissolved (ng/cm² d)] |
|---|---|
| Crocidolite | 5 |
| Sepiolite | 45 |

It is worth mentioning that a correlation exists between initial dissolution rates and the rates obtained in open-system experiments, the values often being identical.

## Conclusions

The retention of fibrous particles in the lung is controlled by several factors; one important factor is the solubility of the inhaled fibres. Solution properties of fibrous materials can be compared by means of the new exponential model. The initial dissolution rates may perhaps be useful in the classification of fibrous materials according to their solubility, from the point of view of their health effects.

## References

Feck, G. (1984) In vitro *Untersuchungen zur Löslichkeit von Glasfasern und einigen Mineralfasern in schwach alkalischer wässriger Lösung*. Ph.D. thesis, Aachen College of Technology

Forster, H. & Feck, G. (1985) *In vitro* studies of artificial mineral fibres [in German]. *Zentralbl. Arbeitsmed.*, 35, 130-135

Lasaga, A.C. (1981) Rate laws of chemical reactions. *Rev. Miner.*, 8, 1-110

Stumm, W. & Morgan, J.J. (1981) *Aquatic Chemistry. An Introduction Emphasizing Chemical Equilibria in Natural Waters*, New York, Chichester, John Wiley

# ADSORPTION OF POLYCYCLIC AROMATIC HYDROCARBONS ON TO ASBESTOS AND MAN-MADE MINERAL FIBRES IN THE GAS PHASE

### P. Gerde & P. Scholander

*Department of Chemical Engineering, Royal Institute of Technology, Stockholm, Sweden*

*Summary.* The adsorption of the aromatic hydrocarbons naphthalene and phenanthrene on to 4 types of asbestos and 2 types of man-made mineral fibres in the gas phase was studied. The asbestos types were chrysotile, anthophyllite, amosite, and crocidolite and the man-made mineral fibres were rock wool and glass wool. The influence of the gas humidity on this adsorption was also studied. The experiments were performed in an open system with the continuous generation of a gas stream of constant flow rate, humidity and hydrocarbon concentration. The results show that chrysotile asbestos is an extremely good adsorbent of polycyclic aromatic hydrocarbons (PAHs) in dry gas. This material adsorbs about 100 000 times more PAH than does glass wool, which has the lowest capacity of the fibres tested. The amphibolic asbestos types lie in the upper half and the rock wool sample in the lower half of the range. However, if the gas phase is humidified to typical ambient air values, there is a dramatic decrease in the adsorption on to the highly adsorbing materials. This means that, at relevant gas-phase humidities, there will be a fairly weak adsorption of PAH of about the same order of magnitude for all the materials tested. Any enhanced biological effect of inhalation of PAH adsorbed on fibres is thus likely to be connected with properties of the fibres other than the mere amounts of PAH adsorbed.

## Introduction

It is well established that asbestos exposure increases the already high risk that smokers run of contracting lung cancer (Selikoff *et al.*, 1968). One hypothesis to explain this synergistic effect is that carcinogenic compounds in the cigarette smoke, in particular polycyclic aromatic hydrocarbons (PAHs), are adsorbed on the surfaces of the asbestos fibres. This may increase the retention of these substances in the lung and thus also the risk of cancer developing (Harington, 1973). Another question of importance is whether the same effect can occur with man-made mineral fibres (MMMF), which are now replacing asbestos in many fields of use. Apparently this adsorption may occur either before inhalation in the ambient air, or after inhalation, probably in the liquid phase of the lining of the lung. This paper deals exclusively with the former possibility.

A wide variety of PAHs are present in the ambient air. The vapour pressures of these compounds decrease drastically with increasing molecular size (Murray *et al.*, 1974). In addition, the larger the molecules the greater is their tendency to be adsorbed on to airborne particles (Eiceman & Vandiver, 1983). Medical research has paid the greatest attention to the low-volatile, heavier members of the group. Benzo[a]pyrene (BP), the reference substance in the group, has a very low vapour pressure which, at 25°C, can be estimated to be about $1 \times 10^{-11}$ atmosphere (Murray et al., 1974; Sonnefelt et al., 1983). The saturated vapour of this hydrocarbon then contains around 100 ng/m³. Tar from tobacco smoke contains BP at a concentration of about 2 µg/g (US Department of Health, Education, and Welfare, 1979). Typically, the content of BP in urban air is in the range 0.6–3.5 ng/m³ of air or 5–50 µg/g of airborne particles (Katz et al., 1978). In urban air, in addition, BP occurs at a concentration of less than 10% in its gaseous state, and this figure should represent a true equilibrium value (Yamasaki et al., 1982). At these very low gas-phase concentrations, it is necessary to consider not only the adsorption equilibria on airborne particles, but also the kinetics of those mass-transfer processes that will lead to such equilibria. This is particularly true for particles such as asbestos fibres, where adsorption is likely to occur at ambient air temperatures. It is known that some types of asbestos are already contaminated with small amounts of PAH when they are mined, and it seems that the fibres are further contaminated with PAH during processing. BP concentrations of up to $3.7 \times 10^{-8}$ g/g asbestos have been measured (Pylev & Shabad, 1973). However, it is not clear whether these PAHs are already associated with other particles when they contaminate the asbestos fibres or whether they are adsorbed directly from the gaseous state. The purpose of this study was to estimate the adsorption of a typical PAH on asbestos fibres and MMMF in the gas phase. The influence of the humidity of the gas phase on these adsorption equilibria was also studied.

## Materials and methods

The very low gas-phase concentrations of BP give rise to experimental problems. Weighing a portion of the material to be studied with sufficient accuracy requires something of the order of $10^{10}$ micron-size particles. One of these particles, when airborne, may have a contact time fully sufficient to reach adsorption equilibrium in a gas phase of such a low concentration, but the experimental bulk sample of particles requires far too long a contact time even with forced convection. It has therefore been necessary to use the more volatile aromatics naphthalene and phenanthrene as model substances for the heavier members of the group. Using 2 different-sized molecules permits a crude extrapolation of adsorption data from these substances to the less volatile PAH. The experiments were performed with 4 types of asbestos — chrysotile, crocidolite, amosite and anthophyllite — and with two types of MMMF, rock wool and glass wool. Before use, all fibrous materials were washed in toluene for 8 h by means of Soxhlet extraction and dried at 120°C for 2 h. The specific surface areas of the washed materials were determined by measuring their B.E.T. isotherms (Adamson, 1976) (Table 1). [¹⁴C]Naphthalene (1.45 MBq/mg, Amersham) and [¹⁴C]phenanthrene (4.0 MBq/mg, Amersham) were used as adsorbates, and when necessary these were diluted with their respective non-labelled analogues (Merck, scintillation grade, and Serva Feinbiochimica, analytical grade).

**Table 1. Specific surface areas of the fibrous materials studied**

| Fibre type | Specific surface area (m$^2$/g) |
|---|---|
| Chrysotile asbestos (UICC) | 40.1 |
| Anthophyllite asbestos (UICC) | 19.6 |
| Amosite asbestos (UICC) | 12.8 |
| Crocidolite asbestos (UICC) | 12.5 |
| Rock wool (Rockwool AB, Skövde, Sweden) | 0.275 |
| Glass wool (AB Gullfiber, Billesholm, Sweden) | 0.717 |

The experiments were performed using a flow-through system with continuous generation of a gas stream of constant flow rate, PAH concentration, and humidity (Figure 1). Nitrogen was used as carrier gas instead of air. A constant concentration of PAH was achieved by passing one gas stream through a bed of crystals of the PAH studied. This almost saturated stream of gas was then diluted with another gas stream to the desired concentration; this procedure has been described elsewhere (Pella, 1976; Westcott et al., 1981). Flow regulation of the saturated gas stream was accomplished by supplying a capillary with a constant pressure from a pressure regulator (Brooks 8601-B). When greater dilutions were desired, a self-diffusion tube was used instead. The diluting stream was regulated by means of a pressure regulator (Brooks 8601-B) and a constant-flow regulator (Brooks 7144). The total flow rate was measured with a soap-bubble meter. The adsorption column and bypass were designed according to Miguel et al. (1979), with a length of 50 mm and an inside diameter of 9 mm. The humidifier unit consisted of 2 glass chambers installed in the line of the diluting gas stream. The gas first enters a humidifier chamber with a water reservoir kept at the same temperature as the adsorption unit. The chamber has a volume of about 50 ml and is equipped with 2 vertical porous glass slabs in order to increase the contact area between gas and water. The nearly saturated effluent from this chamber is passed through a dew-point chamber kept at a lower temperature corresponding to the dew point of the desired humidity at the adsorption temperature. After excess humidity has been precipitated in this vessel, the gas flows back again to the bath with the adsorption unit. The ducts connecting the water baths are jacketed and kept at a higher temperature in order to prevent condensation.

The concentrations of the $^{14}$C-labelled compounds were measured by liquid scintillation counting. The gas phase was then sampled by absorbing the substance from a known volume of gas (500 ml) in toluene. The amounts of PAH absorbed by the fibres were measured by means of a thermal wet-oxidation process (van Slyke & Folch, 1940), wherein all organic matter, including the labelling isotopes, on a weighed portion of fibres is degraded to $CO_2$. This gas is then adsorbed in Protosol tissue solubilizer (New England Nuclear). The weight of fibres used in the experiments varied between 0.02 and 1.5 g, depending on the adsorptivity of the material.

**Fig. 1. Diagram of the adsorption apparatus.**

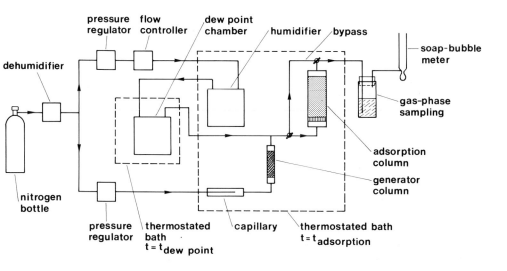

A total flow rate of 100 ml/min was used and, after start up, the gas-phase concentration of the generator was normally stable within less than an hour. Variations in the flow rate and in the aromatic concentration of the bypassed gas stream had standard deviations of ±0.4% and ±4% respectively. Connecting the adsorption column gave break-through curves of the effluent stream varying from undetectable for the low-adsorbing materials to well-defined curves stretching over several weeks for the high-adsorbing materials. At equilibrium, the fibre portion was quickly divided and transferred to 2 bottles, which were immediately sealed to avoid desorption losses. These bottles were sample containers for the wet-oxidation process. A mass balance for naphthalene on a break-through curve from chrysotile asbestos showed that 94% of the substance removed from the gas stream was recovered in the wet-oxidation process.

## Results

### Dry gas

The adsorption of naphthalene and phenanthrene in dry gas on the various fibrous materials at 25°C is shown in Figure 2. The concentration on the fibres in g/g is shown as a function of the relative pressure of the aromatic substances. This is the gas-phase concentration expressed as its fraction of the concentration of the compound in the saturated vapour under same conditions. The concentrations of naphthalene and phenanthrene in the saturated vapour at 25°C are 537 mg/m³ and 1.16 mg/m³, respectively (Sonnefelt et al., 1983). In dry gas, the fibrous materials tested cover a wide range of adsorptivities, with the asbestos types falling in the upper half and the MMMF in the lower half of the range. The differences in the adsorptivities are, of course, partly due to the varying specific surface areas of the materials (Table 1). If the adsorption is expressed per unit of surface area, the ranking of the fibrous materials is somewhat different. At a 1% relative pressure of naphthalene, the following results for the different materials are obtained, in $\mu g/m^2$: chrysotile 150, anthophyllite 50, amosite 10, crocidolite 7, rock wool 10 and glass wool 0.3. Chrysotile is thus an extremely good adsorbent of PAHs in dry gas, anthophyllite a somewhat weaker one, while rock wool, crocidolite and amosite are all weaker still. Finally, glass wool seems to have a very low adsorption capacity for PAHs.

**Fig. 2. Adsorption of naphthalene and phenanthrene in dry gas on to various fibrous materials at 25° C**

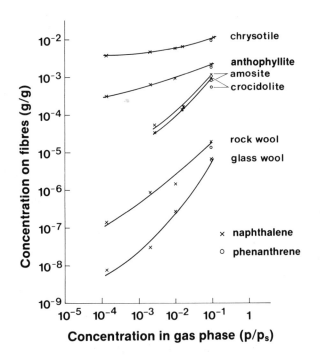

## Humid gas

The humidity of the gas phase has both a varying and in some cases a very dramatic influence on the adsorption of PAHs on to the fibrous materials studied (Figure 3). In general, the greater the adsorptivity of a material in dry gas, the greater the decrease in this capacity with increasing humidity of the gas phase. Chrysotile asbestos adsorbs about 4000 times more naphthalene at a relative pressure of 1% in dry gas than in gas with a relative humidity of 79%. For the rock wool sample, adsorption is only 6 times greater in dry gas than in gas at 79% relative humidity. The rate at which gas-phase humidity affects adsorption is very rapid, as demonstrated by the following simple experiment. Fibres are equilibrated in a gas phase with naphthalene at a relative pressure of 1%. If the relative humidity of the gas stream is suddenly increased to 79%, this will lead to a sharp pulse of naphthalene in the effluent stream from the fibre bed which, in the case of chrysotile asbestos, may reach a concentration 40 times that in the incoming gas stream (Figure 4).

**Fig. 3. Adsorption in the gas phase of naphthalene and phenanthrene on to various fibrous materials at 25° C as a function of relative humidity. The relative pressures of naphthalene and phenanthrene were 0.010 and 0.090, respectively**

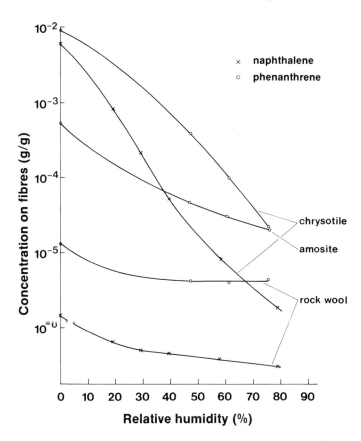

**Fig. 4.** Concentration of naphthalene downstream of the adsorption column when the relative humidity of the incoming gas is suddenly raised from 0% to 79%. The fibres were previously equilibrated with the constant naphthalene concentration of the incoming gas phase of 5.6 $\mu$/litre

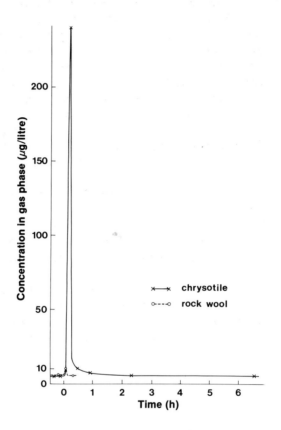

## Discussion

It would seem that the adsorption of aromatic hydrocarbons in dry gas increases with increasing polarity of the adsorbent. Chrysotile asbestos, in particular, with its high surface density of hydroxyl groups, has a very high adsorptivity. As suggested elsewhere (Fournier & Pezerat, 1982), this is probably due to the polar interaction between the charged surface of the chrysotile fibre and the mobile $\pi$ electrons of the aromatic hydrocarbon structure. This view is further supported by the extreme sensitivity of this adsorption of PAHs to the humidity of the gas phase. The aromatic hydrocarbons and the polar water molecules seem to be attracted to the same sites on the surfaces. The weaker adsorption forces between PAHs and the MMMF tested, however, are not nearly as sensitive to the humidity of the gas phase. This is an indication of a more non-polar type of interaction between the hydrocarbons and the amorphous fibre surfaces with a lower surface charge. Similar behaviour has been reported when various benzene derivatives are adsorbed in the gas phase on to soil samples (Chiou & Shoup, 1985). A decrease in the adsorption of these organics with increasing humidity of the gas phase was attributed to competition between the water and the organic molecules for the liquid phase when phenanthrene was adsorbed on to various types of asbestos in organic solvents (Fournier & Pézerat, 1982). The adsorption was considerably reduced if the solvent had not been dried prior to use. Just as in our investigation, this effect was more pronounced with chrysotile asbestos than with the amphibolic asbestos types crocidolite and amosite. The almost identical adsorptivity of naphthalene and phenanthrene on the same fibrous materials at the same relative pressure indicates that the adsorption of PAH of lower volatility can be estimated from adsorption measurements using lighter and more volatile members of this group of substances.

In the range of typical ambient air humidities, the fibrous materials studied adsorb PAHs to roughly the same degree. At a relative pressure of 1% for a model PAH and a relative humidity of 60%, this adsorption will be in the range of $5 \times 10^{-7} - 1 \times 10^{-6}$ g/g of fibres. This implies that the fibres may adsorb more PAH once they have become airborne and released from the kinetic constraints of belonging to a bulk sample. The adsorption also implies an extremely large accumulation of PAHs within the volume element occupied by the fibre as compared with the surrounding gas phase. This effect is by no means unique for fibres as compared with other ambient air particles, however. Any enhanced biological effect on inhalation of PAHs adsorbed on fibres must thus be connected with properties of the fibres other than the mere magnitude of the PAH adsorbed.

## Acknowledgements

This work was supported by the Swedish Work Environment Fund through grant No. 81-0929.

## References

Adamson, A.W. (1976) *Physical Chemistry of Surfaces*, 3rd ed., New York, John Wiley, pp. 565-571

Chiou, C.T. & Shoup, T.D. (1985) Soil sorption of organic vapors and effects of humidity on sorptive mechanism and capacity. *Environ. Sci. Technol.*, *19*, 1196-1200

Eiceman, G.A. & Vandiver, V.J. (1983) Adsorption of polycyclic aromatic hydrocarbons on fly ash from a municipal incinerator and a coal-fired power plant. *Atmos. Environ.*, *17*, 461-465

Fournier, J. & Pezerat, H. (1982) Study of the mode of adsorption of polycyclic aromatic hydrocarbons on asbestos. The case of phenanthrene [in French]. *J. Chim. Phys.*, *79*, 589-596

Harington, J.S. (1973) Chemical factors (including trace elements) as etiological mechanisms. In: Bogovski, P., Gilson, J.C. & Timbrell, V., eds, *Biological Effects of Asbestos (IARC Scientific Publications No. 8)*, Lyon, International Agency for Research on Cancer, pp. 304-311

Katz, M., Sakuma, T. & Ho, A. (1978) Chromatographic and spectral analysis of polynuclear aromatic hydrocarbons — quantitative distribution in air of Ontario cities. *Environ. Sci. Tech.*, *12*, 909-914

Miguel, A.H., Korfmacher, W.A., Wehry, E.L., Mamantov, G. & Natusch, D.F.S. (1979) Apparatus for vapor-phase adsorption of polycyclic organic matter onto particulate surfaces. *Environ. Sci. Technol.*, *13*, 1229-1232

Murray, J.J., Pottie, R.F. & Pupp, C. (1974) The vapor pressures and enthalpies of sublimation of five polycyclic aromatic hydrocarbons. *Can. J. Chem.*, *52*, 557-563

Pella, P.A. (1976) Generator for producing trace vapor concentrations of 2,4,6-trinitrotoluene, 2,4-dinitrotoluene, and ethylene glycol dinitrate for calibrating explosives vapor detectors. *Anal. Chem.*, *48*, 1632-1637

Pylev, L.N. & Shabad, L.M. (1973) Some results of experimental studies in asbestos carcinogenesis. In: Bogovski, P., Gilson, J.C. & Timbrell, V., eds, *Biological Effects of Asbestos (IARC Scientific Publications No. 8)*, Lyon, International Agency for Research on Cancer, pp. 99-105

Selikoff, I.J., Hammond, E.C. & Churg, J. (1968) Asbestos exposure, smoking and neoplasia. *J. Am. Med. Assoc.*, *204*, 104-110

van Slyke, D.D. & Folch, J.J. (1940) Manometric carbon determination. *J. Biol. Chem.*, *136*, 509-541

Sonnefelt, W.J., Zoller, W.H. & May, W.E. (1983) Dynamic coupled-column liquid chromatographic determination of ambient temperature vapor pressures of polynuclear aromatic hydrocarbons. *Anal. Chem.*, *55*, 275-280

US Department of Health, Education, and Welfare (1979) *Smoking and Health: A Report of the Surgeon General (DHEW Publication No. (PHS)79-50066)*, Rockville, MD, Chapter 14, p. 54

Westcott, J.W., Simon, C.G. & Bidleman, T.F. (1981) Determination of polychlorinated biphenyl vapor pressures by a semimicro gas saturation method. *Environ. Sci. Tech.*, *15*, 1375-1378

Yamasaki, H., Kuwata, K. & Miyamoto, H. (1982) Effects of ambient air temperatures on aspects of airborne polycyclic aromatic hydrocarbons. *Environ. Sci. Technol.*, *16*, 189-194

# COMPARATIVE STUDY OF THE EFFECT OF CHRYSOTILE, QUARTZ AND RUTILE ON THE RELEASE OF LYMPHOCYTE-ACTIVATING FACTOR (INTERLEUKIN 1) BY MURINE PERITONEAL MACROPHAGES *IN VITRO*

D. Godelaine & H. Beaufay

*International Institute of Cellular and Molecular Pathology, Brussels, Belgium*

*Summary*. The release of lymphocyte-activating factor (LAF) into the medium of cultured mouse resident peritoneal macrophages was estimated from the effect of this medium on the proliferation of mouse thymocytes in the presence of phytohaemagglutinin. Untriggered macrophages released little LAF into their culture medium. Upon addition of UICC chrysotile A (25-50 $\mu$g/$10^6$ macrophages), LAF appeared in the medium, and release continued for at least 20 h. DQ12 quartz was a more potent inducer of LAF production, while rutile was nearly inactive. Although chrysotile and quartz caused cell damage (as estimated by the release of lactate dehydrogenase), it was established that LAF release was not attributable to leakage of preformed intracellular LAF. The lymphoproliferative activity detected in macrophage media was stable at 56°C and non-dialysable, which corresponds to the properties of interleukin 1. These biochemical observations are consistent with the non-specific stimulation of the immune system found in asbestotic and silicotic patients.

## *Introduction*

Pulmonary fibrosis caused by the inhalation of mineral dusts such as asbestos and silica is often accompanied by immunological abnormalities, e.g., increased serum immunoglobulins, autoantibodies and circulating immune complexes (Kagan *et al.*, 1977; Lange *et al.*, 1974; Pernis *et al.*, 1965; Vigliani & Pernis, 1963), and elevated helper:suppressor T-cell ratios among blood lymphocytes (Miller *et al.*, 1983). Studies on experimental exposure to asbestos have also drawn attention to the stimulatory effects of mineral dusts on the immune system (Miller & Kagan, 1981; Rola-Pleszczynski *et al.*, 1981).

A reasonable explanation for some aspects of this immune stimulation resides in an effect of the mineral dusts on the macrophages, resulting in the production of interleukin 1 (Hartmann *et al.*, 1984; Pernis & Vigliani, 1982). Indeed, this monokine has many effects on cells of the immune system, one of them being the activation of T-cells (Oppenheim *et al.*, 1982).

The present study was undertaken to assess whether exposure of mouse peritoneal macrophages to asbestos *in vitro* is associated with augmented release of a lymphocyte-activating factor (LAF) with properties of interleukin 1. The effects of quartz and rutile were examined in parallel because these mineral particles are often used as fibrogenous and non-fibrogenous references, respectively.

## Materials and methods

### Macrophage cultures

Resident peritoneal cells were collected from female NMRI mice, washed once, and resuspended in Dulbecco's modified Eagle medium (Gibco) supplemented with 10% fetal calf serum (FCS, Gibco). The peritoneal cells were plated in aliquots of 1 ml containing $2 \times 10^6$ cells in 24-well multidishes (Nunc, Gibco). After incubation for 2 h under a 10% $CO_2$/air atmosphere, non-adherent cells were removed by thorough washing and the adherent cells were further incubated in duplicate in 1 ml of medium with 5% FCS and the following agents: UICC chrysotile A from Zimbabwe (obtained from Dr J.C. Wagner, MRC Pneumoconiosis Unit, Penarth, UK), standard DQ12 quartz and rutile (gifts from Dr K. Robock, Institute for Applied Fibrous Dust Research, Neuss, Federal Republic of Germany), sterilized by UV irradiation and suspended in medium by sonication for 2 min. Following incubation for up to 24 h with the added agents, the media were collected and centrifuged to remove the particles. The supernatants are referred to as conditioned media. Cell monolayers were washed, covered with 1 ml of medium with 5% FCS, frozen at $-20°C$, thawed, scraped off with a rubber policeman, sonicated, and centrifuged. The supernatants are referred to as cell lysates.

### Lymphocyte-activating factor assay

LAF was assayed by its capacity to enhance the mitogenic response of murine thymocytes to phytohaemagglutinin, essentially as described by Gery *et al.* (1981). In brief, thymocytes from male BALB/c mice in RPMI medium 1640 (Gibco) supplemented with 5% FCS and $3.3 \times 10^{-5}$ M mercaptoethanol were mixed with phytohaemagglutinin P (Difco) at a final concentration of 1.3 $\mu$l/ml. They were immediately plated in flat-bottomed 96-microwell plates (Nunc, Gibco) in aliquots of 0.15 ml containing $1.5 \times 10^6$ cells. Various serial dilutions of macrophage-conditioned media were added in triplicate in 0.05-ml aliquots. After incubation for 66 h, the cultures were pulsed for 6 h with 1 $\mu$Ci/well of [$^3$H]thymidine (2 Ci/mmol, New England Nuclear) and collected on filter paper with a cell harvester. LAF activity is given by the [$^3$H]thymidine incorporated by thymocytes and expressed as disintegrations per min (dpm).

### Lactate dehydrogenase

Lactate dehydrogenase was assayed as described by (Canonico *et al.* 1978).

## Results

Figure 1 shows the results of a typical experiment in which the LAF activity of conditioned media of macrophages exposed to chrysotile, quartz or rutile, was

Fig. 1. LAF activity of conditioned media of macrophages cultured with various particles. Macrophages were incubated for 20 h: (a) in the absence of any particles; (b) in the presence of UICC chrysotile A; (c) in the presence of DQ12 quartz; (d) in the presence of rutile. LAF activity was then measured by the thymocyte proliferation assay, as described under Materials and methods. ● = 50 µg/ml; o = 100 µg/ml

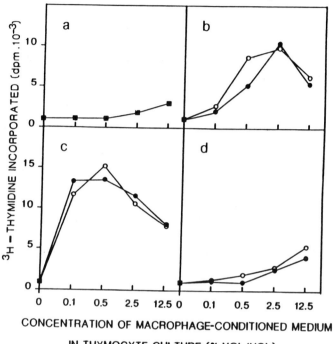

compared with that of conditioned media of unexposed macrophages. Whereas rutile appeared to be essentially inactive, chrysotile and quartz stimulated LAF release by macrophages, quartz being a more potent inducer than chrysotile at the same dose. This effect, which seems to approach a maximum at a dose of 50 µg, was not detected at a dose of 10 µg (result not shown).

The fact that thymocyte proliferation reaches a maximum and decreases at high concentrations of conditioned media derived from macrophages exposed to chrysotile or quartz suggests that a component present in these media adversely affects proliferation. Dialysis of the conditioned media before the thymocyte proliferation assay suppressed this inhibition (result not shown). Prostaglandins $E_2$ are among the macrophage products that can inhibit lymphocyte proliferation (Goodwin et al., 1977, 1979). Figure 2 shows that conditioned media of macrophages cultivated in the presence of $10^{-5}$ M indomethacin induced proliferation of thymocytes without inhibitory effects when used at high concentrations.

At the doses of chrysotile and quartz required for LAF production, macrophages released lactate dehydrogenase, which is indicative of cell damage. We therefore

**Fig. 2.** LAF activity of conditioned media of macrophages cultured with various particles in the absence or presence of indomethacin. Macrophages were incubated for 24 h: (a) in the absence of any particles; (b) in the presence of 100 µg/ml of UICC chrysotile A; (c) in the presence of 100 µg/ml of DQ12 quartz; (d) in the presence of 100 µg/ml of rutile, without (●) or with (o) $10^{-5}$ M indomethacin. LAF activity was then measured by the thymocyte proliferation assay, as described under Materials and methods.

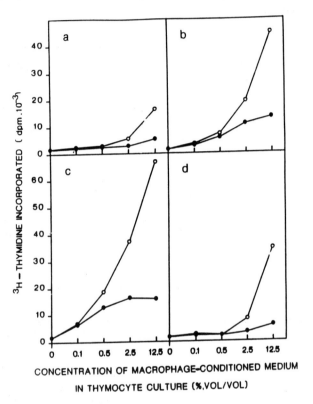

addressed the question whether LAF production by these cells does not simply reflect escape of preformed intracellular LAF. For this purpose, LAF and lactate dehydro-dehydrogenase were assayed in cell lysates and in conditioned media (Table 1). Macrophages that had not been exposed to particles exhibited a high content of LAF after 4 h of culture, which markedly decreased after 20 h; LAF activity in conditioned media remained low after 4 h and 20 h. Because the decline of intracellular LAF does not correlate with its release in the extracellular medium, LAF must have been internally degraded or somewhat changed, as already suggested by Gery et al. (1981). Exposure of macrophages to chrysotile or quartz moderately increased the intracellular pool of LAF, but this pool tended to disappear when time of culture was prolonged, as in control cells; in contrast to such cells, LAF activity in the medium was higher and continued to increase with time. It is important to note that the total activity of LAF was increased by exposure of macrophages to chrysotile or quartz. This is in contrast with the behaviour of lactate dehydrogenase; an increased release of the enzyme was accompanied by a corresponding decrease in the level of intracellular activity, leaving the total activity practically unchanged.

**Table 1. Activity of lymphocyte-activating factor (LAF) and lactate dehydrogenase in conditioned media and lysates derived from macrophages cultured with various particles[a]**

| Particles | Incubation time (h) | LAF activity (dpm × $10^{-3}$)[b] | | | Lactate dehydrogenase activity (mU/ml) | | |
|---|---|---|---|---|---|---|---|
| | | Media | Lysates | Total | Media | Lysates | Total |
| None | 4 | 1.9 | 6.9 | 8.8 | - | - | - |
| | 20 | 2.0 | 2.9 | 4.9 | 2.4 | 57.4 | 59.8 |
| UICC chrysotile A | 4 | 3.0 | 8.1 | 11.1 | - | - | - |
| | 20 | 5.1 | 3.8 | 8.9 | 16.9 | 41.2 | 58.1 |
| DQ12 quartz | 4 | 9.3 | 10.5 | 19.8 | - | - | - |
| | 20 | 11.0 | 3.8 | 14.8 | 42.5 | 12.5 | 55.0 |
| Rutile | 4 | 2.0 | 6.3 | 8.3 | - | - | - |
| | 20 | 2.1 | 3.0 | 5.1 | 4.7 | 56.4 | 61.1 |

[a]At a dose of 50 µg/ml.

[b]LAF activity is expressed as dpm of [$^3$H]thymidine incorporated by thymocytes cultured with media or lysates derived from macrophages at a final dilution of 0.5%. Control thymocyte cultures, incubated with medium alone, incorporated $1.1 \times 10^3$ dpm.

The LAF activity assayed in macrophage-conditioned media shared two properties of interleukin 1: it resisted heating for 30 min at 56°C and was not dialysable through a membrane with a molecular weight cut-off of 10 000 (Oppenheim et al., 1982).

## Discussion

This study has shown that *in vitro* exposure of mouse peritoneal macrophages to UICC chrysotile A stimulates them to release a lymphocyte-activating factor which is most probably interleukin 1. At equal doses, DQ12 quartz is a more potent inducer than chrysotile, while rutile has little or no effect.

Although we observed a direct correlation between the release of lactate dehydrogenase and that of LAF, injury caused by chrysotile or quartz to macrophages does not merely cause a leakage of preformed LAF from the cells, since the intracellular levels of the mediator are increased rather than decreased upon exposure to these particles. Gery *et al.* (1981) reported that exposure of cultured murine peritoneal macrophages to quartz induces a marked increase in LAF release with a small increment in the intracellular activity, and also concluded from the net stimulatory effect that silica particles do not just cause the release of preformed LAF. Similar observations were made with human monocyte cultures (Lepe-Zuniga & Gery, 1984).

Inhalation studies conducted by Kagan and his co-workers (Hartmann *et al.*, 1984) demonstrated that alveolar macrophages from rats exposed to crocidolite or chrysotile asbestos release interleukin 1 during culture. However, this interleukin 1 production (assayed by the enhancement of thymocyte and fibroblast proliferation) was observed only in media from co-cultures of alveolar macrophages and autologous splenic lymphocytes obtained from FCS-immunized animals. The effects of asbestos inhalation therefore appear to be expressed in the context of antigen-directed lymphoid activation, and indeed correlate with I-region-associated (Ia) antigen expression (Kagan *et al.*, 1985).

Our cellular model, in which cultured mouse peritoneal macrophages respond to chrysotile and quartz exposure by increased LAF production, has the merit of being simple and direct, and is consistent with the non-specific stimulation of the immune system of asbestotic and silicotic patients. Despite the uncertainties surrounding the possibility of extrapolating the results to humans, the *in vitro* system described in this paper could be used in the screening of asbestos substitutes that are potentially hazardous to human health.

## Acknowledgements

The authors would like to express their gratitude to Dr C. Peeters-Joris for helpful suggestions. These studies were supported by the Belgian Institut pour l'Encouragement à la Recherche Scientifique dans l'Industrie et l'Agriculture (IRSIA) (Convention No. 4396).

## References

Canonico, P.G., Beaufay, H. & Nyssens-Jadin, M. (1978) Analytical fractionation of mouse peritoneal macrophages: physical and biochemical properties of subcellular organelles from resident (unstimulated) and cultivated cells. *J. Reticuloendothel. Soc.*, 24, 115-138

Gery, I., Davies, P., Derr, J., Krett, N. & Barranger, J.A. (1981) Relationship between production and release of Lymphocyte Activating Factor (Interleukin 1) by murine macrophages. I. Effects of various agents. *Cell. Immunol.*, 64, 293-303

Goodwin, J.S., Bankhurst, A.D. & Messner, R.P. (1977) Suppression of human T-cell mitogenesis by prostaglandin. Existence of a prostaglandin-producing suppressor cell. *J. Exp. Med.*, 146, 1719-1734

Goodwin, J.S., Wiik, A., Lewis, M., Bankhurst, A.D. & Williams, R.C. (1979) High-affinity binding sites for prostaglandin E on human lymphocytes. *Cell. Immunol.*, 43, 150-159

Hartmann, D.P., Georgian, M.M., Mala, M., Oghiso, Y. & Kagan, E. (1984) Enhanced interleukin activity following asbestos inhalation. *Clin. Exp. Immunol.*, 55, 643-650

Kagan, E., Georgian, M.M. & Hartmann, D.P. (1985) Enhanced interleukin production and alveolar macrophage Ia expression after asbestos inhalation. In: Beck, E.G. & Bignon, J., eds, In vitro *Effects of Mineral Dusts (NATO ASI Series, Series G: Ecological Sciences)*, Berlin, Heidelberg, Springer-Verlag, pp. 149-157

Kagan, E., Solomon, A., Cochrane, J.C., Kuba, P., Rocks, P.H. & Webster, I. (1977) Immunological studies of patients with asbestosis. II. Studies of circulating lymphoid cell numbers and humoral immunity. *Clin. Exp. Immunol., 28*, 268-275

Lange, A., Nineham, L.J., Garncarek, D. & Smokik, R. (1983) Circulating immune complexes and antiglobulins (IgG and IgM) in asbestos-induced lung fibrosis. *Environ. Res., 31*, 287-295

Lange, A., Smokik, R., Zatonski, W. & Szymanska, J. (1974) Autoantibodies and serum immunoglobulin levels in asbestos workers. *Int. Arch. Arbeitsmed., 32*, 313-325

Lepe-Zuniga, J.L. & Gery, I. (1984) Production of intra- and extracellular interleukin 1 (Il-1) by human monocytes. *Clin. Immunol. Immunopathol., 31*, 222-230

Miller, K. & Kagan, E. (1981) Manifestations of cellular immunity in the rat after prolonged asbestos inhalation. II. Alveolar macrophage-induced splenic lymphocyte proliferation. *Environ. Res., 26*, 182-194

Miller, L.G., Sparrow, D. & Ginns, L.C. (1983) Asbestos exposure correlates with alterations in circulating T-cell subsets. *Clin. Exp. Immunol., 51*, 110-116

Oppenheim, J.J., Stadler, M.B., Siraganian, R.P., Mage, M. & Mathieson, B. (1982) Lymphokines: their role in lymphocyte responses. Properties of interleukin 1. *Fed. Proc., 41*, 257-262

Pernis, B. & Vigliani, E.C. (1982) The role of macrophages and immunocytes in the pathogenesis of pulmonary diseases due to mineral dusts. *Am. J. Ind. Med., 3*, 133-137

Pernis, B., Vigliani, E.C. & Selikoff, I.J. (1965) Rheumatoid factor in serum of individuals exposed to asbestos. *Ann. N.Y. Acad. Sci., 132*, 112-120

Rola-Pleszczynski, M., Massé, S., Sirois, P., Lemaire, I. & Begin, R. (1981) Early effects of low-dose exposure to asbestos on local cellular immune responses in the lung. *J. Immunol., 127*, 2535-2538

Vigliani, E.C. & Pernis, B. (1963) Immunological aspects of silicosis. *Adv. Tuberc. Res., 12*, 230-279

# BIOLOGICAL EFFECTS OF ASBESTOS FIBRES ON RAT LUNG MAINTAINED *IN VITRO*

F.M. Michiels, G. Moëns, J.J. Montagne & I. Chouroulinkov

*Laboratoire de Recherches Appliquées aux Cancérogènes Chimiques,*
*Institut de Recherches Scientifiques sur le Cancer,*
*Villejuif, France*

*Summary.* The *in vitro* systems used to determine whether asbestos acts as an initiator or as a promoter have failed to give definitive answers. We studied the effect of chrysotile and crocidolite in an initiation-promotion model on the Fischer rat embryo lung. Two assay systems were used in succession: organ culture of the lung cultured for 24 days and epithelial cell culture derived from treated or untreated explants cultured for 25 passages. Apart from the control groups, three major groups were analysed: (1) fibres with complete carcinogenic potency: explants and/or cells treated with fibres alone; (2) fibres with initiating potency; short treatment with fibres, followed by treatment with the classical promoter TPA; (3) fibres with promoting potency: short benzo[*a*]pyrene treatment followed by treatment with the fibres. In organ culture, fibres alone induce only cytotoxic lesions; in the 'fibres with promoting potency' group, precancerous lesions were observed. In epithelial cell culture, several transformation criteria are analysed. Our results with the cell system confirm that fibres act as a promoter, but also as a complete carcinogen. However, for equal doses, crocidolite needs a longer treatment time than chrysotile. These different assays failed to demonstrate any initiating activity of the fibres. The use of organ and cell culture in succession makes it possible to demonstrate the *in vitro* promoting effect of chrysotile and crocidolite.

## *Introduction*

Up to the present, the *in vitro* systems used to determine whether asbestos acts as an initiator or as a promoter have failed to give any definitive answers. In this study, we attempted to determine the mode of action of chrysotile and crocidolite in an initiation-promotion model on the Fischer rat lung in 2 *in vitro* assay systems: organ culture alone and a combined organ culture-cell culture system. In our experiments, treatment is performed in organ culture, and may or may not be followed by continuous treatment in cell culture. This method is used to preserve tissue integrity and the relationship between different tissue components, and maintain proliferation and differentiation. It is capable of producing precancerous lesions similar to these found *in vivo* (Michiels *et al.*, 1981).

## Materials and methods

### Fibres and chemicals

The UICC chrysotile A and crocidolite asbestos used were generously provided by Dr Greffard (Bureau des recherches géologiques et minières, Orleans, France). In organ culture, the fibres were administered at 10 µg/ml of medium and were present throughout the entire duration of the experiment. The cell cultures were treated 24 h after each seeding with 1 µg/cm² crocidolite or 0.4 µg/cm² chrysotile. The fibres were sonicated and suspended in complete medium. After 3 days, the medium was replaced with fresh medium.

Benzo[a]pyrene (BP) (Schuchardt, Munich, Federal Republic of Germany) was only used in organ culture; it was dissolved in acetone and added to the medium as initiator at 1 µg/ml of medium for 24 h, or as a complete carcinogen at 5 µg/ml of medium and was then present throughout the entire experiment.

12-O-Tetradecanoylphorbol 13-acetate (TPA) (LC Service Corporation, Woburn, USA) was only used in cell culture; it was dissolved in acetone and added to the medium at a concentration of 0.25 µg/ml of medium throughout the entire experiment.

### Culture methods

#### Organ culture

Lungs isolated from 17-day-old Fischer rat embryos (IFFA-CREDO, L'Arbresle, France) were cultivated on a Millipore membrane using a rocking system developed in our laboratory in which they were exposed alternately to air and medium. The medium consisted of M 199 supplemented with 5% fetal calf serum, penicillin, streptomycin and amphotericin B; the explants were incubated at 37°C, aerated with 5% of $CO_2$ and the medium was changed every 3 days. Histological analysis was performed between days 1 and 24.

#### Cell culture

Explants cultivated for 24 days were dissociated for 2 h in a 0.02% collagenase solution at 37°C. Cells were then cultivated using Dulbecco's H21 medium supplemented with 10% fetal calf serum and the usual antibiotics. They were reseeded at $5 \times 10^4$ cells per 60-mm Primaria dish weekly for 25 passages.

*Isolation and epithelial characterization.* At the first seeding after the organ culture, the fibroblast-like cells were scraped with a mini-rubber policeman. The epithelial character of the cells was determined by immunofluorescent staining with antikeratin antibodies.

*Cloning efficiency and clone morphology.* A total of 300 cells were seeded per 60-mm culture dish and were fixed and stained with Giemsa 8 days later. To determine the cloning efficiency (CE), the colonies were counted and the results expressed as a percentage of the number of seeded cells. On the same dishes, any abnormal morphological characteristics of the colonies were noted.

*Tumorigenicity.* Suspensions of $2 \times 10^6$ cells were injected intraperitoneally into isogenic 4-week-old rats. At least 6 animals of each group were injected.

*Transformation criteria procedure.* CE and clone morphology were determined after every 2 passages. The tumorigenicity *in vivo* was determined after every 4 passages.

## Results

### In organ culture

*Complete carcinogenic potency*

In total, 425 explants were analysed. After preliminary cytotoxic assays, a dose of 10 µg of fibres per ml of medium was used for both fibres. At this dose, chrysotile and crocidolite were of approximately equal toxicity (as measured by the survival rate expressed as the percentage of healthy explants; at 24 days cultivation, the figures were: chrysotile 83%; crocidolite 89%). The surviving explants displayed pyknotic cells, accumulation of macrophages and regenerative proliferation. No precancerous lesions were found in this group.

*Initiating or promoting potency*

In total, 320 explants were analysed. Initiating potency was not tested in organ culture. In the promoting group, apart from atypical proliferation, precancerous lesions were observed, such as hyperplasia, squamous metaplasia and adenopapillary structure, depending on the type of fibre but essentially on the duration of the treatment. Hyperplasia was seen in 27% at day 24 for both fibres. Adenopapillary structure and squamous metaplasia were observed after 18 days treatment with crocidolite (25 and 27% respectively) and 10 days treatment with chrysotile (25 and 36% respectively).

### In cell culture

*Cell culture groups*

Apart from the negative (untreated) and positive (BP or TPA treatment) control groups, three main groups were analysed: (1) *Complete carcinogen potency groups*: cells derived from fibre-treated explants, some of which were continuously treated in cell culture. (2) *Initiating potency group*: cells derived from fibre-treated explants, followed by TPA treatment in cell culture. (3) *Promoting potency group*: cells derived from explants treated for 24 h with BP followed by fibre treatment, possibly continued in cell culture.

*Cloning efficiency*

The CE of the untreated cells was relatively stable during the whole experiment (±30%). The CE of TPA- or BP-treated cells increased with the number of passages (at passage 3, 28 and 37% respectively and at passage 23, 43 and 73%). Up to passage 12, no difference could be noted between the fibre-treated groups and the untreated group. From passage 13 onwards, the CE of the treated groups exceeded that of the controls and increased with the number of passages. Analysis of the CE at passage 23 demonstrated that: (1) the CE in the fibre initiating groups did not exceed that of the TPA-treated cells; (2) in the complete carcinogen group, the CE of the crocidolite-treated cells was the same as that of the TPA-treated cells; but repeated chrysotile treatment induced an increased CE; (3) in the fibre promoting group, for both fibres, the CE was increased and reached or exceeded that of the BP-treated cells. For both fibres, the CE is higher in the promoting potency group than in the complete carcinogen group and, with regard to these criteria, chrysotile seems to have a higher promoting potency than crocidolite. Only treatment with chrysotile in the complete carcinogen potency group resulted in morphological transformation. Clones defined as type II (overgrowth) were present from passage 13 onwards (3.7%) and increased with the number of passages (passage 23: 14%).

*Tumorigenicity* in vivo

A total of 312 rats were injected with cells derived from the different groups every 4 passages from passage 4 to passage 24. In the untreated group, 20% of tumours appeared up to passage 16, with a long latency period which decreased with the number of passages (passage 16: 83 weeks; passage 24: 21 weeks). The BP-treated cells induced tumours in all rats, with a short latency period at passage 4 and subsequent passages (passage 4: 13 weeks; passage 24: 8 weeks).

The results in the various fibre-treated groups were as follows:

(1) Initiating potency group: Injection of chrysotile/TPA-treated cells gave a relatively constant tumour incidence rate (56%) from passage 4 onwards with a constant latency period (mean: $42 \pm 3$ weeks). Crocidolite/TPA-treated cells induced tumours from passage 12 onwards (16%) with a decreasing latency period (passage 12: 64; passage 24: 35 weeks).

(2) Complete carcinogen potency group: With the chrysotile-treated cells, a constant tumour incidence of 73% was observed from passage 8 onwards (latency period: mean: $32 \pm 3$ weeks). Apart from one tumour at passage 4 and another at passage 12, 83% of tumours appeared in the crocidolite-treated cells from passage 24 onwards, with a latency period of 38 weeks.

(3) Promoting potency group: With the BP/chrysotile treated cells, apart from one tumour at passage 8, tumours appeared from passage 16 onwards (36%) with a decreased latency period (passage 16: 61 weeks; passage 24: 33 weeks). Rats injected with BP crocidolite-treated cells developed tumours from passage 4 (20%) onwards, 100% being reached at passage 20, but with a long latency period (mean: $57 \pm 5$ weeks). Carcinomas and adenocarcinomas were the predominant types of tumour obtained in all the groups. In some cases, polymorphic tumours (at a rate of 12–20%) were also present.

## Discussion

Depending upon the *in vitro* system, some authors have reported the existence of a synergistic effect of asbestos and chemical carcinogens (Mossman & Craighead, 1981) while others have failed to find such an effect (Jaurand *et al.*, 1984), but the promoting or initiating potency of fibres has not yet been demonstrated in such systems. The combination of treatment of lung tissue in organ culture followed by treatment of the explant-derived cells would seem to be a good initiation-promotion model.

In organ culture, both of the fibres, when administered after BP initiation, were able to produce precancerous lesions, whereas they were unable to do so without initiation. These findings are in agreement with those of Topping and Nettesheim (1980), who found no metaplastic lesions in grafted tracheal explants treated with chrysotile.

In cell culture, to avoid problems caused by the high toxicity of chrysotile and to compare the activity of both fibres, we determined the dose giving a minimum of 90% cell survival. In our cell system, this was 1 $\mu$g/cm$^2$ with crocidolite and 0.4 $\mu$g/cm$^2$ with chrysotile. This showed that chrysotile was 2.5 times as potent as crocidolite and is in line with the data of other investigators working with other cell lines (Hesterberg & Barrett, 1984). The results obtained with regard to CE indicated that the cells were in a preneoplastic stage and showed that both fibres are more active as promoters than as complete carcinogens and that the promoting activity of chrysotile is higher than that of crocidolite.

Injection of the cells confirmed that the preneoplastic stage develops into a neoplastic one. Nevertheless, crocidolite cell tumorigenicity appeared earlier in the promotion group than in the complete carcinogen group, a fact not observed with the chrysotile fibres. The latency period in both chrysotile groups is shorter than that in the corresponding crocidolite groups.

These results lead us to the conclusion that this model can be used to determine the complete carcinogen and promoting potency of fibres but not the initiating potency, if any. Other relevant transformation criteria should be used for that purpose, such as chromosomal changes or oncogenic activation. The karyotypic analysis of the various groups is currently under way.

## Acknowledgements

We are grateful to Françoise Maugain for typing the manuscript. This work was supported by a grant from the Programme Interdisciplinaire de Recherche sur l'Environnement (PIREN), Centre National de la Recherche Scientifique (CNRS).

## References

Hesterberg, T.W. & Barrett, J.C. (1984) Dependence of asbestos and mineral dust-induced transformation of mammalian cells in culture on fiber dimension. *Cancer Res.*, *44*, 2170-2180

Jaurand, M.C., Patérour, M.J.P., Bastie-Sigeac, I. & Bignon, J. (1984) *In vitro* transformation of rat pleural mesothelial cells by chrysotile fibers and/or benzo-3,4-pyrene. *Am. Rev. Resp. Dis.*, *129*, A 146

Michiels, F.M., Duverger, A. & Chouroulinkov, I. (1981) Correlation between preneoplastic lesions in rat embryo lung treated with B(a)P or CSC in organ culture and tumour production *in vivo*. *Carcinogenesis*, *2*, 885-896

Mossman, B.T. & Craighead, J.E. (1981) Mechanisms of asbestos carcinogenesis. *Environ. Res.*, *25*, 269-280

Topping, D.C. & Nettesheim, P. (1980) Two-stage carcinogenesis studies with asbestos in Fischer 344 rats. *J. Natl Cancer Inst.*, *63*, 627-630

# TRANSLOCATION OF SUBCUTANEOUSLY INJECTED CHRYSOTILE FIBRES: POTENTIAL COCARCINOGENIC EFFECT ON LUNG CANCER INDUCED IN RATS BY INHALATION OF RADON AND ITS DAUGHTERS

G. Monchaux[1], J. Chameaud[2], J.P. Morlier[2], X. Janson[3], M. Morin[1] & J. Bignon[4]

[1]*Commissariat à l'Energie Atomique, Institut de Protection et de Sûreté Nucléaire, Département de Protection Sanitaire, Fontenay-aux-Roses, France*

[2]*Compagnie Générale des Matières Nucléaires, Laboratoire de Pathologie Pulmonaire Expérimentale, Razes, France*

[3]*Laboratoire d'Etudes des Particules Inhalées, Paris, France*

[4]*INSERM U139, CHU Henri Mondor, Creteil, France*

*Summary.* Exposure to radon 222 and its daughters has been shown to induce lung cancer in rats. The cocarcinogenic effect of intrapleurally injected mineral fibres in rats which have previously inhaled radon has also been established. The aim of this work was to establish whether a similar process could be induced at a distance from the lungs by subcutaneous injection of chrysotile fibres. Three groups of animals were used: (1) 109 rats which inhaled radon only (dose: 1600 working-level months (WLM)); (2) 109 rats given a subcutaneous injection in the sacrococcygeal region of 20 mg of chrysotile fibres after inhalation of the same dose of radon; and (3) 105 rats injected with fibres only. No mesotheliomas occurred in any of the 3 groups. The incidence of lung cancer was 55% in group 2, 49% in group 1 and 1% in group 3. Statistical analysis using Pike's model showed that the carcinogenic insult was slightly higher in group 2 than in group 1. Electron microscopy analysis of fibre translocation from the injection site showed that less than 1% of injected fibres migrated to the regional lymph-nodes and only about 0.01% to the lungs. After injection, the mean length of the fibres recovered in lung parenchyma increased with time, suggesting that short fibres are cleared by pulmonary macrophages whereas long fibres are trapped in the alveolar walls. Although the high tumour incidence observed in group 1 might have masked the cocarcinogenic effect induced by the fibres, it is possible that this effect can occur only at short distances. The translocation of long, thin fibres to the lung and the chronic inflammatory reaction are considered as possible promoters of pulmonary carcinogenesis.

## Introduction

It has been shown that lung cancer can be induced in rats by inhalation of $^{222}$Rn and its daughters at cumulative doses similar to those to which uranium miners are exposed. The tumours so induced were identical to those observed in humans and consisted mainly of epidermoid carcinomas and adenocarcinomas (Chameaud *et al.*, 1982). The cocarcinogenic effect of intrapleurally injected fibres in rats which had previously inhaled radon has also been established (Bignon *et al.*, 1983). The aim of this study was to ascertain whether or not a cocarcinogenic process could also be induced at some distance from the lungs by subcutaneous injection of chrysotile fibres in rats previously exposed to radon. In this connection, the translocation of fibres from the injection site to the regional lymph-nodes and lungs was also studied at various times after injection.

## Materials and methods

### Animals
The experiments were carried out on male SPF Sprague-Dawley rats which were 3 months old at the beginning of radon inhalation.

### Study of carcinogenesis
Rats were divided into the following 3 groups:
Group 1 (G1): 109 rats which inhaled radon only.
Group 2 (G2): 109 rats which were exposed to radon and one month later were given a single subcutaneous injection, in the sacrococcygeal region, of 20 mg of UICC A chrysotile fibres, suspended in 1 ml of saline.
Group 3 (G3): 105 rats injected subcutaneously with fibres only.

### Study of translocation
A total of 25 additional rats were used for this study. Of these, 5 inhaled radon only, 10 inhaled radon and were then injected subcutaneously with UICC A chrysotile fibres, and 10 were injected subcutaneously with these fibres only.

### Inhalation of radon
Rats were exposed to a cumulative dose of 1600 WLM of $^{222}$Rn at about 40% equilibrium with decay products according to a technique previously described (Chameaud *et al.*, 1974).

### Asbestos fibres
Type A chrysotile fibres obtained from UICC were used. The size characteristics of these fibres have already been described (Monchaux *et al.*, 1981).

### Tissue analysis

*Study of carcinogenesis*
Rats were allowed to survive for their full lifespan or until they became moribund, when they were killed. The lungs and thoracic tumours were processed by standard histological methods. The bronchopulmonary lesions found were graded according to a staging system derived from the tumour-node-metastasis (TNM) classification (Bignon *et al.*, 1983).

*Study of translocation*

The rats used were killed at days 1, 7, 15, 30, and 60 after fibre injection, and their lungs and sacroiliac lymph-nodes removed. These organs were analysed by transmission electron microscopy according to a previously described technique (Monchaux *et al.*, 1982) in order to determine quantitative fibre translocation from the injection site to the lungs and sacroiliac lymph-nodes. Two rats from the G2 and G3 groups in the carcinogenesis study, killed at days 588 and 743 respectively, were also examined for their lung fibre content.

## Statistical analysis of carcinogenic data

Mean survival times in the different treatment groups were compared by means of Student's *t*-test. The values obtained for survival time and tumour incidence were analysed using the model developed by Pike (1966) relating the tumour induction rate to the survival time. For the model used:

$$R = 1 - \exp[-b(t-w)^k]$$

where

$R$ = tumour induction rate
$b$ = carcinogenicity factor of Wagner *et al.* (1973)
$t$ = age-specific death rate of animals dying of lung cancer $t$ days after the beginning of treatment
$w$ = latency period for tumour development
$k$ = constant.

## General inflammation

The presence of a general inflammatory reaction in rats injected with fibres was tested by determining serum haptoglobin. The rats in the translocation study were bled by carotid catheterization and their serum was collected. Serum haptoglobin content was determined by an immunoelectrophoretic technique.

# Results

## Carcinogenesis

*Histological findings*

Neither injection-site tumours nor pleural mesotheliomas occurred in any of the 3 experimental groups, and only one peritoneal mesothelioma was observed in G1. Table 1 shows the incidence of lung cancer in the different groups according to histological type.

According to the TNM-derived classification used, no difference was observed between G1 and G2 in tumour size, pleural spreading or lymph-node involvement, but there were more M2 and M3 metastases in G2 than in G1 (52 as compared with 37%).

**Table 1. Incidence of lung tumours in the 3 experimental groups according to histological type**

| Histological type | G1 | G2 | G3 |
|---|---|---|---|
| Epidermoid carcinoma | 15 | 17 | 0 |
| Adenocarcinoma | 32 | 39 | 0 |
| Bronchioloalveolar carcinoma | 7 | 4 | 1 |
| Total | 54/109 (49%) | 60/109 (55%) | 1/105 (1%) |

*Survival time*

There were no statistically significant intergroup differences between the mean survival times of rats with lung tumours (774±140 days in G1, 787±113 days in G2 and 783 days in G3).

*Statistical analysis*

The best estimates of $k$ and $w$ obtained by the method of maximum likelihood were $k=3$ and $w=415$. The values of the carcinogenicity factor $b$ ($\times 10^9$) were, in decreasing order of magnitude, 10.5 for G2, 9.3 for G1 and 0.1 for G3.

## Translocation

It was not possible in this study to measure the absolute quantity of chrysotile fibres translocated from the injection site to the regional lymph-nodes because the total mass of these lymph-nodes was not assessed. Nevertheless, the number of fibres recovered per gram of lymph-node sample increased rapidly during the first month and then remained relatively constant. At day 30 after injection, the number of fibres recovered per gram of dried lymph-node tissue varied between $1.4 \times 10^7$ and $2.4 \times 10^9$. The proportion of fibres translocated from the injection site to the regional lymph-nodes was therefore about 1% or less. The fibres recovered in lymph-nodes were shorter and thicker than those injected. At day 30, only 3% of the fibres recovered in lymph-node tissue were longer than 4 μm.

Most of the fibres recovered in lung tissue were long and thin. For lung parenchyma, fibre characteristics at different times after injection were systematically calculated in relation to the weight of the whole lung. Table 2 shows the proportion of the total number of fibres translocated from the injection site to the lungs and the proportion of fibres corresponding to the size criteria defined by Stanton et al. (1981) for assessment of the probability of tumour induction, i.e., fibres longer than 4 μm with a diameter of up to 1.5 μm, and fibres longer than 8 μm and less than 0.25 μm in diameter.

## General inflammatory reaction

At day 1 after injection, the serum haptoglobin content of rats ranged from 1.2 to 2 mg/ml in G2 and from 0.8 to 2 mg/ml in G3 and then decreased with time. Serum

**Table 2. Proportion of the total number of fibres translocated from the injection site in the lungs and proportion of such fibres corresponding to the granulometric classes defined by Stanton**[a]

| Time after injection (days) | % of fibres translocated to the lungs | | % of fibres of length >4 μm and diameter <1.5 μm | | % of fibres of length >8 μm and diameter <0.25 μm | |
|---|---|---|---|---|---|---|
| | G2 | G3 | G2 | G3 | G2 | G3 |
| 1 | 0.005 | ND[b] | 0 | ND | 0 | ND[b] |
| 7 | 0.008 | 0.006 | 0 | 50 | 0 | 50 |
| 15 | 0.01 | 0.002 | 52 | 47 | 44 | 29 |
| 30 | 0.02 | 0.003 | 67 | 87 | 61 | 87 |
| 60 | 0.02 | 0.004 | 71 | 57 | 50 | 29 |
| 588 | - | 0.12 | - | 30 | - | 13 |
| 743 | 0.4 | - | 45 | - | 13 | - |

[a]See Stanton et al. (1981).

[b]ND, not detected.

haptoglobin levels 60 days after injection were between 0 and 0.8 mg/ml in G2 and about 0.2 mg/ml in G3, compared with the normal level of about 0.7 mg/ml.

## Discussion

The results of this study showed that the subcutaneous injection of fibres slightly increased the incidence of lung cancer in rats which had inhaled radon previously. Even though this increase was small, it was confirmed by the carcinogenicity factor $b$ calculated according to Pike's model. It is possible, however, that the dose of radon given to groups 1 and 2 might have been too great and have masked the potential cocarcinogenic effect of the fibres. The TNM method of classification was used for staging purposes. Stage M2 corresponds to the presence of pulmonary metastasis or of several tumours in the lungs. The higher incidence of M2 multiple lung tumours in group 2 as compared with group 1 might be related to a possible synergistic effect of the fibres on radon-induced lung tumours.

Although haptoglobin may be a poorly sensitive marker of inflammation, a chronic inflammatory reaction estimated on the basis of serum haptoglobin content seems unlikely to play a role in such a cocarcinogenic process.

The fact that no mesotheliomas occurred in G2 or G3 appears to indicate that the subcutaneous injection of fibres did not have a cocarcinogenic effect on the pleura. In any case, it might only be possible to detect such an effect of foreign bodies at a short distance from the injection site.

The translocation data showed that only a few fibres migrated from the injection site to the regional lymph-nodes and lungs. The proportion of fibres translocated to the lungs increased with time to 0.1–0.2% at day 60. The higher proportion of fibres recovered in lung tissue at days 588 and 743 might be related to the possible breaking up of fibres as a result of the chemical digestion of histological blocks previously embedded in paraffin. However, it is worth noting that the proportion of long, thin fibres recovered in lung tissue and corresponding to the carcinogenicity criteria defined by Stanton et al. (1981) increased with time and was much higher than in the injected fibres. These findings therefore suggest that short fibres are cleared by pulmonary macrophages, whereas long, thin fibres are trapped in the alveolar walls. The role of these fibres as a possible synergistic factor in lung carcinogenesis warrants further investigation.

## Acknowledgements

This work was supported in part by grant No. 83-C/17 from the Comité National contre les Maladies Respiratoires et la Tuberculose, Paris, France.

## References

Bignon, J., Monchaux, G., Chameaud, J., Jaurand, M.C., Lafuma, J. & Masse, R. (1983). Incidence of various types of thoracic malignancy induced in rats by intrapleural injection of 2 mg of various mineral dusts after inhalation of $^{222}$Rn. *Carcinogenesis*, 4, 621-628

Chameaud, J., Perraud, R., Lafuma, J., Masse, R. & Pradel, J. (1974) Lesions and lung cancers induced in rats by inhaled radon 222 at various equilibriums with radon daughters. In: Karde, E. & Park, J., eds, *Experimental Lung Cancer. Carcinogenesis and Bioassays*, Berlin, Heidelberg, New York, Springer-Verlag, pp. 411-421

Chameaud, J., Perraud, R., Lafuma, J. & Masse, R. (1982) Cancers induced by 222 radon in the rat. In: Clemente, G.F., ed., *Assessment of Radon and Daughter Exposure and Related Biological Effects*. Salt Lake City, R.D. Press, pp. 198-209

Monchaux, G., Bignon, J., Jaurand, M.C., Lafuma, J., Sébastien, P., Masse, R., Hirsch, A. & Goni, J. (1981) Mesotheliomas in rats following inoculation with acid-leached chrysotile asbestos and other mineral fibres. *Carcinogenesis*, 2, 229-236

Monchaux, G., Bignon, J., Hirsch, A. & Sébastien, P. (1982) Translocation of mineral fibres through the respiratory system after intrapleural injection into the pleural cavity of rats. *Ann. Occup. Hyg.*, 26, 309-318

Pike, M.C. (1966) A method of analysis of a certain class of experiments in carcinogenesis. *Biometrics*, 22, 142-161

Stanton, M.F., Layard, M., Tegeris, A., Miller, E., May, M., Morgan, F. & Smith, A. (1981) Relation of particle dimension to carcinogenicity in amphibole asbestoses and other fibrous minerals. *J. Natl Cancer Inst.*, 67, 965-976

Wagner, J.C., Berry, G. & Timbrell, V. (1973) Mesotheliomata in rats after inoculation with asbestos and other materials. *Br. J. Cancer*, 28, 173-185

# TRANSPLANTATION AND CHROMOSOMAL ANALYSIS OF CELL LINES DERIVED FROM MESOTHELIOMAS INDUCED IN RATS WITH ERIONITE AND UICC CHRYSOTILE ASBESTOS

### L.D. Palekar & J.F. Eyre

*Northrop Services, Inc., Environmental Sciences, Research Triangle Park, NC, USA*

### D.L. Coffin

*US Environmental Protection Agency, Health Effects Research Laboratory, Research Triangle Park, NC, USA*

*Summary.* The histological lesions, chromosomal characteristics and transplantability of 6 erionite-induced and 7 UICC chrysotile-induced rat mesotheliomas are compared. The tumours were of 4 types: tubulopapillary, fibrosarcomatous, mixed fibrosarcomatous and tubulopapillary, and mixed fibrosarcomatous and chondrosarcomatous. Cell lines derived from these tumours displayed heterogeneous chromosome anomalies, but none were unique either to chrysotile or erionite treatment. Six of 7 erionite-induced and 4 of 6 UICC chrysotile-induced tumours had various anomalies of chromosome No. 1. When 7 cell lines were transplanted into syngeneic rats, all produced tumours that were pathologically similar to the original tumour, regardless of the route of injection. Cytogenetically, the cell lines derived from tumours after intrapleural transplantation resembled the injected cell line. The cytogenetic analysis of the cell lines derived from the tumours after subcutaneous transplantation is in progress. The induction period for transplanted tumours was 28–80 days.

## Introduction

Clinical and epidemiological studies have shown that occupational exposure to asbestos causes pleural mesotheliomas. It is also reported that non-occupational exposure to asbestos in the environment, specifically from old buildings (Nicholson *et al.*, 1979), asbestos waste dumps (Nicholson & Pundsack, 1973) and building demolitions (Nicholson *et al.*, 1979), may also pose a health hazard. With the exception of erionite, little is known regarding the potential health hazard of non-asbestos mineral fibres. Erionite, an aluminosilicate of the zeolite family, is known to produce mesotheliomas in human beings (Artvinli & Baris, 1979; Baris *et al.*, 1978).

Animal studies have also indicated that rats exposed to erionite either by inhalation (Wagner *et al.*, 1985) or by intrapleural injection (Palekar *et al.*, 1985; Wagner *et al.*, 1985) produce a higher incidence of mesothelioma with a shorter induction period than after similar exposure to asbestos. It is of great interest to know whether the biological reactions to erionite are similar to those observed for asbestos.

Attempts have been made to evaluate the mechanism of tumorigenesis of erionite. Morphological transformation of C3H 10T½ cells exposed to erionite has been reported by Poole et al. (1983a). Similar exposure to asbestos has had negative results (Poole et al., 1983b). Cytogenetic analysis indicates that both asbestos and erionite alter ploidy and produce clastogenesis in exposed mammalian cells (Kelsey et al., 1986; Palekar et al., 1987).

In this investigation, we examined 7 rat mesotheliomas induced with erionite and 6 induced with UICC chrysotile. A comparison is made between their histological appearances, chromosomal anomalies and transplantabilities into syngeneic hosts.

## Materials and methods

### Induction of tumours

Erionite or UICC chrysotile (20 mg) was injected into the pleural cavities of barrier-sustained, male Fischer 344 (F344) rats at 6 weeks of age. The animals were given food and water *ad libitum* and necropsied after death. Standard histological preparations were made for pathological evaluations.

### Preparation of cell lines

Seven mesotheliomas from UICC chrysotile-treated animals and 6 from erionite-treated animals were excised and primary cell cultures were established. The tumours were excised aseptically and cells were dissociated with 0.25% trypsin (GIBCO). The cells were maintained in DMEM (GIBCO) supplemented with 1 mg $D$-glucose per ml, 100 $\mu$g sodium pyruvate per ml, 20% (v/v) fetal bovine serum, 2.5 $\mu$g fungizone per ml and 50 $\mu$g gentamicin per ml. The cells were incubated at 37°C in 10% $CO_2$ in a humidified atmosphere. Subsequently, confluent cultures were trypsinized and carried through various passages.

### Cytogenetic analysis

A total of 13 tumour cell lines were studied using conventional Giemsa-stained chromosomes. In all but one cell line, T1, 100 metaphase cells were examined. Fifty cells were examined in T1. Standard karyotypes of cells with the modal number of chromosomes were constructed from 2-10 cells from each tumour cell line. Chromosomes were arranged in groups A-D according to Levan (1974). With tumour cell line T3, karyotypes with G-banded chromosomes were prepared, since its modal number was 42. G-bands were elicited with trypsin according to Arrighi and Hsu (1974). Slides for G-band analysis were prepared according to Yunis (1981). G-banded chromosome analysis was also performed on some selected tumour cell lines (T1, T3-T5, T8-T9 and T11) in order to verify abnormalities in chromosome No. 1.

## Transplantation

Except for T11, the tumour cell lines were subcultured to passages (P) 10-11 and then injected. T11 was at P19 when injected.

Male syngeneic F344 rats 6-8 weeks old were treated with $2 \times 10^6$ cells per 0.5 ml saline either by intrapleural (i.pl.) or subcutaneous (s.c.) injection. Control rats received only 0.5 ml saline. In addition, there were 4 untreated control animals. Animals receiving s.c. injections were palpated at the injection site, and the time required for the first appearance of a palpable tumour was recorded. The animals were maintained until the tumour was 13 mm in diameter and then euthanized. When animals receiving i.pl. injections died, they were necropsied. The experiment was terminated after one year, when all animals were necropsied. The tumours were removed from the animals aseptically and divided into 2 parts. One part was preserved in 10% buffered formalin and standard pathological observations were made, the other was processed to establish a cell line.

## *Results*

Mesotheliomas induced in F344 rats with UICC chrysotile and erionite were morphologically similar. The pathological evaluations indicated that there were 4 different types: (1) epithelioid (tubulopapillary); (2) mesenchymal (fibrosarcomatous); (3) biphasic (mixture of epithelioid and fibrosarcomatous); and (4) biphasic (mixture of cartilage or osteoid tissue and fibrosarcomatous). The cell lines derived from the epithelioid tumours appeared epithelial-like *in vitro*. Those derived from fibrosarcomatous tumours were fibroblast-like *in vitro* and those from the biphasic tumours were a mixture of both *in vitro*.

Karyotypes were evaluated for differences or patterns that might be unique and characteristic of either chrysotile- or erionite-induced tumour cell lines, but none was readily identifiable. Many different chromosomes were involved and the karyotypes of the 13 cell lines revealed a heterogeneous display of chromosome anomalies. A detailed description of these karyotypes will be published elsewhere. The modal number and anomalies in chromosome No. 1 are presented in Table 1. Most of the cell lines were either aneuploid (T1, T8 and T9) or polyploid (T4-T7, T10-T13). Two cell lines (T2 and T7) had a population of cells that could be divided into several subpopulations and identified according to their number of chromosomes, and/or anomaly in chromosome No. 1, or by a unique marker chromosome.

G-banded karyotyping of the T3 cell line revealed that it was pseudodiploid and had many structural and numerical abnormalities (-5, +14, -16, plus one marker chromosome) along with a normal pair of chromosome No. 1. Chromosome No. 1 was frequently abnormal, either structurally (T1 and T8), numerically (T2, T4, T6, T7, T10 and T13), or in both ways (T9 and T11).

The 7 tumour lines injected in 26 animals produced tumours in all of them, regardless of the route of injection. The saline control and untreated control animals had no tumours. The results are presented in Table 2. The induction period of transplanted tumours was 28-80 days. The transplanted pleural tumours were fatal; a

**Table 1. Analysis of chromosome No. 1 in rat mesothelioma tumour cell lines**

| Mineral fibre | Cell line and passage (P) number | Chromosome modal number | Chromosome No. 1[a] |
|---|---|---|---|
| UICC | T1P1[b] | 40 | 1q⁻ |
| Chrysotile | T2P3 | 40, 42[c] | −1(40), N(42) |
| | T3P1[b] | 42 | N |
| | T4P3[b] | 53–55 | −1 |
| | T5P1[b] | 76 | N |
| | T6P2 | 92 | +1 |
| Erionite | T7P1 | 42, 84[d] | N(42), +1(84) |
| | T8P4[b] | 43 | 1p+q⁺, 1q⁻ |
| | T9P2 | 48 | +1q⁻ |
| | T10P3 | 63–66 | +1 |
| | T11P7 | 72 | +1q⁻ |
| | T11P19[b] | 70 | +1q⁻ |
| | T12P4 | 74 | N |
| | T13P2 | 82–84 | +1 |

[a]N, normal pair of chromosome No. 1; +1=trisomy 1; −1=monosomy; etc.
[b]Cell lines transplanted into syngeneic hosts.
[c]Bimodal T2 cell line had two subpopulations identifiable as cells with either 40 or 42 chromosomes.
[d]Bimodal T7 cell line had two subpopulations identifiable as cells with either 42 or 84 chromosomes.

substantial weight loss and death occurred in these animals after a short period. The necropsies indicated no apparent lesions except small tumours on the visceral pleura which were identified as mesotheliomas. The animals receiving s.c. injections appeared healthy before and after a tumour was noted at the injection site. These tumours appeared morphologically similar to those induced with i.pl. injections.

The transplanted tumours that resulted from an i.pl. injection of tumour cells yielded *in vitro* cell lines that were cytogenetically similar to the injected cell lines (parent cell lines). The cytogenetic evaluation of the s.c. tumours is in progress.

## *Discussion*

The morphological types of the mesotheliomas were similar in the erionite- and UICC chrysotile-treated animals. This is in agreement with the data presented for animals by Wagner *et al.* (1985). Tubulopapillary and fibrosarcomatous mesotheliomas have also been described in human beings (Suzuki *et al.*, 1976).

The 13 mesothelioma cell lines examined had numerous chromosomal changes, but none was specific to either of the 2 groups. Chromosomal heterogeneity has also been demonstrated in human mesotheliomas (Gibas *et al.*, 1986). The relatively high incidence of abnormalities in chromosome No.1 (6/7 from erionite-induced and 4/6 from UICC chrysotile-induced mesotheliomas) is of interest. These data suggest that

## Table 2. Transplantation of rat mesothelioma cell lines

| Test material | Tumour cell line | Type of primary tumour[a] | Tumour cell type type | Type of transplanted tumour | Incidence of transplanted tumours | |
|---|---|---|---|---|---|---|
| | | | | | Intrapleural | Subcutaneous |
| UICC chrysotile | T1 | TP | E | TP | 2/2 | 2/2 |
| | T3 | FS & CS | E & F | FS & CS | 2/2 | 2/2 |
| | T4 | FS | F | FS | 2/2 | 2/2 |
| | T5 | FS | F | FS | 2/2 | ND[b] |
| Erionite | T8 | FS | F | FS | 2/2 | 2/2 |
| | T9 | FS | F | FS | 2/2 | 2/2 |
| | T11 | FS & TP | E & F | FS & TP | 2/2 | 2/2 |
| Saline control | - | - | - | - | 0/4 | 0/4 |
| Untreated control | - | - | - | - | 0/4 | 0/4 |

[a]TP, tubulopapillary; FS, fibrosarcomatous; CS, chondrosarcomatous; E, epithelial; F, fibroblast.
[b]Not done because of insufficient number of cells at the time of injection.

such abnormalities may be common in rat mesothelioma cells, but the significance of this is uncertain at the present time; however, it is noteworthy that an H-*ras* oncogene has been identified on rat chromosome No.1 (Szpirer et al., 11985).

All cell lines produced tumours in syngeneic hosts. The induction period for the transplanted tumours was 28–80 days, and their histological characteristics were similar to those of the primary tumours. Chahinian et al. (1980) have shown that cell lines from human mesotheliomas injected into nude mice produced tumours morphologically similar to the original tumours in a relatively short time. While the data presented in this paper give no explanation for the difference in the tumorigenic potency of the two mineral fibres, it is conceivable that this difference in potency may be a manifestation of their initial interaction with the mesothelioma rather than connected with the later stages of promotion and/or tumour growth. This can only be confirmed by further research on the early stages of tumorigenesis, starting from the first exposure to mineral fibres.

## Acknowledgements

These studies were funded by the US Environmental Protection Agency (Contract No. 68-02-4032), but the contents of this paper do not necessarily reflect the views and policies of that Agency, nor does the mention of trade names or commercial products constitute endorsement or recommendation for use.

## References

Arrighi, F.E. & Hsu, T.C. (1974) Staining constitutive heterochromatin and Giemsa crossbands of mammalian chromosomes. In: Yunis, J.J., ed., *Human Chromosome Methodology*, 2nd ed., New York, Academic Press

Artvinli, M. & Baris, Y.I. (1979) Malignant mesotheliomas in a small village in the Anatolian region of Turkey: an epidemiological study. *J. Natl Cancer Inst.*, *63*, 17-22

Baris, Y.I., Sahin, A.A., Özesmi, M., Kerse, I., Özen, E., Kolaçan, B., Altinörs, M. & Göktepeli, A. (1978) An outbreak of pleural mesothelioma and chronic fibrosing pleurisy in the village of Karain/Ürgüp in Anatolia. *Thorax*, *33*, 181-192

Chahinian, A.P., Baranek, J.T., Suzuki, Y., Bekesi, T.G., Wisnie, L., Selikoff, J. & Holland, J.F. (1980) Transplantation of human malignant mesothelioma in nude mice. *Cancer Res.*, *40*, 181-185

Gibas, Z., Li, F.P., Antman, K.H., Bernal, S., Stahel, R. & Sandberg, A.A. (1986) Chromosome changes in malignant mesotheliomas. *Cancer Genet. Cytogenet.*, *20*, 191-201

Kelsey, K.T., Yano, E., Liber, H.L. & Little, J.B. (1986) The *in vitro* genetic effects of fibrous erionite and crocidolite asbestos. *Br. J. Cancer*, *54*, 107-114

Levan, G. (1974) Nomenclature for G-bands in rat chromosomes. *Hereditas*, *77*, 37-52

Nicholson, W.J. & Pundsack, F.L. (1973) Asbestos in the environment. In: Bogovski, P., Gilson, J.C., Timbrell, V. & Wagner, J.C., eds, *Biological Effects of Asbestos (IARC Scientific Publications No. 8)*, Lyon, International Agency for Research on Cancer, pp. 126-130

Nicholson, W.J., Sevoslowski, E.J., Jr., Porohl, A.N., Tadaro, J.D. & Adams, A. (1979) Asbestos contamination in the U.S. state schools from use of asbestos surfacing material. *Ann. N.Y. Acad. Sci.*, *330*, 587-596

Palekar, L.D., Eyre, J.F., Price, H.C., Brown, B.G. & Coffin, D.L. (1985) Biological activities of zeolite (erionite) in comparison with asbestos. *Toxicologist*, *5*, 38

Palekar, L.D., Eyre, J.F., Most, B.M. & Coffin, D.L. (1987) Metaphase and anaphase analysis of V79 cells exposed to erionite, UICC chrysotile and UICC crocidolite. *Carcinogenesis*, *8*, 553-560

Poole, A., Brown, R.C. & Fleming, G.T.A. (1983a) Study of the cell transforming ability of amosite and crocidolite asbestos and the ability to induce changes in the metabolism and macromolecular binding of benzo[*a*]pyrene in $C_3H10T\frac{1}{2}$ cells. *Environ. Health Perspect.*, *51*, 319-324

Poole, A., Brown, R.C., Turner, G.J., Skidmore, J.W. & Griffiths, D.M. (1983b) *In vitro* genotoxic activities of fibrous erionite. *Br. J. Cancer*, *47*, 697-705

Suzuki, Y., Churg, J. & Kannerstein, M. (1976) Ultrastructure of human malignant diffuse mesothelioma. *Am. J. Pathol.*, *85*, 241-262

Szpirer, J., Defeo-Jones, D., Ellis, R.W., Levan, G. & Szpirer, C. (1985) Assignment of three rat cellular *ras* oncogenes to chromosomes 1, 4 and X. *Somat. Cell Genet.*, *11*, 93-97

Wagner, J.C., Skidmore, J.W., Hill, R.J. & Griffiths, D.M. (1985) Erionite exposure and mesotheliomas in rats. *Br. J. Cancer*, *51*, 727-730

Yunis, J.J. (1981) Mid-prophase human chromosomes: the attainment of 2000 bands. *Human Genet.*, *56*, 292-298

# CARCINOGENICITY STUDIES ON NATURAL AND MAN-MADE FIBRES WITH THE INTRAPERITONEAL TEST IN RATS

F. Pott[1], M. Roller[1], U. Ziem[1], F.-J. Reiffer[1], B. Bellmann[2], M. Rosenbruch[1] & F. Huth[3]

[1] *Medical Institute for Environmental Hygiene at the University of Düsseldorf, Düsseldorf, Federal Republic of Germany*

[2] *Fraunhofer Institute of Toxicology and Aerosol Research, Hannover, Federal Republic of Germany*

[3] *Municipal Hospital, Hildesheim, Federal Republic of Germany*

*Summary.* Female Wistar rats were injected intraperitoneally (i.p.) with a suspension of 11 fibrous and 3 granular dusts. A dose of 0.25 mg actinolite or UICC chrysotile induced tumours of the peritoneum in more than 50% of the animals. Even 0.05 and 0.01 mg proved to be carcinogenic, although no adhesions of the abdominal organs could be observed. The findings are in conflict with the hypothesis that a scar is always the morphological precondition for the development of an asbestos-induced tumour.

Actinolite injected i.p. in a solution of polyvinylpyridine-$N$-oxide gave a lower tumour incidence than when suspended only in saline, possibly due to inactivation of the fibre surface. Persistent glass fibres were less effective than actinolite having a similar fibre size distribution. On the other hand, relatively thick basalt fibres and ceramic fibres gave higher tumour incidences than expected. Wollastonite fibres were not carcinogenic, probably because of their low durability. Large amounts of polyvinylchloride, $\alpha$-ferric oxide hydrate and wood dust also led only to adhesions of the abdominal organs and fibrosis; a definite carcinogenic effect was not detected.

## *Introduction*

The results of numerous animal experiments have led to the conclusion that the carcinogenic effect of asbestos is caused by its fibrous particle shape. Moreover, the fibres must persist in the bronchial wall or in the serosa tissue for a certain time in order to induce an alteration that can lead to the development of a tumour. Surface properties may also play a role, but precise data are not available.

The most sensitive model for studying the carcinogenic activity of fibres is the intraperitoneal test. Using this method, the effect of a number of fibrous dusts was examined with the aim of finding answers to the following questions:

(1) Can extremely low doses of asbestos induce tumours in a sensitive test model?

(2) Is fibrosis a precondition for tumour development caused by asbestos?

(3) Does polyvinylpyridine-*N*-oxide, a well-known inhibitor of the fibrogenicity of quartz, also inhibit the carcinogenic activity of asbestos?

(4) To what extent do the carcinogenic potencies of wollastonite, certain man-made mineral fibres, and plastic fibres differ from that of asbestos?

## Materials and methods

The fibrous dusts were prepared by cutting and milling the bulk materials. For some samples, the fraction of thinner fibres was separated by sedimentation. Information on the origin of the substances used and the method of dust preparation used has already been published (Pott *et al.*, 1987). The method of fibre measurement has also been described (Bellman *et al.*, 1987). The fibre size distributions of the dusts and the numbers of fibres injected per rat are shown in Table 1.

**Table 1. Fibre length and diameter distribution and number of fibres injected**

| Fibrous material | Fibre length[a] ($\mu$m) | | | Fibre diameter[a] ($\mu$m) | | | No. of fibres injected[b] (millions) | Fibre mass[c] (%) |
|---|---|---|---|---|---|---|---|---|
| | 10%< | 50%< | 90%< | 10%< | 50%< | 90%< | | |
| Actinolite | 0.36 | 1.10 | 4.2 | 0.05 | 0.10 | 0.20 | 102 | 50 |
| UICC Canadian chrysotile | 0.28 | 0.67 | 2.1 | 0.03 | 0.05 | 0.12 | 202 | 55 |
| Glass 104/475 | 0.74 | 2.6 | 9.6 | 0.06 | 0.15 | 0.38 | 680 | 91 |
| Basalt | 3.5 | 17 | 55 | 0.46 | 1.1 | 3.4 | 59 | 15 |
| Ceramic wool (Fiberfrax) | 3.6 | 13 | 51 | 0.36 | 0.89 | 2.2 | 150 | 61 |
| Ceramic wool (MAN) | 4.2 | 16 | 81 | 0.50 | 1.4 | 4.1 | 21 | 6 |
| Wollastonite | 4.1 | 8.1 | 19 | 0.62 | 1.1 | 2.2 | 430 | 25 |
| $\gamma$-Ferric oxide hydrate | - | ~0.5[d] | - | - | ~0.03[d] | - | - | - |
| $\alpha$-Ferric oxide hydrate | - | ~0.1[d] | - | - | ~0.01[d] | - | - | - |
| Kevlar[e] | 2.2 | 4.9 | 12 | 0.28 | 0.48 | 0.76 | 1260 | 88 |
| Polypropylene | 4.7 | 10 | 28 | 0.58 | 1.1 | 2.2 | 409 | 65 |

[a] A fibre is defined as a particle with an aspect ratio >5 : 1.

[b] A fibre is defined as a particle of length >5 $\mu$m, diameter <3 $\mu$m, and aspect ratio >5 : 1. Total dust mass injected: actinolite and chrysotile 250 $\mu$g; for the others, see Table 2.

[c] Mass of fibres injected as % of total mass injected (fibre definition, see footnote[b]).

[d] More precise measurement is planned.

[e] An aramid fibre (E.I. du Pont de Nemours, Newark, DE, USA).

Female Wistar rats (age 8 weeks) were injected intraperitoneally (i.p.) with suspensions of the dusts listed in Table 2. The very small amounts of asbestos were administered in a single injection of 1 ml 0.9% sodium chloride solution; the large amounts of the other fibre types were divided between 5 weekly injections administered in 2 ml sodium chloride solution. Wood dust was injected 3 times (50 + 100 + 150 mg, 50 mg per ml).

Some animals died spontaneously or were killed when in bad health. The surviving animals were sacrificed 130 weeks after the first treatment (the wood dust group after 140 weeks). A post mortem examination was made of the abdominal cavity. Parts of tumours or organs with macroscopically suspected tumour tissue were fixed in formalin and prepared for histological examination on paraffin-embedded H & E stained sections. The rats examined (listed in Table 2) included all autopsied rats, together with those which died relatively early, but excluding those lost through cannibalism. The percentage of dead rats also includes those lost through cannibalism and is related to the number of rats treated.

## Results

The preliminary tumour incidence, latency period, grading of adhesions and survival time are shown in Table 2. The macroscopic appearance and the method of grading adhesions have been described elsewhere (Pott *et al.*, 1987).

## Discussion

The results show that the i.p. test is a very sensitive model for detecting and grading the carcinogenic activity of fibres. However, the influence of particle deposition in the airways and of lung clearance cannot be simulated by the injection route. The spontaneous incidence of mesothelioma and sarcoma in the abdominal cavity is low when those animals which simultaneously have a malignant tumour of the uterus are excluded. In the Wistar rats used, a malignant tumour of the uterus seems to metastasize fairly frequently into the abdominal cavity; in the group injected only with sodium chloride solution (102 rats), 6 of 15 animals with a malignant tumour of the uterus also showed mesothelioma or sarcoma of the peritoneum, but in only 2 rats was a mesothelioma or sarcoma found without a malignant uterine tumour. On the other hand, it is possible that, in the groups treated with dusts, a tumour was classified as a metastasis of a spontaneous uterine tumour even though it was induced by the substance administered and occurred accidentally alongside a malignant neoplasm of the uterus.

Both actinolite and chrysotile induced tumours after injection of an extremely low dose. This confirms the findings of Bolton *et al.* (1984). The dose-response relationship in rats found with the i.p. test obviously cannot be extrapolated to the lung of humans; however, the results confirm, in principle, the hypothesis that there is no threshold for the carcinogenicity of asbestos, even at an extremely low dose. No adhesions and signs of fibrosis were observed macroscopically. Any histologically detectable fibrosis must

Table 2. Tumour incidence,[a] adhesions and life span after intraperitoneal injection of various fibrous and granular dusts in female Wistar rats (8 weeks of age)

| Dust | Dose[b] (mg or ml) | | No. with tumours/ No. examined (percentage) | Grade of adhesions[c] | Lifespan (weeks) after first treatment | | | | Rats with tumours | |
|---|---|---|---|---|---|---|---|---|---|---|
| | | | | | All rats | | | | First to die | Average |
| | | | | | 20%< | 50%< | 80%< | 100%< | | |
| Actinolite, | 0.01 | | 8/35 (23) | 0 | 79 | 103 | 121 | 130 | 103 | 116 |
| FR Germany | 0.05 | | 15/36 (42) | 0 | 75 | 101 | 122 | 130 | 64 | 99 |
| | 0.25 | | 20/36 (56) | 0 | 69 | 90 | 112 | 130 | 58 | 86 |
| in 0.4% PVNO | 0.25 | | 8/35 (23) | 0 | 77 | 96 | 113 | 130 | 80 | 99 |
| in 2% PVNO | 0.25 | | 12/36 (33) | 0 | 95 | 114 | 126 | 130 | 78 | 112 |
| UICC Canadian | 0.05 | | 12/36 (33) | 0 | 81 | 105 | 128 | 130 | 61 | 102 |
| chrysotile | 0.25 | | 23/34 (68) | 0 | 80 | 90 | 115 | 130 | 58 | 95 |
| | 1.00 | | 30/36 (83) | 0-1 | 52 | 63 | 72 | 130 | 35 | 62 |
| Glass fibres, 104/475 | 5 | (5×1) | 34/53 (64) | 3 | 53 | 67 | 85 | 126 | 37 | 65 |
| Basalt wool (Grunzweig and Hartmann) | 75 | (5×15) | 30/53 (57) | 3 | 64 | 79 | 92 | 111 | 54 | 83 |
| Ceramic wool (Fiberfrax) | 45 | (5×9) | 33/47 (70) | 3 | 38 | 51 | 63 | 95 | 30 | 54 |
| Ceramic wool (Manville Corporation) | 75 | (5×15) | 12/54 (22) | 3 | 71 | 91 | 106 | 128 | 64 | 96 |
| Wollastonite (India) | 100 | (5×20) | 0/54 (0) | 1-2 | 88 | 107 | 126 | 130 | - | - |
| | | (5×50) | 8/49 (16) | 2 | 67 | 98 | 120 | 130 | 88 | 107 |
| γ-Ferric oxide hydrate | 250 | (5×50) | 2/51 (4) | 2 | 83 | 112 | 130 | 130 | 61 | 95 |
| α-Ferric oxide hydrate | 250 | | | | | | | | | |
| Kevlar fibres | 20 | (5×4) | 3/53 (6) | 2 | 78 | 106 | 126 | 130 | 35 | 62 |
| | 50 | (5×10) | 2/51 (4) | 2-3 | 86 | 108 | 130 | 130 | 126 | 128 |
| Polypropylene fibres | 500 | (5×100) | 5/51 (10) | 2 | 83 | 107 | 125 | 130 | 62 | 104 |
| Polyvinylchloride[d] | 100 | (5×20) | 2/53 (4) | 0 | 82 | 109 | 130 | 130 | 38 | 79 |
| Titanium dioxide anatase[d] | | | | | | | | | | |
| Wood dust, beech[d] | 250 | (3 injections) | 0/52 (0) | 2 | 104 | 125 | 140 | 140 | - | - |
| Sodium chloride solution | 10 | (5×2) | 2/102 (2) | 0 | 85 | 111 | 127 | 130 | 93 | 107 |

[a]Mesothelioma or sarcoma in the abdominal cavity; tumours of the uterus are excluded; [b]Subdivided as shown in parentheses; [c]For method of grading, see

therefore be extremely small, and it is not usual to call such a small, macroscopically undetectable fibrosis a scar. In earlier studies, mesotheliomas were also observed without fibrosis after i.p. injection of a low dose of fibres (Huth & Pott, 1979). A tumour-promoting effect resulting from chronic foreign body reaction is characterized by extensive collagenous fibrosis (Brand, 1986). Our observations are therefore in conflict with the hypothesis that a scar is always a precondition for tumour development caused by asbestos (Kuschner, 1987). This hypothesis is based on experiments in rats and observations in humans, where large doses of asbestos caused marked fibrosis in all cases and a tumour response in a certain percentage of cases. It should be remembered that exposure to asbestos for a short time or at a low concentration can induce lung carcinomas or mesotheliomas without any detectable asbestosis.

When actinolite was injected together with a solution of polyvinylpyridine-$N$-oxide, the latency period was increased and the tumour incidence decreased in comparison with the group treated with actinolite suspended in saline alone. This result confirms those of an earlier experiment which showed the inhibiting effect more markedly (Pott et al., 1987). It remains to be determined whether the polymer neutralizes a carcinogenic agent on the surface of the actinolite fibres by adsorption, or whether some side-effect, present only under the experimental conditions, slows down tumour development.

A comparison of the dose-response relationship of asbestos fibres and man-made mineral fibres is difficult, because the number of fibres in each size category, their durability and possibly also unknown surface properties must be taken into account, and information on these characteristics is still inadequate. Some results are shown in Table 1 and Figure 1. On the basis of the available data, the fibre size distributions of actinolite and glass fibres 104/475 (Manville Corporation, Denver, CO, USA) are rather similar and both fibre types can be regarded as durable in rats. However, the number of fibres which induced tumours at a level of 60% was much greater for the glass fibres used than for the actinolite fibres. This finding confirms earlier results on the difference between these two dusts (Pott et al., 1987). On the other hand, the relatively thick basalt fibres and one type of ceramic fibre gave high tumour incidences, although the number of fibres injected per rat was much smaller than in the glass fibre group. It was found that 0.25 mg actinolite and 75 mg basalt fibres contained a similar number of fibres longer than 5 $\mu$m and gave a similar tumour incidence. This means that the carcinogenic potency does not decrease with increasing diameter as has been supposed up to now (Pott, 1978; Stanton & Layard, 1978). The relatively high percentage of very long (>20 $\mu$m) basalt and ceramic fibres (see Table 1) or some unknown surface properties may perhaps be responsible for this unexpected effect. The maximum carcinogenic potency of fibres is perhaps not reached at lengths of less than about 20 $\mu$m.

A rather large number of wollastonite fibres of Indian origin did not induce tumours. The durability of these fibres in water was low. However, Stanton et al. (1981) found some tumours after intrapleural administration of wollastonite from a Canadian mine. The differences in the results underline the experience gained with other mineral fibres (attapulgite, sepiolite, glass fibres, etc.), namely, that the

**Fig. 1. Numbers of fibres in each size category injected i.p. per rat. A fibre is defined as a particle of length/diameter ratio > 5:1.**

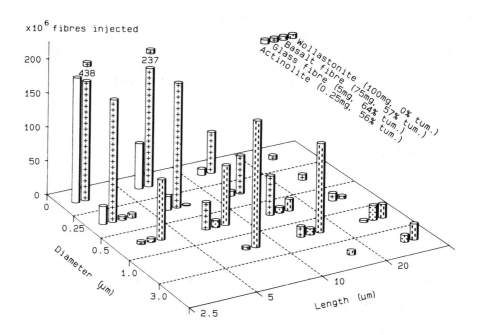

mineralogical name or trade name of a fibrous material generally does not give a sufficient indication of its carcinogenic potency. Up to now, however, it is not possible to define unequivocally the borderline between carcinogenic and non-carcinogenic fibres. The effect of durability or the half-life of fibres *in vivo* on the carcinogenic effect may be different in rats as compared with humans (Pott, 1987). For practical purposes, a standard for the definition of carcinogenic fibres should be developed because the use of inhalable non-asbestos fibres is growing.

The negative results with plastic fibres are harder to evaluate because these fibres have a greater tendency to aggregate than mineral fibres. Nevertheless, a low positive effect at least was expected but could not be detected.

All fibres of γ-ferric oxide hydrate seemed to be shorter than 2 μm; they gave a relatively low tumour incidence, as did a similar sample in an earlier experiment (Pott *et al.*, 1987). The fibres of α-ferric oxide hydrate were much shorter and were not found to be carcinogenic. It appears unlikely that γ-ferric oxide hydrate possesses a specific chemical carcinogenicity in contrast to the α-form. It can therefore be concluded that the carcinogenic activity of fibres starts at a fibre length of 1–2 μm and increases with increasing length. However, it will be realized that a low tumour response to an extremely large number of fibres 1-2 μm in length in a very sensitive test can hardly be considered as relevant for humans.

Titanium dioxide was injected into a control group as a granular dust. The low tumour incidences in the groups treated with large amounts of polyvinylchloride, wood dust, and α-ferric oxide hydrate demonstrate that the Oppenheimer effect (Oppenheimer *et al.*, 1948) is not detectable under such conditions, although large dust deposits in the abdominal cavity and significant adhesions of the organs were observed. It can be concluded that the mechanism of fibre carcinogenicity differs essentially in some way from the tumorigenesis induced by large foreign bodies.

## Acknowledgements

This work was supported in part by the Bundesanstalt für Arbeitsschutz, Dortmund, Federal Republic of Germany.

## References

Bellmann, B., Muhle, H., Pott, F., König, H., Klöppel, H. & Spurny, K. (1987) Persistence of man-made mineral fibres (MMMF) and asbestos in rat lungs. *Ann. Occup. Hyg.*, 31, 693-709

Bolton, R.E., Davis, J.M.G., Miller, B., Donaldson, K. & Wright, A. (1984) The effect of dose of asbestos on mesothelioma production in the laboratory rat. In: *VIth International Pneumoconiosis Conference 1983. VI. Internationale Pneumokoniose-Konferenz 1983, Bochum, 20-30 September 1983*, Vol. 2, Geneva, International Labour Organisation, pp. 1028-1046

Brand, K.G. (1986) Fibrotic scar cancer in the light of foreign body tumorigenesis. In: Goldsmith, D.F., Winn, D.W. & Shy, C.M., eds, *Silica, Silicosis and Cancer. Controversy in Occupational Medicine (Cancer Research Monographs, Vol. 2)*, New York, Praeger Scientific Publications, pp. 281-286

Huth, F. & Pott, F. (1979) Fibrosis following intraperitoneal application of fibrous dusts (asbestos, glass fibres), an obligatory precondition of carcinogenesis? [in German] *Verh. Dtsch. Ges. Pathol.*, 63, 437-439

Kuschner, M. (1987) A review of experimental studies on the effects of MMMF on animal systems. *Ann Occup. Hyg.*, 31, 791-797

Oppenheimer, B.S., Oppenheimer, E.T. & Stout, A.P. (1948) Sarcomas induced in rats by implanting cellophane. *Proc. Soc. Exp. Biol. Med.*, 67, 33-34

Pott, F. (1978) Some aspects on the dosimetry of the carcinogenic potency of asbestos and other fibrous dusts. *Staub-Reinhalt. Luft*, 38, 486-490

Pott, F. (1987) Problems in defining carcinogenic fibres. *Ann. Occup. Hyg.*, 31, 799-802

Pott, F., Ziem, U., Reiffer, F.-J., Huth, F., Ernst, H. & Mohr, U. (1987) Carcinogenicity studies on fibres, metal compounds, and some other dusts in rats. *Exp. Pathol. (Jena)*, 32, 129-152

Stanton, M.F. & Layard, M. (1978) The carcinogenicity of fibrous minerals. In: Gravatt, C.C., LaFleur, P.D. & Heinrich, K.F.J., eds, *Proceedings of Workshop on Asbestos: Definitions and Measurement Methods (NBS Special Publication 506)*, Washington DC, US Government Printing Office, pp. 143-151

Stanton, M.F., Layard, M., Tegeris, A., Miller, E., May, M., Morgan, E. & Smith, A. (1981) Relation of particle dimension to carcinogenicity in amphibole asbestoses and other fibrous minerals. *J. Natl Cancer Inst.*, 67, 965-975

# TOXICITY OF AN ATTAPULGITE SAMPLE STUDIED *IN VIVO* AND *IN VITRO*

A. Renier[1], J. Fleury[1], G. Monchaux[2], M. Nebut[1],
J. Bignon[1] & M.C. Jaurand[1]

[1]*INSERM U139, CHU Henri Mondor, Créteil, France*
[2]*Commissariat à l'Energie Atomique, Fontenay aux Roses, France*

*Summary.* Conflicting data are found in the literature concerning the carcinogenic potency of attapulgite. We tested the carcinogenic potency of French attapulgite in rats, and compared it with 2 chrysotile samples: Rhodesian UICC (Ch A) and short Canadian fibres (Ch C). The mean length of the fibres was 0.77 $\mu$m (attapulgite), 3.21 $\mu$m (Ch A) and 1.25 $\mu$m (Ch C). The mean diameter was 0.06 $\mu$m in the 3 samples.

The particles (20 mg) in saline were inoculated into the pleural cavity of Sprague-Dawley rats allowed to survive for their full lifespan. The incidence rates of mesothelioma were: 0% (saline controls), 0% (attapulgite), 19% (ChC) and 48% (Ch A).

*In vitro* studies were carried out using cultures of rat pleural mesothelial cells (RPMC). Attapulgite and Ch C did not modify cell growth except at high doses of 10 $\mu$g/cm$^2$. Unscheduled DNA synthesis (UDS) was detected using [$^3$H]thymidine incorporation in confluent RPMC (GoG1 arrested) and a scintillation method. UDS was stimulated with either Ch A or Ch C at doses ranging from 2 to 10 $\mu$g/cm$^2$. In contrast, attapulgite did not significantly enhance [$^3$H]thymidine incorporation at doses ranging from 2 to 20 $\mu$g/cm$^2$.

The results show that the attapulgite tested here had no carcinogenic potency. The *in vivo* and *in vitro* reactivity of the fibres used in this experiment might perhaps be related to the fibre size; however, other parameters may also be important.

## Introduction

Attapulgite is a fibrous clay widely used in industry and commerce. Because of its fibrous nature, its carcinogenic potency has been debated. Epidemiological data do not show any relationship between exposure to attapulgite and neoplasia but animal data are conflicting. Several authors have reported the occurrence of mesothelioma following intrapleural or intraperitoneal inoculation of some attapulgite samples (Pott *et al.*, 1974) while others found no carcinogenic effect (Stanton *et al.*, 1981). In this paper, we report data on the effect of French attapulgite. *In vivo*, attapulgite was inoculated in the pleural space of rats. *In vitro*, the effects of attapulgite on rat pleural mesothelial cell cultures was determined, firstly, by measuring cell growth, and secondly, by testing for the stimulation of unscheduled DNA synthesis (UDS), a method used for evaluating the genotoxicity of chemicals.

## Materials and methods

### Fibres and chemicals

Attapulgite was obtained from a deposit at Mormoiron, France. Rhodesian chrysotile (Ch A) was a gift from the UICC and short Canadian chrysotile (Ch C) from J.F. Kimmerle (Quebec, Canada). Mean fibre lengths were 0.77 $\mu$m (attapulgite), 3.21 $\mu$m (Ch A) and 1.25 $\mu$m (Ch C). The mean diameter was 0.06 $\mu$m for the 3 samples. The percentages of Stanton fibres (length $\geqslant 4$ $\mu$m, diameter $\leqslant 1.5$ $\mu$m) were 0, 25.7 and 3.5 respectively. Fibres were dispersed by sonication in culture medium for 5 min (Jaurand et al., 1983).

Hydroxyurea was purchased from Sigma. [$^3$H]Thymidine, specific activity 29 Ci/mole, was purchased from the Commissariat à l'Energie Atomique. The phosphate-buffered saline (PBS) used contained no magnesium and calcium.

### Rats

*In vivo* experiments were carried out with 2-month-old male Sprague-Dawley rats which, slightly anaesthetized, were given an intrapleural injection of 20 mg of the different fibre samples in 1 ml of saline (0.9% NaCl). Each group comprised 36 rats; a control group received saline only.

All rats were housed throughout their lifespan. Lungs and thoracic tumours were fixed and processed for histological analysis according to standard methods.

### Rat pleural mesothelial cells

Rat pleural mesothelial cells (RPMC) were obtained from rat parietal pleura and cultured with standard methods in CEB flasks (CML, France). The procedure for cell isolation and routine subculturing have been described elsewhere (Jaurand et al., 1981). RPMC were between 7 and 15 passages and cultured with Ham F 10 (Seromed, Biopro, France) supplemented with 10% fetal calf serum (Flow) and 2 antibiotics, namely, penicillin (100 U/ml) and streptomycin (50 $\mu$g/ml) (Biomérieux). Cells were grown in a humidified atmosphere of 5% $CO_2$ in air at 37°C.

### Growth analysis

Twenty-four hours after subculturing, RPMC were treated with the various fibres. After 48 h, they were washed with PBS, incubated in fresh medium and counted *in situ* in an inverted phase-contrast microscope (Jaurand et al., 1983).

### Unscheduled DNA synthesis

RPMC were cultured to confluence with Ham F 10 in complete medium on glass coverslips (1 cm²) in multi-well plates (24 wells, CEB). Ten coverslips were used for each fibre concentration. The medium was replaced with RPMI 1640 medium containing 1% fetal calf serum and 5 mM hydroxyurea for 24 h. Cells were then treated with fibres in RPMI 1640 + 1% fetal calf serum + 5mM hydroxyurea containing [$^3$H]thymidine for 20 h; uptake was stopped by 3 washings with phosphate-buffered saline, and acid-soluble material was removed by rinsing 3 times in 5% cold trichloroacetic acid for 60 min.

Thereafter, the coverslips were placed in vials with scintillation fluid to determine the radioactivity. The cell protein content was determined according to Oyama and Eagle (1956).

Results were expressed as cpm/µg of protein. Statistical analysis was carried out using tests.

In addition, the ratio ($R$) of thymidine incorporation in the control culture to thymidine incorporation in the treated culture was calculated.

## Results

### *In vivo* experiments

Animals treated with attapulgite did not develop mesothelioma. Their survival time (788±155 days) was very similar to that of control rats (809±110 days). In contrast, 48% of the animals treated with Ch A developed mesothelioma and 19% of those treated with Ch C. The survival time was 591±130 days in the Ch A-treated group and 788±138 in the Ch C group.

### *In vitro* experiments

#### Cell growth

Attapulgite at 1, 2 and 4 µg/cm² was not cytotoxic; however, at 10 µg/cm², cell growth was inhibited. The same pattern was observed with Ch C. In contrast, Ch A was highly cytotoxic, since a 40% decrease in cell number was observed after 48 h of incubation with 1 µg/cm² (Figure 1).

#### Unscheduled DNA synthesis

UDS was not modified by attapulgite at 2 and 4 µg/cm². At 10 µg/cm² there was a significant increase in one experiment.

Ch A and Ch C were significantly more potent in stimulating DNA repair. An increase of about 20% was observed with 2 µg/cm² (Tables 1 and 2).

## Discussion

The attapulgite used in the present experiments was not carcinogenic following intrapleural inoculation in rats. The *in vitro* data reported here may be relevant to this result. Attapulgite was poorly cytotoxic to RPMC, in agreement with data reported on other cell types (Reiss *et al.*, 1980; Woodworth *et al.*, 1983), and its low UDS enhancement indicates that this sample, unlike chrysotile, may be only a very weak DNA-damaging agent. Previously, we reported no induction of sister chromatid exchanges by another sample of attapulgite (Achard *et al.*, 1987) and UDS was not induced in hepatocytes (Denizeau *et al.*, 1985).

The lower rate of mesothelioma obtained with Ch C when compared with Ch A might be related to a size effect, as reported by Stanton *et al.* (1981), and this might explain why this attapulgite sample was not carcinogenic following intrapleural inoculation. However, a more complete study has been reported elsewhere using

Fig. 1. Relative viability (RV) of RPMC treated with different doses of CH A(■), Ch C (o) or Att (Δ). RV represents the ratio of the number of cells in the treated culture to the number of cells in the control culture.

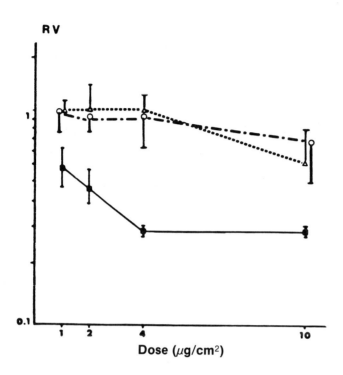

Table 1. Unscheduled DNA synthesis in rat pleural mesothelial cells (RPMC) (cpm/µg of protein ± standard deviation)

| Mineral | RPMC strain | Dose (µg/cm²) | | | |
|---|---|---|---|---|---|
| | | 0 | 2 | 4 | 10 |
| Attapulgite | D18 | 18±2 | 18±2 | 19±1 | 20±2[a] |
| | D18 | 61±5 | 59±7 | 51±11 | 59±10 |
| | A28 | 15±3 | 14±1 | 15±2 | 16±2 |
| Canadian chrysotile | F12 | 35±3 | 56±10[b] | 54±5[b] | 58±15[b] |
| Rhodesian chrysotile | D18 | 18±2 | 18±2 | 23±4[b] | 22±2[b] |
| | F12 | 59±9 | 80±11[b] | 111±22[b] | 85±7.8[b] |
| | F12 | 35±3 | 51±6[b] | 51±11[b] | 47±9[b] |
| | A 28 | 38±5 | 47±7[b] | 52±6[b] | 62±7[b] |

[a] $p<0.01$
[b] $p<0.005$

### Table 2. Unscheduled DNA synthesis in rat pleural mesothelial cells: values of $R^a$

| Mineral | Dose ($\mu g/cm^2$) | | | |
|---|---|---|---|---|
| | 0 | 2 | 4 | 10 |
| Attapulgite | 1 | 0.99±0.04 | 0.96±0.11 | 1.05±0.08 |
| Canadian chrysotile | 1 | 1.6 | 1.5 | 1.7 |
| Rhodesian chrysotile | 1 | 1.21±0.15 | 1.53±0.29 | 1.45±0.14 |

[a] $R$, ratio of thymidine incorporation (cpm/$\mu$g protein) in the control culture to thymidine incorporation (cpm/$\mu$g protein) in the treated culture (mean ± standard deviation)

9 samples of fibres (Jaurand et al., 1987). It was observed that other parameters (e.g., chemistry) were also important in carcinogenic potency.

Determination of UDS seems to be a useful means of predicting the carcinogenic potency of mineral dusts but further experiments need to be carried out in order to assess its predictive value.

## References

Achard, S., Perderiset, M. & Jaurand, M.C. (1987) Sister chromatid exchanges in rat pleural mesothelial cells treated with crocidolite, attapulgite, or benzo[3,4]pyrene. *Br. J. Ind. Med.*, 44, 281-283

Denizeau, F., Marrion, M., Chevalier, G. & Cote, M.G. (1985) Absence of genotoxic effects of non asbestos mineral fibres. *Cell Biol. Toxicol.*, 1, 23-32

Jaurand, M.C., Bernaudin, J.F., Renier, A., Kaplan, H. & Bignon, J. (1981) Rat pleural mesothelial cells in culture. *In Vitro*, 17, 98-106

Jaurand, M.C., Bastié-Sigeac, I., Bignon, J. & Stoebner, P. (1983) Effects of chrysotile and crocidolite on the morphology and growth of rat pleural mesothelial cells. *Environ. Res.*, 30, 255-269

Jaurand, M.C., Fleury, J., Monchaux, G., Nebut, M. & Bignon, J. (1987) Pleural carcinogenic potency of mineral fibres (asbestos, attapulgite) and their cytotoxicity on cultured cells. *J. Natl Cancer Inst.*, 79, 797-804

Oyama, V.I. & Eagle, H. (1956) Measurement of cell growth in tissue culture with a phenol reagent (Folin-Ciocalteau). *Proc. Soc. Exp. Biol. Med.*, 91, 305-307

Pott, F., Huth, F. & Friedrichs, K.H. (1974) Tumorigenic effect of fibrous dusts in experimental animals. *Environ. Health Perspect.*, 9, 313-315

Reiss, B., Milette, J.R. & Williams, G.M. (1980) The activity of environmental samples in a cell culture test for asbestos toxicity. *Environ. Res.*, 22, 315-321

Stanton, M.F., Layard, M., Tegeris, A., Miller, E., May, M., Morgan, E. & Smith, A. (1981) Relation of particle dimension to carcinogenicity in amphibole asbestoses and other fibrous minerals. *J. Natl Cancer Inst.*, 67, 965-975

Woodworth, C.D., Mossman, B.T. & Craighead, J.E. (1983) Induction of squamous metaplasia in organ cultures of hamster trachea by naturally occurring and synthetic fibers. *Cancer Res.*, 43, 4906-4912

# EFFECT OF CHOLINE AND MINERAL FIBRES (CHRYSOTILE ASBESTOS) ON GUINEA-PIGS

### A.P. Sahu

*Scientific Commission for Continuing Studies on Effects of Bhopal Gas Leakage on Life Systems, New Delhi, India*

*Summary.* Both short-term and long-term, low and high doses of parenterally administered choline produce pathological lesions in lungs and lymph-nodes of rats and guinea-pigs. Male guinea-pigs given choline by intraperitoneal injection, 40 doses of 50 mg of choline chloride, 5 days per week for 8 weeks, developed lung lesions consisting of peripheral nodules of small cells, neoplastic bronchiolar epithelium, carcinomatous lesions and changes in the pleural surface by the end of the experiment (680 days). In a second group, a single intraperitoneal injection of 15 mg of chrysotile asbestos given after 210 days of choline administration resulted in the early onset of pulmonary lesions at 570 days but there was, in addition, evidence of enhancement of cancerous lesions both in lung and in lymph-nodes at 570 days and 680 days as compared with choline alone. It is clear from the present experiment that the parenteral administration of carcinogenic mineral fibres (chrysotile asbestos) and availability of excess choline act synergistically in producing cancerous lesions in lungs and other organs.

## Introduction

Choline (trimethyl-$\beta$-hydroxyethylammonium) is a quaternary ammonium compound widely distributed throughout the plant and animal kingdoms. It is a naturally occurring substance of relatively low toxicity in comparison with many other quaternary ammonium compounds. However, it has been reported that high doses of choline are rapidly lethal (Hodge, 1944). High intake of choline produces acute gastrointestinal distress, sweating, salivation and anorexia (Wood & Allison, 1982). Methylamines formed after choline ingestion, and which possibly act as substrate for the formation of nitrosamines, showed marked carcinogenic activity (Zeisel et al., 1983). The marked toxic effects of methyl isocyanate (MIC) on the lungs of mice, rats and guinea-pigs may also possibly be due to methylamine, which is the first reaction product of MIC (Andersson et al., 1985).

The purpose of the experiment reported here was to investigate the effects produced by parenteral administration of carcinogenic mineral fibre (chrysotile asbestos) in conditions where organ-specific toxicity was produced by administration of excess choline.

## Materials and methods

A total of 25 male guinea-pigs, obtained from the Animal Breeding Facility of the Industrial Toxicology Research Centre, Lucknow (average body weight, 325 g), were kept on stock laboratory diet (Hindustan Lever Pellets), leafy vegetables and drinking-water *ad libitum*. They were divided into 3 groups, Group I consisting of 10 animals into which 50 mg of choline chloride in 2 ml of sterile distilled water was injected intraperitoneally (i.p.) for 5 days per week for a period of 8 weeks. Group II, also of 10 animals, was treated in the same way but, at day 210, 15 mg of chrysotile asbestos in 2 ml of 0.15 M sodium chloride solution was injected i.p. The remaining 5 animals (Group III) served as controls. After 570 and 680 days of choline chloride and chrysotile asbestos administration, the animals were killed and their lungs and other organs removed and fixed in 10% formol saline. The tracheobronchial lymph-nodes were excised carefully and fixed in Bouin's solution. Paraffin blocks were prepared and 5-$\mu$m thick sections cut. Multiple sections were stained with haematoxylin and eosin, silver impregnation for reticulin, Giemsa's stain for lymph-nodes and Perl's stain for asbestos bodies.

## Results

The histopathological findings are summarized in Table 1. Detailed results for the different groups are given below.

**Table 1. Histopathological grading of lesions in lungs and lymph-nodes of guinea-pigs given choline or choline and chrysotile asbestos**

| Lesion | Group I | | Group II | | Group III | |
|---|---|---|---|---|---|---|
| | 570 days | 680 days | 570 days | 680 days | 570 days | 680 days |
| **Lung** | | | | | | |
| Hyperplasia | + | ++ | ++ | ++ | − | − |
| Lymphatic dilation | + | + | ++ | ++ | − | − |
| Pleural changes | ++ | ++ | +++ | +++ | − | − |
| Reticulin pattern | ++ | ++ | +++ | +++ | − | − |
| Tumours | + | ++ | +++ | ++++ | − | − |
| **Tracheobronchial lymph-nodes** | | | | | | |
| Eosinophils | + | ++ | ++ | +++ | − | − |
| Reticulin pattern | ++ | ++ | ++ | +++ | − | − |
| Tumours | + | ++ | ++ | +++ | − | − |

## Group 1 (choline chloride alone)

At 570 days, a well marked cellular lining of the pleural surface was observed. Near this surface, the bronchioles and lymphatic vessels were seen to be pulled out of the outer limiting contour of the lung lobe, remaining attached to the lung by a connecting cellular band containing hyperchromatic cells (Figure 1). At 680 days, the significant changes seen were dysplasia of the mucosa with occasional parakeratotic koilocytic changes in the bronchiolar epithelium. At 680 days, the medullary region of the lymph-node contained large amounts of pigment.

**Fig. 1. Lung at 570 days after parenteral adminstration of choline. Cellular mass outside the lung lobe containing hyperchromatic cells and attached to the lung by connecting cellular band. Haematoxylin and eosin × 182**

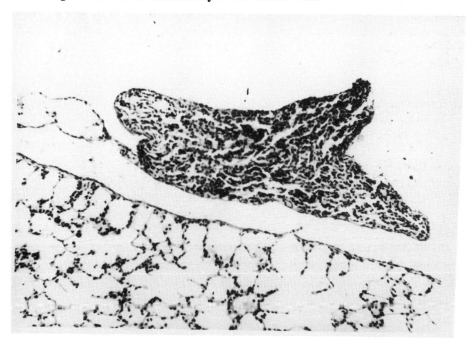

## Group II (choline chloride and chrysotile asbestos)

At 570 days, the bronchioles were lined with transformed epithelial cells and the lumen was filled with cellular debris and pigment (Figure 2). At 680 days, bronchiolar carcinoma together with collections of neoplastic cells around bronchioles, blood vessels, and in and around the lymphatics were observed. The visceral pleura showed generalized thickening of the mesothelium, which was more prominent at some places where it appeared neoplastic. In the lymph-nodes at 570 days, granulomatous lesions were found, together with the deposition of pigment and eosinophils in the medullary region. Perls-positive asbestos bodies were not seen at either 570 or 680 days.

**Fig. 2.** Lung at 570 days after parenteral adminstration of choline followed by chrysotile asbestos treatment. Bronchiolar epithelium showing neoplastic changes (arrow) together with cell debris and pigment in the lumen. Haematoxylin and eosin × 240

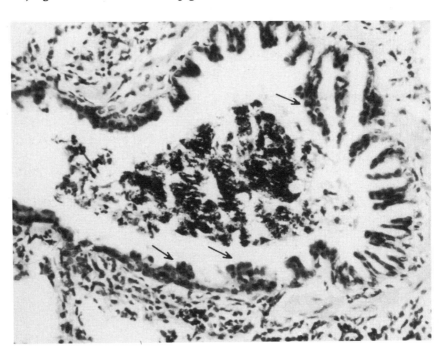

**Group III (control)**

No pathological changes of significance were seen in the lungs and tracheobronchial lymph-nodes of control animals at either 570 or 680 days.

## Discussion

It is clear from these experiments that chronic choline administration produces characteristic pathological lesions in lungs and lymph-nodes. The results were in agreement with those of earlier experiments in rats (Sahu & Shankar, 1984; Sahu *et al.*, 1984, 1986). It is still not clear why choline, after parenteral administration, produced the organ-specific toxic changes in the lungs and associated lymph-nodes. In combination with i.p. administered chrysotile asbestos, it produced not only the early onset of pulmonary lesions but an enhancement of such lesions. Parenteral administration of choline produced the condition described above in guinea-pig lungs and lymph-nodes at 210 days. Choline also acted synergistically with chrysotile asbestos to produce cancerous lesions in lungs and lymph-nodes. The mechanism responsible for the production of such enhanced organ-specific lesions is not known.

## References

Andersson, N., Kerr Muir, M., Salmon, A.G., Wells, C.J., Brown, R.B., Purnell, C.J., Mittal, P.C. & Mehra, V. (1985) Bhopal disaster: eye follow up and analytical chemistry. *Lancet, i,* 761-762

Hodge, H.C. (1944) Acute toxicity of choline hydrochloride administration intraperitoneally to rat. *Proc. Soc. Exp. Biol. Med, 57,* 26-28

Sahu, A.P. & Shankar, R. (1984) Experimental evidence of the possible relationship of choline in pulmonary carcinogenesis. In: *Symposium on Tumor Promotion and Enhancement of Human and Experimental Respiratory Tract Carcinogenesis,* Williamsburg, VA, US Environmental Protection Agency, p. 52

Sahu, A.P., Shukla, L.J. & Krishna Murti, C.R. (1984) Effect of mica dust and choline on the lymph nodes of rats. *Br. J. Pathol., 65,* 533-541

Sahu, A.P., Saxena, A.K., Singh, K.P. & Shankar, R. (1986) Effect of chronic choline administration in rats. *Ind. J. Exp. Biol., 24,* 91-96

Wood, J.L. & Allison, R.G. (1982) Effect of consumption of choline and lecithin on neurological and cardiovascular systems. *Fed. Proc., 41,* 3015-3021

Zeisel, S.H., Wishnok, J.S. & Blusztajn, J.K. (1983) Formation of methylamines from ingested choline and lecithin. *J. Pharmacol. Exp. Ther., 225,* 320-324

# CYTOTOXICITY AND CARCINOGENICITY OF CHRYSOTILE FIBRES FROM ASBESTOS-CEMENT PRODUCTS

### F. Tilkes & E.G. Beck

*Institute of Hygiene, Justus-Liebig University of Giessen, Federal Republic of Germany*

*Summary.* Fibres from weathered asbestos-cement products have little or no haemolytic activity, as compared with UICC chrysotile; this is probably the result of magnesium leaching during the weathering process. Weathered samples of asbestos cement are cytotoxic, but the release of lactic dehydrogenase (LDH) by guinea-pig alveolar macrophages caused by low and intermediate dust concentrations of UICC chrysotile is greater than that of such samples. The influence of serum is different as between UICC chrysotile and asbestos cement. In the former, LDH release by macrophages is enhanced, whereas it is reduced in the latter. Cytotoxicity is length-dependent in respect of LDH release from macrophages and proliferating cells, as well as cell proliferation.

In all test systems, the sample from the unweathered core of an asbestos-cement plate is less toxic; only in the haemolysis system using an unbuffered solution can erythrocyte destruction be observed. This may be because this sample contains fewer single fibres than the others and because the specific surface of those fibres is smaller. The carcinogenicity of the weathered asbestos-cement chrysotile fibres is comparable to that of standard chrysotile fibres following intraperitoneal (i.p.) application.

## Introduction

Asbestos-cement products contain about 15-20% chrysotile. As a result of meteorological factors and air pollution, their surfaces become weathered and fibres are exposed. From a biological point of view, the question arises as to the magnitude of the adverse effects of these fibres on health, and in particular their carcinogenic potential, as compared with that of the standardized original material.

In this study, suitable physically and chemically well-characterized fibre samples from a highly corroded and weathered 20-year-old asbestos-cement plate taken from a building roof were subjected to biological testing (Tilkes & Beck, 1986).

## Materials and methods

**Fibres**

Three well-characterized samples were prepared by Spurny *et al.* (1986) from a weathered asbestos-cement plate. Two fibre fractions were prepared from the weathered plate surface, one with a fibre length of up to 18 $\mu$m (AC-long) and the second with a fibre length of up to 13 µm (AC-short). A third sample was prepared from the core of the corroded plate, which had never been in contact with the ambient air. For purposes of comparison, a sample of UICC standard chrysotile (Rhodesia) was employed which had approximately the same fibre size distribution as the samples derived from the weathered asbestos-cement plate surface. A cement fraction served as a non-fibrous control (Table 1).

### Table 1. Physical properties of dust samples

| Property[a] | UICC chrysotile | Long asbestos-cement fibres | Short asbestos-cement fibres | Core | Cement |
|---|---|---|---|---|---|
| $L_F$ | 4.1±3.7 | 3.3±3.0 | 3.1±2.4 | 4.5±4.8 | - |
| $D_F$ (µm) | 0.3±0.1 | 0.3±0.1 | 0.3±0.1 | 0.3±0.1 | - |
| $L_F/D_F$ | 12.8±11.5 | 14.0±12.1 | 10.8±9.2 | 13.1±11.3 | - |
| Fibres/µg (millions) | 1.6 | 1.86 | 2.1 | 1.4 | - |
| Specific surface (m²/g) | 24.92 | 26.43 | 26.4 | 19.54 | 15.31 |

[a]$L_F$, mean fibre length; $D_F$, mean fibre diameter.

### Test systems

1. Haemoglobin release of rat erythrocytes — haemolysis.
2. LDH release of guinea-pig alveolar macrophages.
3. Phagocytic capacity of rat alveolar macrophages after exposure to dusts.
4. Proliferation and LDH release of phagocytosing tumour cells (mesothelioma induced in Wistar rats after i.p. administration of nemalite) after exposure to the described dusts.

### Animal experiment

In this experiment, 2 mg and 10 mg of each fibre fraction and 25 mg of cement suspended in saline solution were injected i.p. into female Wistar rats (3-4 months old).

## Results and discussion

### Haemolysis

The results show that, in phosphate-buffered saline and in minimal essential medium with HEPES, only the UICC chrysotile sample causes haemolysis. In 0.9% sodium chloride and veronal buffer, the core and cement samples also show a dose-dependent release of haemoglobin (Figure 1).

**Fig. 1.** Haemolysis of rat erythrocytes 2 h after incubation with different dusts in veronal buffer (concentration in mg/ml). COR, corundum; $DQ_{12}$, Doerentruper quartz; LONG AC, long asbestos-cement chrysotile; SHORT AC, short asbestos-cement chrysotile; CORE AC, chrysotile from the core of the asbestos-cement plate; UICC, UICC chrysotile (Rhodesia).

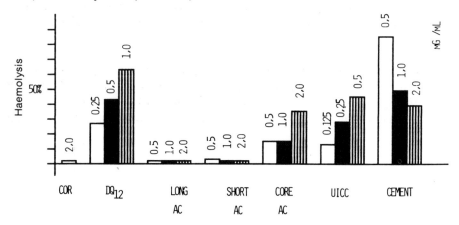

The pH measurements on suspensions of exposed material help to explain these results. Cement-containing samples show a higher pH (8.8–10.4) than the adjusted physiological value of 7.2–7.3. It seems that, in this case, only the surface pH of the dusts accounts for the haemolytic activity demonstrated. With higher buffering capacity, the haemolysis decreases. Chrysotile fibres from asbestos-cement dusts lacking non-fibrous cement particles exhibit only a weak erythrocyte destruction, as compared with UICC chrysotile. In the case of cement, the haemolytic activity decreases at the higher concentrations of 1 and 2 mg/ml. A high absorption capacity for free haemoglobin is probably the reason for these results.

## Macrophages

### Enzyme release

The specific cytotoxicity was calculated using the values for the negative control corundum (=0%) and the positive control Doerentruper quartz ($DQ_{12}$) (=100%) and expressed as percent toxicity relative to these.

$$\text{Percent toxicity (X)} = \frac{X - \text{corundum}}{DQ_{12} - \text{corundum}} \times 100$$

The results obtained were as follows (see also Figure 2):

1. When high concentrations are employed, the long and short asbestos-cement fibre samples and the UICC sample are of similar toxicity.

2. At lower concentrations, UICC chrysotile is the most toxic, followed by the long and short asbestos-cement samples derived from the surface of the asbestos-cement product.
3. The samples derived from the core of the product give rise to significant enzyme release only at the highest concentration.
4. The effect of adding 5% fetal calf serum varies. Whereas the toxicity of all asbestos-cement samples is reduced, the enzyme release of UICC chrysotile is enhanced.

Fig. 2. Release of lactic acid dehydrogenase (LDH) by guinea-pig alveolar macrophages exposed for 20 h to 3 different concentrations of the test samples. 1, 50 $\mu$g/$10^6$ cells; 2, 100 $\mu$g/$10^6$ cells; 3, 150 $\mu$g/$10^6$ cells. – FCS, without fetal calf serum; 5% FCS, with 5% fetal calf serum; CO, control without exposure. For other abbreviations, see Figure 1.

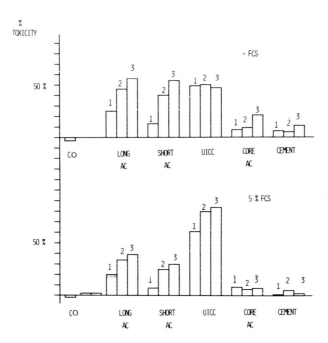

The above-mentioned enhancing effect with UICC chrysotile can perhaps be explained by an increase in the phagocytic capacity of the macrophages in the presence of serum. The reason for the protective effect of serum observed in all samples prepared from the asbestos-cement plate is unclear. Possibly in this case the coating of the fibre surfaces by serum proteins plays both a greater and a different role than with UICC chrysotile. In the past, we have observed a similar effect with coal mine dusts and kaolinite.

## Phagocytosis

The determination of the phagocytic activity of rat alveolar macrophages after exposure to the dusts at a concentration of 150 µg/$10^6$ cells gave the following results. Asbestos-cement dust from the surface causes a depression of macrophage activity similar to that produced by UICC chrysotile, whereas the cement and the core fraction have no significant effect (Figure 3).

**Fig. 3. Phagocytic activity of rat alveolar macrophages 20 h after exposure to different asbestos-cement dusts (150 µg/$10^6$ cells), measured by luminol-enhanced chemiluminescence.**

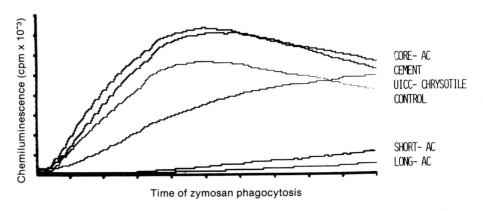

## Cell line

### Proliferation

The results show the same toxicity ranking as enzyme release in macrophage cultures, but the differences between the UICC chrysotile, on the one hand, and the asbestos-cement sample, on the other, are much greater. Whereas the long fraction of asbestos-cement fibres reduces the cell number to 60% of the original figure 92 h after exposure, the short fraction reduces it to 83%.

### Enzyme release

A comparison of LDH release at 92 h yields similar results (Figure 4). Enzyme release after exposure to UICC chrysotile is extremely high, the next highest being that produced by the long asbestos-cement sample. Once again, the short fraction from the surface of the weathered plate shows a significantly lower cytotoxicity with respect to this parameter as well. The sample from the core of the asbestos-cement product and the pure cement sample cause only weak cell damage in terms of both parameters.

## Carcinogenicity

The incidence of abdominal tumours observed after i.p. injection of the various dusts is shown in Table 2.

Fig. 4. Release of lactic dehydrogenase (LDH) by proliferating tumour cells 92 h after incubation with test samples. CO, control; COR, corundum; LONG, long asbestos-cement fraction; SHORT, short asbestos-cement fraction; UICC, UICC chrysotile (Rhodesia); CORE, core asbestos-cement fraction. 1, 150 $\mu$g/2×$10^6$ cells; 2, 300 $\mu$g/2×$10^6$ cells

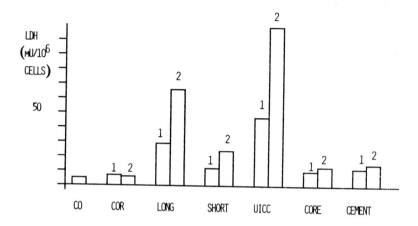

Table 2. Incidence of abdominal tumours after intraperitoneal injection

| Dust | Dose (mg) | No. of animals per group | Tumour incidence (%) |
| --- | --- | --- | --- |
| Phosphate-buffered saline control | 0 | 50 | 1 |
| Cement | 25 | 50 | 6 |
| UICC chrysotile (100% fibres) | 2 | 50 | 50 |
|  | 10 | 25 | 54 |
| Long asbestos-cement chrysotile | 2 | 50 | 38 |
| (30% fibres) | 10 | 25 | 32 |
| Short asbestos-cement chrysotile | 2 | 50 | 40 |
| (approx. 30% fibres) | 10 | 25 | 32 |
| Core | 2 | 50 | 22 |
| (approx. 10% fibres) | 10 | 25 | 48 |

As can be seen from the Table, the carcinogenicity of the weathered asbestos-cement chrysotile fibres is comparable to that of the standard chrysotile fibres following i.p. injection and to the results of Muhle et al. (1986).

## References

Muhle, H., Pott, F., Spurny, K., Fuhst, R. & Takenaka, S. (1986) Carcinogenic effect of weathered asbestos cement products in rats. In: Gesellschaft für Aerosolforschung, American Association for Aerosol Research, ed., *Aerosol — Formation and Reactivity*, Oxford, New York, Pergamon Press, pp. 309-310

Spurny, K., Marfels, H., Pott, F. & Muhle, H. (1986) *Investigation on Corrosion and Weathering of Asbestos Cement Products as well as on the Carcinogenic Effect of the Weathering Products.* Berlin, Umweltbundesamt, Report No.104 08 n 314, pp. 1-172

Tilkes, F. & Beck, E.G. (1986) Fibres from asbestos-cement products in the cell culture. In: Gesellschaft fur Aerosolforschung, American Association for Aerosol Research, ed., *Aerosol — Formation and Reactivity*, Oxford, New York, Pergamon Press, p. 335

# QUALITATIVE AND QUANTITATIVE EVALUATION OF CHRYSOTILE AND CROCIDOLITE FIBRES WITH INFRARED SPECTROPHOTOMETRY: APPLICATION TO ASBESTOS-CEMENT PRODUCTS

F. Valerio & D. Balducci

*Environmental Chemistry Laboratory,
National Institute for Cancer Research, Genoa, Italy*

*Summary.* Infrared (IR) spectrophotometry allows simple and rapid qualitative and quantitative evaluations of different types of asbestos, as well as of other inorganic particles. In particular, chrysotile and crocidolite have characteristic IR spectra, and optical density measurements in the 2710 nm band for chrysotile and the 12 820 nm band for crocidolite permit the quantitative evaluation of each fibre either alone or in mixtures. IR spectra also provide information on changes in fibre structure and in chemical composition as the result, for example, of thermal treatment or acid leaching.

The analytical method that we have developed can detect amounts as small as 0.1 mg of fibre in a 300-mg disk of potassium bromide using a low-cost IR spectrophotometer. The use of a Fourier transform IR spectrophotometer dramatically improves the sensitivity and selectivity. Computer-assisted analysis of spectra offers the possibility of reducing matrix interference and of comparing different spectra. The application of the IR technique to asbestos-cement products and insulating materials is described.

## *Introduction*

The analytical techniques most widely used for the identification of asbestos fibres require expert and well-trained personnel (polarized light microscopy) (MacCrone, 1980) or expensive equipment, and the analysis is long and tedious (electron microscope, X-ray microanalysis) (Middleton & Jackson, 1982; Malami *et al.*, 1981). In general, these techniques also require very careful standardization (Anderson & Long, 1980).

Some authors have reported the use of infrared (IR) spectrophotometry for this purpose (Luoma *et al.*, 1982; Coates, 1977; Jolicoeur & Duchesne, 1981; Beckett *et al.*, 1975; Marconi, 1983; Hlavay *et al.*, 1984). Based on our experience also, this method deserves greater attention. In fact, although it does have some limitations, mainly as a result of possible matrix interference, it is capable of quickly providing both qualitative and quantitative (Kimmerle *et al.*, 1984) information on the type of asbestos contained in the sample material.

In our laboratory, a simple IR spectrophotometric method has been developed, using a low-cost instrument, which allows the quantitation of chrysotile and crocidolite fibres present in a sample mixture in about 90 min. Using a Fourier transform IR spectrophotometer (FTIR) and computer-assisted analysis of the spectra, we also applied the IR technique to asbestos-cement samples. In this case, the problem of the matrix effect was easily overcome and the sensitivity was improved.

## Materials and methods

Crocidolite and Rhodesian chrysotile and amosite were provided by UICC.

Samples of asbestos-cement products and the raw materials for their preparation (asbestos fibres and cement) were kindly supplied by Eternit.

Potassium bromide for IR spectrophotometry and analytical grade ethanol were obtained from Merck.

A Perkin Elmer 710-B double-beam, and a Perkin Elmer 1700 FTIR spectrophotometer were used and spectra were analysed using a program produced by Perkin Elmer (PE 983) and a PE 3600 data station.

All the fibre samples were analysed by the potassium bromide disk method, the potassium bromide being ground and dried in an oven at 110°C for 12 h.

A suitable amount of each sample (asbestos, cement and their mixtures) was weighed and ground in an agate mill for 3 min with two drops of ethanol. As soon as most of the ethanol had evaporated, 300 mg of potassium bromide were added. The sample was mixed until homogeneous and then heated to 110°C for 1 h. The mixture was then placed in a press and a pressure of 12 tons was applied, under vacuum, for 10 min, after which the potassium bromide disk was submitted to IR analysis.

Spectrophotometric analysis was carried out by scanning in the range 2500–16 000 nm. Absorbance in the selected analytical bands was calculated and adjusted according to the width of the disk. This procedure was used both with disks containing a single type of fibre and with disks containing a mixture of two types of fibres.

Disks contained from 0.3 to 1 mg of a single type of fibre, while the total amount of fibre in the disks containing a mixture ranged from 1.1 to 2 mg. Each measurement was made in duplicate.

## Results

The results obtained with asbestos standards have shown that, under our experimental conditions, 0.3 mg is the minimum measurable amount using a PE 710-B IR spectrophotometer.

The main IR absorption peaks for chrysotile, crocidolite and amosite and the corresponding wave numbers are shown in Table 1 (Hodgson, 1979). Quantitative determinations were based on the absorbance in characteristic bands for each type of fibre.

For chrysotile, the band at 2710 nm, corresponding to the OH stretching vibration, was chosen. For crocidolite, the band at 12 820 nm was used. Absorbance was linearly correlated with the amount of each fibre in the potassium bromide disk (Table 2). This result was practically unchanged when a fixed amount of a particular fibre was mixed with a variable amount of another. The results were adequately reproducible. The standard deviation of the absorbance values for the disks with the smallest amount of fibre was 11% of the value of the mean.

**Table 1. Wavenumbers ($cm^{-1}$) of principal IR absorption peaks for asbestos minerals**

| Item | Chrysotile | Crocidolite | Amosite |
|---|---|---|---|
| O-H: |  |  |  |
| stretching vibration | 3686 | 3636 | 3656 |
|  | 3640 | 3619 | 3637 |
|  | - | - | 3618 |
| Si-O: |  |  |  |
| stretching vibration | 1078 | 1143 | 1128 |
|  | 1020 | 1110 | 1082 |
|  | 960 | 989 | 996 |
|  |  | 897 | 981 |
| Silica chain and ring | - | 778 | 775 |
| vibration | - | 725 | 703 |
|  | - | 694 | 638 |
|  | - | 668 | - |
|  | - | 636 | - |
| Cation-oxygen: |  |  |  |
| stretching vibration | 615 | 541 | 528 |
|  |  | 504 | 498 |
|  |  |  | 481 |
|  |  |  | 426 |

**Table 2. Equations of calibration curves for pure and mixed asbestos fibres**

| Equation |  | Correlation coefficient |
|---|---|---|
| Chrysotile[a] | $= (Abs^b + 0.0136)/0.491$ | 0.960 |
| Crocidolite[a] | $= (Abs^b + 0.0097)/0.192$ | 0.997 |
| Chrysotile[a] and crocidolite[c] | $= (Abs^b - 0.0006)/0.505$ | 0.969 |
| Crocidolite[a] and chrysotile[d] | $= Abs^b - 0.0051)/0.196$ | 0.995 |

[a]The concentration was expressed as mg of fibre in 300 mg of potassium bromide disk.

[b]Absorbance (Abs) was measured at 2710 nm for chrysotile and at 12 820 nm for crocidolite.

[c]Variable amounts of chrysotile with 0.8 mg of crocidolite.

[d]Variable amounts of crocidolite with 1 mg of chrysotile.

### Fibre identification in insulating material and asbestos-cement products

The analytical method described was used to identify inorganic insulating materials of unknown composition.

The IR spectrum obtained from a sample of insulating rope used in aluminium casting and from insulating materials sprayed on the frames of a public building are shown in Figures 1 and 2.

**Fig. 1. IR spectrum (PE 710B) of insulating rope. The sample was identified as chrysotile**

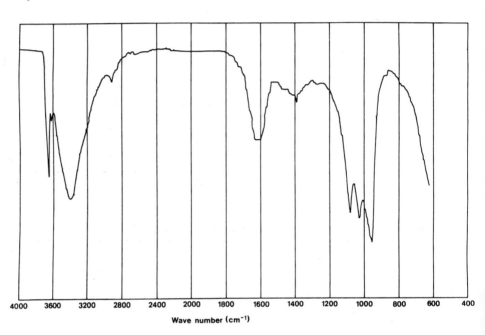

The spectrum of the first sample closely matched that of chrysotile and the absence of interference made it possible to determine the percentage of asbestos fibre in the sample (25%).

In the second sample, a preliminary examination by phase-contrast light microscopy excluded the presence of crocidolite, and the IR spectrum showed a good correlation with that of amosite.

In more complex situations, as in analysing asbestos-cement products, the use of computer-assisted analysis made it possible to subtract the spectrum of pure cement (Figure 3) from that of a sample of asbestos-cement tile (Figure 4). The spectrum obtained (Figure 5) showed a pattern very similar to that of chrysotile.

**Fig. 2. IR spectrum (PE 1700 FTIR) of insulating material. The fibre was identified as amosite.**

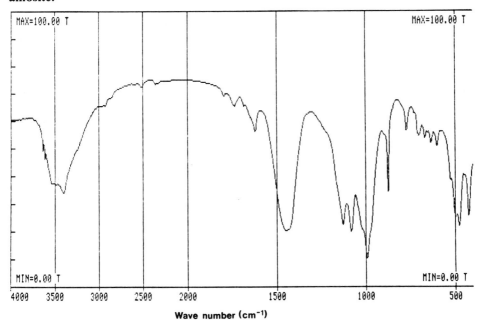

**Fig. 3. IR spectrum of a sample from an asbestos-cement tile**

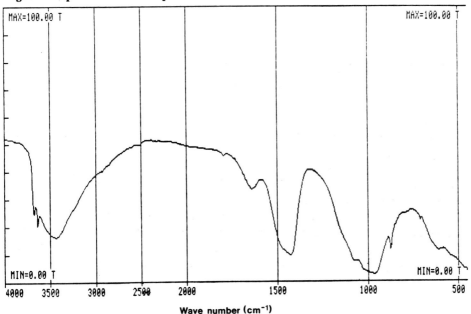

**Fig. 4. IR spectrum of cement used to prepare asbestos-cement tiles**

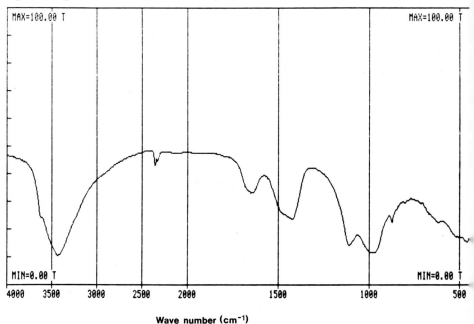

**Fig. 5. Difference spectrum between Figures 3 and 4.**
The PE 983 program for IR analyses was used, with a PE 3600 data station. The arrow points out the characteristic band of chrysotile with maximum at 3686 cm$^{-1}$

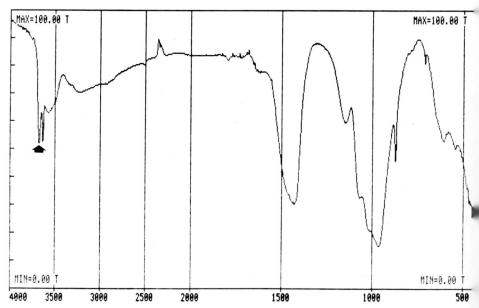

## Discussion

**Possible interference with chrysotile**

Kaolin and antigorite (non-fibrous serpentines), absorbing at 3770 and 3650 cm$^{-1}$, can interfere with chrysotile. In this case, useful indications are given by the spectrum between 400 and 300 cm$^{-1}$, where chrysotile has a characteristic band at 300 cm$^{-1}$. The presence of an absorption peak at 1040 cm$^{-1}$ may confirm the absence of antigorite in the sample (Marconi, 1983).

**Possible interference with crocidolite**

Some difficulties may arise in interpreting IR spectra when two amphiboles are present in the sample at the same time, for example, crocidolite and amosite, since their spectra between 4000 and 600 cm$^{-1}$ are quite similar. Examination of the spectra between 600 and 300 cm$^{-1}$ may help in overcoming this problem, as crocidolite has a typical band at 315 cm$^{-1}$ that is not present with amosite.

To avoid this problem, some authors have suggested that the mixture should be subjected to different physical and chemical treatments (Jolicoeur & Duchesne, 1977; Chen, 1977). Thus, thermal treatment affects the stretching vibration of OH groups in chrysotile fibres, which disappear at temperatures above 650°C. Chrysotile is also very sensitive to acid treatment, while amosite and crocidolite are less affected. When doubts exist as to the true composition of a sample, therefore, thermal pretreatment or acid leaching may give useful additional information.

From the results reported here, the following scheme may be used for the identification by IR spectrometry of asbestos fibres used as lagging in buildings:

(1) Examination by optical microscopy to verify the presence of fibrous material in the sample.
(2) If necessary, destruction of the organic matrix by heating. If a cement matrix is present, the fibres can be separated by sedimentation (Bagioni, 1975) or by treatment with 0.1M hydrochloric acid.
(3) Determination of IR spectra in the range 4000–200 cm$^{-1}$.
(4) Identification of fibrous material in the sample by comparison of the IR spectra with a set of reference spectra.
(5) If confirmation is necessary, subjection of fibres to thermal and/or chemical treatment (acid leaching) and comparison of the resulting changes in the IR spectra with those of reference samples after similar standardized treatment.

## Acknowledgements

The authors are grateful to Eternit for providing asbestos-cement products, and to Italian Perkin Elmer for its valuable assistance.

## References

Anderson, C.H. & Long, J.M. (1980) *Interim Method for Determining Asbestos in Water*, US Environmental Protection Agency (600/4-80-005)

Bagioni, R.P. (1975) Separation of chrysotile asbestos from minerals that interfere with its infrared analysis. *Environ. Sci. Technol.*, *3*, 262-263

Becket, S.T., Middleton, A.P. & Dodgson, J. (1975) The use of infra-red spectrophotometry for the estimation of small quantities of single varieties of UICC asbestos. *Ann. Occup. Hyg.*, *18*, 313-319

Chen, J.T. (1977) Infrared studies of the effects of acid, base, heat and pressure on asbestos and structurally related substances. *J. Assoc. Off. Anal. Chem.*, *60*, 1266-1276

Coates, J.P. (1977) IR analysis of toxic dusts. Analysis of collected samples of asbestos. *Am. Lab.*, *12*, 57-65

Hlavay, J., Antal, I., Gyorgy-Pozsanyi, I. & Inczedy, J. (1984) Determination of chrysotile content of asbestos cement dusts by IR-spectroscopy. *Fresenius Z. Anal. Chem.*, *319*, 547-551

Hodgson, A.A. (1979) Chemistry and physics of asbestos. In: Michaels, L. & Chissick, S.S., eds, *Asbestos, Properties, Applications and Hazards*, Vol. 1, New York, John Wiley & Sons, p. 67

Jolicoeur, C. & Duchesne, D. (1981) Infrared and thermogravimetric studies of thermal degradation of chrysotile asbestos fibers: evidence for matrix effects. *Can. J. Chem.*, *59*, 1521-1526

Kimmerle, F.M., Noel, L. & Khorami, J. (1984) Quantitative IR-ATR spectrometry of asbestos fibres on membrane filters. *Can. J. Chem.*, *62*, 441-451

Luoma, G.A., Yee, L.K. & Rowald, R. (1982) Determination of microgram amount of asbestos in mixtures by infrared spectrometry. *Anal. Chem.*, *54*, 2140-2142

MacCrone, W.C. (1980) Evaluation of asbestos in insulation. *Int. Lab.*, *10* (1), 35-48

Malami, C., Hoke, E. & Grasserbauer, M. (1981) Contribution to the quantitative X-ray analysis of individual submicrometre particles: asbestos fibres. *Mikrochim. Acta*, *1*, 141-158

Marconi, A. (1983) Application of infrared spectroscopy in asbestos mineral analysis. *Ann. Ist. Super. Sanità*, *19* (4), 620-638

Middleton, A.P. & Jackson, E.A. (1982) A procedure for the estimation of asbestos collected on membrane filter using transmission electron microscopy (TEM). *Ann. Occup. Hyg.*, *25*, 381-391

# III. FIBRE LEVEL MEASUREMENTS

# FIBRE LEVELS IN LUNG AND CORRELATION WITH AIR SAMPLES

**B.W. Case**

*Dust Disease Research Unit, School of Occupational Health and Department of Pathology, McGill University, Montreal, Canada*

**P. Sébastien**

*Dust Disease Research Unit, School of Occupational Health, McGill University, Montreal, Canada, and Groupe Pneumopathies Professionnelles, Verneuil en Halatte, France*

## Introduction

In this paper we summarize current knowledge of environmental exposures to fibrous dust. We present first what is known from studies using air fibre measurements, then the data from lung burden studies. Very few non-occupational environments exist for which both types of data are available, but these are particularly useful in extrapolating from exposure to intrapulmonary deposition. The biological significance of lung fibre content at the levels considered here is in most cases unknown, but reported effects are mentioned.

Five categories of 'environmental' exposure can be distinguished: (1) exposure to urban atmospheric fibre pollution; (2) exposure in buildings to asbestos and man-made mineral fibres (MMMF); (3) exposure arising from natural or geological sources; (4) exposure near industrial or mining 'point sources' of fibre emission, as well as from brake emissions; and (5) domestic exposure.

## Environmental exposure as reflected by fibre levels in air samples

### Urban atmospheric fibre pollution

Mineral fibres have been identified in urban air samples from Canada (Chatfield, 1983; Sébastien et al., 1986), France (Sébastien et al., 1983), South Africa (Holt & Young, 1973), the Netherlands (Lanting & van Boeft, 1983), the United Kingdom (Burdett et al., 1984), the United States (Nicholson, this volume, pp. 239-261), Federal Republic of Germany (Friedrichs, 1979; Spurny et al., 1979), Italy (Vigliani et al., 1976), Switzerland (Litistorf et al., 1985), and Japan (Kohyama, this volume, pp. 262-276). The fibre type most commonly identified in these studies has been chrysotile, usually short (less than 5 µm in length) and present in low concentration as a homogeneous 'background'. Sulfates such as calcium sulfate are also frequently seen (Middleton, 1978; Spurny et al., 1979). 'Glass' fibres — structures containing silica and probably corresponding to varieties of MMMF — have also been identified in urban atmospheres (Gaudichet et al., this volume, pp. 291-298) as have a variety of miscellaneous fibres.

## Asbestos and MMMF exposure in buildings

Evidence of specific exposure of building occupants is still controversial. Environmental studies conducted in buildings have either been systematically positive or systematically negative. It is clear, on closer examination, that measurement method is the determining factor. Roughly speaking, studies using the so-called 'direct method' have not reported levels above background (Burdett & Jaffrey, 1986; Chatfield, 1983). Our own view is that the direct method effectively raises the detection limit, and is therefore less appropriate as well as less reliable (Sébastien et al., 1986). When indirect methods are used, asbestos-containing floor tiles, insulating and other materials appear to increase concentrations to up to 1000 times the background level (Constant et al., 1983; Nicholson et al., 1975; Sébastien et al., 1982, 1983). Not all buildings containing asbestos materials are polluted. To determine whether a given building is contaminated, air measurements must be performed: there is no predictable relationship between architectural configuration and air asbestos concentration.

Recent studies in France (Gaudichet et al., this volume, pp. 291-298) have also identified increased levels of MMMF in some buildings in which these materials were used.

## Atmospheric fibre pollution arising from natural (geological) sources

Naturally occurring soil fibre contamination has been reported in several parts of the world. Studies in central Europe (Hillerdal, 1980), and in Corsica (Boutin et al., 1986), have noted pleural plaques and calcifications, suggesting the possibility of such exposure.

Within Europe, there is an apparent 'pleural arc' extending from Finland through the USSR, Austria, Czechoslovakia, Bulgaria, Turkey and Greece within which increased observations of pleural plaques and calcifications have been attributed to fibre exposure from natural deposits of tremolite, anthophyllite or chrysotile. To our knowledge, air measurements in this area have been performed only in Corsica, where low levels of chrysotile and tremolite have been recorded (Steinbauer et al., 1987). There are also reports of environmental air pollution from natural sources in the United States (Cooper et al., 1979; Rohl et al., 1977) and in the 'mesothelioma villages' of Cappadocia, Turkey.

The situation in Turkey requires special comment. Erionite is the fibre type most probably implicated in certain villages of Cappadocia, where mesothelioma prevalence and mortality figures are markedly increased (Artvinli & Baris, 1979; Baris et al., 1978, 1987; Simonato et al.. this volume, pp. 398-405). While environmental exposure to erionite in these villages was suggested by early bulk analysis of local rock samples (Ataman, 1978; Pooley, 1979), results of air measurements have been equivocal (Baris et al., 1981, 1987).

## Asbestos fibre pollution near industrial and mining sites and from brake emissions

No data are available for non-asbestos fibres (MMMF, clay, etc.) contaminating the air in the neighbourhood of industrial sources of these materials. 'Neighbourhood exposure' has been documented only for asbestos in studies, *inter alia*, of mining sites (Sébastien *et al.*, 1986), manufacturing plants and dumping sites (Burdett *et al.*, 1984; Harwood *et al.*, 1975; Lanting & den Boeft, 1983). In all of these studies an increase was found in neighbourhood air fibre concentrations. For brake emissions, divergent results have been noted (Alste *et al.*, 1973; Sébastien *et al.*, 1983). The study reported by Kohyama (this volume, pp. 262-276) suggests that brake emissions are a source of serious asbestos air pollution near main roads in Japan.

As an example of industrial/mining neighbourhood asbestos air pollution, we report here some of the results of our recent studies in the mining towns of Quebec (Sébastien *et al.*, 1986). Throughout 1984, a network of 9 sampling stations in the mining towns and in control urban (Montreal) and rural environments was used to obtain monthly data. Using transmission electron microscopy (TEM), we found measurable levels of chrysotile at all stations. These proved in some months to exceed those in the control areas by a factor of up to 250. There was marked seasonal variation, which can be almost entirely explained by weather conditions and by the month-to-month differences in activity of the Quebec mining industry throughout 1984.

Geometric mean annual chrysotile concentrations were 0.7 fibres per litre (f/l) in control areas, 33.3 in Asbestos, 39.9 in Thetford Mines and 110.4 in the nearby town of Black Lake. Tremolite concentrations were generally very low, being unmeasurable in control areas and 0.2 f/l in Asbestos. While still low in absolute terms, tremolite concentrations in Thetford Mines were almost one order of magnitude higher than those in Asbestos, at 1.5 f/l. As described below, this has important consequences for lung tremolite accumulation.

**Air fibre levels resulting from domestic exposures**

Surprisingly, few studies have systematically reported domestic or household exposures for family members of workers exposed to asbestos and other fibres. To our knowledge, only Nicholson *et al.* (1980) have reported air measurements in such environments. In that study, levels as high as 2000 ng/m$^3$ were recorded in the homes of asbestos workers.

**Summary**

Current knowledge of environmental exposure to mineral fibres can be summarized as follows:

(1) Few or no data exist on actual measurements of exposure to the two main types of industrially significant non-asbestos fibres: MMMF and clay fibres.

(2) A number of asbestos air fibre measurement studies have been carried out in contaminated buildings and for industrial/mining neighbourhood fibre-emission sources. However, in most cases there are no accompanying biological data.

(3) For some areas where neighbourhood and domestic exposures to asbestos have occurred, we have good epidemiological evidence of asbestos-related

disease but few corresponding air fibre measurements. Examples include the classic mesothelioma study of Wagner *et al.* in South Africa (1960), other instances of mesothelioma associated with household exposure (Hammond *et al.*, 1979; McDonald & McDonald, 1980) and observations of pleural plaques in several areas in the 'pleural arc' (Kiviluoto, 1960; Boutin *et al.*, 1986).

(4) There are currently only 3 environmental situations in which we have air measurement, epidemiological and lung fibre burden data. These are the studies in Corsica (Boutin *et al.*, 1986), in Cappadocia (Artvinli & Baris, 1979; Baris *et al.*, 1978, 1987; Sébastien *et al.*, 1984; Simonato *et al.*, this volume, pp. 398-405) and in the Quebec mining area (Case & Sébastien, 1987, 1988a; Sébastien *et al.*, 1986).

## *Environmental exposure as reflected by fibre levels in lung samples*

### Sources of variation in lung fibre analytical studies

The division of environmental air sampling studies into 5 categories is relatively straightforward. While the same scheme is followed in the presentation of the lung fibre analytical data, the situation is much more complex. As an example, consider a child going to school in a town near an asbestos mine. He or she may be exposed to asbestos or other fibres at school (building exposure) and domestically exposed if the father or mother works in the mines or mills; and is certainly exposed in varying degrees to both geological and industrial sources, as well as to the normal 'background' pollution.

The most common deficiency of tissue analytical studies of 'environmentally exposed' individuals is poor definition of the target population. It is often difficult or impossible to exclude the possibility of occupational exposure, and misclassification inevitably occurs. Interpersonal variation in smoking habits, inhaled fibre dose, and respiratory clearance patterns further complicate the picture. Lung sample site may affect the results. Differences have been found between lung lobes (Morgan & Holmes, 1983, 1984), between peripheral and central parenchyma (Churg & Wiggs, 1987; Morgan & Holmes, 1983; Sébastien *et al.*, 1977), between subpleural and parenchymal areas (Sébastien *et al.*, 1980) and even between immediately adjacent sites (Churg & Wood, 1983).

Once sampling is completed, a plethora of technical factors may further complicate the analysis. Sample size, lung disease (Case & Sébastien, 1987b; Churg & Wood, 1983) and prior preparation (embalming, choice of fixative, paraffin embedding) may lead to modifications in final fibre concentrations (Vallyathan & Green, 1985). Methods used to remove organic constituents (chemical digestion versus low-temperature ashing) (Gylseth & Skaug, 1986); filtration method; and filter load, type and pore size (Erlich & Suzuki, 1987; Gylseth & Skaug, 1986; O'Sullivan *et al.*, 1987) all produce measurable variation. Differences in instrumentation, identification criteria, counting rules, final magnification and detection limit lead to marked interlaboratory differences (Gylseth *et al.*, 1985).

Given the degree of error possible, it is important that reporting criteria be standardized and that studies have sufficient case numbers. Controls are often poorly chosen, unmatched, or selected from autopsy series having nothing in common with those from which cases were derived.

Despite these difficulties, a number of lung analytical studies have produced results which correspond well to the environmental air sampling data. Typical 'asbestos bodies' are usually counted via optical microscopy (Churg, 1982). Individual fibres are best identified using TEM at high (generally × 6600 or greater) magnification. A wide variety of inorganic mineral fibres are best quantitatively assessed (type, concentration and dimensions) using TEM coupled to elemental identification systems, such as energy-dispersive X-ray analysis (EDXA).

## Urban and rural residence and lung fibre content

The occurrence of typical asbestos bodies in lung tissue has been known since 1927 (Page, 1935). Most early studies concentrated on sputum analysis of asbestos industry workers (Page, 1935; Simson & Strachan, 1931), but Simson reported one case in which asbestos bodies were present in lung tissue after an exposure limited to 2 months (Simson & Strachan, 1931). Thomson et al. (1963) found asbestos bodies in the lungs of 25% of Cape Town residents, in the first studies to specifically assess urban environments. In 1969, Selikoff and Hammond reported a higher incidence (males 51%; females 39%) among 2000 New York City autopsies. In 1977, Churg and Warnock (1977a,b, 1979, 1980, 1981) began an important series of studies outlining the frequency and nature of asbestos bodies in general autopsy populations. They showed that asbestos bodies can, with proper methodology, be found in the lung tissue of 96% or more of autopsies (Churg, 1982; Churg & Warnock, 1977a). In a study of asbestos bodies of non-occupationally exposed individuals with low counts of such bodies, they found that almost all were formed on amphibole fibre cores (Churg & Warnock, 1979).

Lung burden studies have confirmed the presence of asbestos bodies in most urban and rural residents, with excesses in the former (Breedin & Buss, 1976; Case et al., 1988b). Asbestos bodies have been found in infant lungs (Haque et al., 1985), but concentrations are significantly correlated with increasing age (Case et al., 1988b), implying a relationship with cumulative amphibole exposure from all sources.

TEM lung fibre analytical studies confirm the predominance of short chrysotile fibres, as found in air measurements in urban environments. Langer and Selikoff (1971) found chrysotile in 24 of 28 New York City residents. Churg and Warnock (1980) assessed 21 autopsies from urban dwellers with low (<100 per g dry lung) counts of of asbestos bodies. Of the asbestos fibres identified, 80% were chrysotile, and 90% of chrysotile fibres were less than 5 $\mu$m in length. Most amphiboles present were of the non-commercial variety, suggesting environmental exposure. In analyses of 65 male forensic autopsies across Canada, chrysotile fibre (longer than 5 $\mu$m) concentrations have been found to be unrelated to large city residence or to age (Case et al., 1987c). Conversely, commercial amphibole and tremolite concentrations peak in the 40-60-year-old age-group in males and are highest in lung tissue from residents of large cities. It appears, then, that long amphibole fibres, present in air-sampling studies and in lung in much smaller numbers than chrysotile fibres, accumulate in tissue, accounting for the presence of asbestos bodies and their correlation with increasing age and urban residence.

While asbestos has received most attention, other fibres account for approximately half the lung burden in general populations (Case et al., 1988b; Case & Sébastien, 1987; Churg, 1983). Relative proportions vary from study to study, depending on mineralogical identification criteria and analytical methodology, but fibrous talc, titanium (rutile), mica and other silicates have been reported as having the highest concentrations (Case & Sébastien, 1987). MMMF of 3 chemical types have also been identified in low concentration in a number of subjects in ongoing work in our laboratory.

### Exposure in buildings and lung fibre content

Given the great concern with regard to asbestos contamination in buildings and resulting from asbestos removal, it is surprising that not a single systematic tissue analytical study has been performed in this area. Churg and Warnock (1977a) did study asbestos body counts in 28 patients who had 'insulated or built their own homes' (not necessarily asbestos-exposed), finding no excess.

### Sources of geological exposure and lung fibre content

Lung burden studies have confirmed excesses in several areas where asbestos or other fibrous minerals occur naturally. Since such minerals have commercial value, they are also mined in these areas, and it is difficult to separate 'natural' sources from those associated with economic activity. Tremolite, usually accompanied by chrysotile, has been isolated from the lungs of individuals who lived near mineral deposits containing that fibre in north-western Greece (Langer et al., 1987), Cyprus (McConnochie et al., this volume, pp. 411-419), north-eastern Corsica (Magee et al., 1986) and a farming community 30 km from the Quebec serpentine belt (Case & Sébastien, unpublished data). In every instance, however, domestic or neighbourhood exposure to tremolite or chrysotile in current or former mining operations or in products used in housing (e.g., tremolite in whitewash (Langer et al., 1987)) are likely to have contributed. Epidemiological studies are not yet complete, but preliminary findings suggest an excess of pleural pathology (plaques or mesothelioma) in Greece (Langer et al., 1987), Cyprus (McConnochie et al., this volume, pp. 411-419) and Corsica (Boutin et al., this volume, pp. 406-410).

An important series of studies, including air and bulk sampling data, lung burden, and epidemiological surveys, has related erionite exposure to mesothelioma and lung cancer incidence in certain villages of Cappadocia, Turkey (Baris et al., 1987; Simonato et al., this volume, pp. 398-405). Although erionite is only slightly increased in air samples, sputum analysis has clearly demonstrated an age-related excess of ferruginous bodies in residents of 2 affected villages as compared with controls

(Sébastien et al., 1984). Studies on lung tissue from sheep grazing near control and 'mesothelioma' villages also show an excess of erionite in the latter (Baris et al., 1987). The superior ability of tissue analysis to differentiate between case and control populations suggests that lung burden reflects cumulative rather than current exposure to airborne erionite.

**Lung fibre burden following industrial neighbourhood or domestic exposures**

Separation of non-occupationally exposed individuals with a history of household exposure from those exposed only to mine or plant emissions is extremely difficult. Churg and Warnock (1977a) were unable to find excess asbestos bodies in autopsy lung from 25 women whose husbands worked in the steel or construction industries, but asbestos exposure history was not specifically ascertained. Bianchi et al. (1983), in an autopsy study performed near an Italian shipyard, found 8 of 27 women domestically exposed and with pleural plaques to have asbestos body counts exceeding 1000 per g dry lung.

The only set of studies in which both environmental and lung tissue analytical data are available for a population exposed both domestically and via neighbourhood pollution is that in the Quebec chrysotile mining region (Case & Sébastien, 1987, 1988a; Sébastien et al., 1986). The serpentine belt of Quebec can be divided for practical purposes into 2 distinct subregions, the first of which is the single, large open-pit mine at Asbestos, Quebec. About 70 km distant, separated by a range of hills, is the Thetford Mines/Black Lake/East Broughton area (henceforth designated 'Thetford'). The latter comprises many mines and tailings which sometimes surround the co mmunities, as in Black Lake, where the highest current air levels (0.027 f/ml) have been recorded (Sébastien et al., 1986). As indicated above, chrysotile accounts for the majority of the fibres present in air in both areas, but tremolite, while low in absolute terms, is one order of magnitude higher in Thetford.

Our previous work in Asbestos demonstrated a clear excess of asbestos bodies in the lungs of 22 long-term residents without a history of work in the mining or milling industry, as compared to controls from the same autopsy population (Case & Sébastien, 1988a). TEX/EDXA analyses (Case & Sébastien, 1987) showed lungs from these men and women to contain excess chrysotile, with fibre size distributions closer to those in chrysotile miners and millers than to controls. Tremolite levels were indistinguishable from those in control autopsies.

Recently, we completed a 5-year tissue analytical study in the Thetford area. During that period, all autopsies ($N=200$) underwent an initial screening performed by M Christian Pratte and Dr M. Poulin. Those known to have worked in the mines or associated mills or factories were excluded and for the remaining 25 individuals, personal interviews were conducted with next-of-kin to determine whether there was a history of occupational or domestic exposure to asbestos. Three men and one woman missed in the screening procedure were found to have had occupational exposure in the chrysotile industry. Two subjects had lived on farms in a community 47 km from the mining area throughout their lives. Of the remaining 19, nine were found to have had domestic exposure of 15 years or more, and 10 had had only neighbourhood environmental exposure. There were no differences in any parameter (age; years lived in region; pleural plaque incidence; asbestos body counts; fibre types, concentrations and length and diameter distributions) between 'domestic' and 'neighbourhood' groups.

Median asbestos body counts and total asbestos fibre concentrations were one order of magnitude lower in the two subjects living farthest from the mines than in the 19 environmentally exposed subjects ($p<0.05$; Wilcoxon rank sum). Asbestos body counts and tremolite, chrysotile and total asbestos concentrations were highest in the 4 occupationally exposed subjects ($p<0.01$).

To illustrate the relationship between current air sampling levels of chrysotile and tremolite and lung fibre burden (Table 1), we compared the 19 environmentally exposed Thetford subjects with our previously published data on 19 controls and 22 environmentally exposed individuals in Asbestos (Case & Sébastien, 1988a). We also compared environmental groups with a group of miners and millers from Thetford. As expected, intrapulmonary chrysotile levels are higher in Asbestos and Thetford than in the control area, but somewhat lower than those in Thetford miners and millers. Retained chrysotile fibres were longer and thinner in both Thetford groups.

### Table 1. Lung fibre burden in two Quebec chrysotile mining communities

| Item | Controls[a] | Asbestos: environmental[a] | Thetford | |
|---|---|---|---|---|
| | | | Environmental[a] | Occupational[a] |
| No. analysed | 19 | 22 | 19 | 22 |
| Asbestos bodies[b] | 80[c] | 480[c] | 1880[c] | 35000[c] |
| Chrysotile (f/μg)[d] | 0.08[c] | 0.28 | 0.33[e] | 2.2[c] |
| Tremolite (f/μg)[d] | 0.06 | 0.08 | 0.39[c] | 17.2[c] |
| All asbestos (f/μg)[d] | 0.26[c] | 0.57[c] | 1.16[c] | 24.6[c] |
| Tremolite/all fibres (%)[f] | 14 | 13 | 37 | 74 |
| Asbestos/all fibres (%)[f] | 43 | 60 | 75 | 91 |

[a]Environmental groups include those exposed non-occupationally either at home (domestic) or through residence near mines (neighbourhood exposure). Controls are from the same autopsy population as for the Asbestos environmental group (Case & Sébastien, 1987, 1988a). The Thetford occupational group includes unselected miners and millers from the same autopsy population as the Thetford environmental group.

[b]Median typical asbestos body count per g dry lung tissue.

[c]$p<0.05$ or less as compared with every other group (Wilcoxon rank sum for medians; independent sample $t$-test for geometric means).

[d]Geometric mean fibre concentration per μg dry lung tissue

[e]$p<0.05$ or less, except as compared with Asbestos (Wilcoxon rank sum for medians; independent sample $t$-test for geometric means).

[f]Arithmetic mean percentage.

Tremolite was present at the level of detection in controls and in Asbestos, one order of magnitude higher in Thetford, and another one to two orders of magnitude higher in Thetford workers. Tremolite fibre lengths were similar in all areas, but again, both Thetford groups had lower mean fibre diameters.

These results parallel almost exactly the air sampling data for 1984 (Sébastien *et al.*, 1986), although it is reasonable to assume that environmental levels of both chrysotile and tremolite were much higher in the period during which subjects in both mining areas acquired most of their lung burden. Lung tremolite concentration is probably of greatest importance in both fibre accumulation in tissue and the production of biological effects. We found a significant correlation between lung tremolite concentration in the environmentally exposed in Thetford and years lived in the mining region (Spearman rank $R=0.52$; $p<0.05$), and no correlation for chrysotile ($R=0.09$). Further, although no asbestos-related lesions had been noted in the environmentally exposed in Asbestos, 4 of the 19 Thetford residents had pleural plaques, and one (with domestic exposure) had a mesothelioma confirmed by histology review.

## Conclusions

In summary, non-occupational lung burden studies are influenced by choice of cases and controls, interpersonal and technical factors, and in most cases by a number of possible sources of exposure. The results of such studies are unlike air sampling data since they largely reflect lifetime accumulation rather than a current cross-section. Chrysotile may be an exception, in that its rapid clearance from lung makes it a better indicator of recent rather than cumulative exposure. Conversely, lung tremolite and possibly erionite content appear to accurately reproduce lifelong exposure histories, even at non-occupational levels.

## Acknowledgements

This work was supported by grant MA-8578 from the Medical Research Council of Canada.

## References

Alste, J., Watson, D. & Bagg, J. (1976) Airborne asbstos in the vicinity of a freeway. *Atmos. Environ.*, *10*, 583-589

Artvinli, M. & Baris, Y.I. (1979) Malignant mesothelioma in a small village in the Anatolian region of Turkey: An epidemiologic study. *J. Natl Cancer Inst.*, *63*, 17-22

Ataman, G. (1978) Zeolite outcrops of Cappadocia and their possible relationship with certain types of lung cancer and with pleural mesothelioma [in French]. *C.R. Acad. Sci. Paris*, *287*, 207-210

Baris, Y.I., Sahin, A.A., Özesmi, M., Kerse, I., Özen, E., Kolaçan, B., Altinörs, M. & Göktepeli, A. (1978) An outbreak of pleural mesothelioma and chronic fibrosing pleurisy in the village of Karain/Ürgüp in Anatolia. *Thorax*, *33*, 181-192

Baris, Y.I., Saracci, R., Simonato, L., Skidmore, J.W. & Artvinli, M. (1981) Malignant mesothelioma and radiological chest abnormalities in two villages in central Turkey. *Lancet, i*, 984-987

Baris, I., Simonato, L., Artvinli, M., Pooley, F., Saracci, R., Skidmore, J. & Wagner, J.C. (1987) Epidemiological and environmental evidence of the health effects of exposure to erionite fibres: a four-year study in the Cappadocian region of Turkey. *Int. J. Cancer, 39*, 10-17

Bianchi, C., Brolio, A. & Bittesini, L. (1983) Asbestos exposure in the Monfalcone shipyard area (Italy): A study based on a necropsy series. In: *Proceedings of the Tenth International Social Security Association World Congress on the Prevention of Occupational Accidents and Diseases*, pp. 81-85

Boutin, C., Viallat, J.R., Steinbauer, J., Massey, D.J., Charpin, D. & Mouries, J.C. (1986) Bilateral pleural plaques in Corsica: a non-occupational asbestos exposure marker. *Eur. J. Respir. Dis., 69*, 4-9

Breedin, P.H. & Buss, D.H. (1976) Ferruginous (asbestos) bodies in the lungs of rural dwellers, urban dwellers and patients with pulmonary neoplasms. *South. Med. J., 69*, 401-404

Burdett, G.J., Le Guen, J.M.M. & Riod, A.P. (1984) Mass concentrations of airborne asbestos in the non-occupational environment. A preliminary report of UK measurements. *Ann. Occup. Hyg., 28*, 31-38

Burdett, G.J. & Jaffrey, S.A.M.T. (1986) Airborne asbestos concentrations in buildings. *Ann. Occup. Hyg., 30*, 185-199

Case, B.W. & Sébastien, P. (1987) Environmental and occupational exposures to chrysotile asbestos: A comparative microanalytic study. *Arch. Environ. Health, 42*, 185-191

Case, B.W. & Sébastien, P. (1988a) Biological estimation of environmental and occupational exposure to asbestos. In: McCallum, R., ed., *Proceedings of the VIth International Symposium on Inhaled Particles (Cambridge, 1985)* (in press)

Case, B.W. & Sébastien, P. (1988b) Lung fiber analysis in accident victims: A biological assessment of general environmental exposures. In: *Proceedings of the First International Symposium on Environmental Epidemiology, Pittsburgh, 1987. Arch. Environ. Health* (in press)

Chatfield, E.J. (1983) Short mineral fibres in airborne dust. In: *Short and Thin Mineral Fibres. Identification, Exposure and Health Effects, Proceedings of a Symposium of the National Board of Occupational Safety and Health Research Department, Solna, Sweden*, pp. 9-81

Churg, A. (1982) Fiber counting and analysis in the diagnosis of asbestos-related disease. *Hum. Pathol., 13*, 381-392

Churg, A. (1983) Nonasbestos pulmonary mineral fibres in the general population. *Environ. Res., 31*, 189-200

Churg, A. & Warnock, M.L. (1977a) Correlation of quantitative asbestos body counts and occupation in urban patients. *Arch. Pathol. Lab. Med., 101*, 629-634

Churg, A. & Warnock, M.L. (1977b) Analysis of the cores of ferruginous (asbestos) bodies from the general population. I. Patients with and without lung cancer. *Lab. Invest., 37*, 280-286

Churg, A. & Warnock, M.L. (1979) Analysis of the cores of ferruginous (asbestos) bodies from the general population. III. Patients with environmental exposure. *Lab. Invest., 40*, 622-626

Churg, A., & Warnock, M.L. (1980) Asbestos fibers in the general population. *Am. Rev. Resp. Dis., 122*, 669-678

Churg, A. & Warnock, M.L. (1981) Asbestos and other ferruginous bodies: their formation and clinical significance. *Am. J. Pathol., 102*, 447-456

Churg, A., & Wood, P. (1983) Observations on the distribution of asbestos fibers in human lungs. *Environ. Res., 31*, 374-380

Churg, A. & Wiggs, B. (1987) Accumulation of long asbestos fibers in the peripheral upper lobe in cases of malignant mesothelioma. *Am. J. Ind. Med., 11*, 563-569

Constant Jr, P.C., Bergman, F.J., Atkinson, D.R., Rose, D.R., Watts, D.L., Logue, E.E., Hartwell, T.D., Price, B.P. & Ogden, J.S. (1983) *Airborne Asbestos in Schools*, Environmental Protection Agency (*EPA Report No. 560/5-83-003*)

Cooper, W.C., Murchio, J., Popendorf, W. & Wenk, H.R. (1979) Chrysotile asbestos in a California recreational area. *Science*, *206*, 685-688

Erlich, A. & Suzuki, Y. (1987) A rapid and simple method of extracting asbestos bodies from lung tissue by cytocentrifugation. *Am. J. Ind. Med.*, *11*, 109-116

Friedrichs, K.H. (1979) Morphological aspects of fibers. In: Lemen, R. & Dement, J.M., eds, *Dust and Disease*, Park Forest South, IL, Pathotox Publishers, pp. 51-64

Gylseth, B., Churg, A., Davis, J.M.G., Johnson, N., Morgan, A., Mowe, G., Rogers, A. & Roggli, V. (1985) Analysis of asbestos fibers and asbestos bodies in tissue samples from human lung: An international laboratory trial. *Scand. J. Work Environ. Health*, *11*, 107-110

Gylseth, B. & Skaug, V. (1986) Relation between pathological grading and lung fibre concentration in a patient with asbestosis. *Br. J. Ind. Med.*, *43*, 754-759

Hammond, E.C., Selikoff, I.J., Garfinkel, L. & Nicholson, W.J. (1979) Mortality experience of residents in the neighbourhood of an asbestos factory. *Ann. N.Y. Acad. Sci.*, *330*, 417-422

Haque, A.K., Hernandez, J.C. & Dillard III, E.A. (1985) Asbestos bodies found in infant lungs. *Arch. Pathol. Lab. Med.*, *109*, 212

Harwood, C.F., Ostrich, P.K., Siebert, P. & Stockholm, J.D. (1975) Asbestos emissions from baghouse controlled sources. *J. Am. Ind. Hyg. Assoc.*, *36*, 595-603

Hillerdal, G. (1980) *Pleural Plaques: Occurrence, Exposure to Asbestos, and Clinical Importance*. Dissertation No. 363, Faculty of Medicine, Uppsala University

Holt, P.F. & Young, D.K. (1973) Asbestos fibres in the air of towns. *Atmos. Environ.*, *7*, 481-483

Kiviluoto, R. (1960) Pleural calcification as a roentgenologic sign of non-occupational endemic anthophyllite-asbestosis. *Acta Radiol., Suppl.*, *194*, 1-67

Langer, A.M. & Selikoff, I.J. (1971) Chrysotile asbestos in the lungs of persons in New York City. *Arch. Environ. Health*, *22*, 348-361

Langer, A.M., Nolan, R.P., Constantopoulos, S.H. & Moutsopoulos, H.M. (1987) Association of Metsovo lung and pleural mesothelioma with exposure to tremolite-containing whitewash. *Lancet*, *i*, 965-967

Lanting, R.W. & den Boeft, J. (1983) Atmospheric pollution by asbestos fibres. In: Reinisch, D., Schneider, H.W. & Birkner, K.-F., eds, *Fibrous Dusts – Measurement, Effects, Prevention*, Düsseldorf, VDI-Verlag (*VDI Berichte 475*), pp. 123-128

Litistorf, G., Guillemin, M., Buffat, P. & Iselin, F. (1985) Ambient air pollution by mineral fibres in Switzerland (A brief survey of outdoor concentrations). *Staub-Reinhalt. Luft*, *45*, 302-307

Magee, F., Wright, J.L., Chan, N., Lawson, L. & Churg, A. (1986) Malignant mesothelioma caused by childhood exposure to long-fiber low aspect ratio tremolite. *Am. J. Ind. Med.*, *9*, 529-533

McDonald, A.D. & McDonald, J.C. (1980) Malignant mesothelioma in North America. *Cancer*, *46*, 1650-1656

Middleton, A.P. (1978) On the occurrence of fibres of calcium sulphate resembling amphibole asbestos in samples taken for evaluation of airborne asbestos. *Ann. Occup. Hyg.*, *21*, 91-93

Morgan, A. & Holmes, A. (1983) Distribution and characteristics of amphibole asbestos fibres, measured with the light microscope, in the left lung of an insulation worker. *Br. J. Ind. Med.*, *40*, 45-50

Morgan, A. & Holmes, A. (1984). The distribution and characteristics of asbestos fibres in the lungs of Finnish anthophyllite mine-workers. *Environ. Res.*, *33*, 62-75

Nicholson, W.J., Rohl, A.N. & Weisman, I. (1975) *Asbestos Contamination of the Air in Public Buildings*. (*EPA Report No. 450/3-76-004*) Research Triangle Park, NC, US Environmental Protection Agency, Office of Quality Planning and Standards

Nicholson, W.J., Rohl, A.N., Weisman, I. & Selikoff, I.J. (1980) Environmental asbestos concentrations in the United States. In: Wagner, J.C., ed., *Biological Effects of Mineral Fibres (IARC Scientific Publications No. 30)*, Lyon, International Agency for Research on Cancer, pp. 823-827

O'Sullivan, M.F., Corn, C.J. & Dodson, R.F. (1987) Comparative efficiency of Nucleopore filters of various pore sizes as used in digestion studies of tissue. *Environ. Res., 43*, 97-103

Page, R.A. (1935) A study of the sputum in pulmonary asbestosis. *Am. J. Med. Sci., 189*, 44-55

Pooley, F.D. (1979) Evaluation of fibre samples taken from vicinity of two villages in Turkey. In: Lemen, R. & Dement, J.H., eds, *Dust and Disease*, Park Forest South, IL, Pathotox Publishers, pp. 41-44

Rohl, A.N., Langer, A.M. & Selikoff, I.J. (1977) Environmental asbestos pollution related to use of quarried serpentine rock. *Science, 196*, 1319-1322

Sébastien, P., Fondimare, A., Bignon, J., Monchaux, G., Desbordes, J. & Bonnaud, G. (1977) Topographic distribution of asbestos fibres in human lung in relation to occupational and nonoccupational exposure. In: Walton, W.H. & McGovern, B., eds, *Inhaled Particles IV, Part 2*, New York, Pergamon, pp. 435-444

Sébastien, P., Janson, X., Gaudichet, A., Hirsch, A. & Bignon, J. (1980) Asbestos retention in human respiratory tissues: comparative measurements in lung parenchyma and in parietal pleura. In: Wagner, J.C., ed., *Biological Effects of Mineral Fibres (IARC Scientific Publications No. 30)*, Vol. 1, Lyon, International Agency for Research on Cancer, pp. 237-246

Sébastien, P., Bignon, J. & Martin, M. (1982) Indoor airborne asbestos pollution: from the ceiling and the floor. *Science, 216*, 1410-1413

Sébastien, P., Billon-Galland, M.A. & Gaudichet, A. (1983) Biometric data on urban atmospheric pollution with asbestos [in French]. In: Reinisch, D., Schneider, H.W. & Birkner, K.-F., eds, *Fibrous Dusts — Measurement, Effects, Prevention*, Düsseldorf, VDI-Verlag (*VDI-Berichte 475*), pp. 105-108

Sébastien, P., Bignon, J., Baris, Y.I., Awad, L. & Petit, G. (1984) Ferruginous bodies in sputum as an indication of exposure to airborne mineral fibres in the mesothelioma villages of Cappadocia. *Arch. Environ. Health, 39*, 18-23

Sébastien, P., Plourde, M., Robb, R., Ross, M., Nadon, B. & Wypnuk (1986) *Ambient Air Asbestos Survey in Quebec Mining Towns: Part II — Main Study*, Montreal, Supply and Services Canada (*Environment Canada Report No. EPS 5/AP/RQ-2E*), pp. 52

Selikoff, I.J. & Hammond, E.C. (1969) Asbestos bodies in the New York City population in two periods of time. In: Shapiro, H., ed., *Pneumoconiosis: Proceedings of the International Conference, Johannesburg*, London, Oxford University Press, pp. 99-105

Selikoff, I.J., Nicholson, W.J. & Langer, A.M. (1972) Asbestos air pollution. *Arch. Environ. Health, 25*, 1-13

Simson, F.W. & Strachan, A.S. (1931) Asbestosis bodies in the sputum: A study of specimens from fifty workers in an asbestos mill. *J. Pathol. Bacteriol., 34*, 1-4

Spurny, K.R., Stober, W., Opiela, H. & Weiss, G. (1979) On the evaluation of fibrous particles in remote ambient air. *Sci. Total Environ., 11*, 1-40

Steinbauer, J., Boutin, C., Viallat, J.R., Dufour, G., Gaudichet, A., Massey, D.G., Charpin, D. & Mouries, J.C. (1987) Pleural plaques and environmental asbestos exposure in North Corsica [in French]. *Rev. Fr. Mal. Respir., 4*, 23-27

Thomson, J.G., Kaschula, R.O.C. & MacDonald, R.R. (1963) Asbestos as a modern urban hazard. *South Afr. Med. J., 27*, 77

Vallyathan, V. & Green, F.H.Y. (1985) The role of analytical techniques in the diagnosis of asbestos-associated disease. *CRC Crit. Rev. Clin. Lab. Sci., 22*, 1-42

Vigliani, E.C., Patroni, M., Ocella, E., Rendal, R.E., Skikne, M. & Ellis, P. (1976) Presence and identification of fibres in the atmosphere of Milan. *Med. Lav., 67*, 551-567

Wagner, J.C., Sleggs, C.A. & Marchand, P. (1960) Diffuse pleural mesothelioma and asbestos exposure in the North Western Cape Province. *Br. J. Ind. Med., 17*, 260-271

# NON-OCCUPATIONAL MALIGNANT MESOTHELIOMAS

### A.R. Gibbs

*Department of Pathology, Llandough Hospital,
Penarth, South Glamorgan, UK*

### J.S.P. Jones

*Department of Pathology, City Hospital, Nottingham, UK*

### F.D. Pooley

*Institute of Materials, University College, Cardiff, UK*

### D.M. Griffiths & J.C. Wagner

*MRC External Staff Team on Occupational Lung Diseases,
Llandough Hospital, Penarth, South Glamorgan, UK*

*Summary.* The mineral content of the lungs from 84 cases of malignant pleural mesothelioma was estimated by electron microscopy and energy-dispersive X-ray analysis. These cases were chosen because the history of asbestos exposure was absent, indirect or ill-defined. The occupational exposures were classified according to the method of Zielhuis, and the results indicated that this classification is unnecessarily complicated. The chrysotile counts in the lungs from these mesothelioma cases were similar to those in controls and in a previous series of mesotheliomas in which the majority had had direct exposure to asbestos. Amphibole counts were intermediate between those in controls and in the previous series of mesotheliomas with direct asbestos exposure. These findings confirm those of previous studies indicating that amphiboles are more important than chrysotile in the causation of malignant mesothelioma. The results confirm that some mesotheliomas develop in the absence of asbestos exposure.

## *Introduction*

The association between asbestos exposure and the development of malignant mesotheliomas of the pleura and peritoneum is well recognized (Wagner *et al.*, 1960). The majority of series of malignant mesotheliomas reported in the literature have found evidence of asbestos exposure in 68–99% of cases (Borow *et al.*, 1973; Cochrane & Webster, 1978; Tagnon *et al.*, 1980; Whitwell & Rawcliffe, 1971). However, in some series, the association has been found in less than 50% of cases (Brenner *et al.*, 1982; Oels *et al.*, 1971). This has led to the suggestion that other agents may cause malignant mesothelioma (Greenberg & Lloyd Davies, 1974; Peterson *et al.*, 1984). Agents which have been implicated in the causation of malignant mesothelioma include radiation

(Hirsch et al., 1982; Maurer & Egloff, 1975), beryllium (Oels et al., 1971), chronic inflammation (Brenner et al., 1982; Roggli et al., 1982) and non-asbestos mineral fibres (Wagner & Pooley, 1986).

There are several possible explanations of the differing proportions of cases with associated asbestos exposure reported in the various series, including misdiagnosis of malignant mesothelioma, unreliable occupational histories and different geographical populations. Histopathological diagnosis of malignant mesothelioma can be extremely difficult, and it is for this reason that expert panels have been set up (Jones et al., 1985; Planteydt, 1980; Spirtas et al., 1986). Since exposure to asbestos may precede the appearance of malignant mesothelioma by several decades, the exposure history may be unreliable; equally, if the persons involved are influenced by considerations of compensation in particular cases, then exposure to asbestos may be over-estimated.

In this study we have examined the mineral fibre content of lungs from persons who have died from malignant mesothelioma of the pleura in whom the history of asbestos exposure was absent, indirect or vague. The objectives of the study were to:

(1) correlate lung mineral content with the Zielhuis groupings of occupational exposure to asbestos;

(2) determine whether any mesotheliomas were unrelated to asbestos exposure; and

(3) compare the role of amphiboles and chrysotile in the causation of mesothelioma.

## Materials and methods

The mineral content of the lungs was evaluated in 84 cases diagnosed between 1979 and 1986 by A.R. Gibbs, J.S.P. Jones and/or J.C. Wagner as definite or probable malignant mesothelioma of the pleura in which exposure to asbestos was absent, indirect or ill-defined. Exposure histories were classified according to Zielhuis (1977): Group Ia = direct exposure to asbestos; Group Ib = indirect exposure to asbestos, e.g., non-asbestos workers in the shipbuilding industry; Group II = paraoccupational exposure, e.g., the wives of males working with asbestos; Group III = neighbourhood exposure, e.g., people living in the vicinity of asbestos-processing factories; Group IV = exposure to asbestos in ambient air; and Group V = no exposure to asbestos.

The mineral content was analysed electron microscopically (Pooley & Clarke, 1979). Blocks of formalin-fixed lung were selected from the apex of the upper lobe, the apex of the lower lobe and the base of the lower lobe, pooled and dried at 8°C, digested in sodium hydroxide, washed and ashed in an oxygen atmosphere. The final extract was suspended and filtered on to a Nuclepore filter. The filter preparation holding the known weight of dried tissue was coated with carbon, the filter dissolved with chloroform and the carbon film mounted on to a gold electron microscope grid. These preparations were examined electron microscopically for fibrous particles and their numbers counted, the values obtained being related to the original weight of lung tissue. A total of 100-200 fibres were analysed by the energy-dispersive X-ray technique to determine the percentage of the different types of fibres.

## Results

All mineral fibre counts are expressed in numbers of fibres $\times 10^6$ per g of dried lung tissue.

Of the total number of mesotheliomas, 61 occurred in males and were of the following types: epithelial 23; mixed 18; connective tissue 20. The 23 which occurred in females were of the following histological types: epithelial 11; mixed 9; connective tissue 3.

The numbers in the different Zielhuis groups, together with age and sex, are shown in Table 1; mean total and differential fibre counts are shown for each of the Zielhuis groups in Table 2, while Tables 3-5 show the total and differential fibre counts for Zielhuis groups 2, 3 and 5.

**Table 1. Numbers of mesotheliomas in different Zielhuis groups, age and sex**

| Zielhuis group | Total no. of mesotheliomas | Mean age and age range (years) at death | M:F |
|---|---|---|---|
| Ia  | 16 | 68.2 (42-82) | 16:0 |
| Ib  | 28 | 65.1 (44-83) | 27:1 |
| II  | 13 | 62.1 (47-72) | 1:12 |
| III | 5  | 55.4 (31-77) | 0:5  |
| IV  | 1  | 69.0          | 0:1  |
| V   | 21 | 64.4 (41-83) | 17:4 |

**Table 2. Mean total and differential fibre counts for different Zielhuis groups**

| Zielhuis group | Total no. of cases | Fibre count (range in parentheses) | | | |
|---|---|---|---|---|---|
| | | Total no. | Amosite | Crocidolite | Chrysotile |
| Ia  | 16 | 37.6 (0-255)        | 4.5 (0-64.6)  | 1.8 (0-15.7)   | 19.6 (0-229)     |
| Ib  | 28 | 24.9 (2.6-345)      | 0.5 (0-25.0)  | 0.5 (0-18.3)   | 14.8 (0.5-438)   |
| II  | 13 | 277.8 (5.6-2507)    | 1.5 (0-6.1)   | 31.8 (0-251.1) | 218.9 (1.9-2507) |
| III | 5  | 174 (14.6-779.1)    | 1.7 (0-7.9)   | 5.2 (0.8-9.1)  | 147.9 (6.0-692.5)|
| IV  | 1  | 6.4                 | 0.1           | 0.9            | 1.5              |
| V   | 21 | 42.5 (0-188.3)      | 0.7 (0-4.6)   | 5.5 (0-101.7)  | 19.6 (0-76.5)    |

**Table 3. Total and differential fibre counts for Zielhuis Group II (paraoccupational cases)**

| Case no. | Fibre count | | | | |
| --- | --- | --- | --- | --- | --- |
| | Total | Amosite | Crocidolite | Chrysotile | Mullite |
| 45 | 51.1 | 1.0 | 4.3 | 26.1 | 19.7 |
| 46 | 151.8 | 4.6 | 7.6 | 135.1 | 4.6 |
| 47 | 2563.5 | 0.0 | 29.5 | 2507.0 | 0.0 |
| 48 | 79.1 | 2.5 | 9.0 | 28.5 | 39.1 |
| 49 | 307.0 | 0.0 | 251.1 | 31.0 | 24.9 |
| 50 | 61.5 | 1.2 | 1.2 | 9.2 | 49.9 |
| 51 | 251.2 | 0.0 | 108.8 | 62.2 | 80.3 |
| 52 | 5.6 | 0.0 | 0.0 | 1.9 | 3.4 |
| 53 | 48.3 | 2.2 | 0.0 | 2.9 | 33.4 |
| 54 | 20.7 | 6.1 | 1.6 | 6.7 | 4.1 |
| 55 | 26.1 | 0.3 | 0.3 | 2.0 | 8.3 |
| 56 | 58.3 | 2.2 | 0.0 | 25.8 | 30.3 |
| 57 | 43.2 | 0.0 | 0.0 | 7.2 | 33.9 |

**Table 4. Total and differential fibre counts for Zielhuis Group III (neighbourhood exposure)**

| Case no. | Total | Amosite | Crocidolite | Chrysotile | Mullite |
| --- | --- | --- | --- | --- | --- |
| 58 | 779.1 | 7.9 | 7.9 | 692.5 | 70.8 |
| 59 | 36.1 | 0.4 | 3.2 | 23.8 | 8.7 |
| 60 | 16.9 | 0.0 | 5.1 | 8.2 | 3.6 |
| 61 | 23.2 | 0.0 | 9.1 | 6.0 | 8.1 |
| 62 | 14.6 | 0.0 | 0.8 | 8.9 | 4.9 |

Tremolite was found in only 8 cases, but in only 2 did the number of fibres exceed 1 million, and these were both in Group Ia. The amounts of chrysotile were similar for controls and mesotheliomas (Figure 1a). The amounts of amphiboles in the group of mesothelioma cases studied fell between those for controls and for known asbestos-exposed mesotheliomas (Figures 1b–d).

## Discussion

One of the major problems encountered by the epidemiologist when assessing the relationship of environmental exposure to a particular mineral is the unreliability of the occupational history. Because relevant events will have occurred many years ago, the patient or relative may have been unaware of exposure to a particular mineral or that the important mineral may have only accounted for a relatively small proportion of the total mineral exposure. Exposure may also have been brief. Assigning

**Table 5. Total and differential fibre counts for Zielhuis Group V (no exposure)**

| Case no. | Total | Amosite | Crocidolite | Chrysotile | Mullite |
|---|---|---|---|---|---|
| 64 | 188.3 | 3.7 | 101.7 | 49.8 | 33.2 |
| 65 | 90.5 | 0.9 | 0.9 | 75.9 | 12.8 |
| 66 | 17.7 | 0.2 | 0.2 | 3.1 | 14.2 |
| 67 | 78.3 | 0.0 | 0.0 | 72.6 | 5.7 |
| 68 | 66.0 | 0.0 | 2.9 | 2.9 | 60.2 |
| 69 | 23.9 | 0.5 | 0.5 | 2.2 | 20.7 |
| 70 | 21.5 | 0.5 | 0.5 | 6.6 | 14.0 |
| 71 | 8.0 | 1.0 | 0.3 | 0.5 | 5.1 |
| 72 | 8.0 | 0.6 | 0.2 | 1.1 | 5.1 |
| 73 | 11.6 | 0.2 | 1.9 | 0.7 | 5.8 |
| 74 | 16.8 | 0.3 | 0.6 | 5.9 | 8.4 |
| 75 | 23.3 | 0.2 | 0.7 | 9.6 | 12.6 |
| 76 | 4.7 | 0.0 | 0.0 | 1.3 | 2.7 |
| 77 | 0.0 | 0.0 | 0.0 | 0.0 | 0.0 |
| 78 | 7.8 | 0.4 | 0.1 | 2.2 | 4.9 |
| 79 | 94.1 | 0.0 | 1.8 | 76.5 | 15.0 |
| 80 | 101.7 | 0.0 | 1.0 | 47.8 | 37.6 |
| 81 | 56.2 | 4.6 | 0.7 | 4.6 | 42.2 |
| 82 | 9.9 | 0.0 | 0.0 | 4.5 | 4.7 |
| 83 | 2.1 | 0.1 | 0.0 | 0.6 | 0.0 |
| 84 | 62.0 | 1.2 | 0.6 | 43.0 | 16.0 |

individuals to the different Zielhuis occupational groups is difficult and sometimes arbitrary: the history is often not precise or the case can be assigned to more than one group. Lung fibre analysis provides additional information concerning mineral exposure, as illustrated in the data shown here. Our results suggest that it would be advantageous to reduce the number of occupational asbestos exposure groups to three: (1) direct exposure; (2) indirect exposure, which would include paraoccupational, neighbourhood and ambient exposures; and (3) no exposure.

Mesotheliomas caused by paraoccupational exposure have been described by Newhouse and Thompson (1965) and Greenberg and Lloyd Davies (1974), but there is little information on lung fibre burden in these cases. In the group-II cases from this study, 45–51 were from the Nottingham mesothelioma registry and 52–57 from the Cardiff mesothelioma registry. We consider it highly unlikely that cases 46, 47, 48, 49 and 51 were only exposed paraoccupationally, despite their histories, since the mineral fibre analysis of these cases is very similar to that obtained in a unique group of Nottingham gas-mask workers (Table 6) who were exposed to considerable quantities of crocidolite (Jones et al., 1980b). In group III, it seems probable from lung fibre analysis that case 58 had some direct exposure to asbestos. In group V, case 64 is similarly likely to have had direct industrial exposure. In a study of 36 mesotheliomas, Hirsch et al. (1982) considered that mineral fibre measurement indicated asbestos exposure in 9 cases where occupational questionnaires were inconclusive or negative.

**Fig. 1.** Quantities of different asbestos fibres in lung tissues of the present series of mesothelioma cases compared with controls and with a 1976 series of occupational mesothelioma cases. (From Jones *et al.*, 1980a).

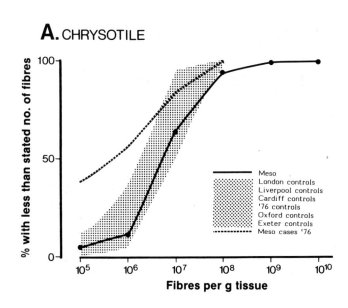

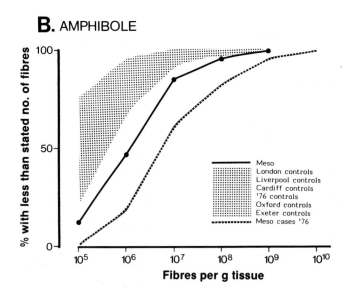

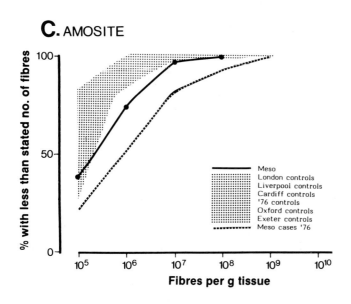

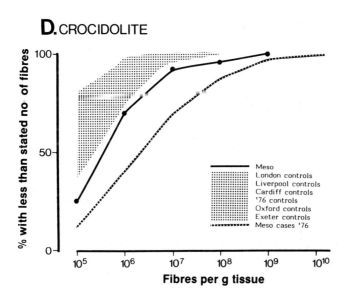

**Table 6. Total and differential fibre count for Nottingham gas-mask workers**[a]

| Total | Amosite | Crocidolite | Chrysotile | Mullite |
|---|---|---|---|---|
| 99.6 | - | 68.6 | 31.0 | - |
| 52.0 | - | 29.0 | 23.0 | - |
| 143.0 | - | 128.0 | 15.0 | - |
| 238.1 | - | 235.9 | 2.2 | - |
| 204.0 | - | 195.0 | 9.0 | - |
| 993.0 | - | 977.0 | 16.0 | - |
| 46.0 | - | 26.0 | 20.0 | - |
| 1494.0 | - | 498.0 | 996.0 | - |
| 110.0 | - | 80.0 | 30.0 | - |
| 297.0 | - | 295.0 | 2.0 | - |
| 219.26 | - | 181.23 | 38.03 | - |
| 742.70 | - | 65.3 | 677.40 | - |
| 94.0 | - | 93.0 | 1.0 | - |
| 210.0 | - | 50.0 | 160.0 | - |
| 38.0 | - | 29.0 | 9.0 | - |
| 1140.0 | - | 1120.0 | 20.0 | - |
| 17.62 | - | 1.12 | 16.5 | - |
| 34.39 | - | 33.16 | 1.23 | - |
| 469.0 | - | 455.0 | 14.0 | - |
| 51.0 | - | 50.0 | 1.0 | - |
| 248.2 | - | 245.5 | 2.7 | - |
| 7.35 | - | 2.30 | 5.05 | - |

[a]From Jones et al., 1980b.

We did not find any significant difference in the histological typing of the mesotheliomas as between males and females or as between the different occupational groups. Hirsch et al. (1982) found only the epithelial types of mesothelioma in the non-asbestos-related group.

The mean age at death was similar in the various occupational groups, except for group III, where it was lower; 3 of these cases were thought to have had their neighbourhood exposure during childhood.

Previous studies have indicated that amphiboles rather than chrysotile cause mesothelioma (Churg et al., 1984; Jones et al., 1980; Wagner et al., 1982). In the present study, the mesothelioma cases show amphibole levels greater than those of controls but lower than those in a series of previous mesotheliomas with known asbestos exposure (Jones et al., 1980). This raises the possibility that this is a mixed group of cases, some related to and some unrelated to asbestos. Although open to argument, if we take the cut-off value for amphibole fibres as 1 million fibres per g of dried lung tissue as significant, 39 of these cases would be considered to be unrelated to asbestos. However, previous studies have shown that the dimensions of the amphibole fibres appear to be critical in the causation of disease, and we are carrying out further studies to determine the sizes of these fibres.

The role of chrysotile in the causation of mesothelioma is disputed. There is a considerable amount of evidence to show that mesothelioma occurring after exposure to chrysotile ore or its products is due to contamination by amphiboles (Churg et al., 1984; Pooley & Mitha, 1986; Wagner et al., 1982). In the present series, the amounts of chrysotile appeared similar for both mesotheliomas and controls, and only 3 cases showed chrysotile counts greater than those for controls without raised amphibole counts. It is possible that chrysotile might potentiate the effects of amphiboles, but we believe that it has either no potential (or a very low one) for mesothelioma induction on its own.

These preliminary findings illustrate the value of electron microscopic mineral fibre analysis in detecting exposure to asbestos in cases where careful occupational histories are negative.

## References

Borow, M., Conston, A., Livornese, L. & Schalet, N. (1973) Mesothelioma following exposure to asbestos: a review of 72 cases. *Chest, 64*, 641-646

Brenner, J., Sordillo, P.P., Magill, G.B. & Golbet, R.B. (1982) Malignant mesothelioma of the pleura: a review of 123 patients. *Cancer, 49*, 2431-2435

Churg, A., Wiggs, B., Depaoli, L., Kampe, B. & Stevens, B. (1984) Lung asbestos content in chrysotile workers with mesothelioma. *Am. Rev. Respir. Dis., 130*, 1042-1045

Cochrane, J.C. & Webster, I. (1978) Mesothelioma in relation to asbestos exposure: a review of 70 serial cases. *South Afr. Med. J., 54*, 279-281

Greenberg, M. & Lloyd Davies, T.A. (1974) Mesothelioma register 1967-1968. *Br. J. Ind. Med., 31*, 91-104

Hirsch, A., Brochard, P., DeCremoux, H., Erkan, L., Sébastien, P. & Bignon, J. (1982) Features of asbestos exposed and unexposed mesothelioma. *Am. J. Ind. Med., 3*, 413-422

Jones, J.S.P., Pooley, F.D., Clark, N.J., Owen, W.G., Roberts, G.H., Smith, P.G., Wagner, J.C., Berry, G. & Pollock, D.J. (1980a) The pathology and mineral content of lungs in cases of mesothelioma in the United Kingdom in 1976. In: Wagner, J.C., ed., *Biological Effects of Mineral Fibres*, Vol. 1 (*IARC Scientific Publications No. 30*), Lyon, International Agency for Research on Cancer, pp. 187-200

Jones, J.S.P., Pooley, F.D., Sawle, G.W., Madely, R.J., Smith, P.G., Berry, G., Wignall, B.K. & Aggarwal, A. (1980b) The consequences of exposure to asbestos dust in a war-time gas mask factory. In: Wagner, J.C., ed., *Biological Effects of Mineral Fibres*, Vol. 2 (*IARC Scientific Publications No. 30*), Lyon, International Agency for Research on Cancer, pp. 637-653

Jones, J.S.P., Lund, C. & Planteydt, H.T. (1985) *Colour Atlas of Mesothelioma*, Lancaster, MTP Press

Maurer, R. & Egloff, B. (1975) Malignant peritoneal mesothelioma after cholangiography with thorotrast. *Cancer, 36*, 1381-1385

Newhouse, M.L. & Thompson, H. (1965) Mesothelioma of pleura and peritoneum following exposure to asbestos in the London area. *Br. J. Ind. Med., 22*, 261-269

Oels, H.C., Harrison, E.G., Carr, D.T. & Bernatz, P.E. (1971) Diffuse malignant mesothelioma of the pleura: a review of 37 cases. *Chest, 60*, 564-570

Peterson, J.T., Greenberg, S.D. & Buffler, P.A. (1984) Non-asbestos related malignant mesothelioma. *Cancer, 54*, 951-960

Planteydt, H.T. (1980) Experiences with observer variation in mesothelioma panels. In: Wagner, J.C., ed., *Biological Effects of Mineral Fibres*, Vol. 1 (*IARC Scientific Publications No. 30*), Lyon, International Agency for Research on Cancer, pp. 211-216

Pooley, F.D. & Clarke, N.J. (1979) Quantitative assessment of inorganic fibrous particles in dust samples with an analytic transmission electron microscope. *Ann. Occup. Hyg.*, *22*, 253-271

Pooley, F.D. & Mitha, R. (1986) Fibre types, concentrations and characteristics found in lung tissues of chrysotile exposed cases and controls. *Accompl. Oncol.*, *1*, 1-11

Roggli, V.L., McGavran, M.H., Subach, J., Sybers, H.D. & Greenberg, S.D. (1982) Pulmonary asbestos body counts and electron probe analysis of asbestos body cores in patients with mesothelioma: a study of 25 cases. *Cancer*, *50*, 2423-2432

Spirtas, R., Keehn, R.J., Beebe, G.W., Wagner, J.C., Hochholzer, L., Davies, J.N.P., Ortega, L.G. & Sherwin, R.P. (1986) Results of a pathology review of recent US mesothelioma cases. *Accompl. Oncol.*, *1*, 144-152

Tagnon, I., Blot, W.J., Stroube, R.B., Day, N.E., Morris, L., Peace, B.B. & Fraumeni, J.F. (1980) Mesothelioma associated with the shipbuilding industry in coastal Virginia. *Cancer Res.*, *40*, 3875-3879

Wagner, J.C. & Pooley, F.D. (1986) Mineral fibres and mesothelioma. *Thorax*, *41*, 161-166

Wagner, J.C., Sleggs, C.A. & Marchand, P. (1960) Diffuse pleural mesothelioma and asbestos exposure in the North Western Cape Province. *Br. J. Ind. Med.*, *17*, 260-271

Wagner, J.C., Berry, G. & Pooley, F.D. (1982) Mesotheliomas and asbestos type in asbestos textile workers: a study of lung contents. *Br. Med. J.*, *285*, 603-606

Whitwell, F. & Rawcliffe, R.M. (1971) Diffuse malignant pleural mesothelioma and asbestos exposure. *Thorax*, *26*, 6-22

Zielhuis, R.L. (1977) *Public Health Risks of Exposure to Asbestos*. Oxford, Pergamon Press

# INCIDENCE OF FERRUGINOUS BODIES IN THE LUNGS DURING A 45-YEAR PERIOD AND MINERALOGICAL ANALYSIS OF THE CORE FIBRES AND UNCOATED FIBRES

S. Shishido, K. Iwai & K. Tukagoshi

*The Research Institute of Tuberculosis, Japan Anti-Tuberculosis Association, Tokyo, Japan*

*Summary.* In order to determine the level of asbestos pollution in the lungs of members of the general population in and around Tokyo, the incidence of ferruginous bodies in autopsied or resected lungs during 5 periods over the 45 years from 1937 to 1981 was studied under a light microscope. Core fibres, after removal of their ferruginous coatings with oxalic acid, and uncoated fibres were analysed using a scanning electron microscope equipped with a Kevex energy-dispersive X-ray spectrometer.

The incidence of ferruginous bodies in 5 g (wet) of digested lung tissue was shown to be 10% in period I (1937-1941), 18% in period II (1947-1951), 70% in period III (1958-1963), 74.4% in period IV (1970-1973) and 81.0% in period V (1980-1981).

The major types of core fibres of ferruginous bodies were found to be asbestos, including amosite, crocidolite, chrysotile and the tremolite-actinolite series, but a small number of fibres of materials other than asbestos were also detected.

In contrast, a large number of short fibres less than 5 $\mu$m in length in 1 g of wet lung tissue were classified as belonging to the Mg+Si group (the ratio of Mg to Si components being 30% or over) and presumed to be chrysotile.

Thus an annual increase in asbestos deposition in the lungs of people living in and around Tokyo has been demonstrated and fine chrysotile fibres less than 5 $\mu$m in length seem to be the main type of deposited fibres.

## Introduction

Asbestos is not produced in Japan but is imported from Canada, the United States of America, the Soviet Union, South Africa, etc., and a vast number of manufactured asbestos products are widely used. The amount of asbestos imported annually has been increasing year by year, and there is considered to be a high probability that asbestos fibres originating from the many different asbestos industries and asbestos products will be inhaled by the general population.

In order to determine the level of asbestos deposition in the lungs of people living in and around Tokyo, we studied the incidence of ferruginous bodies, using the digestion method, in lungs which had been autopsied or operated on, over 5 periods between 1937 and 1981. The core fibres removed by oxalic acid treatment as well as the uncoated fibres found in the lungs of the general population were examined using a scanning electron microscope equipped with a Kevex energy-dispersive X-ray spectrometer.

## *Materials and methods*

### Detection of ferruginous bodies

*Materials*

Five groups of specimens from lungs which had been autopsied or operated on, from persons ranging in age from 20 to 59 years, were examined. Almost all of them resided in and around Tokyo. In the first group, the lung specimens were obtained during the period 1937-1941, and 40 cases of tuberculosis (32 male and 8 female) were included. The second group (1947-1951) consisted of 50 cases of tuberculosis (36 male and 14 female). The third group (1958-1961) consisted of 100 cases of tuberculosis (67 male and 33 female). The fourth group (1970-1973) consisted of 95 cases of tuberculosis (59 male and 36 female). The fifth group (1980-1981) consisted of 105 cases other than primary lung cancer and gastrointestinal cancer (65 male and 40 female).

*Methods*

The technique for extracting ferruginous bodies from 5-g wet weight lung specimens followed Smith and Naylor's digestion method (Smith & Naylor, 1972) with sodium hypochlorite, chloroform and ethanol. The ferruginous bodies remaining on a Millipore filter with a pore size of 5 $\mu$m were counted under a light microscope.

### Identification of the core fibres of ferruginous bodies

*Materials*

A total of 35 subjects chosen at random from positive cases showing ferruginous bodies in the 5-g wet weight lung specimens were studied for identification of core fibres. The 35 subjects were made up of 8 with 1-10 ferruginous bodies in the 5 g of lung tissue, 10 with 11-20, 5 with 21-40, 5 with 51-100, 3 with 101-200 and 5 with over 200.

*Methods*

We used Churg and Warnock's method (Churg & Warnock, 1980), as modified by us, for extracting ferruginous bodies in the lung. In this method, a lung sample of 1-2 g wet weight was cut into small pieces and dissolved overnight in 30 ml of 10% sodium hypochlorite. To this sample, 15 ml of 30% hydrogen peroxide was carefully added and it was then placed in an incubator at a temperature of 60°C for 4 h to remove foreign matter and carbonaceous material. The digest was filtered through a Millipore filter with a pore size of 3 $\mu$m.

For the determination of the core fibres of the ferruginous bodies, Dodson's oxalic acid treatment method (Dodson *et al.*, 1983) was used, as follows. The portion containing a ferruginous body remaining on a Millipore filter was marked with a circle under a light microscope. The filter was then fixed in a filter holder and immersed in 40 ml of 8% oxalic acid in a water bath and maintained at 75°C for 4 h, after which it was removed and the portion previously marked with a circle was cut out and placed on Formvar-carbon coated, nickel electron microscope grids and treated with acetone. The core fibres thus obtained by removing the coating material from a ferruginous body were observed and analysed using a scanning electron microscope equipped with a Kevex energy-dispersive X-ray spectrometer.

### Enumeration of uncoated fibres

*Materials*

A group of 25 was chosen at random from the specimens used for the study of ferruginous bodies, made up as follows: 6 without detectable ferruginous bodies in 5 g of lung tissue, 6 with 1-10 ferruginous bodies, 6 with 11-20, 3 with 51-100 and 4 with over 100.

*Methods*

The lung tissue (0.5 g in wet weight) was cut up and dissolved with sodium hypochlorite and hydrogen peroxide as described above. The digest was filtered through Millipore filters with pore sizes of 0.45 (10 subjects) and 1.2 $\mu$m (15 subjects). After acetone treatment, the fibrous materials on 10 meshes out of 300 of this grid were analysed by X-ray spectrometer, and the number of fibres presumed to be asbestos were counted.

## *Results*

### Detection of ferruginous bodies

The secular trends in the prevalence of ferruginous bodies in lung tissue of the general population by sex and by age during the 45 years from 1937 to 1981 are shown in Tables 1 and 2. The incidence during and after the period 1958-1961 showed a significant increase as compared with the previous two periods. A significant difference by sex was noted in the period from 1947 to 1951 ($p<0.05$), while no such differences were found in the other periods. The incidence by age showed a greater tendency to increase in older age-groups than in younger ones during 1980-1981, but no such tendencies were observed in the other periods.

Table 3 shows the number of ferruginous bodies found in the 5 g of lung tissue in each study period. The majority of the positive cases showed from 1 to 10 bodies and none had over 21 in the periods before 1951. The cases which had more than 21 bodies in the lung tissue increased in number after 1958, when there were 5 with over 200 bodies.

Figure 1 shows the amount of asbestos imported into Japan annually and the trends in the prevalence of ferruginous bodies. The marked increase in the prevalence of ferruginous bodies in the lungs of the general population clearly reflects the increase of asbestos imports into the country.

**Table 1. Incidence of ferruginous bodies in 5-g lung tissue specimens by sex in 5 different periods**

| Period | Incidence | | | | | |
|---|---|---|---|---|---|---|
| | Male | | Female | | Total | |
| | No. | % | No. | % | No. | % |
| 1937-1941 | 3/32 | 9.4 | 1/8 | 12.5 | 4/40 | 10.0 |
| 1947-1951 | 9/36 | 25.0 | 0/14 | 0 | 9/50 | 18.0 |
| 1958-1961 | 52/67 | 77.6 | 18/33 | 54.5 | 70/100 | 70.0 |
| 1970-1973 | 46/59 | 78.0 | 25/36 | 69.4 | 71/95 | 74.9 |
| 1980-1981 | 58/65 | 89.2 | 27/40 | 67.5 | 85/105 | 81.0 |

**Table 2. Incidence of ferruginous bodies in 5-g lung tissue specimens by age in 5 different periods**

| Period | Incidence by age in years | | | | | | | | | |
|---|---|---|---|---|---|---|---|---|---|---|
| | 20-29 | | 30-39 | | 40-49 | | 50-59 | | Total | |
| | No. | % | No. | % | No. | % | No. | % | No. | % |
| 1937-1941 | 2/13 | 15.4 | 1/16 | 6.3 | 0/5 | 0 | 1/6 | 16.6 | 4/40 | 10.0 |
| 1947-1951 | 3/14 | 21.4 | 2/17 | 11.8 | 3/14 | 21.4 | 1/15 | 20.0 | 9/50 | 18.0 |
| 1958-1961 | 18/27 | 66.7 | 23/30 | 76.7 | 20/30 | 66.7 | 9/13 | 69.2 | 70/100 | 70.0 |
| 1970-1973 | 16/23 | 69.6 | 22/28 | 78.6 | 22/26 | 84.6 | 11/18 | 61.1 | 71/95 | 74.9 |
| 1980-1981 | 12/18 | 66.7 | 19/27 | 70.4 | 25/29 | 86.2 | 29/31 | 93.6 | 85/105 | 81.0 |

## Identification of the core fibres of ferruginous bodies

*Effects of digestion, acetone and oxalic acid treatment on the chemical analysis of asbestos fibres*

Three kinds of UICC standard asbestos fibres, chrysotile, amosite and crocidolite, were examined for their chemical constituents before and after treatment with 10% sodium hypochlorite, 30% hydrogen peroxide, acetone and 8% oxalic acid. No change in their chemical constituents was detected in any of the types of asbestos fibres tested after treatment with 10% sodium hypochlorite, 30% hydrogen peroxide and acetone. In the oxalic acid treatment, no leaching was noted in amosite and crocidolite, but a decrease in the magnesium component was demonstrated in chrysotile fibres in various degrees.

Table 3. Number of ferruginous bodies in 5-g lung tissue specimens in 5 different periods

| No. of ferruginous bodies | Number of cases | | | | |
|---|---|---|---|---|---|
| | 1937-1941 | 1947-1951 | 1958-1961 | 1970-1973 | 1980-1981 |
| 0 | 36 | 41 | 30 | 24 | 20 |
| 1-10 | 3 | 8 | 47 | 54 | 54 |
| 11-20 | 1 | 1 | 5 | 3 | 17 |
| 21-50 | 0 | 0 | 12 | 8 | 7 |
| 51-100 | 0 | 0 | 4 | 4 | 4 |
| 101-200 | 0 | 0 | 1 | 0 | 1 |
| Over 200 | 0 | 0 | 1 | 2 | 2 |
| Total | 40 | 50 | 100 | 95 | 105 |

Fig. 1. Incidence of ferruginous bodies and annual amount of asbestos imports

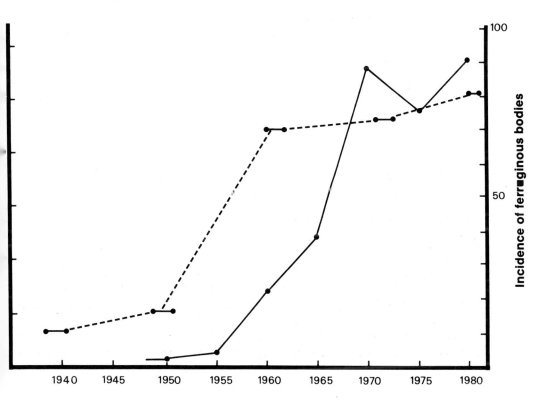

Asbestos imports———; incidence ---------

*Identification of core fibres*

Figure 2 shows the morphology of a ferruginous body and its analysis; many elements are present with iron predominant. It was therefore impossible to identify the core fibre of this body. Figures 3 and 4 show the EDX spectra of fibres of chrysotile and of the amosite-crocidolite series respectively, obtained after removal of the coating material from ferruginous bodies by oxalic acid treatment. Apart from these commercial asbestos fibres, tremolite-actinolite series asbestos, cores consisting of magnesium, silicon and calcium, cores of silicon alone, cores containing conspicuous amounts of aluminium and cores containing a complex of the constituents, phosphorus, sodium, chlorine and potassium, were observed in various numbers. The numbers and types of each type of asbestos or non-asbestos core fibres are shown in Table 4 according to the number of ferruginous bodies in the 5-g wet weight lung tissue. Ferruginous bodies which had core fibres containing silicon alone tended to be found in the group having chrysotile cores in ferruginous bodies. The majority of the core fibres were asbestos, but non-asbestos cores were also found.

**Table 4. Identification of core fibres after removal of coating material from ferruginous bodies**

| No. of ferruginous bodies/5 g wet lung | No. of subjects | Chrysotile | Amosite or crocidolite | Tremolite-actinolite series | Mg+Si+Ca+ group | Si component only | Including Al component | Nothing detected | Others |
|---|---|---|---|---|---|---|---|---|---|
| 1-10 | 8 | 4 | 3 | 1 | 0 | 1 | 0 | 0 | 0 |
| 11-20 | 9 | 9 | 11 | 1 | 6 | 4 | 4 | 0 | 0 |
| 21-50 | 5 | 9 | 13 | 2 | 7 | 4 | 2 | 0 | 1 |
| 51-100 | 5 | 12 | 11 | 7 | 4 | 9 | 4 | 1 | 3 |
| 101-200 | 3 | 10 | 11 | 0 | 6 | 3 | 2 | 1 | 2 |
| Over 200 | 5 | 11 | 37 | 3 | 5 | 4 | 4 | 1 | 2 |
| Total no. of fibres | | 55 | 86 | 14 | 28 | 25 | 16 | 3 | 8 |
| No. of subjects | 35 | 19 | 24 | 11 | 12 | 12 | 11 | 3 | 5 |

*Enumeration of uncoated fibres*

Almost all particles deposited in lungs were less than 5 $\mu$m in length, although a small number of fibres over 5 $\mu$m length, such as chrysotile and the amosite-crocidolite series, were found. Of a large number of these particles of various shapes, fibrous particles less than 5 $\mu$m in length containing 2 components, namely Mg and Si, with a content of 30% or more of Mg were classified as the Mg+Si group and presumed to be chrysotile. The number of Mg+Si group fibres per 1 g of wet lung tissue was determined under the electron microscope and the results were related to the number of ferruginous bodies in the 5 g wet weight lung tissue found under the light

**Fig. 2. Ferruginous body of club-like shape and its EDX spectrum**

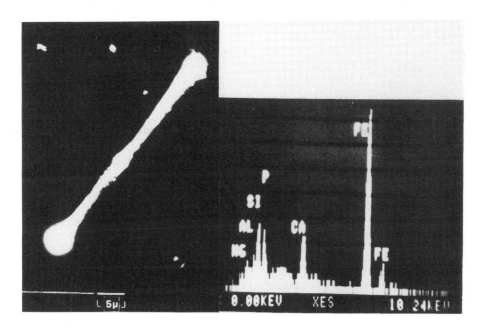

**Fig. 3. A chrysotile fibre after removal of ferruginous coating with oxalic acid and its EDX spectrum**

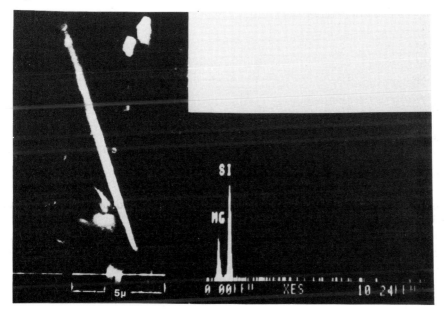

**Fig. 4. Amosite or crocidolite fibre after removal of the ferruginous coating oxalic acid and its EDX spectrum**

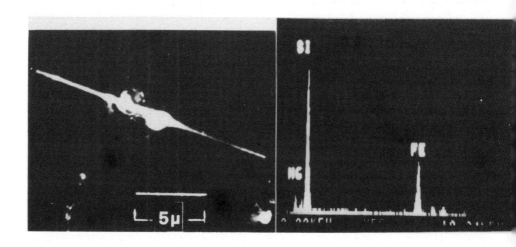

microscope (Table 5). A large number of Mg+Si group fibres were found, and there was a conspicuous increase among the subjects containing larger numbers of ferruginous bodies.

**Table 5. Number of fibres of Mg + Si group less than 5 μm in length per 1 g wet lung by number of ferruginous bodies in 5 g wet lung**

| No. of ferruginous bodies/5 g wet lung[a] | Average number of fibres of Mg + Si group less than 5 μm per 1 g wet lung[b] |
|---|---|
| 0 ($N = 6$) | $41 \times 10^3$ <br> ($8 \times 10^3 - 77 \times 10^3$) |
| 1-20 ($N = 12$) | $46 \times 10^3$ <br> ($0 - 172 \times 10^3$) |
| 51-100 ($N = 3$) | $117 \times 10^3$ <br> ($94 \times 10^{-3} - 137 \times 10^3$) |
| Over 100 ($N = 4$) | $217 \times 10^3$ <br> ($137 \times 10^3 - 361 \times 10^3$) |

[a]Number of cases ($N$) in parentheses.
[b]Range in parentheses.

## Discussion

Thomson et al. (1963), using the lung smear method, found 132 cases of asbestos bodies among 500 consecutive autopsies from the general population in Cape Town. As a result of this report, many subsequent studies were carried out in various parts of the world, using a number of different methods, to investigate asbestos deposition in the lungs of members of the general population. By the digestion method, Bignon et al. (1970) detected ferruginous bodies at a rate of 100% in France, and Breedin and Buss (1976) found them at a rate of more than 90% in North Carolina. These rates are higher than our results and asbestos pollution is believed to be higher in Europe and America. On the other hand, Bhagavan and Koss (1976) showed that the incidence of asbestos bodies in necropsy lung tissue 5 g in wet weight in New York was 40.9 in 1940, 61.7 in the 1950s and 91.1% in the 1970s. In Japan, secular trends in the prevalence and concentration of ferruginous bodies during the 45 years from 1937 to 1981 also showed an increase and this closely reflected the increase in the amount of asbestos imports into Japan. The results of these two studies lead us to suspect that air pollution by asbestos fibres has been increasing during the period covered by them.

It remains uncertain whether the cores of these bodies are truly of asbestos or of some other kinds of fibres. Gross et al. (1967, 1968) demonstrated experimentally that bodies similar to asbestos bodies in shape could be formed by aluminium silicate, silicon carbide whiskers and glass fibres in the lungs of hamsters. They proposed that these bodies should be called ferruginous bodies, defined as ferritin-like protein-coated bodies, unless the cores were chemically identified. In our study a large number of asbestos cores were found after the coating material had been removed from ferruginous bodies. On the other hand, a considerable number of cores contained the 3 components, magnesium, silicon and calcium, or the silicon component alone. However, we suggest that the majority of the cores containing the above-mentioned 3 elements were tremolite-actinolite series and that the cores containing silicon alone were chrysotile, because the EDX pattern of magnesium, silicon and calcium is similar to that of the tremolite-actinolite series. Moreover, the cores with the silicon component alone had a tendency to be found among the subjects having chrysotile cores; it was therefore highly probable that these cores were chrysotile, the magnesium component of which had been leached out by oxalic acid treatment. The level of asbestos deposition high enough to form bodies may therefore reasonably be estimated from the number of ferruginous bodies in the lung.

The use of oxalic acid for dissolving the ferruginous coat proved to be an excellent method for detecting amphibole asbestos fibres because of their high resistance to acid. On the other hand, it should be pointed out that the magnesium of chrysotile was easily leached out by oxalic acid treatment. If, however, the acidity, temperature and effective treatment time are controlled to avoid leaching out a significant amount of magnesium, this method of detecting chrysotile is also useful.

Using a combination of morphology with a scanning transmission electron microscope and analysis with an EDX spectrometer and electron diffraction, Churg and Warnock (1980) found a very large number of uncoated asbestos fibres in the lungs of 21 urban dwellers who had fewer than 100 ferruginous bodies in 1 g of lung tissue; 80% of these fibres were chrysotile and 90% were less than 5 μm in length. In our study, a large number of fine uncoated fibres less than 5 μm in length having the two components Mg and Si and an Mg content of over 30% were found even in lungs which had no detectable ferruginous bodies in 5 g wet weight of lung tissue. Though talc or some similar material may be present in this group, the majority of these short fibres were thought to be chrysotile and considerable chrysotile deposition in the lungs of the general population in and around Tokyo was therefore suspected. Further morphological studies by transmission electron microscope and analysis by electron diffraction are needed to confirm our results.

## References

Bhagavan, B.S. & Koss, L.G. (1976) Secular trends in prevalence and concentration of pulmonary asbestos bodies — 1940-1972. *Arch. Pathol. Lab. Med.*, *100*, 539-541

Bignon, J., Goni, J., Bonnaud, G., Jaurand, M.C., Dufour, G. & Pinchon, M.C. (1970) Incidence of pulmonary ferruginous bodies in France. *Environ. Res.*, *3*, 430-442

Breedin, P.H. & Buss, D.H. (1976) Ferruginous (asbestos) bodies in the lungs of rural dwellers, urban dwellers, and patients with pulmonary neoplasms. *South. Med. J.*, *69*, 401-404

Churg, A. & Warnock, M.L. (1980) Asbestos fibers in the general population. *Am. Rev. Respir. Dis.*, *122*, 669-678

Dodson, R.F,, Williams, M.G., Jr & Hurst, G.A. (1983) Method for removing the ferruginous coating from asbestos bodies. *J. Toxicol. Environ. Health*, *11*, 959-966

Gross, P., Cralley, L.J. & de Tréville, R.T.P. (1969) 'Asbestos' bodies: their nonspecificity. *Am. Ind. Hyg. Assoc. J.*, *28*, 541-542

Gross, P., de Tréville, R.T.P., Cralley, L.J. & Davis, J.M.G. (1968) Pulmonary ferruginous bodies. *Arch. Pathol.*, *85*, 539-546

Smith, M.J. & Naylor, B. (1972) A method for extracting ferruginous bodies from sputum and pulmonary tissue. *Am. J. Clin. Pathol.*, *58*, 250-254

Thomson, J.G., Kaschula, R.O.C. & MacDonald, R.R. (1963) Asbestos as a modern urban hazard. *South Afr. Med. J.*, *19*, 77-81

# AIRBORNE MINERAL FIBRE LEVELS IN THE NON-OCCUPATIONAL ENVIRONMENT

W.J. Nicholson

*Division of Environmental and Occupational Medicine,
Mount Sinai School of Medicine of the City University of New York,
New York, NY, USA*

*Summary.* Numerous sources of asbestos exist that may contribute to non-occupational exposures, among the important ones being building surfacing materials that have been damaged or allowed to deteriorate. Even more important is the potential exposure from improperly controlled maintenance activities in buildings. Evidence exists suggesting that vehicle braking makes a significant contribution to ambient asbestos levels, but more data are required to establish its extent. Many asbestos materials are present in homes, and fibres may be released during home renovations or repairs. Little information exists on the levels of other mineral fibres in the non-occupational environment or on the relative contributions from potential sources.

## Introduction

A wide variety of potential sources of asbestos and other fibres exist that may contribute to non-occupational exposures (see Table 1). Virtually all products containing asbestos currently manufactured incorporate the fibre in a matrix from which release is difficult. However, the incorporation of asbestos in the past in highly friable material provides current sources of exposure. The most important of these materials are located in commercial and public buildings and, to a lesser extent, in private dwellings. The greatest potential for the release of fibres from these materials occurs during building maintenance and renovation. To the extent that such activities are identified and procedures established to prevent fibre release, non-occupational exposures can be minimized.

**Table 1. Potential sources of asbestos in the non-occupational environment**

| | |
|---|---|
| *Industrial activity* | Mine, mill and factory operations; building renovation and demolition |
| *Building operations* | Inadvertent damage or deterioration of asbestos materials, surfacing materials, thermal insulation and cement products; maintenance activities |
| *Vehicle braking* | Automobiles; subway cars |
| *Home materials* | Heating systems; air supply systems; appliances; wallboard; filler; textured paint |

## Measurement of environmental asbestos concentrations

**Analytical techniques**

Asbestos air concentrations encountered in buildings and in the general environment are virtually always below the limit of sensitivity of optical microscopy of 0.001–0.01 fibres >5 $\mu$m in length per ml. The ineffectiveness of optical microscopy below this level is the consequence of the presence of non-asbestos inorganic or organic fibres that cannot be distinguished from the asbestos minerals with phase-contrast or polarizing-light microscopy. At concentrations below 0.01 fibres per ml (f/ml) the fibres seen may largely consist of such non-asbestos fibres. An exception to this is the analysis of amosite fibres, where their size, density and configuration allow most non-asbestos fibres to be rejected from consideration, with a concomitant reduction in the limit of sensitivity. However, if confounding fibres, such as gypsum, are present, special precautions must be taken. Additionally, the practical resolution of the optical microscope precludes observation of fibres <0.3–0.4 $\mu$m in diameter.

Recommended analytical methods specify the use of transmission electron microscopy to enumerate and size asbestos fibres (Burdett, 1984; Samudra et al., 1977; Yamate et al., 1984). Samples for such analysis are usually collected either on a Nuclepore® (polycarbonate) filter with a pore size of 0.4 $\mu$m or less or on a Millipore® (cellulose ester) filter with a pore size of 0.8 $\mu$m or less. Samples collected on Nuclepore filters are prepared for direct analysis by carbon coating the filter to entrap the collected particles. A segment of the coated filter is then mounted on an electron microscope grid, which is placed on a filter paper saturated with chloroform, the vapours of which dissolve the filter material. Special precautions must be exercised during and following sample collection to ensure that fibres are not lost from the filter before reaching the analytical laboratory. The fibres of Millipore filters can also be directly transferred to electron microscope grids. In this case, an initial treatment with acetone vapour serves to partially collapse the filter. The filter is then partially etched in a low-temperature asher to expose fibres within the filter matrix and coated with carbon to entrap the fibres as above (Burdett & Rood, 1983). After coating, the remainder of the filter is dissolved with acetone. Without the etching step, as much as 90% of the fibres on an 0.8-$\mu$m filter can be lost in the dissolution process. The loss is greater for the thinner fibres and smaller losses occur with smaller pore size filters.

An alternative to the above direct analyses is to prepare samples by ashing a portion of the filter in a low-temperature oxygen furnace. This removes the membrane filter material and all organic material collected in the sample. The residue is recovered in a liquid phase, dispersed by ultrasonification, and filtered on a Nuclepore filter. Here, one can select the amount of material to be filtered to achieve the desired material loading. The refiltered material is coated with carbon and mounted on a grid as above. Earlier electron microscopic analysis utilized a rub-out technique in which the ashed residue was dispersed by grinding in a nitrocellulose film on a microscope slide and a portion of the film was then mounted on an electron microscope grid for scanning (Nicholson & Pundsack, 1973). A disadvantage of the indirect method is that fibres are broken apart by the dispersal techniques with a concomitant increase in the number of smaller fibres. However, light ultrasonic treatment does not break apart most fibres $>5$ µm in length (Sébastien, 1985). In contrast, physical grinding or prolonged ultrasonic treatment is likely to significantly alter the number of both short and long fibres. This results, however, in a more uniform distribution of the asbestos material, allowing greater precision in the measured concentrations, but all information on size distribution is lost.

The grids prepared by either the direct or indirect methods can be scanned at both high and low magnification to identify all asbestos fibres within the scanned area. Chrysotile asbestos is identified on the basis of its morphology in the electron microscope and amphiboles are identified by their selected area electron-diffraction patterns, supplemented by energy-dispersive X-ray analysis. Both scanning (SEM) and transmission electron microscopic (TEM) techniques can be utilized. However, SEM analysis usually identifies only fibres thicker than 0.2 µm, because of contrast limitations, while TEM can visualize fibres of all sizes. Both SEM and TEM analysis can identify most fibres, particularly the larger ones, but the X-ray or electron-diffraction patterns of some fibres may not be completely clear. Fibre concentrations expressed as fibres per unit volume are calculated based on sample volume and filter area counted. Mass concentrations are reported using fibre volume and density relationships. Mass concentrations can be converted into equivalent *optical* fibre concentrations using data from the parallel analysis of samples by phase-contrast microscopy and TEM. The approximate relationship is that 3 ng/m$^3$ is equivalent to 0.1 fibres $>5$ µm (optical) per ml. The uncertainty in this relationship, resulting from differing fibre size distributions, is about a factor of 5 (US Environmental Protection Agency, 1986a).

A disadvantage of the direct-transfer processes is that filter-loading requirements are crucial. Too little material requires excessive scanning time or results in low fibre counts of high statistical uncertainty; too much material results in obscuration of fibres by other debris and support-film breakage during electron microscopic analysis. The obscuration effect can be substantial for both short and long fibres. Results of parallel analyses by Chatfield (1985a) using both direct and indirect preparations indicated that in 8 samples the mass was 12 times greater (geometric mean), the number of fibres of all sizes was 6 times greater (geometric mean) and the number of fibres greater than 5 µm in length was 2.5 times greater (arithmetic mean) when indirect techniques were utilized. The increases in mass and in the numbers of both long and short fibres cannot be explained by fibre break-up, but are probably due to obscuration effects or losses in the direct preparations.

**Analytical considerations**

Measurements of ambient air contaminants should satisfy two criteria for sensitivity. They should be able to measure either: (1) typical ambient air concentrations; or (2) levels commensurate with lifetime risks of the order of $10^{-6}$, whichever is greater. The latter criterion follows from the levels at which US environmental regulatory agencies typically act to control exposures when the potential number of cancer deaths in the US population per year exceeds 1. Figure 1, adapted from Travis et al. (1987) shows the history of US environmental regulatory action. Of 23 regulatory decisions by US agencies (Environment Protection Agency, Food and Drug Administration, Consumer Product Safety Comission) where the population risk per year exceeded 1, exposure reduction was mandated in 15. An additional 5 positive actions were taken for risks greater than those depicted in Figure 1. The 8 cases considered for which regulation has not yet been mandated were for exposures to formaldehyde, aflatoxin, saccharin and polycyclic organic matter. Risks with fewer than 1 potential cancer death per year in the US population or where the individual risk was less than $10^{-6}$ were typically not regulated. In the case of asbestos, ambient background concentrations are of the order of 1 ng/m$^3$ or 0.1 fibres >5 µm per litre. Four estimates of lifetime risk from such exposures are shown in Table 2. As these are all well above an individual risk level of $10^{-6}$, most even above $10^{-5}$, any widespread source of asbestos is of regulatory concern. Thus, analytical procedures should be

**Fig. 1. Regulatory action as a function of individual and population risk for lifetime exposure to various environmental agents; ● regulatory action taken; o regulatory action not taken or deferred**

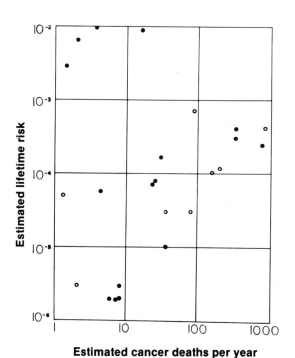

Data from Travis et al. (1987)

**Table 2. Lifetime risks of death per 100 000 from mesothelioma and lung cancer from a continuous lifetime exposure to 0.1 f/l**

| Source | Lifetime risk per 100 000 |
| --- | --- |
| US Consumer Product Safety Commission (1983) | 2.3-3.4 |
| Doll and Peto (1985) | 0.6 (approx.)[a] |
| US Environmental Protection Agency (1986) | 2.4-4.2 |
| US National Academy of Sciences (Breslow et al., 1986) | 4.2-11.2 |

[a]For males.

utilized that would, at least, quantify ambient background exposures equivalent to about 0.1 fibres >5 µm per litre. This places a substantial burden on the analyst. At typical loadings of 0.5 m³ of air per cm² of filter, a concentration of 0.1 f/l yields an average deposition density of 1 fibre per 200 grid squares. Since at least 4 fibres should be counted to establish a meaningful concentration value, the difficulty of counting fibres greater than 5 µm at typical ambient air concentrations can easily be appreciated.

A problem with any of the ambient asbestos air measurement methods is that none of the variables used to measure asbestos concentrations corresponds to a measure of increased cancer risk. The least meaningful measure is the concentration of fibres of all sizes. In many circumstances such values reflect the presence of a vast number of tiny single chrysotile fibrils, 1 µm long and 0.04 µm in diameter, which have substantially less carcinogenic potential than fibres 10 times longer and thicker. Because all estimates of risk in occupational settings are associated with measures of exposure in terms of fibres >5 µm, the enumeration of such fibres is a useful component of any environmental asbestos evaluation. It should be recognized, however, that even perfect enumeration of fibres >5 µm does not give a direct indication of carcinogenic risk, because the carcinogenicity of fibres varies according to their length and diameter, the longer and thinner fibres being the more carcinogenic (Pott, 1980; Stanton et al., 1981). Furthermore, there is no evidence that the carcinogenicity of fibres vanishes at the arbitrary counting cut-off value of 5 µm. Because of their much greater number, fibres <5 µm may, in fact, be the dominant contributors to the cancer risk of a particular aerosol. Were all asbestos aerosols to have the same fibre size distribution, the different carcinogenic potentials of fibres of different sizes would not be important. A count of all fibres >5 µm would always count both the same fraction of the asbestos present and the same fraction of the carcinogenic potential. Unfortunately, fibre size distributions vary tremendously, even for aerosols of the same fibre type. There is thus a highly variable relationship between fibres <5 µm and carcinogenic potential.

Evidence for a differing unit fibre exposure risk, part of which may be attributed to fibre size distribution effects, is seen in the lung cancer risk calculated from mortality and exposure data in 14 studies of asbestos-exposed workers (US Environmental Protection Agency, 1986a). Figure 2, which summarizes all those studies for which dose-response information is available, shows the percentage increase in carcinogenic risk for an exposure to one fibre (optical) per ml for one year. The horizontal line is the best estimate of risk for a particular study. The wide vertical bar is the uncertainty in the study associated with the number of lung cancer deaths and with epidemiological uncertainties in the data (e.g., effects of incomplete tracing). The thin vertical line is an additional 2-5-fold estimated uncertainty in the average measure of exposure for the study. As can be seen, the uncertainty of the risk estimated from the data of a single study is enormous, often exceeding an order of magnitude. There is no evidence that fibre type plays a particularly important role in these differences; the greatest differences are between exposures to the same fibre type. Exposures to virtually pure chrysotile aerosols differ by as much as a factor of 30 (e.g., in the studies of McDonald et al., 1980 as compared with those of Dement et al., 1983). Differences by as much as a factor of 10 exist between risk estimates from studies of the same process (e.g., Finkelstein et al., 1983; Weill et al., 1979).

Fig. 2. Percentage increase in lung cancer for a 1-year exposure to 1 f/ml in 14 asbestos-exposed cohorts

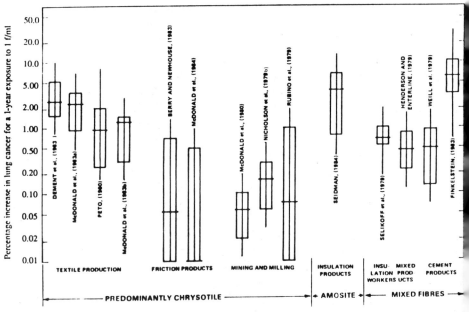

Epidemiological study and predominant exposure

The open bar reflects the estimated 95% confidence limits associated with measures of response. The vertical line represents the uncertainties associated with measures of exposure, generally ± a factor of 2. The horizontal line is the best estimate of risk for the study concerned.

The geometric mean risk from all studies, with the exception of those on chrysotile mining and milling, each weighted by the reciprocal of its estimated variance, is 1% per year of exposure to 1 f/ml. The calculated uncertainty in that value is a factor of 3; the calculated uncertainty in the risk associated with an unmeasured exposure is a factor of 10. The extrapolation of these risks estimated from conditions in occupational settings to exposures 2 or more orders of magnitude less adds even greater uncertainties to any risk estimation process.

It is within the framework of these uncertainties that we must consider the various measurements of airborne levels of mineral fibres in the non-occupational environment. Fibre concentrations must be interpreted cautiously and meaningful measures of concentrations in terms of fibres >5 µm at ambient air levels are extremely difficult to obtain. Mass measurements, while suffering from the loss of important information on fibre size, may relate best to carcinogenic risk. Mass certainly accounts better for the length dependence of carcinogenic risk than any fibre count. For the fibres >8 µm it also accounts well for the relation between carcinogenic risk and diameter (Bertrand & Pezerat, 1980; see also Peto, this volume, pp. 457-470). Ideally, of course, the best measurement of fibre aerosols is one in which every fibre is identified and sized, allowing any of the various measures of exposure to be utilized. It remains to be seen whether this can be achieved on a practical basis. These various analytical considerations are summarized in Table 3.

## General ambient air

Asbestos of the chrysotile variety has been found to be a ubiquitous contaminant of ambient air. Table 4 lists fibre concentration data from a variety of studies in both urban and rural areas over the period 1969-1983. As can be seen, arithmetic mean air concentrations were of the order of 1-3 ng/m$^3$ in 1970, a time when asbestos was extensively used in construction, in particular in relatively uncontrolled spray applications for fire-proofing purposes in the United States. During daytime hours, shorter-term sampling indicated that higher levels were present than those found in 24-h samples. Earlier data are of doubtful accuracy because analytical techniques were in process of development and the control of asbestos contamination in laboratory processing was less stringent than it is now. In recent years, no systematic study has been made of asbestos in urban air. However, virtually all studies of outside air, taken in conjunction with studies of building air, show mean concentrations to be less than 1 ng/m$^3$. Because of the above-mentioned limitations on the counting of fibres >5 µm, however, most such measurements have only provided upper limits of exposure. (See, however, Rödelsperger et al., this volume, pp. 361-366).

Few data exist on the fibre size distribution of asbestos in the general ambient air. From ratios of fibres >5 µm to total fibres, it would appear to be similar to that of chrysotile in occupational environments, where the percentage of fibres >5 µm is between 5 and 0.5% (Gibbs & Hwang, 1980; Nicholson et al., 1972). Data from Kohyama (this volume, pp. 262-276) indicate that 2% of fibres in the ambient air of Japan are >5 µm in length.

## Buildings with asbestos-containing surfacing materials

Considerable data have accumulated in recent years on the air concentrations of asbestos fibres associated with the use of asbestos-containing surfacing materials. The

Table 3. Advantages and disadvantages of preparation, electron microscopy and counting methods

| Method | Advantages | Disadvantages |
|---|---|---|
| **Preparation** | | |
| Direct | Fibre size maintained | Some fibres obscured<br>Sampling conditions may be critical<br>Possible fibre loss |
| Indirect | No obscuration of fibres<br>Losses less likely<br>Sampling conditions less stringent | Information on fibre size distribution lost |
| **Electron microscopy** | | |
| SEM[a] | Preparation techniques less critical<br>Scanning time less per area scanned | No standard methodology available<br>Only fibres thicker than 0.2 $\mu$m usually counted |
| TEM[a] | Fibres of all sizes can be counted<br>Standard methodologies available | Extremely long scanning times may be required<br>Sample preparation critical |
| **Counting** | | |
| All fibres | None | Small fibres overwhelm count<br>Relates least to biological hazard<br>Counts affected by instrument resolution<br>Time consuming |
| Fibres >5 $\mu$m | Relates best to optical counts | Problems with low counts<br>Does not consider clusters or fibre bundles |
| Mass | Preparation techniques are less stringent<br>May relate best to carcinogenic risk<br>Improved sensitivity | May be dominated by a few fibres or clusters<br>Difficult to relate to optical counts |

[a]SEM, scanning electron microscopy; TEM, transmission electron microscopy.

data, however, represent information about a moving target. Before the early 1970s neither the hazard nor the extent of use of asbestos in buildings was fully appreciated. Conditions in some buildings deteriorated to a point that was unacceptable by any standards and extremely high air concentrations occurred. For example, phase-contrast microscopic measurements in excess of 10 f/ml were measured for short

## Table 4. Summary of ambient air asbestos concentrations, 1969-1983

| Sample set | Collection period | No. of samples | Arithmetic mean concentration[a] (ng/m³) | (fibres >5 µm per litre) |
|---|---|---|---|---|
| Quarterly composites of 5-7 24-hour US samples (US Environmental Protection Agency, 1971; Nicholson & Pundsack, 1973) | 1969-1970 | 187 | 3.3 C | - |
| Quarterly composites of 5-7 24-hour US samples (US Environmental Protection Agency, 1974) | 1969-1970 | 127 | 3.4 C | - |
| 16-hour samples of 5 US cities (US Environmental Protection Agency, 1974) | 1974 | 34 | 13 C | - |
| 5-day samples in Paris, France (Sébastien et al., 1980) | 1974-1975 | 161 | 0.96 C | - |
| 5-day, 7-hour control samples for US school study (Constant et al., 1983) | 1980-1981 | 31 | 6.5 (6C, 0.5A) | - |
| 12-hour samples in Toronto, Ontario (Chatfield, 1983) | 1980-1981 | 24 | 0.83[b] | ND[c] |
| 12-hour samples in Southern Ontario (Chatfield, 1983) | 1980-1981 | 48 | 0.20[b] | ND[c] |
| UK urban and rural background (Le Guen et al., 1983) | 1979-1981 | 8 | <1-5 | - |
| Long-term samples in 3 cities in the Federal Republic of Germany (Friedrichs et al., 1983) | 1982 | 6 | - | 2.8 C[d] 2.6 A[d] |
| Urban Switzerland (Litistorf et al., 1985) | 1977 | 10 | 0.74 | 0.4 |
| Rural Switzerland (Litistorf et al., 1987) | 1981-1983 | 10 | 0.23 | ND |
| Rural Austria (Felbermayer, 1983) | 1978-1980 | 143 | - | <0.1 |

[a]C, chrysotile; A, amphibole.
[b]An exceptionally high sample contributed all the mass.
[c]Not detected, i.e., less than a detection limit of approximately 4 f/l.
[d]Fibres of all sizes.

periods of time in a school library during book replacement as a result of inadvertent contact with ceiling material (Sawyer, 1977). Increasing awareness of the presence of asbestos in buildings and public concern led to the elimination of the worst exposure hazards and contamination was then reported less frequently.

A variety of studies, using transmission electron microscopy, were undertaken to evaluate the quality of air in buildings under several different circumstances; the results are shown in Table 5. Different methods of analysis were used by different investigators, but the location and conditions of sampling were likely to be a far more important factor than differences in analytical methods in the results that were obtained. Most of the earlier studies focused on the potentially more severe exposures and were thus not representative of all building circumstances. In other studies, buildings were also chosen for sampling by non-random criteria and similarly do not provide a representative sample of all buildings. Overall, the studies present a reasonably consistent picture. In buildings with evidence of severe damage or deterioration, the probability of detecting contamination was high. On the other hand, if the surfacing material or thermal insulation was undamaged, had suffered only minor damage or the surface had been sealed to prevent dusting, excess air concentrations were rarely detected.

Nicholson *et al.* (1975, 1976) analysed 116 samples of indoor and outdoor air collected in and around 19 commercial and public buildings in 5 cities in the USA. The buildings studied were chosen by local or federal air pollution control agencies solely on the basis of ease of access (they belonged to the Government or the owners were willing to allow the tests to be carried out). The choice was thus not random, but neither was any building selected because of a perceived hazard. After collection, the samples were coded by the EPA so that the sites from which they were collected were not known when they were analysed. The results provided no evidence of contamination of buildings with cementitious or plaster-like surfacing material, but the air concentrations in some buildings with surfacing material consisting of a loosely bonded mat were greater than those of control samples and samples collected in buildings with cementitious/plaster material. In this set of samples, and that considered below, open-face filters were utilized. The possibility that some non-respirable asbestos material contributed to the mass cannot be excluded.

Nicholson *et al.* (1978, 1979a) collected 25 samples in primary and secondary schools. The sampling was conducted so as to reflect the general ambient background of schools with substantially damaged surfacing material. Sample collection was observed in order to ensure that the material collected did not reflect an unusual release of fibres near the sampler. However, the schools were in operation during the sampling and normal student activity (except for vandalism) took place during the course of sampling. Two short-term samples of custodial sweeping showed even higher concentrations than those listed, but the results were uncertain because of low sample volume.

Sébastien *et al.* (1980) published data from samples collected in 25 buildings in Paris. Over half the samples were collected in one large school building in which sprayed material containing both amphibole and chrysotile fibres was present.

## Table 5. Summary of ambient air asbestos concentrations in buildings, 1974–1985

| Sample set | Collection period | No. of samples | Arithmetic mean concentration[a] | |
|---|---|---|---|---|
| | | | (ng/m³) | (fibres >5 μm per litre) |
| US buildings with friable asbestos in plenums or as surfacing materials (Nicholson et al., 1975, 1976) | 1974 | 54 | 48 C | - |
| US buildings with cementitious asbestos material in plenums or as surfacing materials (Nicholson et al., 1975, 1976) | 1974 | 28 | 15 C | |
| New Jersey schools with damaged asbestos surfacing materials in pupil use areas (Nicholson et al., 1978, 1979a) | 1977 | 27 | 217 C | - |
| Buildings with asbestos materials in Paris, France (Sébastien et al., 1980) | 1976-1977 | 135 | 35 (25 C, 10 A) | - |
| US school rooms/areas with asbestos surfacing material (Constant, 1983) | 1980-1981 | 54 | 183 (179 C, 4 A) | - |
| US school rooms/areas in buildings with asbestos surfacing material (Constant, 1983) | 1980-1981 | 31 | 61 (53 C, 8 A) | - |
| Ontario buildings with asbestos surfacing materials (Ontario Royal Commission, 1984) | 1982 | 63 | 2.1 | ND[b] |
| Ontario office and school buildings (Chatfield, 1985) | 1977-1982 | 55 | 1.1[c] | ND[d] |
| UK schools, laboratories and factories (Burdett & Jaffrey, 1986) | 1983-1985 | 114 | 1.5 | <0.1-2[e] |

[a] C, chrysotile; A, amphibole.

[b] Not detectable, i.e., less than a detection limit of approximately 4 f/l.

[c] Two further samples had concentrations of 640 and 360 ng/m³, the latter being from a single fibre.

[d] Not detectable, i.e., less than a detection limit of approximately 4 f/l, except for one sample which had a concentration of 20 f/l.

[e] Most samples had an average concentration less than a detection limit of 0.3 f/l.

Substantial damage was reported to have occurred in many of the buildings sampled, including those areas of the school building from which most samples were taken. In some cases visible dust was reported on tables and other furniture. As with the previous school study, the buildings were not selected randomly, but were those brought to the attention of the researchers through written requests for help from building owners or occupants.

A sampling of 25 schools, chosen by a random procedure from those in a large city, was reported by the US Environmental Protection Agency (1983). The results of the survey indicated a substantially increased air concentration (a population-weighted mean concentration of 179 ng/m$^3$) in building areas with sprayed surfacing materials, while 31 outside samples averaged 6 ng/m$^3$. Of special concern were samples collected in schools in which asbestos material was present, but not on the floor or in the room in which the samples were taken. These showed an average concentration of 53 ng/m$^3$, indicating dispersal of fibres from the source. Additionally, numerous small respirable clumps and bundles of asbestos were noted, but not included in the listed mass concentrations. One problem with this study, which indicates concentrations as high as those measured in buildings selected because of the presence of extensive damage, is that sample taking was not routinely monitored. In some cases, the samples, which were collected over a 5-day period, were in classrooms under the control of the teacher; in other cases, they were in corridors and sampling conditions were not kept under continuous observation. The possibility of atypical activities occurring during sampling cannot be excluded.

A report on 63 samples taken in 19 buildings was published by the Royal Commission of Matters of Health and Safety Arising from the Use of Asbestos in Ontario (Ontario Royal Commission, 1984). All of the buildings selected contained asbestos surfacing material. However, qualitative exposure evaluation (algorithms) indicated minimal problems in all but 6 buildings. A major problem with this study is that of analytical sensitivity. The count of a single fibre would appear to correspond to a concentration of 1 f/l. Thus, the analytical threshold (the concentration at which 4 fibres would be counted) is at a concentration 40 times that of typical ambient air. In only 5 buildings were asbestos fibres $>5$ $\mu$m detected, but in those, concentrations substantially exceeded the typical background (where few fibres of any length were detected). There was a poor correlation between either mass or fibre levels with algorithm estimates of hazard.

Chatfield (1985b) has reported on concentrations measured in 8 Ontario office and school buildings. Concentrations ranged from undetectable levels to 17 ng/m$^3$, the average being 1.1 ng/m$^3$, except for 2 samples which had concentrations of 640 and 360 ng/m$^3$, the mass (360 ng) in one case being contributed by one fibre. Counts of fibres $>5$ $\mu$m were severely limited by the analytical sensitivity of the counting protocol. In 57 samples only 12 such fibres were observed, 6 being in one sample. The statistical significance of either the mass or $>5$ $\mu$m fibre concentration is difficult to assess.

The results obtained from 15 commercial, school or domestic buildings in the United Kingdom (Burdett & Jaffrey, 1986) indicated relatively low fibre concentrations. The analytical sensitivity of the protocol for the average of a sample test (up to

26 samples for an individual building) ranged from 1 fibre >5 μm per litre (2 building sample sets) to 0.1 fibres >5 μm per litre (3 building sample sets). The analytical sensitivity of most sample sets was about 0.3 fibres >5 μm per litre; the analytical sensitivity of a typical single sample, however, was only 2–3 fibres >5 μm per litre or 20–30 times the value for typical ambient air. Detectable results (an average of more than 4 fibres counted) were observed in 4 of the 15 buildings. The highest measured concentration was in a school with 'some damage' to sealed spray materials; asbestos fibre concentrations ranged from <35 to 250 fibres of all sizes per litre and from <3 to 12 fibres >5 μm per litre (average 2 f/l), levels substantially above background. Average mass concentrations in the buildings were generally <0.1 ng/m$^3$; the highest average concentration measured was 15 ng/m$^3$. However, as direct-preparation techniques were utilized, the possibility of fibre obscuration must be noted.

All the above studies were carried out under conditions such that surfacing materials were not disturbed. However, measurements indicate that routine maintenance activities can be a substantial source of short-term building air contamination, if proper precautions are not taken. Thus, the Ontario Royal Commission reported increased air levels from inspections above suspended ceilings and maintenance work therein. In one case the exposure to a worker above the ceiling was 12 fibres >5 μm per ml. Earlier, Sawyer (1977) had reported optical fibre counts during routine building maintenance ranging from 0.2 to 17.1 f/l.

Thermal insulation materials are also readily damaged during the course of the maintenance or repair of high-temperature equipment or during the course of building activities. In many cases these materials are not located in the public areas of buildings. However, debris can be carried into such areas by maintenance personnel if proper precautions are not taken.

Asbestos abatement work is also a serious potential source of non-occupational asbestos exposure. While procedures have been specified that should minimize building contamination following renovation, removal, enclosure or encapsulation of asbestos materials, these may not always be followed (see e.g., Burdett et al., this volume, pp. 277–290). The EPA has monitored the efficacy of controls and clean-up procedures in 4 schools undertaking removal of asbestos materials and in one school encapsulating surfacing material. The results are shown in Table 6. As can be seen, no measurable contamination was present in the schools after completion of the work. Some escape of fibres during removal work occurred, but it was successfully dealt with.

## Use of air monitoring for hazard evaluation

The principal source of asbestos contamination in virtually all buildings is the deliberate or inadvertent dislodgement of the material by building occupants during the course of their use of the building or by maintenance activities undertaken by building personnel. Control of the former is often difficult and, where damage caused by building users is common, the only recourse may be removal of the material or enclosure by other building products so as to prevent future damage. In contrast to

**Table 6. Geometric mean of chrysotile fibre and mass concentrations before, during and after asbestos abatement**[a]

| Sampling location | Concentration | | | | | | | |
|---|---|---|---|---|---|---|---|---|
| | Before abatement | | During abatement[b] | | Immediately after abatement | | After school resumed | |
| | (f/l)[c] | (ng/m³) | (f/l) | (ng/m³) | (f/l) | (ng/m³) | (f/l) | (ng/m³) |
| *Encapsulation* | | | | | | | | |
| Rooms with unpainted asbestos | 1423.6 | 6.7 | 117.2 | 0.6 | 13.7 | 0.1 | 248.1 | 1.2 |
| Rooms with painted asbestos | 622.9 | 2.7 | - | - | 0.8 | 0.0 | 187.2 | 0.8 |
| Asbestos-free rooms | 250.6 | 1.2 | 0.5 | 0.0 | 9.3 | 0.0 | 30.7 | 0.2 |
| Outdoors | 3.5 | 0.0 | 0.0 | 0.0 | 6.5 | 0.0 | 2.8 | 0.0 |
| *Removal* | | | | | | | | |
| Rooms with asbestos | 31.2 | 0.2 | 1736.0 | 14.4 | 5.6 | 0.1 | 23.9 | 0.2 |
| Asbestos-free rooms | 6.1 | 0.1 | 12.0 | 0.1 | 1.6 | 0.0 | 18.1 | 0.1 |
| Outdoors | 12.6 | 0.1 | 1.3 | 0.0 | 20.0 | 0.1 | 7.9 | 0.0 |

[a]Source: US Environmental Protection Agency (1985, 1986b).
[b]Measurements outside work containment areas.
[c]Fibres of all lengths.

surfacing material, where control of damage may be difficult, control of thermal insulation is much easier. Encapsulation of pipe coverings by appropriate covering material and establishment of an appropriate building maintenance programme will go far to eliminate exposures to building occupants and maintenance personnel. The extent to which the various abatement options are utilized in a building will, of course, depend on the potential for continued fibre release and for appropriate control. Future building use and the frequency of renovation work also play important roles in any decision to remove or treat asbestos materials in a building.

Air monitoring is generally not a satisfactory means of evaluating whether a specific control activity should be adopted since it only gives a value at one particular point in time and the circumstances are usually artificial. Absence of fibre contamination when no building activity is taking place does not provide any information on what may occur in the future. Furthermore, to accurately establish an average level for a building, from which one might make a highly uncertain risk assessment, would involve such extensive sampling as to make it totally impractical. In general, abatement activities must be undertaken on the basis of the observed conditions and building circumstances rather than air monitoring.

One important fact is that no generalization applicable to all buildings can be made from measurements made in buildings selected according to various sampling criteria. Buildings with intact asbestos provide no information on the hazard in buildings that are severely damaged. The fact that most buildings have material in good condition, with little evidence of release, does not mean that no action should be taken to control emissions in buildings with severe problems. Thus, average building air concentrations cannot be used to justify inaction in a particular building. It is well to keep in mind that many people have drowned in rivers having an average depth of 30 cm. The converse is also true. The finding of substantially elevated concentrations in a building with severe problems does not indicate the likelihood of fibre release in other buildings with quite different materials and conditions.

## Other building materials

Weathering of asbestos-cement wall and roofing materials has been shown to be a source of asbestos air pollution (Nicholson, 1978). Seven samples taken after heavy rain in a school constructed of asbestos-cement panels showed asbestos concentrations of 20–4500 ng/m$^3$ (arithmetic mean = 780 ng/m$^3$); all but 2 samples exceeded 100 ng/m$^3$. It was suggested that asbestos was released from the asbestos-cement walkways and roof panels and entrained into the school air when dry conditions returned. No significantly elevated concentrations were observed in a concurrent study of houses constructed of similar material. Here roof-water run-off landed on the ground and was not re-entrained, while that of the schools fell on to a smooth walkway, allowing easy re-entrainment when dry. Contamination from asbestos-cement products has also been documented by Spurny *et al.* (1980).

Air contamination by fibres released from vinyl-asbestos tile as a result of wear has been reported by Sébastien *et al.* (1982), who found concentrations of asbestos of up to 170 ng/m$^3$ in a building with worn asbestos floor tiles.

## Private dwellings

Asbestos is commonly found in a wide variety of products in the home environment. Most notable among the potential sources of airborne asbestos are thermal insulation products on boilers and high-temperature water lines, insulation material in various space heaters, textured paints, old wallboard and joint compounds, and air-supply duct materials. Additionally, a variety of household products has been found in past years to contain asbestos materials. Depending on the region, these may continue to be produced, and include appliances such as hair dryers, toasters, electrical cords, and portable heaters. Vinyl-asbestos floor tile is commonly used in homes. However, the potential for release in home circumstances is very limited unless the tile is sanded or physically abraded during home renovations. Of greater concern are subflooring materials containing asbestos. Generally, release of asbestos fibres from such architectural materials is limited; however, during building alterations, by either the owner or outside contractors, contamination may occur.

Burdett and Jaffrey (1986) measured air concentrations in 24 buildings, mostly private houses, having warm-air heaters containing asbestos. All but 3 buildings had concentrations less than the analytical sensitivity of the counting protocol (1 fibre >5 µm per litre). In 2 buildings, average concentrations of 1 fibre >5 µm per litre were measured.

Nicholson (1988) measured asbestos air concentrations in homes in which asbestos-paper air ducts were used for air-conditioning systems. Most of the homes also had chrysotile-containing textured paint in all living areas. The results are shown in Table 7. There was no statistical difference between air concentrations in homes with asbestos paint and those without, nor between samples collected while the air-conditioning system was in operation and those taken with it switched off. However, there was a significant difference between the concentrations of all indoor samples and those taken outdoors for control purposes. The difference is small and would not warrant any abatement action. However, a problem exists in that homeowners, unaware that paint may contain asbestos, may disturb the painted surfaces during renovations.

Table 7. Asbestos concentrations in homes containing asbestos products

| Sampling circumstances[a] | No. of samples | Asbestos concentration[b] (ng/m³) |
|---|---|---|
| *Asbestos in textured paint* | | |
| A/C fans off | 12 | 6.1 (2.0-12.7) |
| A/C fans on | 12 | 2.9 (0.0-11.0) |
| Average | | 4.5 |
| *No asbestos in textured paint* | | |
| A/C fans off | 3 | 4.5 (3.9-5.4) |
| A/C fans on | 3 | 2.2 (0.4-3.8) |
| Average | | 3.3 |
| Outside air | 7 | 0.9 (0.0-4.3) |

[a]A/C, air conditioning.
[b]Ranges in parentheses.

## Degradation of friction materials

One of the more significant remaining sources of environmental asbestos exposure may be emissions from braking of automobiles and other vehicles. Measurements of brake and clutch emissions in the USA reveal that 2.5 tons of unaltered asbestos are released to the atmosphere annually and an additional 68 tons fall on to roadways, where some of the asbestos may be dispersed by passing traffic (Jacko et al., 1973). Air concentrations from such releases, however, were not measured by the authors.

Substantially elevated chrysotile asbestos concentrations have been found in subway systems, the cars of which used chrysotile pads in brake systems (Chatfield, 1983). Analyses of samples collected during 1976 in the Toronto subway indicated concentrations of up to 2.7 fibres of all sizes per ml and, with one exception (a value of 20 000 ng/m³) mass concentrations up to 2500 ng/m³. Lower values were found in a 1980 survey and attributed to the use of direct-analysis techniques. A 1981 study of the Stockholm subway found concentrations of 0.10–0.12 fibres >5 $\mu$m per ml and mass concentrations of 170–430 ng/m³.

The data on the contribution of automobile braking to ambient air asbestos concentrations are extremely limited. Lanting and den Boeft (1983) reported concentrations in a traffic tunnel to be approximately 10 times those in urban air (5 fibres >5 $\mu$m per litre as compared with about 0.7 for urban air), and suggested that vehicle braking was the source of the excess. In an earlier study, Alste et al. (1976) reported average concentrations at a freeway exit of 0.5 fibres of all sizes per ml over 9-hour sampling periods, much higher than at a distant site. Williams and Muhlbaier (1982) measured asbestos emissions from braking in a test dynamometer and estimated that normal driving releases 2.6 $\mu$g of airborne asbestos per km of travel. Using the ratio of this value to total lead emissions per km and measured ambient lead levels, they estimated the contribution of automobile braking to airborne asbestos to be 0.063 ng/m³. This procedure is likely to underestimate the effect of braking, as a significant fraction of the emitted lead will rapidly settle to the ground, while the emitted asbestos will tend to remain in suspension. It remains possible that automobile braking makes the major contribution to the asbestos in the ambient air of cities, but data are lacking either to confirm or refute this possibility. Data supporting this hypothesis are presented by Kohyama in this volume, pp. 262–276.

## Other fibre exposures

Man-made and other mineral fibres have largely replaced the asbestos minerals in products where the potential for their release is high, as in thermal insulation or surfacing materials. Fortunately, most man-made mineral fibres (MMMF) are of non-respirable diameter, so that the use of such fibres carries a much lower risk than that of asbestos minerals. However, in the absence of regulation, the use of other naturally occurring fibrous minerals is increasing rapidly. Some of these, as well as some MMMF, are of the same size and durability as asbestos and may present equal health risks.

There are very few data on the presence of non-asbestos inorganic fibres in the general environment. They are noted by analysts as present, usually in numbers exceeding asbestos, but not quantified. Gypsum fibres are commonly present in the air of buildings during renovation; their confounding effect in air samples can be avoided by special procedures (Burdett, 1985). Friedrichs et al. (1983) published data on the numbers of other fibres in ambient air samples in the Federal Republic of Germany; their results, in terms of fibres of all sizes, are given in Table 8. As can be seen, the concentrations of fibres of the 3 listed non-asbestos minerals were less than those of chrysotile and amphibole, but large numbers of other fibres were not identified. Altree-William and Preston (1985) analysed 193 samples using a SEM and found average concentrations of 0.32, 1.47 and 5.10 fibres of all sizes per litre for asbestos, other minerals and organic fibres, respectively. Spurny and Stober (1981) identified mineral fibres in urban and rural air in the Federal Republic of Germany. Total fibre concentrations were in the range 4–15 f/l, less than 6% of which were asbestos fibres. The percentages of the various fibres and some aspects of analyses of single fibres are shown in Table 9.

**Table 8. Concentrations of fibres of all sizes measured in three cities in the Federal Republic of Germany during 1982**

| City | Air concentrations (f/m³) | | | | | | Sampling period |
|---|---|---|---|---|---|---|---|
| | Chrysotile | Amphibole | Rutile | Iron oxide | Glass | Total | |
| Düsseldorf | 2320 | 1350 | 330 | 630 | 660 | 24 160 | Monday-Friday |
| Dortmund | 2600 | 2390 | – | 630 | 2490 | 31 800 | Monday-Friday |
| Duisburg | 1980 | 2610 | 440 | 480 | 1410 | 35 970 | Monday-Friday |
| Düsseldorf | 750 | 720 | – | 700 | 360 | 16 510 | Friday-Monday |
| Dortmund | 3320 | 1780 | 510 | 760 | 1270 | 23 490 | Friday-Monday |
| Duisburg | 5740 | 6890 | 1460 | 610 | 1220 | 24 070 | Friday-Monday |
| Averages | 2790 | 2620 | 460 | 640 | 1230 | 26 000 | |

Source: Friedrichs et al. (1983).

Fibrous glass materials are common constituents of thermal and acoustic insulation materials. Balzer et al. (1971) sampled building and outside air in order to determine whether fibrous glass-lined ducts could be eroded. Their data indicated concentrations of glass fibres in the ambient air of 0.26–4.5 f/l and concentrations of other fibres of 1.5–3.3. The fibres enumerated were those thicker than 1.6 $\mu$m and longer than 4.8 $\mu$m (glass fibres of these dimensions were distinguishable in a petrographic microscope). They suggested that their study indicated a significant reduction in numbers of such fibres after passage through a standard building air filter and down a fibrous glass-lined duct. However, the data supporting this conclusion were scant and not subject to verification by statistical analysis. Gaudichet et al. (this volume, pp. 291–298) have provided extensive data on air pollution by MMMF.

A variety of non-asbestos minerals are being used at an increasing rate as replacements for asbestos, the most important being attapulgite and wollastonite. Attapulgite fibres, which are generally less than 5 $\mu$m in length, are used in drilling muds, cat litter adsorbents and filters, and as fillers in pesticides, fertilizers,

**Table 9. Percentages of various mineral fibres in samples of urban and rural air in the Federal Republic of Germany**

| Fibre type | Sample | | | | |
|---|---|---|---|---|---|
| | A | B | C | D | E |
| Potential non-contaminated asbestos fibres | 1.02 | 0.82 | 1.94 | 0.52 | - |
| Potential contaminated asbestos fibres | 1.51 | 1.22 | 6.80 | 1.04 | 0.41 |
| Potential asbestos fibres leached of Mg or Fe | 1.02 | 0.82 | 5.83 | 0.52 | 0.41 |
| Other fibrous silicates | 2.54 | 1.22 | 9.71 | 1.04 | - |
| Fibrous gypsum | 24.37 | 38.36 | 46.60 | 53.88 | 27.06 |
| Contaminated fibrous gypsum | 15.74 | 20.40 | 2.91 | 3.11 | 1.22 |
| Fibrous ammonium sulfate | 48.22 | 3.66 | 0.16 | 15.02 | 30.74 |
| Unidentified fibre-like particles | 5.58 | 33.50 | 26.05 | 24.87 | 40.16 |

Source: Spurny and Stober (1981).

paints and filling compounds, pharmaceuticals and cosmetics. Their uncontrolled use in this way may lead to environmental contamination. Particular concern exists with regard to their use as pet waste adsorbents, but no air level data are available. Wollastonite occurs as long thin fibres and is also used as a filler or as a reinforcing fibre in insulation and wall board. Mine and mill concentrations of, respectively, 1 and 20 f/ml have been reported (Shasby *et al.*, 1979), but no data exist for non-occupational exposures.

## References

Alste, J., Watson, D. & Bagg, J. (1976) Airborne asbestos in the vicinity of a freeway. *Atmos. Environ.*, 10, 583-589

Altree-Williams, S. & Preston, J.S. (1985) Asbestos and other fibre levels in buildings. *Ann. Occup. Hyg.*, 29, 357-363

Balzer, L., Cooper, W.C. & Fowler, D.P. (1971) Fibrous glass-lined air transmission systems: an assessment of their environmental effects. *Am. Ind. Hyg. Assoc. J.*, 32, 512-518

Berry, G. & Newhouse, M.L. (1983) Mortality of workers manufacturing friction materials using asbestos. *Br. J. Ind. Med.*, 40, 1-7

Bertrand, R. & Pézerat, H. (1980) Fibrous glass: carcinogenicity and dimensional characteristics. In: Wagner, J.C., ed., *Biological Effects of Mineral Fibres* (IARC Scientific Publications No. 30), Lyon, International Agency for Research on Cancer, pp. 901-912

Breslow, L., Brown, S. & van Ryzin, J. (1986) Risk from exposure to asbestos. *Science*, 234, 923

Burdett, G.J. (1984) Proposed analytical method for determination of asbestos fibres in air. (HSE internal report No. IR/L/DI/84/02 prepared for International Standards Organization Joint Working Group ISO/TC/147/SC2WG18-ISO/TC/146/WG1, Brussels, March 1984)

Burdett, G.J. (1985) Use of membrane-filter, direct-transfer technique for monitoring environmental asbestos releases. In: *Asbestos Fibre Measurements in Building Atmospheres, Proceedings of a Conference: March 1985, Toronto, Ontario*, Toronto, Ontario Research Foundation, pp. 87-110

Burdett, G.J. & Jaffrey, S.A.M.T. (1986) Airborne asbestos concentration in buildings. *Ann. Occup. Hyg.*, 30, 185-199

Burdett, G.J. & Rood, A.P. (1983) Membrane-filter, direct-transfer technique for the analysis of asbestos fibres or other inorganic particles by transmission electron microscopy. *Environ. Sci. Technol.*, 17, 643-648

Chatfield, E.J. (1983) Measurement of asbestos fibre concentrations in ambient atmospheres. (Report prepared for the Royal Commission on Matters of Health and Safety Arising from the Use of Asbestos in Ontario).

Chatfield, E.J. (1985a) Limitations of precision and accuracy in analytical techniques based on fibre counting. In: *Asbestos Fibre Measurements in Building Atmospheres, Proceedings of a Conference: March 1985, Toronto, Ontario*, Toronto, Ontario Research Foundation, pp. 115-131

Chatfield, E.J. (1985b) Airborne asbestos levels in Canadian public buildings. In: *Asbestos Fibre Measurements in Building Atmospheres, Proceedings of a Conference: March 1985, Toronto, Ontario*, Toronto, Ontario Research Foundation, pp. 177-201

Constant, P.C., Bergman, F.J., Atkinson, G.R., Rose, D.R., Watts, D.L., Logue, E.E., Hartwell, T.D., Price, B.P. & Ogden, J.S. (1983) *Airborne Asbestos Levels at Schools*. Washington DC, US Environmental Protection Agency, Office of Toxic Substances (EPA 560/5-83-003)

Dement, J.M., Harris, R.L., Jr, Symons, M.J. & Shy, C.M. (1983) Exposures and mortality among chrysotile asbestos workers. Part II: Exposure estimates. *Am. J. Ind. Med.*, 4, 421-433

Doll, R. & Peto, J. (1985) *Asbestos: Effects on Health of Exposure to Asbestos*, London, Health and Safety Commission

Felbermayer, W. (1983) Abwitterung von Asbestzementprodukten — Immissionsmessergebnisse aus Österreich. In: Reinisch, D., Schneider, H.W. & Birkner, K.F., eds, *Fibrous Dusts — Measurement, Effects, Prevention (VDI-Berichte 475)*, Düsseldorf, VDI-Verlag, pp. 143-146

Finkelstein, M.M. (1983) Mortality among long-term employees of an Ontario asbestos-cement factory. *Br. J. Ind. Med.*, 40, 138-144

Friedrichs, K.-H., Hohr, D. & Grover, Y.P. (1983) Ergebnisse von nicht quellenbezogenen Immissionsmessungen von Fasern in der Bundesrepublik Deutschland. In: Reinisch, D., Schneider, H.W. & Birkner, K.F., eds, *Fibrous Dusts — Measurement, Effects, Prevention (VDI-Berichte 475)*, Düsseldorf, VDI-Verlag, pp. 113-116

Gibbs, G.W. & Hwang, C.Y. (1980) Dimensions of airborne asbestos fibres. In: Wagner, J.C., ed., *Biological Effects of Mineral Fibres (IARC Scientific Publications No. 30)*, Lyon, International Agency for Research on Cancer, pp. 69-78

Henderson, V.L. & Enterline, P.E. (1979) Asbestos exposure: factors associated with excess cancer and respiratory disease mortality. *Ann. N.Y. Acad. Sci.*, 330, 117-126

Jacko, M.G., DuCharme, R.T. & Somers, J.T. (1973) How much asbestos do vehicles emit? *Automob. Eng.*, 81, 38-40

Lanting, R.W. & den Boeft, J. (1983) Ambient air concentration of mineral fibres in the Netherlands. In: Reinisch, D., Schneider, H.W. & Birkner, K.F., eds, *Fibrous Dusts — Measurement, Effects, Prevention (VDI-Berichte 475)*, Düsseldorf, VDI-Verlag, pp. 123-128

Le Guen, J.M.M., Burdett, G.J. & Rood, A.P. (1983) Mass concentration of airborne asbestos in the non-occupational environment — a preliminary report of UK measurements. In: Reinisch, D., Schneider, H.W. & Birkner, K.F., eds, *Fibrous Dusts – Measurement, Effects, Prevention (VDI-Berichte 475)*, Düsseldorf, VDI-Verlag, pp. 137-141

Litistorf, G., Guillemin, M., Buffat, P. & Iselin, F. (1985) Ambient air pollution by mineral fibres in Switzerland. *Staub-Reinhalt. Luft, 45*, 302-307

McDonald, A.D., Fry, J.S., Woolley, A.J. & McDonald, J.C. (1983a) Dust exposure and mortality in an American chrysotile textile plant. *Br. J. Ind. Med., 40*, 361-367

McDonald, A.D., Fry, J.S., Woolley, A.J. & McDonald, J.C. (1983b) Dust exposure and mortality in an American factory using chrysotile, amosite and crocidolite in mainly textile manufacturing. *Br. J. Ind. Med., 40*, 368-374

McDonald, A.D., Fry, J.S., Woolley, A.J. & McDonald, J.C. (1984) Dust exposure and mortality in an American chrysotile asbestos friction production plant. *Br. J. Ind. Med., 41*, 151-157

McDonald, J.C., Liddell, F.D.K., Gibbs, G.W., Eyssen, G.E. & McDonald, A.D. (1980) Dust exposure and mortality in chrysotile mining, 1910-75. *Br. J. Ind. Med., 37*, 11-24

Nicholson, W.J. (1978) *Chrysotile asbestos in air samples collected in Puerto Rico, Report to the Consumer Products Safety Commission (CPSC contract No. 77128000)*, New York, City University of New York, Mount Sinai School of Medicine

Nicholson, W.J. (1988) Asbestos use in private dwellings: potential sources of exposure (in preparation)

Nicholson, W.J., Holaday, D.A. & Heimann, H. (1972) Direct and indirect occupational exposure to insulation dusts in United States shipyards. In: *Safety and Health in Shipbuilding and Ship Repairing: Proceedings of a Symposium Organized by the Government of Finland, 30 August–2 September, 1971, Geneva, Switzerland*, Geneva, International Labour Office, pp. 37-47

Nicholson, W.J. & Pundsack, F.L. (1973) Asbestos in the environment. In: Bogovski, P., Gilson, J.C., Timbrell, V. & Wagner, J.C., eds, *Biological Effects of Asbestos (IARC Scientific Publications No. 8)*, Lyon, International Agency for Research on Cancer, pp. 126-130

Nicholson, W.J., Rohl, A.N. & Weisman, I. (1975) *Asbestos Contamination of the Air in Public Buildings*, Research Triangle Park, NC, US Environmental Protection Agency, Office of Air Quality Planning and Standards (*EPA Report No. 450/3-76-004*)

Nicholson, W.J., Rohl, A.N. & Weisman, I. (1976) Asbestos contamination of building air supply systems. In: *Proceedings, International Conference on Environmental Sensing and Assessment, September 1975, Las Vegas, NV*, New York, Institute of Electrical and Electronics Engineers Inc. (paper No. 29-6)

Nicholson, W.J., Rohl, A.N., Sawyer, R.N., Swoszowski, E.J. & Todaro, J.D. (1978) *Control of Sprayed Asbestos Surfaces in School Buildings: A Feasibility Study (NIEHS contract No. N01-ES-7-2113)*, New York, Mount Sinai School of Medicine, Environmental Sciences Laboratory

Nicholson, W.J., Swoszowski, E.J., Jr, Rohl, A.N., Todaro, J.D. & Adams, A., (1979) Asbestos contamination in United States schools from use of asbestos surfacing materials. *Ann. N.Y. Acad. Sci., 330*, 587-596

Nicholson, W.J., Selikoff, I.J., Seidman, H., Lilis, R. & Formby, P. (1979b) Long-term mortality experience of chrysotile miners and millers in Thetford Mines, Quebec. *Ann. N.Y. Acad. Sci., 330*, 11-21

Ontario Royal Commission (1984) *Report of the Royal Commission on Matters of Health and Safety Arising from the Use of Asbestos in Ontario*, Toronto, Ontario Ministry of the Attorney General

Peto, J. (1980) Lung cancer mortality in relation to measured dust levels in an asbestos textile factory. In: Wagner, J.C., ed., *Biological Effects of Mineral Fibres (IARC Scientific Publications No. 30)*, Lyon, International Agency for Research on Cancer, pp. 829-836

Pott, F. (1980) Animal experiments on biological effects of mineral fibres. In: Wagner, J.C., ed., *Biological Effects of Mineral Fibres, (IARC Scientific Publications No. 30)* Lyon, International Agency for Research on Cancer, pp. 261-272

Rubino, G.F., Piolatto, G., Newhouse, M.L., Scansetti, G., Aresini, G.A. & Murray, R. (1979) Mortality of chrysotile asbestos workers at the Balangero Mine, northern Italy. *Br. J. Ind. Med., 36,* 187-194

Samudra, A.V., Harwood, C.F. & Stockham, J.D. (1977) *Electron Microscopic Measurement of Airborne Asbestos Concentrations: A Provisional Methodology Manual (EPA report No. 600/2-77-178/REV)* Research Triangle Park, NC, US Environmental Protection Agency, Environmental Sciences Research Laboratory

Sawyer, R.N. (1977) Asbestos exposure in a Yale building (analysis and resolution). *Environ. Res., 13,* 146-169

Sébastien, P. (1985) Assessing asbestos exposure in buildings. In: *Asbestos Fibre Measurements in Building Atmospheres, Proceedings of a Conference, March 1985, Toronto, Ontario,* Toronto, Ontario Research Foundation, pp. 139-151

Sébastien, P., Billon-Gallard, M.A., Dufour, G. & Bignon, J. (1980) *Measurement of Asbestos Air Pollution Inside Buildings Sprayed with Asbestos (EPA report No. 560/13-80-026)* Washington DC, US Environmental Protection Agency, Survey and Analysis Division

Sébastien, P., Bignon, J. & Martin, M. (1982) Indoor airborne asbestos pollution from the ceiling and the floor. *Science, 214,* 1410-1413

Seidman, H. (1984) Short-term asbestos work exposure and long-term observation. In: *Docket of Current Rule Making for Revision of the Asbestos (Dust) Standard,* Washington DC, US Department of Labor, Occupational Safety and Health Administration

Selikoff, I.J., Hammond, E.C. & Seidman, H. (1979) Mortality experience of insulation workers in the United States and Canada, 1943-1976. *Ann. N.Y. Acad. Sci., 330,* 91-116

Shasby, D.M., Petersen, M., Hodous, T., Boehlcke, B. & Merchant, J. (1979) Respiratory morbidity of workers exposed to wollastonite through mining and milling. In: Lemen, R., & Dement, J.M. eds, *Dust and Disease,* Park Forest South, IL, Pathotox Publishers, pp. 251-256

Spurny, K.R., Stober, W., Weiss, G. & Opiela, H. (1980) Some special problems concerning asbestos fibre pollution in ambient air. In: Benarie, M.M., ed., *Atmospheric Pollution 1980: Proceedings of the 14th International Colloquium: May, Paris, France (Studies in Environmental Science: 8)* Amsterdam, Elsevier, pp. 315-322

Spurny, K.R. & Stober, W. (1981) Some aspects of analysis of single fibers in environmental and biological samples. *Int. J. Environ. Anal. Chem., 9,* 265-281

Stanton, M.F., Layard, M., Tegeris, A., Miller, E., May, M., Morgan, E. & Smith, A. (1981) Relation of particle dimension to carcinogenicity in amphibole asbestos and other fibrous minerals. *J. Natl Cancer Inst., 67,* 965-975

Travis, C.C., Richter, S.A., Crouch, E.A., Wilson, R. & Klema, E.D. (1987) Cancer risk management. *Env. Sci. Technol., 21,* 415-420

US Consumer Product Safety Commission (1983) *Report to the US Consumer Product Safety Commission by the Chronic Hazard Advisory Panel on Asbestos,* Washington DC, Directorate for Health Sciences

US Environmental Protection Agency (1971) *Measurement of Asbestos in Ambient Air (Contract No. CPA 70-92),* Washington DC, National Air Pollution Control Administration

US Environmental Protection Agency (1974) *A Preliminary Report on Asbestos in the Duluth, Minnesota Area,* Duluth, MN, Office of Enforcement and General Counsel, Office of Technical Analysis

US Environmental Protection Agency (1983) *Airborne Asbestos Levels in Schools, (EPA-560/5-83-003),* Washington DC, Office of Toxic Substances

US Environmental Protection Agency (1985) *Evaluation of Asbestos Abatement Techniques. Phase I: Removal (EPA-560/5-85-019),* Washington DC, Office of Toxic Substances

US Environmental Protection Agency (1986a) *Airborne Asbestos Health Assessment Update (EPA-600/8-84-003F),* Washington DC, Office of Health and Environmental Assessment

US Environmental Protection Agency (1986b) *Evaluation of Asbestos Abatement Techniques. Phase II: Encapsulation with Latex Paint (EPA-560/5-86-016)*, Washington DC, Office of Toxic Substances

Weill, H., Hughes, J. & Waggenspack, C. (1979) Influence of dose and fiber type on respiratory malignancy risk in asbestos cement manufacturing. *Am. Rev. Respir. Dis.*, 120, 345-354

Williams, R.L. & Muhlbaier, J.L. (1982) Asbestos brake emissions. *Environ. Res.*, 29, 70-82

Yamate, G., Agarwal, S.C. & Gibbons, R.D. (1984) *Methodology for the Measurement of Airborne Asbestos by Electron Microscopy. Draft Report (Contract No. 68-02-3266)*, Washington DC, Office of Research and Development, US Environmental Protection Agency

# AIRBORNE ASBESTOS LEVELS IN NON-OCCUPATIONAL ENVIRONMENTS IN JAPAN

N. Kohyama

*National Institute of Industrial Health, Ministry of Labour
Kawasaki, Japan*

*Summary.* Airborne asbestos levels in non-occupational environments in Japan were determined by analytical transmission electron microscopy (ATEM) for about 100 air samples from various outdoor settings. Asbestos fibres (chrysotile) were found in almost all samples. The fibre (mass) concentrations were in the range of 4-367 fibres per litre (0.02-47.2 ng/m$^3$) with a geometric mean of 18 f/l (0.3 ng/m$^3$). The mass concentrations were similar to the earlier data reported from other countries. Samples from main roads showed extremely high asbestos concentrations and short fibre lengths compared with those of the other samples. This strongly suggested that braking of vehicles was a significant emission source of airborne asbestos. Laboratory experiments using a brake testing machine demonstrated that asbestos fibres were released during braking. In addition, the present study found high levels of airborne asbestos in some highly polluted areas, such as a serpentine quarry, a town adjacent to an asbestos mine, and factories making asbestos slate-board. On the other hand, chrysotile fibres were also found in air samples from a small isolated island in the Pacific Ocean as well as in ice samples from ten thousand years ago in Antarctica. These facts suggest that chrysotile fibres have been liberated both by industrial activities and natural weathering, and have circulated around the earth.

## *Introduction*

Asbestos has been widely used as an industrial raw material in Japan as in other countries. The asbestos used in Japan has mostly been imported, since a small chrysotile mine in Hokkaido produces only a few thousand tons of low-grade asbestos annually. The amount consumed annually increased markedly in Japan in the 1960s and reached 300 000 tons in the early 1970s. This consumption level has been maintained in clear contrast to the situation in the USA, where asbestos consumption has fallen sharply from 800 000 tons to below 200 000 tons since 1972 or 1973 and has been lower than that of Japan since 1983. Consequently, Japan now comes second only to the USSR in asbestos consumption.

To protect the health of asbestos workers, various governmental regulations have been laid down in most countries. In addition, air pollution levels of asbestos in non-occupational environments have been measured in many countries. Many human activities involving the manufacture of asbestos products, construction work, the demolition of buildings, waste disposal, the braking of vehicles, asbestos mining, the crushing of serpentine rocks, as well as natural weathering, have all been suggested as the source of airborne asbestos.

In Japan, since levels of airborne asbestos in outdoor settings had not been systematically determined, the Environment Agency of Japan set up the Examination Committee of Counter-Asbestos Pollution Sources (ECCAPS) in 1977 to review the methods for determining asbestos concentrations in non-occupational environments and for measuring air pollution. The Committee's first report was published in 1984, the asbestos in all air samples being measured by means of a phase-contrast light microscope (PCM) and, in some of the samples, an analytical transmission electron microscope (ATEM). The author was engaged by the Committee and investigated asbestos air pollution levels with the ATEM. This paper reports the first detailed results on airborne asbestos levels in various environmental areas in Japan, as shown by the ATEM.

## Method and samples

### Air sampling and sample preparation

Ambient air was passed through a membrane filter (47 mm in diameter) mounted in an open-faced filter holder, 9.6 cm² in effective area, with an air flow rate of 10 litres per min for 4 h. About 2.4 m³ of air (0.25 m³/cm²) was passed through the filter. Half of each filter was used for the measurement by the PCM, which enabled fibres more than or equal to 5 $\mu$m in length to be counted. The other half was used for the ATEM measurement, employing an 'indirect transfer' method of sample preparation, slightly modified by the author as follows. The dust side of the half-filter was made to adhere closely to a glass slide by means of a few droplets of acetone and ashed in a low-temperature plasma asher. To resuspend the residue in isopropanol, the surface of the glass slide was wetted with a few droplets of isopropanol and shaved with a new blade.

Both the glass slide and the blade holding the residue were then immersed in about 70 ml isopropanol in a 100 ml conical flask and dispersed ultrasonically for a short time. The suspension was filtered through a Nuclepore filter of effective area 2.0 cm². In this way the air sample of 1.2 m³ (0.25 m³/cm) was condensed to 0.6 m³/cm². The Nuclepore filter was vacuum-coated with carbon and cut into a few pieces 3 mm square. These were placed on 200-mesh nickel TEM grids in a petri dish with chloroform (Jaffe washer) and left overnight.

By means of this process, a carbon film containing dust was left on a TEM grid after the Nuclepore filter had been dissolved with chloroform.

### Asbestos counting and calculation of concentration

Each opening of the nickel TEM grid is a square 100 $\mu$m in size. Some 3-5 such squares were then randomly selected, in which all the asbestos fibres present were counted and their sizes measured by the ATEM at both low and high magnifications, e.g., ×3000 and ×20 000. After the measurements had been made, the fibre concentration was calculated from the following equation:

$$F = (f/n) \times 2.0 \times 10^4 \times (1/L)$$

where  $F$ = fibre concentration (no. per litre)

$f$ = number of fibres

$n$ = number of grid squares examined

$L$ = total air volume sampled (litres)

The lower limit attainable under the present conditions is about 4 fibres per litre (f/l).

The mass concentration was then calculated from the fibre concentration ($F$) and the average fibre length ($A$) in $\mu$m and diameter ($D$) in $\mu$m, by the following equation:

$$M \text{ (ng/m}^3) = C \times F \times A \times D^2$$

where $C$ is a constant of value 2 and 2.75 for chrysotile and amphibole fibres, respectively.

**Standard error in fibre counting**

The standard error depends primarily on the air volume collected on the filter, the number of grid squares observed under the ATEM, and the total number of fibres counted, as well as the errors due to the sample preparation technique and the skill of the operator. The relationship between the standard error and fibre concentration can be expressed as shown in Figure 1 for the present counting conditions, an air volume of 1.2 m³ and 3, 5, and 10 openings. In this calculation, a Poisson distribution was assumed, and the errors caused by both sample preparation and the skill of the operator were ignored. Figure 1 shows that, the higher the count, the lower the standard error, and vice versa. For example, the concentration of 500 f/l has a standard error of 11% and 9% when 3 and 5 openings are examined, respectively, while the figures for 100 f/l are 25% and 19%, for 10 f/l 78% and 61%, and for 4 f/l 245% and 96%, respectively. When the concentration is around the lower detection limit, therefore, 5 or more openings should be observed.

**Sampling locations**

Air sampling and counting by PCM were performed by four prefectoral government institutes plus one private university institute in accordance with the standard specifications previously decided by the Committee. Air sampling locations were chosen locally in various types of area, such as residential areas, shopping districts, beside main roads, etc., as seen in Table 1. These areas were distributed nation-wide, as shown in Figure 2. About 100 specimens out of the 700 collected were examined by the ATEM and the asbestos concentrations and size distributions determined.

**Fig. 1. Diagram for use in estimating the standard error in the concentration of asbestos fibres under the given experimental conditions**

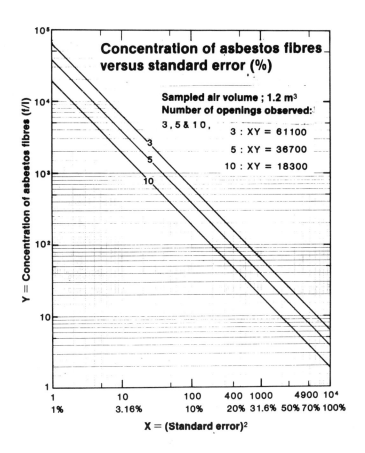

## Results

**Airborne asbestos levels in areas surrounding a serpentine quarry**

In order to determine the reliability of the present method, measurements were first made on air samples collected at and around a serpentine quarry producing crushed stone. The serpentine rock includes a proportion of chrysotile asbestos together with the other serpentine minerals antigolite and lizardite. The crusher and hopper in the centre of the quarry were seen to be the main sources of asbestos dust emission because clouds of dust could sometimes be observed visually around them. The air sampling points were located around the quarry at various positions, as shown in Figure 3. As can be seen from Figure 4, extremely high concentrations of chrysotile fibres of the order of $10^4$ f/l or more could be detected near the crusher and hopper and the concentrations decreased inversely as the square of the distance from

**Fig. 2. Locations where air samples were collected**

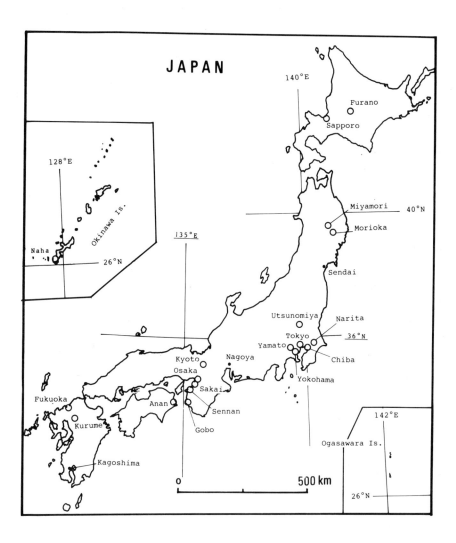

the emission source. They fell to background levels of about 20–30 f/l at points about 500 m to 1 km from the source. The mass concentrations also decreased sharply from 5000 to 10 ng/m$^3$ in moving away from the source within an area of radius 500 m. The nearest dwellings are located about 200 m from the source.

Based on this observation, it was considered that the present method would be able to give fairly reliable data on airborne asbestos concentrations in outdoor environments.

## Table 1. Airborne asbestos levels in various non-occupational environments in Japan[a]

| Environment | Number of samples analysed | Fibre concentration Geometric mean (f/l) | Fibre concentration Range (f/l) | Mass concentration Geometric mean (ng/m³) | Mass concentration Range (ng/m³) |
|---|---|---|---|---|---|
| A. Residential area | 8 | 19.8 | <4–111 | 0.23 | <0.02–9.89 |
| B. Shopping area | 1 | – | 17 | – | 0.16 |
| C. Inland industrial area | 8 | 14.0 | <4–91 | 0.18 | <0.02–10.0 |
| D. Seaside industrial area | 1 | – | 5 | – | 0.02 |
| E. Harbour | 0 | – | – | – | – |
| F. Agricultural area | 4 | 21.8 | 7–47 | 0.17 | 0.08–0.29 |
| G. Area of small factories processing asbestos | 3 | 25.3 | 10–57 | 0.41 | 0.07–3.45 |
| H. Freeways | 15 | 5.0 | <4–10 | 0.03 | <0.02–0.48 |
| I. Main roads | 25 | 29.6 | <4–367 | 0.33 | <0.02–47.2 |
| J. Disposal site for waste materials | 2 | – | 12–14 | – | 0.14–0.20 |
| K. Around buildings being demolished | 0 | – | – | – | – |
| L. Dockyard | 2 | – | 32–54 | – | 0.62–0.65 |
| M. Around automobile repair shops | 3 | 31.6 | 7–104 | 0.47 | 0.07–1.14 |
| N. Inside school room | 2 | – | 18–26 | – | – |
| O & P. Isolated island and in the Pacific Ocean | 19 | 9.7 | 4–48 | 0.05 | <0.01–0.50 |
| Q. Contamination level in sample preparation | 1 | – | <4 | – | – |
| R. Around serpentine quarry | 18 | b | 100–40 000[c] | b | 10–5000 |
| S. In a town adjacent to a chrysotile mine | 6 | 487 | 342–801 | 245 | 32.6–3418 |
| T. Around factories making asbestos slate-board | 5 | 178 | 11–849 | 51.5 | 0.61–618 |

[a]Fibres >5 µm in length: A-P about 2%, R-T about 15%, therefore 1 ng/m³ is equivalent to 2 f/l and 0.3–0.5 f/l, respectively.
[b]See text.
[c]Within an area of radius 500 m.

**Fig. 3. Air-sampling points around a serpentine quarry producing crushed stone.**

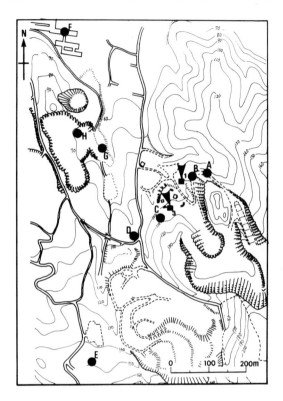

1 and 3: crusher and hopper; A to H: air-sampling points.

### Airborne asbestos levels in various environmental areas

The concentrations of airborne asbestos fibres determined by the ATEM method for samples collected in various environments are shown in Figure 5 and the means and ranges for each area are given in Table 1. Asbestos fibres were found in about 80% of all samples and the fibre concentrations varied from below the lower limit of detection (4 f/l) to 367 f/l, with a geometric mean of 18 f/l (an arithmetic mean of 41 f/l). Almost all the asbestos fibres detected were chrysotile. The distribution of fibre length showed fibres less than 1 $\mu$m in length to be predominant (71.3%); those over 5 $\mu$m accounted for only 2.0% in settings A to P (Table 1). The concentration of asbestos fibres longer than 5 $\mu$m was therefore estimated as 0.36 f/l in ambient air.

Extremely high concentrations were observed along main roads, including crossings where many vehicles were passing, whereas fibre concentrations along freeways were low. The average fibre length was about 0.5 $\mu$m, showing that significant numbers of shorter fibres were present as compared with an average of 1.1 $\mu$m in other areas. Moreover, along main roads, the concentration fell at night and was

**Fig. 4. Concentrations of asbestos (chrysotile) fibres around the serpentine quarry**

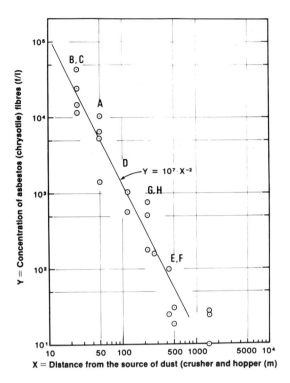

The letters have the same meaning as in Figure 3.

lower at places 50 m from such roads. These facts suggest that the high concentrations found along main roads were due to scatter from braking operations and rescattering by passing vehicles. It is also considered that the low concentrations along freeways could be attributed to the fact that most vehicles rarely brake on freeways, or at least near the sampling locations.

Laboratory experiments on asbestos emissions from a brake test piece using a brake testing machine and both scanning and transmission electron microscopy demonstrated that many asbestos fibres were released during braking (the details will be reported elsewhere).

In the other areas shown in Figure 5, it was remarkable that some samples from residential areas (A), an inland industrial area (C), and around automobile repair shops (M) showed relatively high concentrations of asbestos fibres.

The geometric means of the mass concentrations of airborne asbestos were all under 0.5 ng/m$^3$, except for the last 3 settings (R, S and T) in Table 1. All the mass concentration data are plotted in Figure 6, in which the concentrations are mostly less than 1 ng/m$^3$, except for the data from main roads and the last 3 settings. The asbestos

**Fig. 5. Concentrations of airborne asbestos fibres measured in various non-occupational environments in Japan (air samples collected in 1982–1983)**

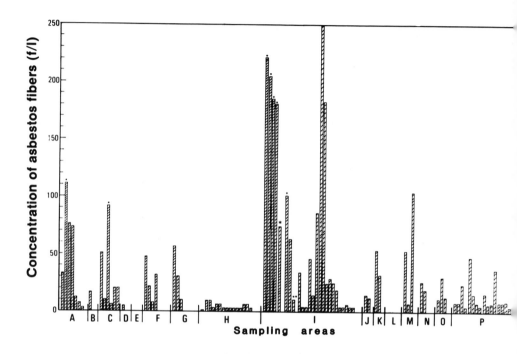

levels by main roads were also higher than those of the other areas when the data were expressed in terms of mass concentration.

### Area around an asbestos mine and factories making asbestos slate-board

High concentrations of airborne asbestos were found in a town adjacent to an asbestos mine and around factories making asbestos slate-board, as noted in Table 1 (S and T). The asbestos fibres detected near factories were both chrysotile and amosite, and longer than those found in the other areas, about 10–20% of the fibres observed being more than 5 $\mu$m in length, as mentioned in the footnote to Table 1.

### Airborne asbestos levels in isolated islands in the Pacific Ocean

Asbestos fibres were also found in air samples taken at Ogasawara Island, a small, isolated island located about 1000 km to the south of the Japanese islands, as shown in Figures 5 and 6 and Table 1 (O and P). The geometric means of 9.7 f/l and 0.05 ng/m³ seem to be slightly lower than those for the main islands of Japan. Moreover, 36 and 11 f/l were also found in air samples collected in Fiji, just to the south of the equator in the central Pacific Ocean. As only two figures have been obtained so far, it is difficult to compare the airborne asbestos level in Fiji with that in Ogasawara Island. It is,

**Fig. 6. Mass concentrations of airborne asbestos measured in various non-occupational environments**

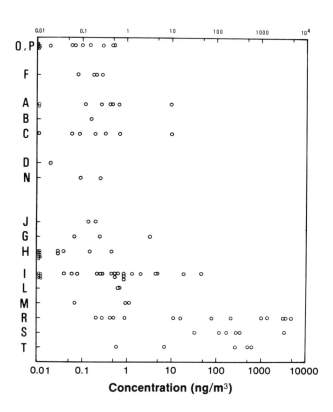

The letters refer to the environmental areas given in Table 1.

however, clear that some airborne chrysotile asbestos fibres are present in the air even in such isolated islands.

## Asbestos concentrations in snow and ice samples from Tokyo and Antarctica

To estimate the levels of airborne asbestos before the material was used in industry, measurements were made of the amount of asbestos fibres in ice samples collected from various depths in Antarctica. The ice samples had been presented by the National Institute for the Polar Region and the date of the original snow fall had already been determined. They were melted and filtered on a Nuclepore filter; subsequent preparation was similar to that used for the air sample. The asbestos concentration was expressed as number of fibres per litre of water. The results are shown in Table 2 in comparison with asbestos concentrations in snow from Tokyo and suburban Kawasaki. In the 2 ice samples formed from snow which fell on the Mizuho base in Antarctica in 1930 and 1970–1973, chrysotile fibres in concentrations of $1.09 \times 10^5$ and $.25 \times 10^5$ f/l, respectively, were found. In contrast, the concentrations of chrysotile fibres in the snow which fell in Kawasaki and Tokyo in 1984 were 10-100 times as great as those of the ice samples from Antarctica mentioned above.

**Table 2. Asbestos concentrations in snow and ice samples**[a]

| No. of samples | Sampling location | Date when snow fell | Asbestos concentration (f/l) |
|---|---|---|---|
| 11 | Mitakashi, Tokyo, Japan | January 1984 | $1.25 \times 10^7$ |
| 12 | Kawasaki, Kanagawa, Japan | January 1984 | $1.65 \times 10^6$ |
| 1 | Mizuho base, Antarctica (70-100 cm below the surface, collected in 1977) | 1970-1973 | $2.25 \times 10^5$ |
| 2 | Mizuho base, Antarctica (5 m below the surface) | ca. 1930 | $1.09 \times 10^5$ |
|  | Yamato Mountains, Antarctica (YM 179) | >10 000 years ago | $5.83 \times 10^5$ |

[a]Detection limit: $1 \times 10^4$ f/l.

## Discussion

### Asbestos emission from braking of vehicles

Airborne asbestos levels were determined by ATEM in the various environmental situations covered in the present study. As shown in Figure 5, asbestos emissions from the braking of vehicles were found. The fibres were relatively short in length and sometimes found as bundles made up of large numbers of fibrils. Stanton and Layard (1978), Stanton et al. (1981) and Pott (1978, 1980) concluded from tests on animals that long, thin mineral fibres produce a higher incidence of mesothelioma than short thick mineral fibres. There are, however, still differences of opinion as to the biological effects, and especially the carcinogenicity, of short chrysotile fibres. Further studies are therefore needed to clarify the nature of these effects since asbestos fibres found in non-occupational environments were short, and the same applies to the high concentrations sometimes detected near main roads.

### Comparison with PCM data

The data shown in Figure 5 are plotted again in Figure 7 together with the PCM data (ECCAPS, 1985). For the measurement of atmospheric asbestos concentrations by the ECCAPS, the PCM was employed as the main method because most of the relevant data on the industrial environment had previously been obtained in this way and very little information on airborne fibrous substances in the non-occupational environment had so far been made available. The PCM is also convenient and popular. Although the PCM measures all fibres and the presence of non-asbestos fibres will give rise to false positives, valuable information was obtained. The results obtained by the PCM seemed to be reasonably comparable with the ATEM counts, except for main roads, as can be seen from Figure 7. The concentrations found by the PCM were in the range 0.1–10 f/l with a geometric mean of 1.0. It was therefore possible to compare the airborne fibre levels in the non-occupational environment with those in the occupational environment. A limitation of the PCM method was noted in the case of main roads, where the asbestos levels measured by ATEM were clearly much higher than those in other areas, but the PCM did not show any such difference. This discrepancy may be due to the fact that the samples from main roads consist of very fine fibres. As has already been pointed out, it is true that only the ATEM can give reliable results for all cases. However, as shown in this study, parallel measurement by the PCM and ATEM is desirable in order to compare the data for non-occupational environments with those for occupational environments already obtained.

**Fig. 7. Comparison of ATEM results with those of PCM**

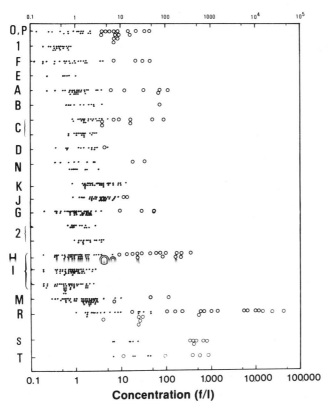

The letters refer to the environmental areas given in Table 1. In addition: 1: remote mountainous area; 2: road crossing. Solid circles, data by PCM; open circles, data by ATEM

## Airborne asbestos levels in highly polluted areas

Many studies have shown that asbestos levels in certain areas, e.g., around an asbestos mine and serpentine quarry, are higher than those in the general environment (Buyce & Dunn, 1980; Rohl *et al.*, 1977; Singh & Thouez, 1985). In Figures 6 and 7, preliminary data on the airborne asbestos levels in a town adjacent to an asbestos mine and the area around factories making asbestos slate-board have also been plotted for purposes of comparison. Further examination of these areas is now in progress. From these limited data, the mean asbestos fibre concentration in the town near the asbestos mine was about 20 times higher than the means of the other areas, and that of the mass concentration was 1000 times higher than those of the other areas, according to the ATEM data. Concentrations in the areas around a number of factories were also similar to those found in the town. The PCM data on these areas showed that the levels were higher than the average of the other areas by a factor of a few tens, but were lower by a factor of 1/100-1/300 than the limit of 2 fibres per ml prescribed by law for the working environment in Japan.

## Comparison with earlier data

The study data were presented in terms of both fibre and mass concentrations. Fibre concentration has many advantages in showing differences in air pollution levels in various environments, as mentioned above. However, as indicated by many authors, artefacts of size distribution will occur during sample preparation, particularly because of the effects of ultrasonication and/or low-temperature ashing. As a result, the number of fibres will be changed, so that the index of mass concentration will be more suitable than the fibre concentration for use in comparisons between the data obtained by different analysts. The study data expressed as mass concentrations are summarized in Figure 8, and show that the airborne asbestos levels observed in various settings in Japan are quite similar to those for outdoor ambient air reported by the US Environmental Protection Agency (1985) and other earlier data (Burdett *et al.*, 1984; Chatfield, 1983; Sébastien *et al.*, 1979).

**Fig. 8. Levels of airborne asbestos in various settings, as shown by this study**

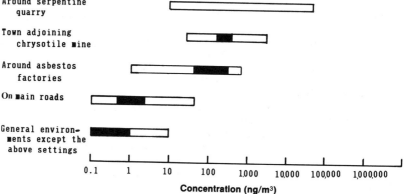

## Global circulation of asbestos fibres

Since asbestos fibres were found in the air of an isolated island in the Pacific Ocean, the long-distance circulation of asbestos fibres all over the earth was considered to be likely. It is generally believed that asbestos fibres contained in snow or ice are derived from airborne asbestos and that the concentrations are comparable with those of airborne asbestos in the past. Although the data obtained so far are limited, they suggest that airborne asbestos levels, at least in urban areas, have been elevated by industrial activities. However, further study will certainly be needed because the data examined were too few to enable this problem to be solved. It was also very interesting that some chrysotile asbestos was found even in snow which fell before the human use of asbestos began. These data may indicate that chrysotile fibres have been liberated both by industrial activities and natural weathering, and have then been circulated around the earth.

## Acknowledgements

This study was supported in part by the Japan Environment Agency. The author wishes to express his sincere thanks to Dr Shigeji Koshi, Chairman of the Committee of Counter-Asbestos Pollution Sources and all the members of that Committee and especially to Dr K. Asakuno, Dr K. Kimura, Mr M. Tanaka, Dr A. Mori and Mr T. Watanabe for air sample collection. The author also thanks Ms Y. Shinbo for her technical assistance.

## References

Burdett, G.J., Le Guen, J.M.M. & Rood, A.P. (1984) Mass concentrations of airborne asbestos in the non-occupational environment — a preliminary report of U.K. measurements. *Ann. Occup. Hyg.*, 28, 31-38

Buyce, M.R. & Dunn, J.R. (1980) Clearing air — A geological approach permits a quick and fair assessment of asbestos(?) hazard in crushed stone. *Ann. Forum Geol. Ind. Miner.*, 14, 34-39

Chatfield, E.J. (1983) Measurement of asbestos fibre concentrations in ambient atmosphere, Study No. 10. In: *Royal Commission on Matters of Health and Safety Arising from the Use of Asbestos in Ontario*, Toronto, Publications Mail Order Service

Examination Committee of Counter-Asbestos Pollution Sources (1985) *Manual for Controlling Asbestos Emission*, Tokyo, Gyousei Co. [in Japanese]

Pott, F. (1978) Some aspects on the dosimetry of the carcinogenic potency of asbestos and other fibrous dust. *Staub-Reinhalt. Luft*, 38, 486-490

Pott, F. (1980) Animal experiments on biological effects of mineral fibres. In: Wagner, J.C., ed., *Biological Effects of Mineral Fibres (IARC Scientific Publications No. 30)*, Lyon, International Agency for Research on Cancer pp. 261-272

Rohl, A.N., Langer, A.M. & Selikoff, I.J. (1977) Environmental asbestos pollution related to use of quarried serpentine rock. *Science*, 196, 1319-1322

Sébastien, P., Billon, M.A., Dufour, G., Gaudichet, A., Bonnaud, G. & Bignon, J. (1979) Levels of asbestos air pollution in some environmental situations. *Ann. N.Y. Acad. Sci.*, 330, 401-415

Singh, B. & Thouez, J.-P. (1985) Ambient air concentration of asbestos fibres near the town of Asbestos, Quebec. *Environ. Res.*, 36, 144-159

Stanton, M.F. & Layard, M. (1978) The carcinogenicity of fibrous materials. In: Gravatt, C.C., LaFleur, P.D. & Heinrichs, K.F.J., eds, *Workshop on Asbestos: Definitions and Measurement Methods (National Bureau of Standards Special Publication No. 506)*, Washington, DC, National Measurement Laboratory, pp. 143-151

Stanton, M.F., Layard, M., Tegeris, A., Miller, E., May, M., Morgan, F. & Smith, A. (1981) Relation of particle dimension to carcinogenicity in amphibole asbestoses and other fibrous minerals. *J. Natl Cancer Inst.*, *67*, 965-975

US Environmental Protection Agency (1985) *Guidance for Controlling Asbestos-containing Materials in Buildings* (EPA 560/5-85-024)

# AIRBORNE ASBESTOS FIBRE LEVELS IN BUILDINGS: A SUMMARY OF UK MEASUREMENTS

### G.J. Burdett, S.A.M.T. Jaffrey & A.P. Rood

*Dust Investigations Section, Occupational Medecine and Hygiene Laboratory, Health and Safety Executive, London, UK*

*Summary.* The UK Health and Safety Executive, in conjunction with the Department of the Environment, has carried out a number of surveys of airborne asbestos fibre concentrations in buildings. All samples have been collected on membrane filters and analysed by analytical transmission electron microscopy.

Four categories of buildings under normal occupation have been investigated; non-domestic buildings containing sprayed or trowelled asbestos, domestic buildings containing sprayed asbestos or asbestos plaster, buildings with warm air heaters containing asbestos and buildings without asbestos materials. A number of buildings have also been surveyed during and after the removal of asbestos materials.

The choice of measurement indices and analytical procedures is reviewed, before measurements are compared in terms of the concentration of asbestos fibres greater than 5 $\mu$m long. The decision whether to remove asbestos from occupied buildings is discussed with reference to the associated cost and risk. In the present survey, management of undamaged asbestos appeared preferable to large-scale removal.

## *Introduction*

This paper provides a summary of recent investigations carried out at the Occupational Medicine and Hygiene Laboratory (OMHL) of the UK Health and Safety Executive on airborne asbestos fibre concentrations in buildings. Two situations have been studied: buildings under normal occupation which contained asbestos materials in their fabric, and the removal of asbestos from buildings. It is hoped that these data can be used to gain an initial estimate of the cost and risk associated with asbestos abatement.

## *Measurement indices and analytical methods*

Fundamental to any assessment of risk is the definition of the disease and the causative agent. By 1958 (Walton, 1982), medical evidence had convinced the British asbestos industry that the fibrogenic effects of asbestos in the lung were caused by the longer fibres. A fibre counting method using optical microscopy was adopted, a fibre being defined as a particle of length >5 $\mu$m, width <3 $\mu$m and an aspect ratio >3:1.

This definition of a fibre has remained intact over the last 30 years and has become the index of exposure for occupational control and epidemiology.

Over this period, the carcinogenic effects of asbestos (e.g., lung cancers and mesothelioma) have also been fully appreciated. Whereas the fibrogenic effect can be controlled by reducing individual exposure, the mechanism leading to the onset of a carcinogenic response some 10–40 years after first exposure is not yet understood. Although laboratory and animal experiments have shown that fibre dimensions are an important factor in both direct administration (Pott, 1978; Stanton et al., 1977) and inhalation of asbestos (Davis et al., 1986), fibre chemistry and durability are also known to be important modifiers. The carcinogenic response is particularly important as no lower threshold of effect has been identified and any exposure to asbestos fibres carries a probability (or risk) of disease. In the public environment, where asbestos materials have been widely used for nearly a century, there are two important consequences: a risk evaluation and management process is necessary, and even a small risk in a large population can give a substantial number of deaths.

The question of the measurement index best suited to assess the relatively low exposures and risks in the public environment remains open. The approach adopted at OMHL has been to measure all the indices available by modern analytical transmission electron microscopy (TEM): fibre number, fibre size, fibre mass and fibre chemistry. A 'direct' method of sample preparation (Burdett & Rood, 1983) has been used to characterize the asbestos aerosol, as present at the entrance to the respiratory system. The method is based on a standard $0.8$-$\mu$m pore size membrane filter, with a collection flow rate of 3–8 l/min for a duration of approximately 4 h. Half-filters are prepared with minimal disturbance to the particulates, for both optical phase-contrast microscopy (PCM) screening of optically defined fibres and TEM analysis of the indices mentioned above.

Results obtained by the direct method differ from most of those obtained previously by indirect methods of sample preparation. Indirect sample preparation methods alter the asbestos fibre concentration and size distribution to varying extents depending on the method used, the type of asbestos in the sample, the amount of fibres held in the matrix and its solubility. Indirect results can therefore usually only be interpreted in terms of the mass of asbestos. Depending on the sample preparation, this may be a measure of the total mass of airborne asbestos, regardless of whether the fibres and agglomerates are respirable, or a partial estimate of mass, ignoring the large agglomerates. A significant overestimate of the mass of respirable fibres is likely, unless size-selective sampling is employed.

It is important to realize that airborne exposure is not a measure of dose; only a portion of the inhaled asbestos will be deposited and much will be removed by macrophage and mucociliary action. The situation is further complicated by the tendency of chrysotile, in particular, to split and divide in lung fluids. In fact, the removal of chrysotile can be so efficient that, in lung autopsies of Canadian chrysotile miners, Pooley (1976) found that a trace contaminant — tremolite asbestos — was more prevalent than chrysotile. Although a more useful measure of dose may be obtained from lung samples, sample preparation techniques alter the fibre number and size distribution and the result will not be specific to a single non-occupational situation. At the present time, the multicharacterization of fibre exposure offers the most meaningful measurement from which dose and risk can be modelled (Burdett et al., 1988).

In this summary, only one index has been considered — the concentration of asbestos fibres >5$\mu$m long, <3 $\mu$m wide and of aspect ratio >3:1, as analysed by the TEM. All reference to fibre concentrations, expressed as fibres per ml (f/ml), is in terms of the above index, unless otherwise indicated. Although this includes a proportion of fibres which would not be visible by PCM, long, thin asbestos fibres are often thought to be the most carcinogenic. Other indices of measurement from the sites have been published elsewhere (Burdett & Jaffrey, 1986; Jaffrey et al., 1988).

As a sample applies only to the immediate vicinity and point in time, a site average for each sampling exercise has been calculated by dividing the sum of the asbestos fibres counted in each individual TEM analysis by the sum of the equivalent volumes of air analysed in each individual TEM analysis. This has the advantage of a multipoint characterization of the site and increased confidence in reporting low concentrations. A limit of quantification (LOQ) of 4 asbestos fibres counted has been used, which, on the assumption of a Poisson distribution, ensures that the limit of detection (one asbestos fibre counted) would be achieved with over 95% confidence on any recount of the same sample. The decimal place to which the result is reported is calculated by dividing 4 by the total volume of air analysed in the TEM analysis, e.g., 40 litres of air would give a concentration of 0.0001 f/ml.

## Asbestos in buildings under normal occupation

Two types of installation have been studied: sprayed/trowelled asbestos coatings on ceilings and asbestos material in warm air circulating heaters. Of the wide variety of asbestos products in buildings, these 2 groups appeared to have the most potential to give a continuous release of airborne fibres. A summary of the results is given in Table 1.

Of the 39 asbestos-containing sites, some 30 (73%) did not exceed the LOQ. At 12 sites where sprayed or trowelled asbestos was still present, only 3 (25%) gave values above the LOQ. Half the buildings showed evidence of slight damage to the asbestos coatings and 2 buildings had more extensive damage. One room at site 7, which had a large area of damaged sprayed asbestos, gave the highest individual sample result of 0.012 f/ml of chrysotile and was the only site to exceed an average of 0.001 f/ml.

Average asbestos concentrations in 24 buildings with warm air heaters containing asbestos did not exceed 0.001 f/ml, except at one site (0.002 f/ml), where a heater had been dismantled and the asbestos insulation left damaged and exposed. Only 5 sites (21%) equalled or exceeded the LOQ. Ambient samples (outside the buildings), 4 control houses without any asbestos materials, and laboratory blank filters, all gave levels below the LOQ.

Follow-up studies at a number of the sites included in Table 1 have been carried out. Four schools (sites 7-10) with sprayed chrysotile and amosite on the ceilings were

**Table 1. Summary of average concentrations of asbestos fibres >5 μm long measured by transmission electron microscopy**

| Site no. | Number of samples analysed | Type of asbestos identified[a] | Average concentration (f/ml) |
|---|---|---|---|
| *Non-domestic buildings containing sprayed or trowelled asbestos insulation* | | | |
| 1 | 3 | C | <0.001 |
| 2 | 4 | A+K | 0.0005 |
| 3 | 6 | C | <0.0002 |
| 4 | 13 | A+C | <0.0001 |
| 5 | 16 | A | <0.0001 |
| 6 | 26 | A | <0.0008 |
| 7 | 6 | A+C | 0.002 |
| 8 | 9 | A | <0.0003 |
| 9 | 8 | A+C | <0.0003 |
| 10 | 10 | C | <0.0005 |
| 11 | 5 | C | <0.0003 |
| 12 | 5 | A+C | <0.001 |
| *Domestic buildings containing sprayed asbestos or asbestos plaster* | | | |
| 13 | 5 | C | <0.0001 |
| 14 | 14 | A+C | 0.0004 |
| 15 | 16 | A+C+K | 0.0007 |
| *Buildings with warm air heaters containing asbestos* | | | |
| 16 | 3 | C | 0.0003 |
| 17 | 9 | A | 0.002 |
| 18 | 5 | C | <0.0001 |
| 19 | 5 | A+C | <0.0001 |
| 20 | 4 | C+A | 0.001 |
| 21 | 8 | C | 0.0001 |
| 22 | 3 | C | <0.001 |
| 23 | 2 | A+C | <0.001 |
| 24 | 3 | ND | <0.001 |
| 25 | 2 | A | <0.0001 |
| 26 | 2 | ND | <0.002 |
| 27 | 5 | A+C | <0.001 |
| 28 | 3 | ND | <0.001 |
| 29 | 2 | ND | <0.001 |
| 30 | 2 | A+C | 0.001 |
| 31 | 2 | A | <0.001 |
| 32 | 4 | A+C | <0.001 |
| 33 | 3 | A | <0.001 |
| 34 | 2 | A+C | <0.001 |
| 35 | 2 | ND | <0.001 |
| 36 | 2 | C | <0.001 |
| 37 | 2 | ND | <0.001 |
| 38 | 2 | ND | <0.001 |
| 39 | 4 | A | <0.001 |
| *Buildings/electric warm air heaters without asbestos materials* | | | |
| 40 | 8 | ND | <0.0003 |
| 41 | 6 | ND | <0.001 |
| 42 | 3 | ND | <0.001 |
| 43 | 2 | C | <0.001 |

[a] A = amosite; C = chrysotile; K = crocidolite; ND = not determined.

resampled one year later. The asbestos insulation from one of the schools (site 7) had been removed some 10 months earlier. As can be seen from Table 2, there is little change as compared with Table 1, but the levels of asbestos fibres were still slightly elevated at site 7, even though the asbestos had been removed.

Table 2. Results obtained in a resurvey of 4 schools with sprayed asbestos on ceilings[a]

| Site no. | No. of samples analysed | Type of asbestos identified[b] | Concentration of asbestos fibres $>5\mu m$ long (f/ml) |
|---|---|---|---|
| 7[c] | 11 | A+C | 0.0008 |
| | (9) | (A+C) | (0.002) |
| 8 | 9 | C | <0.0002 |
| | (9) | (C) | (<0.0003) |
| 9 | 9 | A+C | 0.0004 |
| | (8) | (A+C) | (<0.0003) |
| 10 | 8 | C | <0.0003 |
| | (10) | (C) | (<0.0005) |

[a]Values found 12 months earlier are shown in parentheses.
[b]A, amosite; C, chrysotile.
[c]The asbestos was removed 1 month after the first survey.

Measurements made during normal occupation do not give much information about short-term events and minor maintenance. A limited experiment to measure the release of badly damaged friable amosite asbestos has been reported by Burdett (1987). It was found that moderate (1.5 m/s) air currents were insufficient to suspend fibres and direct physical disturbance was required. Paik et al. (1983), using PCM analysis on personal samples from 127 buildings where maintenance activities were carried out, found geometric mean concentrations of 0.08–0.19 f/l for various types of work. Further work and co-operation with other workers in TEM assessments of fibre releases from maintenance activities are at present under way.

## Removal of asbestos from buildings

Asbestos removal from buildings has been studied in some detail. An example of some of the limitations of dry removal of asbestos is shown in Figure 1, where the position and the measured asbestos fibre concentration (fibres $>5 \mu m$ long per ml) are shown schematically. A sprayed crocidolite insulation was being removed from the walls and ceiling of a large unoccupied industrial building prior to demolition. The building was divided into 3 equal areas by studwork and polythene sheets, the removal taking place in the centre section (Figure 1a). Access to the centre section was from the east end of the building via a 3-stage polythene sheet entrance. An air filtration system was used to create a slight negative pressure inside the centre section. Measured TEM concentrations of crocidolite fibres from static membrane filter samples in 3 phases of the work are shown in Figure 1b–d. Even after close inspection, the addition of a further air filtration unit and a retest of the enclosure with a smoke generator, this trial showed how difficult containment can be in some situations. In this instance, wind pressure on the enclosure was responsible.

**Fig. 1. Average TEM concentrations of asbestos fibres >5 $\mu$m long (f/ml) during the dry removal of sprayed crocidolite asbestos: (a) plan, (b) during removal; (c) no work in progress; (d) during removal after further improvements to the enclosure**

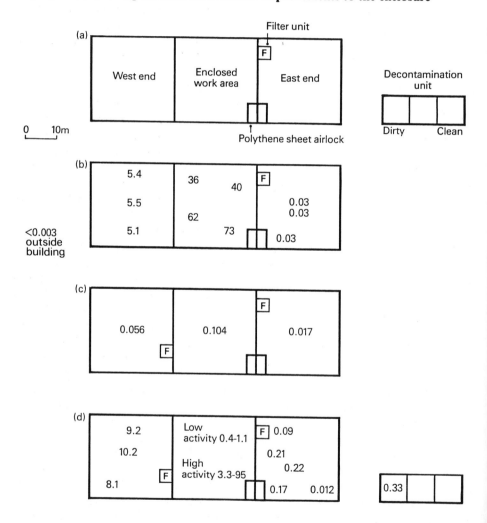

A more detailed history of fibre concentrations in 2 buildings before, during and after asbestos removal is given in Figures 2–5. Both buildings contained sprayed-trowelled amosite insulation behind a false ceiling of non-asbestos perforated tiles. The average concentrations of asbestos fibres $>5\mu$m long are plotted on a log axis with the various stages of monitoring listed on the abscissa. Three post-removal samples with high concentrations (e.g., 10.7 f/ml) were excluded as they could not be explained.

In the first building, a 6-floor teaching block, 2 phases of asbestos removal are shown in Figure 2. Phase 1 saw the removal of asbestos from the top 3 floors during an Easter vacation, refurbishment being carried out at intervals during the following 26 weeks. Phase 2 saw the removal of asbestos from the lower 3 floors during the summer vacation, some 12 weeks later. Refurbishment took place over the following 9 weeks.

Samples taken inside the building during normal occupation, prior to removal, gave an average concentration of $<0.0002$ f/ml, which was not significantly different from the ambient concentration (0.0003 f/ml). Although no samples were taken inside the enclosure during the dry removal of the asbestos, airborne concentrations of the order of 1–50 f/ml were likely. A single sample (0.29 f/ml), taken in the stairwell adjacent to an enclosure, demonstrated that some leakage was occurring. Samples taken after the removal had been completed and the areas cleaned and vacated, gave measurable concentrations of amosite fibres. The levels recorded in both phases of removal decayed with time, although in phase 1 the refurbishment activities initially generated higher airborne concentrations. Samples taken 9 weeks after reoccupation, during normal activities, gave average amosite concentrations of 0.0004 f/ml in both phase 1 and phase 2 areas, slightly above the levels found before any asbestos was removed.

The second building studied in detail consisted of 4 floors and a basement. The building was of steel frame and concrete panel construction and all the steel members (with the exception of the top floor and basement) had a covering of about 30 mm of amosite-based insulation. Results from simulated maintenance activities and dry asbestos removal are shown in Figure 3. Only one sample was taken inside an enclosure during stripping and was very heavily loaded, with an estimated fibre concentration of 10–30 f/ml. As found at other sites, there was measurable leakage from the plastic enclosures during asbestos removal. It was also found that outdoor samples taken close to the building in an approximately downwind position gave increased levels when the ground floor and the first floor were being stripped.

The persistence of airborne amosite fibres after removal on the first and second floors is shown in Figures 4 and 5. As the building was left vacant for some months without refurbishment, the effect of disturbing the floor dust was also monitored. The vinyl floor coverings had been removed during the final cleaning of the building, leaving an unsealed concrete surface. In the course of a detailed visual inspection, several areas were located where traces of amosite material were suspected. In the first exercise, dust on shelves and a small floor area where contamination was expected was moderately disturbed with a hand brush for less than 1 min. On a separate visit, dust was vigorously disturbed by heavy sweeping with a broom for 5 min in suspect areas.

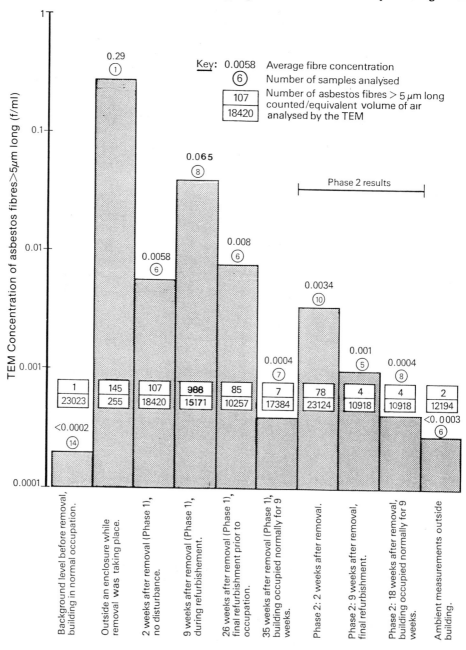

Fig. 2. Average TEM concentrations of asbestos fibres >5 μm long before, during and after two phases of an asbestos-removal programme from a six-storey teaching block

**Fig. 3. Average TEM concentrations of asbestos fibres >5 μm long during disturbance and removal of asbestos from a four-storey building**

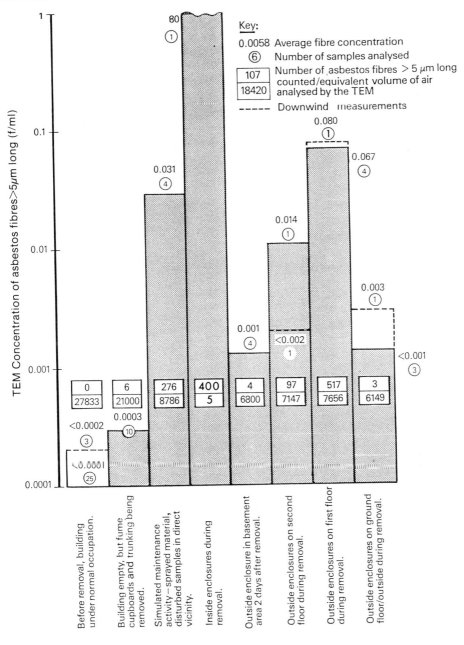

**Fig. 4.** Average TEM concentrations of asbestos fibres >5 μm long, first floor, before, during and after asbestos removal

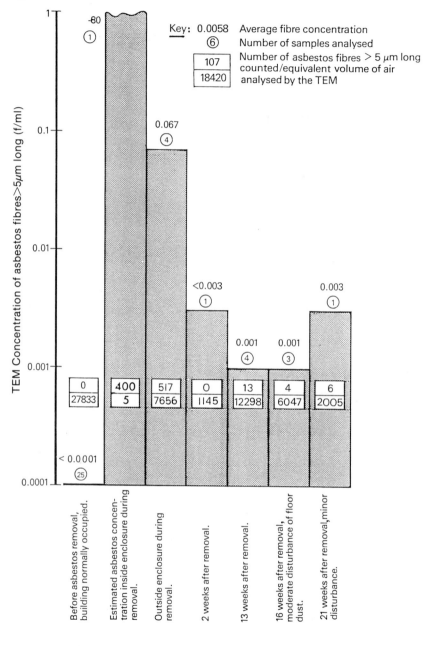

**Fig. 5.** Average TEM concentrations of asbestos fibres >5 μm long, second and third floors, before, during and after asbestos removal

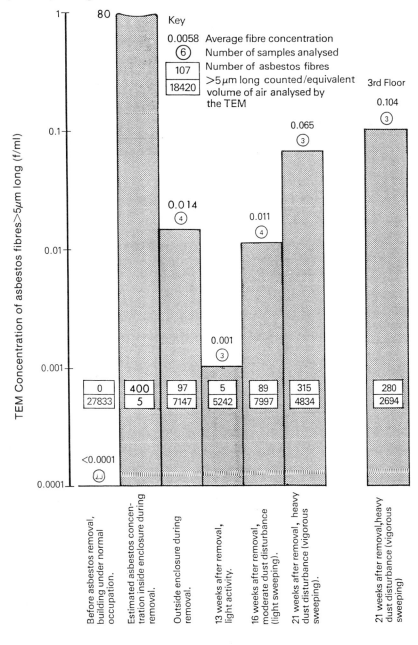

Static samples were collected in the near vicinity for periods of 3–4 h. Considerable resuspension of amosite fibres was found.

## Discussion and conclusions

Under normal occupation, about one-quarter of buildings with sprayed asbestos gave fibre concentrations above the LOQ. This is in agreement with other published work where direct sample preparation was used (Ontario Royal Commission, 1984). The results have also been compared in terms of asbestos mass concentrations with indirect measurements in similar buildings (Burdett & Jaffrey, 1986). A higher proportion of asbestos-polluted buildings (about half to three-quarters) was found. This increase can be explained by the fact that different methods of sample preparation give different measurement indices. Warm air heaters containing asbestos materials did not release fibres in concentrations in excess of 0.001 f/ml during normal usage and only 5 sites of the 24 surveyed exceeded the LOQ.

All 3 asbestos removal sites gave evidence of leakage from the plastic enclosures. If peak concentrations of 100 f/ml are assumed, it can be calculated that the containment must be about 99.9% efficient for leakage to be below UK control limits for crocidolite and 99.999% efficient not to be detected by TEM analysis of a single membrane filter sample. The concentrations shown in Figure 1 indicate that, on both occasions, only 90% containment was achieved. Dry removal of asbestos places stringent demands on the work practices adopted. It has long been known (Sawyer & Spooner, 1978) that, if the asbestos can be made damp, a reduction in fibre concentration during asbestos removal by a factor of about 30 can be achieved.

The environment surrounding a building can be contaminated by a number of other routes: incomplete removal of the asbestos leaving material intact, bulk spillage of material during handling of waste outside the enclosure, fall-off from workers' clothes and footwear, and specific incidents where the precautions were inadequate for one reason or another. The results to date suggest that residual asbestos can continue to present a risk for some months after removal. The extent of the risk is dependent on: (1) the amount and the distribution of the asbestos remaining; (2) the time to reoccupation (if the building is vacated); (3) the ventilation; and (4) the degree of disturbance. Again (Sawyer & Spooner, 1978), it would appear that maintenance and custodial staff may be more at risk than other occupiers of the building.

With the data available, some indication of the cost and risk associated with asbestos removal can be obtained. Doll and Peto (1985), using the above data from buildings under normal occupation, calculated a minimal risk of 1 in 100 000 of developing an asbestos-related cancer. The higher levels associated with damaged friable sprayed asbestos suggest that some action is necessary to minimize exposures. Physical damage is evidence of previous airborne release and a careful assessment of a possible recurrence must be made. This should cover the extent of the damage, the current and future use of the area, the accessibility of the material and any reason for a continuing deterioration of the material (e.g., water penetration or maintenance activities). Some management action will always be required, even if only to place warning signs, and assessment schemes are available (Department of the Environment, 1986).

At some point, removal of asbestos will have a cost-benefit advantage. This might be thought to be the point when the continuous management of a building becomes more expensive than removal, but the decision will invariably depend on a number of unquantifiable factors, such as the perceived risk, the difficulty in selling or letting the building, or the refusal of people to work in it. Each factor will vary with time and from one country to another, the ability to obtain compensation being an important factor.

A risk-benefit analysis is often used to justify a course of action. If removal is recommended, it is usually on the assumption that, on completion, the risk from asbestos ceases. The data previously given suggest that this assumption is invalid. An example of a school used previously (Burdett, 1986), demonstrates the problem involved in deciding whether to manage or remove asbestos. A child's total exposure time in school can be taken as 15 000 hours. If throughout this period an average exposure to asbestos of 0.0005 f/ml is assumed, the cumulative exposure is 7.5 f/ml. The results reported here suggest that cumulative exposures of the same order may be achieved in a few days if pupils are in the vicinity of the enclosure during removal. If a building is reoccupied soon after removal, asbestos concentrations may remain higher than before for some weeks and, in some circumstances, for longer periods. In this situation, the benefits for the occupants do not appear to be very great, especially as the incidence of mesothelioma has been shown by Peto (1983) to have a fourth-power dependence on the age since first exposure.

If the asbestos material is in a condition such that it can be managed with only minor maintenance and repairs, it is unlikely that levels will exceed 0.0005 f/ml except when work is being carried out or damage takes place. At present, very little information on non-occupational exposures during maintenance or damage is available for purposes of comparison with exposures from removal. However, the potential for an increasing release of asbestos remains and, at some time before demolition, removal will be necessary.

From the present results, it must be concluded that asbestos removal cannot be assumed to remove the risk to the occupants. When large areas of asbestos are removed from buildings, it remains difficult, with the existing technology, to eremoval would appear to give the lowest risk at present, especially if the asbestos material is in good condition and has suffered little physical disturbance.

## Acknowledgements

The co-operation and help of Dr P. Corcoran, Dr J. Llewellyn and A. Wilson of the Department of the Environment was instrumental in enabling such a wide range of sites to be covered. Many other people in the Health and Safety Executive also gave their help.

## References

Burdett, G.J. & Rood, A.P. (1983) Membrane-filter, direct-transfer technique for the analysis of asbestos fibres or other inorganic particles by transmission electron microscopy. *Envir. Sci. Technol.*, *17*, 643-648

Burdett, G.J. (1986) Measured asbestos concentrations in buildings. In: *Proceedings of a NSCA Workshop on Asbestos: Policies for the Future, Newcastle, 26-27 March 1986*, Brighton, National Society for Clean Air, Part II, pp. 1-25

Burdett, G.J. & Jaffrey, S.A.M.T. (1986) Asbestos concentrations in public buildings. *Ann. Occup. Hyg.*, *30*, 185-199

Burdett, G.J. (1987) The measurement of airborne asbestos releases from damaged amosite insulation subject to physical attrition. In: *Proceedings, Workshop on Asbestos Fibre Measurements in Building Atmospheres, Ontario Research Foundation, Mississauga, Ontario, Canada, March 1985*, Missisauga, Ontario Research Foundation, pp. 209-228

Burdett, G.J., Firth, J.G., Rood, A.P. & Streeter, R. (1988) Application of fibre retention and carcinogenicity curves to fibre size distributions of asbestos. In: *Inhaled Particles VI* (in press)

Davis, J.M.G., Addison, J., Bolton, R.E., Donaldson, K., Jones, A.D. & Smith, T. (1986) The pathogenicity of long versus short fibre samples of amosite asbestos administered to rats by inhalation and intraperitoneal injection. *Br. J. Exp. Pathol.*, *67*, 415-430

Department of the Environment (1986) *Asbestos Materials in Buildings*, London, Her Majesty's Stationery Office

Doll, R. & Peto, J. (1985) *Asbestos: Effects on Health of Exposure to Asbestos*, London, Her Majesty's Stationery Office

Jaffrey, S.A.M.T., Burdett, G.J. & Rood, A.P. (1988) The measurement of airborne asbestos concentrations in two U.K. buildings: before, during and after asbestos removal. *Int. J. Environ. Stud.* (in press)

Ontario Royal Commission (1984) *Report of the Royal Commission on Matters of Health and Safety Arising from the Use of Asbestos in Ontario*, Vol. 2, Toronto, Ontario Government Bookshop, pp. 557-576

Paik, N.W., Walcott, R.J. & Brogan, P.A. (1983) Worker exposure to asbestos during removal of sprayed material and renovation activity in buildings containing sprayed material. *Am. Ind. Hyg. Assoc. J.*, *44*, 428-432

Peto, J. (1983) Dose and time relationships for asbestos-related cancers and risk assessment in the general population. In: Reinisch, D., Schneider, H.W. & Birkner, K.-F., eds, *Fibrous Dusts—Measurement, Effects, Prevention (VDI-Berichte 475)*, Düsseldorf, VDI-Verlag pp. 309-311

Pooley, F.D. (1976) An examination of the fibrous mineral content of asbestos lung tissue from the Canadian chrysotile mining industry. *Environ. Res.*, *12*, 281-287

Pott, F. (1978) Some aspects on the dosimetry of the carcinogenic potency of asbestos and other fibrous dusts. *Staub-Reinhalt. Luft*, *38*, 486-490

Sawyer, R.N. & Spooner, C.M. (1978) *Sprayed Asbestos Containing Materials in Buildings—A Guidance Document* (EPA 450/2-78-014), Research Triangle Park, NC, US Environmental Protection Agency

Stanton, M.F., Layard, M., Tegeris, A., Miller, E., May, M. & Kent, E. (1977) Carcinogenicity of fibrous glass: Pleural response in the rat in relation to fiber dimension. *J. Natl Cancer Inst.*, *58*, 587-603

Walton, W.H. (1982) The nature, hazards and assessment of occupational exposure to airborne asbestos dust: A review. *Ann. Occup. Hyg.*, *25*, 115-247

# LEVELS OF ATMOSPHERIC POLLUTION BY MAN-MADE MINERAL FIBRES IN BUILDINGS

A. Gaudichet, G. Petit, M.A. Billon-Galland & G. Dufour

*Laboratoire d'Etude des Particules Inhalées,
Département de Paris, Paris, France*

*Summary.* The widespread use of man-made mineral fibres (MMMF) as insulation products in buildings can be a potential source of indoor pollution, but few data are available. Pollution levels were determined at 79 indoor and 18 outdoor locations. The standardized method (membrane filter method) was adapted to the environmental settings: 10 m³ of air was sampled over a period of 5 days; the specimen was prepared by an indirect method (low-temperature ashing); counting was carried out at a magnification of 250; MMMF were identified by their optical properties under the polarizing optical microscope.

Indoor pollution levels were found to be in the range 0–6230 respirable fibres per m³, with a median value of 3. Outdoor airborne pollution levels were lower than 15 respirable fibres per m³, with a median value of 1.

More data are needed in order to make a more precise assessment of the risk associated with environmental MMMF pollution; a standardized method of obtaining such data is also necessary.

## *Introduction*

World production of man-made mineral fibres (MMMF) has been estimated at $4585 \times 10^3$ tons per year in 1973. Of this amount, 80% is accounted for by various types of mineral wool used essentially for thermal or sound insulation, mainly glass wool, rock wool, slag wool and glass fibres. All these materials are amorphous and consist essentially of $SiO_2$, $Al_2O_3$, $B_2O_3$ and alkali. Although their nominal diameter is about 6 $\mu$m, the range of particle size is very large (1–25 $\mu$m), depending on the method of manufacture (Hartung, 1984; Klingholz, 1977; Konzen, 1984). The proportion of respirable fibres emitted during manufacture varies between 66 and 79% (Head & Wagg, 1980).

MMMF are present in public buildings and private houses mainly in the materials used for thermal or sound insulation, either alone or in association with asbestos. Such materials take the form mainly of friable surfacing materials (sprayed insulation), materials for coating pipes, under-ceiling panels or partitions, and air filters used in air-conditioning installations. MMMF can also be used in domestic appliances, such as ventilator hoods, joints in cookers, etc.

Very few data are available on pollution inside buildings. Balzer et al. (1971) and Esmen et al. (1980) showed that MMMF in air-conditioning transmission systems did not make any significant contribution to ambient air pollution. Rindel (1984) estimated that ambient air concentrations in buildings were smaller by a factor of 100–1000 than those in occupational environments. Finally, concentrations in schools and other public buildings were measured by Schneider (1984), who found that respirable fibres were present at levels ranging from undetectable to 84 000 fibres per $m^3$.

The Laboratoire d'Etude des Particules Inhalées (LEPI), which has been measuring pollution by asbestos fibres inside buildings since 1975, has in recent years also undertaken measurements of possible pollution by MMMF. The results are presented below.

## Sampling and analytical methods

### Search for and identification of materials in a building liable to contain MMMF

Registers of both the construction and inspection of buildings were consulted in the search for such materials. In the case of sprayed insulation, the problem of non-uniformity arises; to overcome this, three core samples must be taken at several points in the coating.

Under the polarizing optical microscope, MMMF can be clearly distinguished by means of their optical properties from natural (crystalline) mineral fibres and in particular from asbestos and even from cellulose fibres: they are more or less rectilinear and distortions produced on cooling (drops) may be present; their relief is generally high and, above all, they are isotropic under cross-polarized lighting (Monkman, 1979).

### LEPI method of sampling and analysis

In industrial hygiene, two methods of evaluating atmospheric pollution by MMMF have been used in the European and American research programmes (Walton, 1984): determination of the mass concentration of total respirable dust and more recently the determination of the numerical concentration of MMMF by the membrane filter method (standardized method for the determination of asbestos).

In the absence of a standardized method for environmental situations, the sampling strategy adopted by LEPI for MMMF is similar to that used for the determination of atmospheric pollution in buildings by asbestos fibres (Sébastien et al., 1979):

— passage of air through a cellulose ester membrane filter 47 mm in diameter and 0.45 $\mu$m nominal pore size;

— a flow rate of 5 l/min;

— samples taken from Monday to Friday during working hours, giving a total air volume of about 10 $m^3$. The fact that large amounts of dust which has settled out is redispersed by human activity in buildings is the reason why sampling is restricted to working hours.

The membrane thus obtained is incinerated at low temperature in order to eliminate particles of organic matter. The ash is suspended in liquid and concentrated by microfiltration on a cellulose ester filter which is mounted between slide and cover-slip after clearing with triacetin.

The mineral fibres are counted by scanning the whole area of the membrane at a magnification of 250. Fibres having the optical properties of MMMF (checked under cross-polarized lighting) are counted and measured so that the numerical concentration of so-called respirable fibres (of diameter less than or equal to 3 $\mu$m) can be estimated separately from that of fibres of diameter greater than 3 $\mu$m.

## Results

LEPI has carried out 79 determinations of pollution by MMMF inside buildings since 1981. The materials suspected of causing such pollution were as follows:

— in 31 locations: sprayed under-ceiling surfacing materials consisting entirely of MMMF, of which 28 were in contact with the ambient air and 3 were located inside a plenum;

— in 8 locations: mixed sprayed surfacing materials consisting of MMMF and asbestos, of which 1 was located inside a plenum;

— in 17 locations: surfacing materials associated with ventilation or air-conditioning systems;

— in 21 locations: wall panels;

— in 2 locations: various applications.

The results obtained are given in Table 1 for the respirable fibres and Table 2 for all fibres combined.

A wide range of concentrations was found in the 79 locations, varying from 0.2 to 6778 total fibres per m$^3$ and from 0 to 6230 respirable fibres per m$^3$. This is reflected in the large difference between the mean and median values: 258 and 13 total fibres per m$^3$ respectively and 225 and 8 respirable fibres per m$^3$. As far as the different types of surfacing material are concerned, it would appear that the sprayed insulation is the biggest source of pollution and is associated with a higher proportion of respirable fibres.

For the 31 locations where there was sprayed insulation consisting entirely of MMMF, we compared the pollution levels (expressed as total fibres per m$^3$) with an algorithmic index (see Figure 1) based on a visual inspection of:

— the condition of the surfacing material, in increasing order of deterioration, on the scale 0, 2–5;

— water damage, graded 0, 1 and 2;

— contact with the ambient air (most sprayed insulation was in direct contact) graded 0, 1 and 4;

**Table 1. Characteristics of atmospheric pollution by respirable man-made mineral fibres (MMMF)**

| Source of pollution inside buildings | No. of samples | Numerical concentration (fibres per m$^3$) | | | | |
|---|---|---|---|---|---|---|
| | | Mean | Standard deviation | Median | Minimum | Maximum |
| **Type of surfacing material** | | | | | | |
| Sprayed insulation (MMMF only) | 31 | 225 | 1098 | 8 | 0 | 6230 |
| Sprayed insulation (MMMF + asbestos) | 8 | 116 | 296 | 3 | 1 | 900 |
| Surfacing materials associated with ventilation or air conditioning | 17 | 63 | 141 | 1 | 0 | 470 |
| Wall panels | 21 | 88 | 271 | 1 | 0.1 | 1100 |
| Other | 2 | | | | 0 | 1 |
| **Overall** | | 137 | 715 | 3 | 0 | 6230 |
| **Outdoor background level** | 18 | 2 | 3 | 1 | ND[a] | 15 |

[a]ND, not detected.

— the accessibility, graded 0, 1 and 4 (inaccessible, occasional and easy access);
— the activity in the room concerned, graded 0, 1 and 2;
— the presence or otherwise of ventilation, graded 0 and 1;
— the friability (slight, moderate, high), graded 0, 1 and 4.

The index is given by the sum of the first 6 grades multiplied by the friability; the higher it is, the more the surfacing material is likely to be a source of pollution.

In addition to all these determinations inside buildings, samples were taken, for purposes of comparison, under the same operating conditions, at 18 locations outdoors in Paris (Tables 1 and 2). The outdoor background values found were in the range 0.3–22 total fibres per m$^3$ and 0–15 respirable fibres per m$^3$.

## Discussion

With regard to methodology, our experience with the metrology of mineral fibres and MMMF in the environment, and in particular inside buildings, is the basis for the following comments.

**Table 2. Characteristics of atmospheric pollution by total man-made mineral fibres (MMMF)**

| Source of pollution inside buildings | No. of samples | Numerical concentration (fibres per m$^3$) | | | | | |
|---|---|---|---|---|---|---|---|
| | | Mean | Standard deviation | Median | Minimum | Maximum | No. of respirable fibres |
| **Type of surfacing material** | | | | | | | |
| Sprayed insulation (MMMF only) | 31 | 258 | 1193 | 13 | 2 | 6778 | 87 |
| Sprayed insulation (MMMF + asbestos) | 8 | 119 | 298 | 4 | 1 | 907 | 97 |
| Surfacing materials associated with ventilation or air conditioning | 17 | 171 | 370 | 3 | 0.5 | 1135 | 37 |
| Wall panels | 21 | 135 | 405 | 3 | 0.2 | 1616 | 65 |
| Other | 2 | | | | 4 | 26 | |
| **Overall** | | 184 | 797 | 6 | 0.5 | 6778 | 74 |
| **Outdoor background level** | 18 | 4 | 5 | 2 | 0.3 | 22 | |

As far as sampling methods are concerned, the low levels commonly encountered in the environment call for a duration of sampling greater than in industrial hygiene. Under these conditions, the direct counting of fibres on filters by optical and electron microscopy is made difficult by the total particle load (organic and inorganic). An indirect method of preparation is therefore recommended. This makes possible a better distribution of the fibres on the membrane and the elimination of particles that could interfere with the counting of the fibres (Chatfield, 1984; Sébastien et al., 1985).

As far as the counting of fibres on the membrane is concerned, while electron microscopy, by virtue of its greater resolution, is better adapted to the counting of fibres of small diameter, if account is taken of the characteristics of the MMMF used in the manufacture of insulating materials, optical microscopy remains both the most suitable and the quickest method of evaluating the pollution of the ambient air in the majority of buildings.

In light microscopy, if the standardized method of determining pollution by asbestos fibres is used, counting is effected by phase-contrast microscopy in random fields and at a magnification of 400 (Asbestos International Association, 1982). Nevertheless, for 9 samples, comparative counts were made at two different

**Figure 1. Values of algorithmic index and concentration of total MMF**

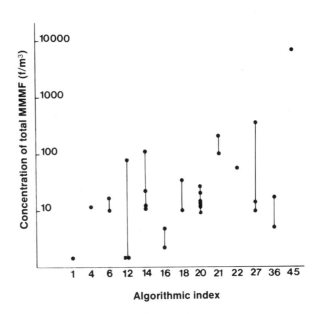

magnifications, namely 250 (entire membrane scanned) and 400 (100 random fields). Comparison of pairs of measurements on the 9 slides showed a high degree of correlation between the two magnifications; values in Spearman's test for the respirable and total fibres were 0.89 and 0.91 respectively ($p<0.05$). Counting can therefore be carried out at a lower magnification ($\times$ 250) without phase contrast; the identification of MMMF can be checked by their isotropy under cross-polarized lighting.

The results of this study show that:

— average background concentrations of MMMF in outdoor locations in Paris are very low (2–4 fibres per m$^3$) in comparison with those found by Friedrichs (1979) in Düsseldorf (820 fibres per m$^3$) and by Balzer (1976) in various towns in California (2200 fibres per m$^3$, of which 860 were glass fibres). Nevertheless, it is difficult to compare these results because of differences in sampling and analytical methods;

— when compared with our outdoor background values, the levels of ambient pollution at our 79 locations are of the same order of magnitude and, in general, very low, as shown by the median values. Nevertheless, the highest levels (0.006 fibres per ml) are 30 times less than that recently called for by Doll (1987) in order to eliminate all risk of lung cancer in those working with MMMF;

— finally, for sprayed surfacing materials, which seem to cause the greatest pollution, the lack of any correlation between the algorithmic index and the level of ambient pollution with MMMF, even though based on only a small number of results, seems similar to that found by LEPI in a co-operative study with the US Environmental Protection Agency on indoor pollution by asbestos fibres (Sébastien et al., 1982).

In conclusion, metrology of MMMF in the environment, and particularly inside buildings, calls for:

— the definition of a standardized method of sampling and analysis adapted to both the outdoor and indoor urban environment;

— the testing of this methodology by means of comparison programmes;

— an increase in measurements of the outdoor and indoor background levels so as to enable doctors and hygienists to assess the possible risk;

— a better evaluation of the capacity of different types of insulating material to emit MMMF into the ambient air as a function of their technical characteristics.

## References

Asbestos International Association (1982) *Recommended Technical Method No. 1: Reference Method for the Determination of Airborne Asbestos Fibre Concentrations at Workplaces by Light Microscopy (Membrane Filter Method)*, London

Balzer, J.L. (1976) Environmental data: airborne concentrations of fibrous glass found in various operations. In: *Proceedings of a Symposium on Occupational Exposure to Fibrous Glass, College Park, Maryland*, Washington, DC, US Department of Health, Education, and Welfare, pp. 82-90

Balzer, J.L., Cooper, W.C. & Fowler, D.P. (1971) Fibrous glass-lined air transmission systems: an assessment of their environmental effects. *Am. Ind. Hyg. Assoc. J.*, 32, 512-518

Chatfield, E.J. (1984) Measurement and interpretation of asbestos fibre concentration in ambient air. In: Baunach, F., ed., *Fifth International Colloquium on Dust Measuring Technique and Strategy*, Johannesburg, pp. 269-296

Doll, R. (1987) Overview and conclusions. In: *Proceedings, WHO/IARC Conference on Man-Made Mineral Fibres in the Working Environment. Ann. Occup. Hyg.*, 31, 805-820

Esmen, N.A., Whittier, D., Kahn, R.A., Lee, T.C., Sheehan, M. & Kotsko, N. (1980) Entrainment of fibers from air filters. *Environ. Res.*, 22, 450-465

Friedrichs, K.H. (1979) Morphological aspects of fibres. In: Lemen, R. & Dement, J.M., eds, *Dust and Disease*, Park Forest South, IL, Pathotox Publishers, pp. 51-64

Hartung, W.J.A. (1984) Technical history of MMMF with particular reference to fibre diameter and dustiness. In: *Biological Effects of Man-Made Mineral Fibres*, Vol. 1, Copenhagen, WHO Regional Office for Europe, pp. 12-19

Head, I.W.H. & Wagg, R.M. (1980) A survey of occupational exposure to man-made mineral fibre dust. *Ann. Occup. Hyg.*, 23, 235-258

Klingholz, R. (1977) Technology and production of man-made mineral fibres. *Ann. Occup. Hyg.*, 20, 153-159

Konzen, J.L. (1984) Production trends in fibre sizes of MMMF insulation. In: *Biological Effects of Man-made Mineral Fibres*, Vol. 1, Copenhagen, WHO Regional Office for Europe, pp. 44-63

Monkman, L.J. (1979) Procedure for the detection and identification of asbestos and other fibres in fibrous inorganic materials. *Ann. Occup. Hyg.*, 22, 127-139

Rindell, A. (1984) Man-made mineral fibres (MMMF) in indoor climate. In: *Proceedings of the Third International Conference on Indoor Air Quality and Climate*, Vol. 2, Stockholm, Swedish Council for Building Research, pp. 221-224

Schneider, T. (1984) Man-made mineral fibres (MMMF) and other fibres in the air and in settled dust. In: *Proceedings of the Third International Conference on Indoor Air Quality and Climate*, Vol. 2, Stockholm, Swedish Council for Building Research, pp. 183-188

Sébastien, P., Billon-Galland, M.A., Dufour, G., Gaudichet, A., Bonnaud, G. & Bignon, J. (1979) Levels of asbestos air pollution in some environmental situations. *Ann. N.Y. Acad. Sci.*, 330, 401-415

Sébastien, P., Billon-Galland, M.A., Dufour, G., Petit, G. & Gaudichet, A. (1982) *Assessment of asbestos exposure in buildings with sprayed materials using both the algorithm method and the ATEM measurement method*, Washington, DC, US Environmental Protection Agency, Office of Toxic Substances (EPA Contract 68-01-5915)

Sébastien, P., Plourde, M., Robbs, R. & Ross, M. (1985) *Ambient Air Asbestos Survey in Quebec Mining Towns - Part I, Methodological Study*, Montreal, Environment Canada (document EPS 3/AP/RQ/IE)

Walton, W.H. (1984) Peer review: Dust measurements in the manufacture and use of MMMF: Present knowledge and future requirements. In: *Biological Effects of Man-made Mineral Fibres*, Vol. 1, Copenhagen, WHO Regional Office for Europe, pp. 264-277

# EXPOSURE TO CERAMIC MAN-MADE MINERAL FIBRES

## J.J. Friar & A.M. Phillips

*Health and Safety Executive, Bootle, UK*

*Summary.* Ceramic fibres (also known as refractory fibres) are regarded here as man-made mineral fibres (MMMF) capable of withstanding temperatures of 1000–1600°C without appreciable distortion or softening. Ceramic fibres are manufactured largely from the aluminosilicate group of minerals but some contain only alumina, zirconia or silica. Simultaneous personal gravimetric and optical fibre count samples were taken throughout the industry. It has not been possible to correlate gravimetric results with fibre counts in any meaningful way. The general conclusions are as follows:

(*a*) gravimetrically, exposures ranged from less than 1 mg/m$^3$ for light tasks to over 10 mg/m$^3$ for some insulation workers. Exposures above 10 mg/m$^3$ were not necessarily associated with correspondingly high fibre counts;

(*b*) fibre counts rarely exceeded 1 f/ml, and it appears that ceramic fibre materials, in company with other MMMF, do not readily produce high airborne fibre counts;

(*c*) control of dust from mineral wools to 5 mg/m$^3$ achieves control to below 1 f/ml. This relationship does not hold for superfine MMMF and does not always hold for ceramic fibres.

## Introduction

Man-made mineral fibres (MMMF) include mineral wools (rock wool, slag wool, glass wool), continuous filament, superfine materials and ceramic fibres. Superfine materials are here regarded as materials which are predominantly made up of fibres of diameter less than 3 $\mu$m, length greater than 5 $\mu$m and a length-to-diameter ratio greater than 3:1, although almost all MMMF materials will contain a proportion of such fibres. The names of these classes of MMMF have different origins and are not necessarily mutually exclusive. Furthermore, classes and fibres within these classes may have different dimensions and/or compositions.

For the purposes of this paper, ceramic fibres are defined as MMMF capable of withstanding temperatures of 1000°C without appreciable distortion or softening. Ceramic fibres are also known as refractory fibres, particularly in the United States. The term 'ceramic' is derived from the fired pottery clay origins of the materials and 'refractory' from the heat-resistant nature of these fibres. Although the term 'refractory' more accurately describes their properties, the term 'ceramic fibres' is in common use in the UK and will be used throughout this paper.

Ceramic fibres are manufactured largely from the aluminosilicate group of minerals traditionally used as pottery clays, but some are made from high-purity alumina, zirconia or silica. There are also rare specialist materials which are compounds of boron and silicon with nitrogen or carbon, and new formulations appear from time to time.

Ceramic fibre products include blankets, felts, yarns, papers, vacuum-formed shapes and boards. These products are generally intended for high-temperature insulation purposes but some have found use in circumstances where the refractory properties are incidental. Some ceramic fibres are used in composites, etc., in the high-technology sector of industry. Ceramic fibres are still relatively specialized materials. There appears to be no single major consumer of such products and much of the exposed work-force is involved with these materials on an infrequent and irregular basis. This may change as the products gain a firmer foothold in the market.

Ceramic fibres containing silica which have been subjected to temperatures in excess of 900°C for prolonged periods are likely to contain a significant proportion of cristobalite (a hazardous form of crystalline silica). Insulation removal workers must therefore take precautions against it in circumstances where it is likely to arise, for example, in furnace wrecking.

## *Exposure limits*

Internationally, there has been little effort to date to deal with MMMF in general or with ceramic fibres separately either in regulations or standard-setting. Many countries, however, have carried out the necessary preparatory work for MMMF in general, and regulatory action could be initiated with little delay. The action under consideration includes labelling, special restrictions on the use of superfine fibres and limit values. In the UK, a control limit has been established for MMMF of 5 mg/m$^3$ total dust and for superfines a recommended limit of 1 f/ml; these are 8-hour time-weighted averages.

Currently, only Sweden specifically regulates MMMF under its Ordinance on synthetic inorganic fibres, which applies to materials containing more than 5% by weight of such fibres. Most countries group MMMF with 'nuisance' or 'non-toxic' dusts and, where limits exist, these are normally in terms of threshold limit values for total or respirable dust. Total dust limits vary from 8 to 10 mg/m$^3$ and respirable dust limits from 4 to 5 mg/m$^3$. Sweden and Poland have a limiting value specific to MMMF of 2 f/ml (full working day). Manufacturers of ceramic fibres in the United States have applied a 2 f/ml in-house standard.

## *Exposure survey*

Over the past 2 years we have carried out a survey of occupational exposure to ceramic MMMF. This was intended to cover a representative sample of such exposures rather than to aim at complete coverage. In addition, exposure data from previous surveys by the UK Health and Safety Executive were pooled with the most recent survey data for the purposes of this paper. Where a number of visits were made to the same site, only the results from the most recent visit are included. This paper contains all such measurements where both gravimetric and fibre count samples were taken at the same time and on the same individual.

Workers in the manufacturing sector have historically been exposed to total dust concentrations of ceramic fibres of up to 70 mg/m$^3$ (and occasionally higher). Several processes are potentially very dusty but dust removal by means of local exhaust ventilation is a well known technique that has been applied to these processes in recent years with considerable success. The group with the highest exposure is that involved in operations requiring the manual handling of ceramic fibre products. Workers engaged in insulation removal are, in general, potentially exposed to far higher concentrations than those encountered elsewhere. Because it is not normally possible to control the exposure of these workers by engineering means, appropriate personal protective equipment is often necessary.

In view of the large variety of ceramic fibre materials and of conditions of use, it is not surprising that it has not been possible to correlate gravimetric sampling results with fibre counts in any meaningful way. Some typical gravimetric and fibre count results are related to particular jobs in Table 1. Graphical representations of the data on simultaneous gravimetric sampling and fibre counts are given in Figures 1 and 2, respectively.

**Table 1. Manufacture and use of ceramic fibres: some typical exposures**

| A. Manufacture | | | B. Use | | |
|---|---|---|---|---|---|
| Process description or job | mg/m$^3$ | f/ml | Process description | mg/m$^3$ | f/ml |
| Needler operator | 2.5 | 0.5 | Wrapping ceramic fibre blanket around pipe weld | 5.2 | 0.8 |
| Baling raw fibre | 1.5 | 0.4 | | | |
| Fibre chopping | 10+ | 0.8 | Stripping and relining furnace panel | 10.5 | 1.2 |
| Product reeling | 2.5 | 0.8 | Kiln building | 13.1 | 1.75 |
| Bagging/chopping raw fibre | 10+ | 1.2 | Handling blanket ceramic fibre | 0.4 | 1.0 |
| Mixing during product formation | 4.5 | 0.4 | Machining and ventilation control of ceramic fireboard | 0.4 | 0.6 |
| Packing of products | 0.7 | 0.02 | Insulation work using blanket | 10.0 | 1.0 |
| | | | Handling operations - manual handling but with little cutting or machining | 2.5 | 0.1 |

**Figure 1. Ceramic fibres — all processes: gravimetric results**

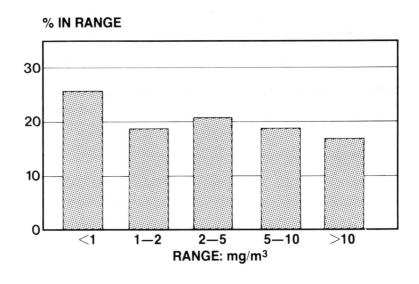

**Fig. 2. Ceramic fibres — all processes: fibre counts.**

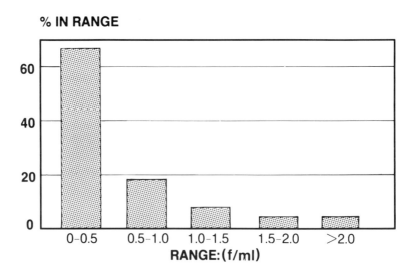

A study of the exposure data suggests the following general conclusions:

(a) gravimetrically, exposures ranged from less than 1 mg/m$^3$ for light tasks to over 10 mg/m$^3$ for some insulation workers. Exposures above 10 mg/m$^3$ were not necessarily associated with correspondingly high fibre counts;

(b) fibre counts rarely exceed 1 f/ml, and it appears that ceramic fibre materials, in company with other MMMF, do not readily produce high airborne fibre counts;

(c) control of dust from mineral wools to 5 mg/m$^3$ achieves control to below 1 f/ml. This relationship does not hold for superfine MMMF and does not always hold for ceramic fibres.

The use of ceramic fibre materials is increasing and such mineral fibres may in the future contribute significantly to fibre loading in the non-occupational environment.

# COMPARATIVE STUDIES OF AIRBORNE ASBESTOS IN OCCUPATIONAL AND NON-OCCUPATIONAL ENVIRONMENTS USING OPTICAL AND ELECTRON MICROSCOPE TECHNIQUES

J. Cherrie, J. Addison & J. Dodgson

*Institute of Occupational Medicine,
Edinburgh, UK*

*Summary.* We have compared asbestos fibre and general fibre counts from scanning electron microscopy (SEM) and transmission electron microscopy (TEM) with fibre counts from phase-contrast optical microscope (PCOM) methods. Three different types of sample have been evaluated: laboratory-prepared samples of different fibre types; chrysotile asbestos textile factory samples; and non-occupational and environmental samples from sites where asbestos might be found.

TEM produced total fibre number assessments which were greater than those found with SEM which, in turn, produced fibre counts greater than those obtained with the PCOM. However, when fibres longer than 5 $\mu$m were alone counted, the two electron microscope (EM) methods provided similar results. This indicates that TEM is advantageous in comparison with SEM when counting or sizing short fibres and has no advantage for fibres longer than 5 $\mu$m.

For fibres longer than 5 $\mu$m in both the laboratory-prepared and asbestos factory samples, the EM asbestos counts were higher than the PCOM fibre counts, the ratios depending on the fibre type in the case of the former. The PCOM fibre counts in samples from the non-occupational situations were shown to be poor predictors of airborne asbestos fibre concentrations determined by EM. This was mainly due to the presence of high and variable proportions of non-asbestos fibres in these samples.

It is concluded that, in order to convert EM asbestos fibre concentrations (>5 $\mu$m) to equivalent PCOM asbestos concentrations, they should be divided by 4.0 for chrysotile and 1.7 for amphibole asbestos.

## Introduction

In order to assess the very low asbestos exposure found outside the workplace, it is necessary to make fibre counts by electron microscopy (EM). This causes difficulties in relating such exposure estimates to industrial experience, where the optical microscope has been used in epidemiological studies and in compliance monitoring.

To be able to translate risk estimates based on workplace epidemiology studies into those applicable to the general environment, it is first necessary to understand the relationship between fibre concentration measurements obtained using the optical and electron microscope methods. In this paper, this relationship is assessed for a range of asbestos samples and environments. Appropriate factors are derived for converting EM asbestos fibre concentrations into equivalent optical fibre concentrations. Fuller details have been reported by Cherrie et al. (1987).

## Materials and methods

Two types of samples were studied, namely, specially prepared samples of pure asbestos and samples taken at the workplace and in ambient air.

The pure asbestos samples were obtained by filtering asbestos fibres from liquid suspension or by collection of airborne dust samples from an experimental chamber (Cherrie et al., 1986). The liquid suspension samples were either collected on cellulose ester membrane filters (Millipore AAOC) and prepared for optical microscopy using the acetone-triacetin technique (Health and Safety Executive, 1983) or on polycarbonate membrane filters (Nuclepore, 0.1 $\mu$m) for EM. The Nuclepore filters were prepared for scanning electron microscopy (SEM) by directly mounting a section of the membrane on to an aluminium stub and coating it with a thin layer of gold (WHO Regional Office for Europe MMMF Technical Committee, 1985) and, for transmission electron microscopy (TEM), by a direct-transfer technique (Chatfield et al., 1978).

The airborne dust samples collected in the experimental chamber were sampled simultaneously on to cellulose ester membranes (Millipore AAOC) for phase-contrast optical microscopy (PCOM), Nuclepore filters (0.4-$\mu$m pore size) for SEM and PVC-acrylic copolymer membrane filters (Gelman DM800) for TEM. Sample preparation for PCOM and SEM were as described above. Samples for TEM were prepared using a direct-transfer technique (Burdett & Rood, 1983).

The PCOM samples were all evaluated in accordance with the European Reference Method rules (Health and Safety Executive, 1983). The EM evaluations were based on those of the Asbestos International Association (1984). Two magnifications were used with the SEM:

1. ×2000 — all fibres longer than 5 $\mu$m were counted;

2. ×10 000 — all fibres were counted and sized regardless of length or diameter.

At the higher magnification the chemical composition of each fibre was assessed using energy-dispersive X-ray analysis (EDXA).

In the TEM evaluations, photomicrographs were taken at a magnification of ×8000. Counting and sizing were done on photographically enlarged images, otherwise the procedures and rules were as for the SEM evaluations.

The airborne dust samples were obtained from 5 locations: inside an asbestos textile factory; outside the same factory; after remedial work on asbestos insulation (clearance samples); in a building with sprayed asbestos fire insulation; and in a general urban location. All of the samples were collected on to Gelman DM800 filters. Each sample was divided into three and the sections were prepared for PCOM, SEM and TEM using the methods described by Le Guen et al. (1980) and Burdett and Rood (1983). The evaluation procedures were as described above.

## Results and discussion

### Scanning electron microscopy

The results obtained in the SEM evaluation of long fibres were different for the two magnifications (×2000 and ×10 000). The differences were particularly apparent for the pure asbestos samples, where the ×2000 results were up to 10 times higher. When these samples were re-evaluated, however, better agreement was obtained between the two scan magnifications and these data were also in agreement with the original ×10 000 evaluations. We have concluded that the ×2000 assessments in the early part of this study were unreliable. Observations of fibres on a SEM is dependent on the magnification. Middleton (1982) has shown that, for magnifications above about ×2000, little improvement in visibility is obtained. However, at these relatively low magnifications, fibres 5 $\mu$m in length are represented on the SEM screen by images of only about 10 mm. When large numbers of fibres occur near a size classification boundary, reliable assessment of fibre length is not possible; relatively high densities were found with the pure asbestos samples, making accurate evaluation difficult. It was probable that the microscopists were incorrectly over-estimating the fibre lengths and hence the fibre density. This finding is consistent with our previous experience in using SEM for evaluating man-made mineral fibres for which a magnification of ×5000 has been recommended (WHO Regional Office for Europe MMMF Technical Committee, 1985). For this reason, further discussion of SEM data in relation to PCOM or TEM is restricted to the ×10 000 data.

### Pure asbestos samples

The results obtained with the pure asbestos samples showed that total fibre concentrations assessed by TEM were greater than those measured by SEM (×10 000) which, in turn, were greater than those determined by PCOM.

Table 1 shows the geometric mean ratios of EM fibre concentrations to PCOM concentrations (standard deviations in parentheses).

The two EM methods were in better agreement when comparison was restricted to fibres longer than 5 $\mu$m. When all the data were pooled, for fibres longer than 5 $\mu$m, from the pure asbestos samples, the differences between TEM and SEM were not significant ($p > 0.05$). In the small number of samples where the PCOM fibre densities were low (<100 fibres per mm²) (f/mm²), the difference between the total EM and PCOM fibre counts were larger; SEM counts were approximately 35 times PCOM counts and TEM counts were about 80 times PCOM counts. The individual ratios of EM to PCOM, for fibres longer than 5 $\mu$m, ranged from 0.8 to 9.4, which is explained partly by the known variability in all three microscope techniques (PCOM: Hayes and Clayton (1980), Ogden (1982); SEM: Cherrie (1983); TEM: Steel and Small (1985)) and partly by difficulties in obtaining homogeneous preparations in the laboratory. As a consequence, and because of the relatively small number of samples evaluated, it is difficult to distinguish differences between the various asbestos varieties. However, there are some indications that the ratio of EM fibre concentration to PCOM count is less for the amphibole varieties than for chrysotile.

**Table 1. Geometric mean ratios and standard deviations of electron microscopy fibre density to phase-contrast optical microscopy fibre density (laboratory prepared samples)**[a]

| Asbestos type | PCOM[b] density (fibres per mm²) | No. of samples | All fibres | | Fibres of length >5 μm | | |
|---|---|---|---|---|---|---|---|
| | | | SEM[b] (×10 000) | TEM[b] | SEM (×10 000) | SEM (×2000) | TEM |
| Chrysotile | >100 | 7 | 7.9 (1.6) | 22.0 (2.1) | 1.6 (1.4) | 4.2 (1.5) | 1.2 (1.5) |
| Amosite | >100 | 4 | 4.7 (1.2) | 8.4 (3.3) | 1.1 (1.2) | 2.1 (1.5) | 1.4 (4.0) |
| Crocidolite | >100 | 5 | 7.5 (1.1) | 22.0 (2.0) | 0.9 (1.2) | 1.7 (1.6) | 1.4 (1.5) |
| Tremolite | >100 | 3 | 2.9 (1.0) | 7.9 (1.7) | 1.1 (1.3) | 1.1 (1.1) | 3.4 (1.5) |
| All fibre types | <100 | 7 | 20.0 (3.1) | 61.0 (2.0) | 2.3 (1.7) | 12.0 (2.8) | 4.1 (1.7) |

[a]Standard deviations in parentheses.
[b]PCOM, phase-contrast optical microscopy; SEM, scanning electron microscopy; TEM, transmission electron microscopy.

## Industrial and environmental samples

The results from samples collected inside the asbestos factory (Table 2) showed good agreement between SEM and TEM. On average, between 16% and 20% of fibres longer than 5 μm were seen by PCOM. Between 63% and 93% of all fibres measured by SEM were identified as asbestos by EDXA, with an average of 94% of all fibres longer than 5 μm as asbestos.

In the 4 sets of environmental samples, the ratios of airborne fibre concentration assessed by EM to that assessed by PCOM were dependent on the source and composition of samples. The fibre composition of the non-occupational samples varied substantially, ranging in asbestos content from 0 to 100% for those taken inside the building with spray insulation, for example, and averaging 35%.

The geometric mean ratio of SEM asbestos fibre density, longer than 5 μm, to that found with the PCOM was 4.0 for the samples from the asbestos factory and 0.4 for those collected inside the building with sprayed asbestos.

In approximately 40% of all the environmental samples, the PCOM concentration was greater than the EM asbestos fibre concentration. It is clear that the PCOM is a poor predictor of the actual asbestos fibre concentration and hence the risk.

## Conclusions

SEM at ×10 000 and TEM give equivalent estimates of the number of asbestos fibres longer than 5 μm, although TEM is able to detect more of the shorter fibres.

From the data obtained from the samples collected in the asbestos factory, it is possible to estimate the equivalent PCOM fibre concentrations from the EM asbestos fibre concentrations by dividing by 4.0. Similarly, for the pure asbestos air samples, the conversion factor for the amphiboles would be 1.7.

**Table 2. Geometric mean ratios and standard deviations of electron microscopy fibre density to phase-contrast optical microscopy fibre density (environmental samples)**[a]

| Location | PCOM[b] fibre density range (fibres per mm²) | No. of samples | Fibres of all sizes | | | Fibres of length >5 μm | | |
|---|---|---|---|---|---|---|---|---|
| | | | SEM | | TEM | SEM | | TEM |
| | | | All | Asbestos | All | All | Asbestos | All |
| Asbestos factory | 170-370 | 11 | 18 (2.0) | 17 (1.7) | 15 (1.5) | 4.2 (2.2) | 4.0 (2.1) | 6.1 (1.4) |
| Outside asbestos factory | 2-6 | 12 | 20 (2.0) | 4.2 (2.2) | A[c] | 3.7 (2.1) | 1.6 (1.9) | A |
| Clearance | 3-39 | 10 | 7.2 (2.3) | 3.3 (2.6) | 30 (1.6) | 2.7 (2.5) | 1.6 (2.5) | 2.5 (2.1) |
| Building with spray insulation | 12-160 | 10 | 5.7 (2.1) | 0.6 (3.8) | 7.8 (2.2) | 1.6 (5.3) | 0.4 (3.5) | 9 (-) |
| Urban environment | 0-3 | 12 | 10 (2.7) | - | A | 2 (1.8) | - | A |

[a]Standard deviations in parentheses.

[b]PCOM, phase-contrast optical microscopy; SEM, scanning electron microscopy; TEM, transmission electron microscopy.

[c]A, samples unsuitable for evaluation due to artefacts.

## Acknowledgements

This work was carried out with the financial support of the Medical Research Council.

## References

Asbestos International Association (1984) *Method for the Determination of Airborne Asbestos Fibres and Other Inorganic Fibres by Scanning Electron Microscopy* (Recommended Technical Method RTM2), London

Burdett, G. & Rood, A. (1983) Sample preparation for monitoring asbestos in air by transmission electron microscopy. *Anal. Chem.*, 55, 1642-1645

Chatfield, E.J. & Dillon, M.J. (1978) Some aspects of specimen preparation and limitations of precision in particulate analysis by SEM and TEM. *Scanning Electron Microsc.*, 487-496

Cherrie, J.W. (1983) Scanning electron microscopy: an investigation of the reproducibility of counting and sizing of asbestos fibres by SEM. In: *Asbestos International Association Fourth International Colloquium on Dust Measuring Technique and Strategy*, London pp. 369-378

Cherrie, J.W., Jones, A.D. & Johnston, A.M. (1986) The influence of fiber density on the assessment of fiber concentration using membrane filter method. *Am. Ind. Hyg. Assoc. J.*, *47*, 465-474

Cherrie, J.W., Dodgson, J., Groat, S. & Carson, M. (1987) *Comparison of Optical and Electron Microscopy for Evaluating Airborne Asbestos*, Edinburgh, Institute of Occupational Medicine, IOM Report TM/87/01

Le Guen, J.M.M., Rooker, S.J. & Vaughan, N.P. (1980) A new technique for the scanning electron microscopy of particles collected on membrane filters. *Environ. Sci. Technol.*, *14*, 1008-1011

Health and Safety Executive (1983) *Asbestos Fibres in Air. Determination of Personal Exposure by the European Reference Version of the Membrane Filter Method* (MDHS 39), London

Hayes, J.R. & Clayton, R. (1980) A routine for the control of the performance of microscopists evaluating airborne asbestos fibre samples on membrane filters by phase contrast microscopy. *Ann. Occup. Hyg.*, *23*, 381-401

Middleton, A.P. (1982) Fibre visibility in the SEM. In: *Asbestos International Association. Fourth International Colloquium on Dust Measuring Technique and Strategy*, London, pp. 362-368

Ogden, T.L. (1982) *The Reproducibility of Asbestos Counts*, London, Health and Safety Executive

Steel, E. & Small, J. (1985) Accuracy of transmission electron microscopy for the analysis of asbestos in ambient environment. *Anal. Chem.*, *57*, 209-213

WHO Regional Office for Europe MMMF Technical Committee (1985) *Reference Methods for Measuring Airborne Man-made Mineral Fibres (MMMF)*, Copenhagen

# ALVEOLAR AND LUNG FIBRE LEVELS IN NON-OCCUPATIONALLY EXPOSED SUBJECTS

G. Chiappino[1], K.H. Friedrichs[2], A. Forni[1], G. Rivolta[1] & A. Todaro[1]

[1] Research Centre on Biological Effects of Inhaled Dusts,
Institute of Occupational Health, University of Milan, Milan, Italy

[2] Medical Institute for Environmental Hygiene at the
University of Düsseldorf, Düsseldorf, Federal Republic of Germany

*Summary.* Mineral fibre concentrations and characteristics were evaluated by the same electron microscope methods in 15 bronchoalveolar lavage samples and in 40 surgical lung tissue samples of subjects with no occupational exposure to asbestos. Both fibre alveolar load and lung burden evaluated by transmission electron microscopy were higher in the groups of industrial workers with no specific asbestos exposure than in the groups of individuals exposed only to general environmental pollution. In both types of samples, the fibre burden consisted of extremely small fibres (mean length $<2$ $\mu$m, mean diameter $<0.1$ $\mu$m), with a trend towards a further reduction in mean length in the lung tissue as compared with the alveolar load. These data suggest that there is a need for a critical reconsideration of the methods of evaluating environmental fibre pollution and of those for assessing exposure-effect relationships.

## Introduction

The 'alveolar load' of mineral fibres determined in bronchoalveolar lavage (BAL) fluid represents at any given time the dynamic resultant of the number of inhaled fibres which reach the alveoli and of clearance via the airways and *via* penetration into the interstitium. In contrast to the evaluation of fibres in lung tissue, the alveolar load can be determined *in vivo* with a less invasive technique, can be controlled over time and can be correlated with indicators of tissue reaction.

The aim of this study was to determine the alveolar load or the lung tissue burden of inorganic fibres in subjects with no occupational asbestos exposure and undergoing BAL or thoracotomy respectively, by using the same methods for sample preparation and fibre counting and analysis.

The investigation was carried out both with scanning (SEM) and transmission (TEM) electron microscopes. However, in a preliminary comparison of results, the SEM data appeared to be incomplete because of the extremely small size of most fibres and the poor analytical resolution of this microscope; only TEM data are therefore reported.

## Materials and methods

### Subjects and materials

Two series of subjects have been investigated, the first comprising 15 adult males undergoing BAL for the assessment of various respiratory disorders. Of these, 10 had been occupationally exposed to industrial dusts, but had no known occupational exposure to asbestos, and 5 had been exposed only to environmental pollution. BAL was performed according to the method currently used at our institution (Forni et al., 1985), and the sediment of 10 ml BAL fluid was frozen for mineralogical studies.

The second series comprised 40 consecutive subjects undergoing thoracotomy for neoplastic or non-neoplastic lung disorders. Of these, 13 were industrial workers with no known occupational asbestos exposure, and 27 had been exposed only to environmental pollution. A piece of macroscopically normal lung tissue was excised from the surgical specimen and frozen.

## Methods

For TEM evaluation of fibre load, both BAL sediments and lung-tissue samples were digested with potassium hydroxide according to Pooley (1972), washed and filtered through Nuclepore filters (pore diameter 0.4 $\mu$m). The filters were carbon coated and transferred on to 200-mesh gold grids using the technique described by Chatfield (1984).

Ten grid openings per sample were evaluated by counting, measuring and analysing fibres at $\times 16\,000$ with a TEM (Philips 301 G) equipped with Kevex 7000 for X-ray microanalysis.

The values obtained were adjusted to 1 ml BAL fluid $\times 10^{-2}$ and to 1 g dry lung, respectively.

## *Results*

The concentrations of mineral fibres in BAL fluid and in lung tissue are reported in Tables 1 and 2, respectively. It will be seen that exposure to general industrial pollution increased both the alveolar and tissue load of asbestos and non-asbestos fibres, even though the increase was not statistically significant in the Wilcoxon signed rank test for unpaired data.

**Table 1. Alveolar load of fibres in 15 subjects with no occupational asbestos exposure (arithmetic mean values and ranges)**

| Subjects | No. | Fibres per ml BAL[a] fluid $\times 10^{-2}$ | |
| --- | --- | --- | --- |
| | | Asbestos | Other |
| Industrial workers | 10 | 164.7 (48-437) | 87.1 (38-288) |
| Not employed in industry | 5 | 70.8 (8-145) | 52.9 (31-83) |

[a]BAL, bronchoalveolar lavage.

**Table 2. Lung tissue burden of fibres in 40 subjects with no occupational asbestos exposure (arithmetic mean values and ranges)**

| Subjects | No. | Fibres per g dry lung $\times 10^6$ | |
|---|---|---|---|
| | | Asbestos | Other |
| Industrial workers | 13 | 6.60 (1.4-23.7) | 7.51 (1.5-26.5) |
| Not employed in industry | 27 | 3.97 (0-16.6) | 4.76 (0.7-28.6) |

The size characteristics of asbestos fibres in both series of subjects are reported in Tables 3 and 4. The data reported in Table 3 show that the alveolar load of asbestos consists predominantly of ultrashort and ultrathin fibres with a very high aspect ratio. A study of the distribution of all fibres in BAL samples by length classes ($<$ and $>5$ $\mu$m) showed that fibres $>5$ $\mu$m long account for an extremely small proportion ($<3\%$) of fibres present in the alveoli.

**Table 3. Size of asbestos fibres in bronchoalveolar lavage of 15 subjects with no occupational exposure to asbestos (arithmetic mean values)**

| Subjects | Chrysotile | | | Amphibole | | |
|---|---|---|---|---|---|---|
| | Length ($\mu$m) | Diameter ($\mu$m) | Length: diameter | Length ($\mu$m) | Diameter ($\mu$m) | Length: diameter |
| Industrial workers | 2.04 | 0.07 | 29:1 | 1.9 | 0.11 | 17:1 |
| Not employed in industry | 1.96 | 0.06 | 33:1 | 2.0 | 0.10 | 20:1 |

In the lung, the fibre size is similar to that found in the alveoli and, moreover, there seems to be a further reduction in mean length (Table 4).

## Discussion

The data reported show good agreement between alveolar load and lung tissue burden for both the concentrations and the size characteristics of mineral fibres.

Both types of samples in subjects with no known occupational exposure to asbestos contain predominantly ultrashort and ultrathin fibres, both asbestos and

**Table 4. Size of asbestos fibres in lung-tissue samples from 40 subjects with no occupational exposure to asbestos (arithmetic mean values)**

| Subjects | Chrysotile | | | Amphibole | | |
|---|---|---|---|---|---|---|
| | Length ($\mu$m) | Diameter ($\mu$m) | Length: diameter | Length ($\mu$m) | Diameter ($\mu$m) | Length: diameter |
| Industrial workers | 1.98 | 0.08 | 25:1 | 1.98 | 0.12 | 17:1 |
| Not employed in industry | 1.27 | 0.08 | 16:1 | 1.57 | 0.14 | 11:1 |

non-asbestos, in similar amounts, in agreement with the results of other studies on lung tissue (Churg & Warnock, 1980; Churg, 1983, 1986).

Our experience with the alveolar load of fibres in a larger series of cases, including workers occupationally exposed to asbestos, has demonstrated that the percentage of fibres longer than 5 $\mu$m in the asbestos-exposed is around 15%, significantly higher than in subjects exposed only to general environmental pollution (unpublished data). In our opinion, the finding of a relatively high percentage of fibres longer than 5 $\mu$m in the alveolar load can be considered a useful indicator of occupational exposure.

The fact that most fibres found in the lung are extremely short and thin suggests that there is a need for a critical revision of the methods currently used for the determination of environmental pollution, which limit the count to fibres longer than 5 $\mu$m. On this matter, we are in complete agreement with the view expressed by McDonald (1984).

Because of the marked numerical predominance of submicroscopic fibres found in the lung, and since the biological effects of fibres are probably more closely linked to number and aspect ratio than to fibre mass, it seems advisable that future epidemiological and experimental investigations should pay particular attention to this component of the lung burden.

## References

Chatfield, E.J. (1984) *Determination of Asbestos Fibre in Air and Water*, Mississauga, Ontario, Ontario Research Foundation (*Document No.14*)

Churg, A. (1983) Non asbestos pulmonary mineral fibers in the general population. *Environ. Res., 31*, 189-200

Churg, A. (1986) Lung asbestos content in long-term residents of a chrysotile mining town. *Am. Rev. Respir, Dis., 134*, 125-127

Churg, A. & Warnock, M.L. (1980) Asbestos fibers in the general population. *Am. Rev. Respir. Dis., 122*, 669-678

Forni, A., Guerreri, M.C. & Chiappino, G. (1985) New methods in the study of occupational lung diseases: bronchoalveolar lavage. Experience of the Institute of Occupational Medicine of the University of Milan [in Italian]. *Med. Lav., 76*, 11-16

McDonald, J.C. (1984) Aspects of the asbestos standard. In: Gee, J.B.L., Morgan, W.K.C. & Brooks, S.M., eds, *Occupational Lung Disease*, New York, Raven Press, pp. 139-149

Pooley, F.D. (1972) Electron microscope characteristics of inhaled chrysotile asbestos fibre. *Br. J. Ind. Med., 29*, 146-153

# FIBRE CONTENT OF LUNG IN AMPHIBOLE- AND CHRYSOTILE-INDUCED MESOTHELIOMA: IMPLICATIONS FOR ENVIRONMENTAL EXPOSURE

A. Churg & J.L. Wright

*Department of Pathology and Health Sciences Centre Hospital, University of British Columbia, Vancouver, BC, Canada*

*Summary.* Using 9 pairs of exposure-period-matched shipyard and insulation workers (amphibole exposure) and chrysotile-industry workers (chrysotile exposure) with mesothelioma, and an additional 9 pairs of workers with asbestosis, we found that the chrysotile workers with mesothelioma had 400 times the median lung fibre burden of the shipyard and insulation workers with mesothelioma. Mesothelioma in the chrysotile workers was associated with a 3 times greater median fibre burden than asbestosis, whereas in the shipyard and insulation workers mesothelioma was associated with only 1/35 the median amphibole burden seen in cases of asbestosis. In the chrysotile workers, the tremolite:chrysotile ratio and the mean fibre sizes were the same for both mesothelioma and asbestosis cases. These data suggest that total fibre load is crucial to the induction of mesothelioma by chrysotile, and that this phenomenon requires, on average, as high a fibre burden as induction of asbestosis by chrysotile. By contrast, for amphibole exposure, mesothelioma appears at a much lower fibre burden than asbestosis. The fibre types appear to differ by at least two orders of magnitude in their potential for inducing mesothelioma. Estimates of risk from environmental exposure must take these differences into account.

## Introduction

The risk of developing mesothelioma after low-level occupational or environmental exposure to chrysotile asbestos is an area of controversy. Attempts to estimate such risks have been confounded by the sparsity of cases of mesothelioma developing after chrysotile (as opposed to amphibole or mixed chrysotile and amphibole) exposure (Hughes & Weill, 1986), and by the use of estimates based on populations with supposed exposure to chrysotile which turn out to have also been exposed to amosite and crocidolite (Peto, 1978; Wagner *et al.*, 1982; Berry & Newhouse, 1983; McDonald, 1980). In this paper we have attempted to approach this problem by analysing lung asbestos burden in workers with mesothelioma and exposure to chrysotile and amphibole.

## Materials and methods

We selected cases for this study from two sets of autopsy lungs: one from shipyard and insulation workers from the Pacific North-West, and the other from long-term miners and millers from the Quebec chrysotile industry. The latter cases, together with detailed occupational histories, were kindly supplied by Dr M. Poulin and Mr C. Pratte of the Hôpital Géneral de la Région de l'Amiante at Thetford Mines, Quebec. Although the shipyard and insulation workers had historically been exposed to both chrysotile and amphibole, preferential clearance of chrysotile (Churg & Wiggs, 1986) has left almost entirely amphibole in their lungs, and we treated them as an amphibole-exposed population (see Discussion).

The cases from each group were matched by disease, years of exposure within 5 years, and year of last exposure within 5 years. Pathological diagnoses were confirmed using standard criteria. Initially, we found 11 mesotheliomas in the chrysotile workers, but analysis revealed that 2 of these lungs contained large amounts of amosite asbestos, hence these cases were excluded. The remaining lungs contained no amosite or crocidolite. Thus we were able to assemble 9 matched cases. To put the data into the context of other asbestos-induced disease, we assembled 9 similarly matched pairs of workers with asbestosis.

Fibres were recovered from lung by our previously published bleach-digestion technique (Churg & Wiggs, 1986). From 200 to 400 fibres were identified, counted and sized in each case. Statistical comparisons were performed by analysis of variance on the original or log-transformed data, with subsequent contrasts and appropriate correction for multiple comparisons.

## Results

Table 1 shows the demographic data. The mean ages and the mean years of exposure are statistically the same in all groups. The mesothelioma cases were exposed slightly more recently than the asbestosis cases.

### Table 1. Demographic data (mean±SD)[a]

| Item (years) | Mesothelioma | | Asbestosis | |
|---|---|---|---|---|
| | Amphibole | Chrysotile | Amphibole | Chrysotile |
| Mean age | 64±9 | 65±7 | 70±9 | 68±8 |
| Mean exposure | 37±7 | 36±11 | 33±7 | 34±8 |
| Mean year of last exposure | 1978±7 | 1978±7 | 1971±5 | 1972±6 |

[a] Amphibole indicates shipyard and insulation workers; chrysotile indicates chrysotile miners and millers.

Table 2 shows the fibre concentration data, using the combination of chrysotile plus tremolite fibres for the chrysotile-exposed workers, and amosite plus crocidolite (virtually all amosite) for the shipyard and insulation workers. For the shipyard and insulation workers, the mean chrysotile plus tremolite level was $0.8 \times 10^6$ fibres per g dry lung in those with asbestosis and $1.7 \times 10^6$ in those with mesothelioma.

**Table 2. Fibre concentrations by disease and exposure (fibres per g dry lung $\times 10^6$)**

| Item | Shipyard and insulation workers (amosite + crocidolite) | Chrysotile industry workers (chrysotile + tremolite) | $p^a$ |
|---|---|---|---|
| **A. Occupational exposure** | | | |
| *Mesothelioma* | | | |
| Mean±SD | 44±11.4 | 540±710 | <0.001 |
| Median | 0.7 | 290 | |
| Range | 0.07-35 | 50-2200 | |
| *Asbestosis* | | | |
| Mean±SD | 30±33 | 330±390 | <0.01 |
| Median | 26 | 110 | |
| Range | 1.0-100 | 50-1200 | |
| **B. General population of Vancouver**[b] | | | |
| Mean | 0.001 | 0.7 | |
| Median | 0 | 0.4 | |
| Range | 0-0.03 | 0-2.5 | |

[a]Concentration of chrysotile/tremolite versus amosite/crocidolite; mesothelioma versus asbestosis cases: $p<0.001$ for shipyard and insulation workers, not significant for chrysotile workers.

[b]From Churg & Wiggs (1986).

Table 3 shows the tremolite:chrysotile ratios for the chrysotile workers with mesothelioma or asbestosis; these were not statistically different. Table 4 shows the mean fibre sizes for the same two groups; these were essentially identical.

**Table 3. Tremolite:chrysotile concentration ratios in chrysotile workers**

| Group | Mean±SD | Median | Range |
|---|---|---|---|
| Mesothelioma | 6.6±3.0 | 5.3 | 3.7-12.6 |
| Asbestosis | 10.8±7.9 | 7.7 | 2.5-28.4 |

Table 4. Geometric mean fibre sizes (geometric SD) in chrysotile workers

| Group | Length | Width | Aspect ratio |
|---|---|---|---|
| Mesothelioma-chrysotile | 2.5 (2.2) | 0.03 (1.4) | 73 (2.3) |
| Asbestosis-chrysotile | 2.7 (2.1) | 0.03 (1.3) | 97 (2.0) |
| Mesothelioma-tremolite | 2.0 (2.0) | 0.16 (2.1) | 13 (2.1) |
| Asbestosis-tremolite | 2.2 (2.0) | 0.14 (2.1) | 16 (2.1) |

## Discussion

Our aim in this paper has been to attempt to use analytical electron microscopy to compare the lung burden of asbestos associated with chrysotile with that associated with amphibole-induced mesothelioma, and to examine the mineral parameters associated with chrysotile-induced mesothelioma. For this purpose, we have used one of the few groups of workers with true chrysotile-induced tumours. This approach has the major advantage of excluding mesotheliomas induced by occult amosite or crocidolite exposure, a serious problem which has probably led to overestimates of the risk of mesothelioma after chrysotile exposure. This technique also allows a fibre-for-fibre comparison of asbestos concentration, and may be more appropriate than attempts to use exposure estimates based on air sampling, since the usual light microscopic air-sampling techniques count a widely differing and entirely unpredictable proportion of amphibole and chrysotile fibres (Pooley & Ranson, 1986).

A major problem with this approach is that such analysis examines only residual fibre content, and no information is gained on fibres which were once present and have now been cleared. We have attempted to deal with clearance effects in a crude way by matching total years of exposure and year of last exposure. Since there is good evidence that chrysotile retention is much less than amphibole retention (Churg & Wiggs, 1986), the effect of this manoeuvre is probably to underestimate differences in the pathogenicity of the two fibre types.

Another problem is that the shipyard and insulation workers used in this study were exposed to both chrysotile and amphibole, but preferential clearance has left them with a residual chrysotile/tremolite level within or close to that seen from background environmental exposure (Churg & Wiggs, 1986). Since there are no populations in North America with pure amphibole exposure, this limitation has to be accepted for the study to be performed. Given the very large lung burden of chrysotile ore components required to produce mesothelioma in the chrysotile workers, it is likely that the role of chrysotile was much less than that of amosite or crocidolite in the shipyard and insulation group.

This study shows that, in workers with exposure only to chrysotile ore components (i.e., chrysotile and its natural contaminant, tremolite), mesothelioma is associated with a median lung fibre burden some 400 times larger than that seen in workers with amphibole exposure and mesothelioma. The data on the cases of asbestosis put these values into useful perspective: in the workers from the chrysotile industry, mesothelioma appears at about a 3 times higher median lung burden than asbestosis, whereas in those with exposure to amphibole, mesothelioma appears at only 1/35 the median burden seen in those with asbestosis.

We have previously suggested that tremolite may be the agent of major importance in mesotheliomas produced by chrysotile ore (Churg *et al.*, 1984). In this study we were not able to find any difference in the ratio of tremolite to chrysotile in the asbestosis as compared with the mesothelioma cases, as might have been expected on the basis of animal experiments. These observations suggest that total lung burden of chrysotile ore may be an important determinant in the genesis of mesothelioma, although clearly other factors must play a role as well. Our data certainly do not rule out a role for tremolite.

Observations of these types may provide a method for estimating risks for amphiboles as compared with chrysotile in persons exposed at low occupational and environmental levels. Certainly the data indicate that mesothelioma induction by chrysotile requires, on average, a very high burden of fibres (roughly as many fibres as are required to induce asbestosis), and also suggest that the relative risks for the two fibre types differ by at least two orders of magnitude and probably more. To ignore such differences, as has been done in some environmental risk assessments and government regulations, is to grossly overestimate the risk of mesothelioma from chrysotile exposure.

## References

Berry, G. & Newhouse, M.I. (1983) Mortality of workers manufacturing friction materials using asbestos. *Br. J. Ind. Med.*, *40*, 1-7

Churg, A. & Wiggs, B. (1986) Fiber size and number in workers exposed to processed chrysotile asbestos, chrysotile miners, and the general population. *Am. J. Ind. Med.*, *9*, 143-152

Churg, A., Wiggs, B., DePaoli, L., Kampe, B. & Stevens, B. (1984) Lung asbestos content in chrysotile workers with mesothelioma. *Am. Rev. Respir. Dis.*, *130*, 1042-1045

Hughes, J.M. & Weill, H. (1986) Asbestos exposure — quantitative assessment of risk. *Am. Rev. Resp. Dis.*, *133*, 5-13

McDonald, A.D. (1980) Malignant mesothelioma in Quebec. In: Wagner, J.C., ed., *Biological Effects of Mineral Fibres (IARC Scientific Publications No.30)*, Lyon, International Agency for Research on Cancer, pp. 673-681

Peto, J. (1978) The hygiene standard for chrysotile asbestos. *Lancet*, *i*, 484-489

Pooley, F.D. & Ranson, D.L. (1986) Comparison of the results of asbestos fibre counts in lung tissue obtained by analytical electron microscopy and light microscopy. *J. Clin. Pathol.*, *39*, 313-317

Wagner, J.C., Berry, G. & Pooley, F.D. (1982) Mesotheliomas and asbestos type in asbestos textile workers: a study of lung contents. *Br. Med. J.*, *285*, 603-606

# LEVELS OF AIRBORNE MAN-MADE MINERAL FIBRES IN DWELLINGS IN THE UK: RESULTS OF A PRELIMINARY SURVEY

S.A.M.T. Jaffrey & A.P. Rood

*Occupational Medicine and Hygiene Laboratories, London, UK*

J.W. Llewellyn & A.J. Wilson

*Department of the Environment, London, UK*

*Summary.* Levels of airborne man-made mineral fibres (MMMF) were measured during the insulation of lofts and after the disturbance of the insulation wools. Transmission electron microscopy was used for the analysis. Generally, the personal samples showed fibre levels of up to 0.7 f/ml, whereas static samples showed fibre levels of 0.05 f/ml in the lofts. Little contamination of living space occurred during these operations.

## *Introduction*

Recent concern over the possibility that fibres other than asbestos could cause lung cancer in man has led to a survey of the airborne levels of man-made mineral fibres (MMMF) in dwellings in the UK. Concentrations of MMMF were measured in 17 houses, 9 of which were already insulated. Air was sampled before and after some disturbance of MMMF insulation in these houses. Six uninsulated houses were insulated with glass fibre/rock wool fibre blankets or rock wool loose-lay products during the exercise. A further 6 houses were subsequently insulated with blown glass fibre or blown rock wool fibre insulation. Air in these houses was monitored 2-3 days before insulation, during insulation and 2-7 days after insulation.

## *Experimental*

Air samples were collected throughout each house 2-3 days before any disturbance in the loft. Similarly, about 30-50 samples of air were collected from each house at various stages of the experiments.

The experiments in the houses that contained insulation comprised: (1) minor disturbance in the loft, e.g., moving household items around without disturbing the insulation for about 10 min; (2) major disturbance of the insulation, which was carried out 2-3 days after the minor disturbance, and lasted for 30 min. Here the MMMF insulation was physically disturbed by beating the blanket with a piece of wood or lifting the blanket from the joists and replacing it.

In houses where new insulation was installed, sampling was carried out: (i) while the insulation was being installed; and (ii) 2-7 days after insulation. In certain cases the new insulation was disturbed in the manner described above and samples of air again collected. Samples for gravimetric analysis were collected on a tared 37-mm or 25-mm diameter glass fibre filter at a rate of 60 l/min. For measurement of fibre concentration by transmission electron microscopy (TEM), air was drawn through a 25-mm diameter Millipore 'MF' filter, pore size 0.8 $\mu$m, using a battery-driven pump at a flow rate of 3-7 l/min.

Filters were cut in half and prepared for TEM analysis using a published method (Burdett & Rood, 1983). A total of 100 fields of view (100 $\mu$m×100 $\mu$m) were examined for each sample at up to ×10 000 magnification. Each fibre was analysed by energy-dispersive X-ray analysis. The majority of the MMMF fell into two categories, namely, glass fibres containing silicon as their major constituent with calcium and sodium as minor constituents, and rock wool fibres made up of silicon and calcium as major components, with aluminium, magnesium, iron and sodium forming the minor constituents.

## Results and discussion

Respirable fibres, i.e., fibres of diameter <3 $\mu$m and length >5 $\mu$m but <100 $\mu$m, are alone considered. Some 30-35% of the MMMF detected were outside the respirable range.

### Disturbance of existing MMMF insulation

After the minor disturbance in the lofts of these houses, the concentration of airborne fibres was found to be generally less than 0.01 fibres per ml (f/ml). After the major disturbance, fibre concentrations in the loft on a time-weighted average of 4 h were found to be up to 0.04 f/ml (see Table 1). Personal samples, taken during the disturbance activity only (approximately 30 min), gave fibre concentrations of up to 0.2 f/ml.

**Table 1. Fibre concentrations in houses with existing insulation after major disturbance**

| Site No. | Concentration of respirable fibres[a] (f/ml) | | | |
|---|---|---|---|---|
| | Loft | First floor | Ground floor | Personal |
| 1 | 0.015 | <0.002 | <0.002 | 0.19 |
| 2 | 0.014 | <0.002 | <0.002 | 0.084 |
| 3 | 0.035 | <0.002 | <0.002 | 0.14 |
| 4 | 0.01 | 0.0006 | <0.002 | 0.07 |
| 5 | 0.007 | 0.003 | <0.002 | 0.20 |
| 6 | <0.006 | 0.002 | <0.002 | 0.04 |
| 7 | 0.002 | 0.002 | <0.002 | <0.009 |
| 8 | 0.008 | <0.002 | <0.002 | 0.06 |
|   | 0.01 | <0.002 | <0.002 | 0.07 |

[a]Defined as those <3 $\mu$m in diameter and 5-100 $\mu$m in length.

## Installation of new insulation

After the preliminary background sampling, air was monitored in these houses while the insulation was being laid (see Table 2). Samples of air were also collected 2-7 days after the installation of the insulation. The insulation was carried out with the following products:

(1)  glass fibre blankets;
(2)  rock wool blankets;
(3)  rock wool loose-lay;
(4)  rock wool blown-fibre insulation;
(5)  glass fibre blown-fibre insulation.

**Table 2. Fibre concentrations after installation of new insulation**

| Site No. | Concentration of respirable fibres[a] (f/ml) | | | | Type of insulation[b] |
|---|---|---|---|---|---|
| | Loft | First floor | Ground floor | Personal | |
| 1  | 0.02    | 0.005   | 0.001   | 0.41 | RWL      |
| 2  | 0.024   | 0.004   | -       | 0.20 | RWB      |
| 3  | 0.04    | 0.008   | <0.002  | 0.085 | RWB     |
| 4  | 0.06    | 0.002   | 0.003   | 0.18 | GFB      |
| 5  | 0.005   | <0.002  | <0.002  | 0.25 | GFB      |
| 6  | 0.15    | 0.06    | 0.016   | 1.77 | GFB      |
| 7  | 0.032   | 0.005   | 0.003   | 0.50 | RW blown |
| 8  | 0.029   | 0.014   | 0.03    | 0.55 | RW blown |
| 9  | 0.053   | 0.006   | <0.001  | 0.67 | GF blown |
| 10 | 0.74    | 0.002   | 0.002   | 0.50 | GF blown |
| 11 | -       | -       | -       | 0.16 | GF blown |
| 12 | <0.0002 | <0.001  | <0.001  | 0.15 | GF blown |

[a]Defined as those <3 μm in diameter and 5-100 μm in length.
[b]RWL, rock wool loose-lay; RWB, rock wool blankets; GFB, glass fibre blankets; RW blown, rock wool blown-fibre insulation; GF blown, glass fibre blown-fibre insulation.

During insulation with rock wool blankets, fibre concentrations in lofts were found to range between 0.02 and 0.04 f/ml, whereas during insulation with glass fibre blankets, fibre concentration in two lofts were 0.005 and 0.06 f/ml and 0.15 f/ml in a third case. Personal samples over short periods (about 30 min) gave fibre concentrations of 0.1-0.4 f/ml during installation of both types of blanket material. In one case, however, where sampling was carried out for 5 and 10 min, fibre levels were 1.7 and 1.85 f/ml, respectively.

Blown-fibre insulation produced slightly higher fibre levels in the loft of 0.03-0.05 f/ml. On one occasion, however, the fibre concentration reached 0.74 f/ml.

The personal samples collected when blown-fibre insulation was installed were mostly in the range 0.5-0.6 f/ml. Fibre levels measured during the follow-up studies

2-7 days after insulation were below the limit of quantification of the TEM method. Major disturbance of the blown-fibre insulation produced fibre levels of 0.04-0.07 f/ml, which were less than the fibre concentrations found in the lofts with existing blankets and loose-lay insulation after similar treatment.

The total dust concentration after the major disturbance in the houses with existing insulation was found to be in the range 0.05-0.5 mg/m$^3$. During installation of new insulation, concentrations of airborne total dust over a 4-h period varied between 1 and 6 mg/m$^3$. It is estimated that less than 10% of this dust consists of MMMF, the rest being the terrestrial/road dust which normally collects in the loft void. The personal samples over short periods of up to 30 min were, on average, 15 mg/m$^3$.

When existing insulation suffered a major disturbance and when insulation was installed in the loft, the fibre levels in other parts of the house remained low even though a trap door remained open throughout the sampling period. On two occasions, however, the level of fibres found on stairways on ground and first floors exceeded 0.01 f/ml. These were: (i) when blown rock wool insulation was used to insulate the loft; and (ii) when one brand of glass-fibre blanket insulation was being laid. The airborne-fibre fraction from this blanket was found to contain very fine fibres (over 80% <1 $\mu$m in diameter), which may have contributed to the high fibre levels found (1.70 f/ml) in personal samples).

## Conclusions

Measurement by TEM of airborne MMMF levels in 9 dwellings having various types of MMMF loft insulation, and in a further 12 dwellings during installation of various types of loft insulation has shown that, during physical disturbance of the insulation, fibre levels in lofts were up to 0.04 f/ml over a 4-h period and personal exposures during disturbance of insulation were up to 0.2 f/ml.

During installation of new insulation, fibre levels in lofts generally ranged from 0.005 to 0.06 f/ml. At two sites levels of 0.15 and 0.74 f/ml were recorded.

Personal exposures during installation of blanket insulation were generally in the range 0.1-0.4 f/ml (at one site, 1.77 f/ml). Personal exposure during installation of blown insulation were in the range 0.5-0.7 f/ml.

During disturbance of existing loft insulation and installation of new insulation, contamination of the living space of the dwelling was minimal, with the exception of one house where the living areas became contaminated with fine MMMF contained in the particular blanket used.

## Reference

Burdett, G.J. & Rood, A.P. (1983) Membrane-filter direct transfer technique for the analysis of asbestos fibres or other inorganic particles by transmission electron microscopy. *Environ. Sci. Technol.*, *17*, 643-648

# SMOKING AND THE PULMONARY MINERAL PARTICLE BURDEN

P.-L. Kalliomäki[1], O. Taikina-aho[2], P. Pääkkö[3], S. Anttila[3], T. Kerola[3], S.J. Sivonen[2], J. Tienari[4] & S. Sutinen[5]

[1]*Institute of Occupational Health, Helsinki, Finland*
[2]*Institute of Electron Optics, University of Oulu, Oulu, Finland*
[3]*Department of Pathology, University of Oulu, Oulu, Finland*
[4]*Department of Applied Mathematics and Statistics, University of Oulu, Oulu, Finland*
[5]*Meltola Hospital, Meltola, Finland*

*Summary.* The total pulmonary mineral particle burden and types of environmental particles were assessed in relation to smoking in 11 unselected autopsy lungs from adult male smokers and paired male non-smokers matched by age and lung. The lungs were fixed intrabronchially with formalin-polyethylene glycol-alcohol solution at a standard pressure and air-dried. A sample of 1-2 $cm^3$ was taken from the posterior or apicoposterior segment of the right/left upper lobe and plasma ashed at low temperature. The mineral particles were identified by scanning transmission electron microscopy (STEM), electron microprobe analysis and electron diffraction. The number, mass and volume were calculated from the STEM image.

The smokers' lung tissue had a lower number ($54\pm15\times10^6$), mass ($5.1\pm3.2$ μg), volume ($183\pm122\times10^{-5}$ $mm^3$) and surface area ($104\pm44$ $mm^2/cm^3$ of lung tissue) of particles than the non-smokers' lung tissue ($68\pm42\times10^6$, $12.6\pm13.4$ μg, $468\pm501 \times10^{-5}$ $mm^3$ and $191\pm167$ $mm^2/cm^3$ of lung tissue, respectively). All mineral types except talc were more numerous in the non-smokers' than in the smokers' lung tissue. The mineral particles were typical of the Finnish bedrock: quartz $15\pm7\%$, plagioclase $8\pm4\%$, microcline $13\pm5\%$, micas $22\pm10\%$, talc $4\pm4\%$ and kaolinite $10\pm5\%$. Fibres were observed in only 2 cases, amounting to 1% in each. The lower mineral particle content of the smokers' lungs probably reflects more active clearance mechanisms caused by cigarette smoke.

## Introduction

Mineral particles in the human lung are of both environmental and occupational origin. Traditionally, only asbestos fibres and quartz have been considered fibrogenic or carcinogenic, but the quantity of asbestos fibres in human lungs is small compared with those of other minerals (Churg & Wiggs, 1985). The number of pure quartz particles is normally much higher, but still only about one-tenth of the number of all mineral particles in lung tissue. Other minerals, such as zeolite (Baris *et al.*, 1987) and clay (Wagner *et al.*, 1986), may also cause fibrosis. We have found it important to examine the character and quantity of the entire mineral particle burden in the human lung. Cigarette smoke also contains mineral particles, such as talc, which may carry toxic compounds and gases on their surfaces, thus promoting their penetration into the lungs. Chronic bronchitis and pulmonary emphysema are clearly related to smoking, but these diseases may at the same time have an effect on the pulmonary clearance of mineral particles. The aim of this work was to assess the total mineral particle burden and the types of environmental particles in lung tissue in relation to smoking.

## Materials and methods

The materials were based on an initial series of 11 unselected autopsy lungs from adult male smokers or ex-smokers and as many paired non-smoker controls matched by age and lung (Table 1). The occupation was obtained from hospital records and the smoking habits were ascertained from the next of kin. The total smoking time in years was estimated and the number of pack years (i.e., the number of years multiplied by the number of packs of cigarettes smoked daily) was calculated when possible. Four cases, a non-smoker and 3 smokers, had to be omitted because of large amounts of endogenous iron which prevented the analyses.

**Table 1. Cases and controls: age, occupation, smoking history and severity of emphysema**[a]

| No. | Case and control | Age (years) | Occupation | Smoking time/ pack years | Severity of emphysema[a] |
|---|---|---|---|---|---|
| 1 | Smoker | 75 | Stoker | 60/120 | 60 |
|   | Non-smoker | 75 | Forester | 0/0 | 0 |
| 2 | Ex-smoker | 82 | Forest worker | NA[b] | 05 |
|   | Non-smoker | 81 | Farmer | 0/0 | 05 |
| 3 | Ex-smoker | 58 | Farmer | NA | 05 |
|   | Non-smoker | 58 | Typographer | 0/0 | 0 |
| 4 | Ex-smoker | 82 | Car mechanic | 30/23 | 10 |
|   | Non-smoker | 84 | Sawmiller | 0/0 | 0 |
| 5 | Smoker | 87 | Worker | 70/140 | 50 |
|   | Non-smoker | 86 | NA | 0/0 | 05 |
| 6 | Smoker | 77 | Farmer | 35/26 | 10 |
|   | Non-smoker | 77 | NA | 0/0 | 05 |
| 7 | Ex-smoker | 77 | NA | 40/NA | 05 |
|   | Non-smoker | 79 | Manager | 0/0 | 0 |
| 8 | Ex-smoker | 71 | Worker | 60/30 | 30 |
|   | Non-smoker[c] | 71 | Farmer | 0/0 | 0 |
| 9 | Smoker[c] | 62 | Tradesman | 32/16 | 05 |
|   | Non-smoker | 62 | Welder | 0/0 | 0 |
| 10 | Ex-smoker[c] | 71 | NA | 50/NA | 70 |
|   | Non-smoker | 72 | Secretary | 0/0 | 05 |
| 11 | Ex-smoker[c] | 69 | Farmer | 45/68 | 50 |
|   | Non-smoker | 69 | NA | 0/0 | 05 |

[a]Grade 0, normal lung; grades 05-20, mild; grades 30-50, moderate; grades 60-100, severe (Sutinen et al., 1981).
[b]NA, not available.
[c]Case omitted because of large amounts of endogenous iron preventing the mineral particle analyses.

One lung from each patient was excised and radiographed during continuous air inflation (Sutinen et al., 1979), fixed intrabronchially with a formalin-polyethylene glycol-alcohol solution, dried by air insufflation and sliced sagittally. The lungs were studied morphologically on the basis of the radiographs, gross specimens and histological sections (Sutinen et al., 1982). A peripheral sample of 1-2 cm$^3$, avoiding the pleural surface, was taken from the posterior or apicoposterior segment of the upper lobe of the right/left lung. The samples were ashed in aluminium cups in a low-temperature asher (100-120°C) with an oxygen plasma. The ash was dissolved in 1 M nitric acid to remove excess salts and then in absolute ethanol. The residue was sonicated and various dilutions were filtered on to a 0.1-$\mu$m pore size, 25-mm diameter Nuclepore filter. A piece of carbon-coated filter containing particulates was cut with a knife, avoiding contamination, transferred to a transmission electron microscope (TEM) grid and dissolved slowly in chloroform vapour. The grid was then coated with carbon again to minimize charging.

At least 100 particles (>0.1 $\mu$m) were identified by scanning transmission electron microscopy (STEM), electron microprobe analysis and electron diffraction using a JEOL 100CX scanning transmission microscope fitted with a PGT 1000 energy-dispersive spectrometer (EDS). Each particle was identified on the basis of its chemical composition as calculated from the EDS spectrum. About 1 out of every 10 mineral identifications was confirmed by electron diffraction. The volumes of the particles were calculated from their dimensions as measured on the STEM image, the thickness being assumed to be equal to the width for all minerals other than sheet silicates (0.2×width). The masses of the minerals were then calculated from their known densities and calculated volumes.

In order to calculate the number of mineral particles per unit weight of formalin-fixed, dry lung tissue and to be able to compare our results with those of previous studies, the volume of excised lung after fixation was measured by water displacement (Sutinen et al., 1979) (mean±SD: 2450±400 ml) and the total weight of a fresh lung was obtained (775±130 g). In our series, 1 g of fresh, wet lung tissue corresponded to 3.2±0.6 cm$^3$ of inflated, fixed lung tissue. In our tests, performed on another collection of 35 lungs inflated and fixed with wet formalin, the ratio of the fresh lung weight (760±200 g) to the formalin-fixed lung weight (1280±290 g) was 0.60±0.15, and for 10 lung samples weighing 0.4–1.1 g, the ratio of formalin-fixed, wet lung tissue weight to formalin-fixed, dry lung tissue weight was 8.5±0.8 (obtained by vacuum drying) (Pääkkö et al., 1987). On the basis of these determinations, 1 g of formalin-fixed, dry lung tissue corresponded to about 16 cm$^3$ of inflated, fixed lung tissue.

## Results

The numbers of particles of the various minerals per unit volume of lung tissue are shown in Table 2. The mean total number of mineral particles in the smokers' lungs was (54±15)×10$^6$, their mass 5.1±3.2 $\mu$g, their volume (183±122)×10$^{-5}$ mm$^3$ and their surface area 104±44 mm$^2$/cm$^3$ of lung tissue; the figures for the non-smokers' lungs were $(68\pm42)\times10^6$, $12.6\pm13.4$ μg, $468\pm501 \times 10^{-5}$ mm³ and $191\pm167$ mm²/cm³ of lung tissue, respectively. The mean total number of mineral particles in the smokers' lungs corresponded to $860\times10^6$ and in the non-smokers' lungs to $1080\times10^6$ per g of formalin-fixed, dry lung tissue. The differences between the smokers and non-smokers were not statistically significant. The mineral particles were typical of the Finnish bedrock: micas $22\pm10\%$, quartz $15\pm7\%$, microcline $13\pm5\%$, plagioclase $8\pm4\%$, kaolinite $10\pm55\%$ and talc $4\pm4\%$ (Figures 1 and 2). Fibrous minerals, namely anthophyllite and crocidolite, were observed in only 2 cases, amounting to 1%. All the mineral types except talc were more numerous in the lung tissue of the non-smokers than in that of the smokers.

### Table 2. Numbers of particles in lung tissue

| Type of mineral and chemical formula | Mean No. of particles ($\times10^6$/cm³) (SD) | |
|---|---|---|
| | Smokers ($n = 7$) | Non-smokers ($n = 7$) |
| Plagioclase [(Na,Ca)Al(Al,Si)Si$_2$O$_8$] | 3.8 (2.6) | 6.9 (6.0) |
| K-feldspar (KAlSi$_3$O$_8$) | 5.8 (3.0) | 10.6 (8.5) |
| Quartz (SiO$_2$) and silica | 6.0 (3.0) | 11.2 (8.4) |
| Kaolinite [Al$_4$Si$_4$O$_{10}$(OH)$_8$] | 5.8 (3.6) | 6.2 (4.9) |
| Mica[a] | 10.4 (5.3) | 18.0 (14.4) |
| Talc [Mg$_6$Si$_8$O$_{22}$(OH)$_4$] | 2.0 (1.5) | 1.4 (1.2) |
| Mullite (Al$_6$Si$_2$O$_{22}$) | 1.2 - | 0.3 - |
| Amphiboles[b] | 0.3 - | 0.8 - |
| Chlorite [(Mg,Al,Fe)$_{12}$(Si,Al)$_8$O$_{20}$(OH)$_{16}$] | 0.3 - | 0.8 - |
| Titanium (Ti) | 3.1 (4.6) | 3.3 (2.9) |
| Aluminium (Al) | 0.7 (1.2) | 1.0 (1.3) |
| Chromium compounds | 1.1 (1.2) | 0.7 (1.1) |
| Sulfur compounds | 0.8 (1.0) | 0.5 (0.6) |
| Iron particles | 2.0 (2.0) | 2.6 (3.4) |
| Various | 4.6 (2.8) | 3.7 (3.3) |
| Total | 53.9 (15.0) | 67.8 (41.8) |

[a]Muscovite [KAl$_2$(Al,Si$_3$)O$_{10}$(OH)$_2$] and biotite [K(Mg,Fe)$_3$(Al,Fe)Si$_3$O$_{10}$(OH,F)$_2$]

[b]Anthophyllite [Mg$_7$(Si$_8$O$_{22}$)(OH)$_2$], crocidolite (Na$_2$Fe$_2$Fe$_3$(Si$_8$O$_{22}$)(OH)$_2$], hornblende [(Ca,Na)$_{2-3}$(Mg,Fe$^{2+}$, Fe$^{3+}$,Al)$_8$O$_{22}$(OH)$_2$] and tremolite-actinolite [Ca$_2$(Mg,Fe)$_5$Si$_8$O$_{22}$(OH)$_2$].

**Fig. 1.** TEM micrograph showing a typical block-like plagioclase mineral. ×28 000

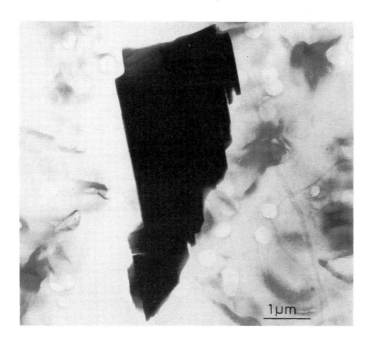

**Fig. 2.** TEM micrograph of a talc particle. These are always plate-like, thin and never fibrous. ×10 000

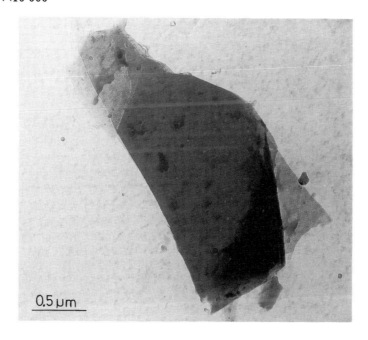

## Discussion

It is not easy to compare our results directly with those of certain previous studies because we prefer to calculate the amount of particles per unit volume. We therefore determined the ratio of fresh lung weight to the volume of excised lung after fixation by water displacement, the ratio of fresh lung weight to the formalin-fixed lung weight, and finally the ratio of formalin-fixed, wet lung tissue weight to formalin-fixed, dry lung tissue weight. From these ratios we were able to express our results as numbers of particles per unit weight of formalin-fixed, dry lung tissue. The total mineral particle burden in the lung tissue in our study seemed to be higher than that found in a previous study (Churg & Wiggs, 1985). The difference may be due to the differences in the digestion methods used — Churg and Wiggs (1985) used a wet digestion method while our method was low-temperature plasma ashing — and to the size of the smallest particles analysed.

The types of mineral particles in the lungs of both the smokers and the non-smokers were typical of the Finnish bedrock, and most of them can be expected to exist in airborne dust. The proportions of kaolinite and talc, especially in the lungs of the smokers, nevertheless exceeded the level that could be expected in mineral dust originating from granitic bedrock. Both talc and kaolinite are used in the paper industry, while talc is also used in many other industries. Fibrous minerals, namely anthophyllite and crocidolite, were found only in the lungs of 2 smokers.

The fact that the smokers' lung tissue contained a smaller number and mean mass of particles than that of the non-smokers may reflect the activation of pulmonary clearance mechanisms by cigarette smoking. Further studies with larger materials and containing cases with different smoking habits and different grades of emphysema would be needed to confirm this finding and to determine the effects of chronic lung disease caused by smoking on the pulmonary mineral particle burden.

## Acknowledgements

This study was supported financially by the Finnish Academy of Science.

## References

Baris, I., Simonato, L., Artvinli, M., Pooley, F., Saracci, R., Skidmore, J. & Wagner, C. (1987) Epidemiological and environmental evidence of the health effects of exposure to erionite fibres: a four year study in the Cappadocian region of Turkey. *Int. J. Cancer*, 39, 10-17

Churg, A. & Wiggs, B. (1985) Mineral particles, mineral fibers, and lung cancer. *Environ. Res.*, 37, 364-372

Pääkkö, P., Kokkonen, P., Anttila, S., Kerola, T., Kalliomäki, P.-L. & Sutinen, S. (1987) Metal concentrations in lung tissue in relation to smoking and pulmonary emphysema. In: *Trace Elements in Human Health and Disease*, Extended Abstracts from the Second Nordic Symposium, Odense, Denmark, 17-21 August 1987, pp. 148-151

Sutinen, S., Pääkkö, P. & Lahti, R. (1979) Post-mortem inflation, radiography, and fixation of human lungs: a method for radiological and pathological correlations and morphometric studies. *Scand. J. Respir. Dis.*, 60, 29-35

Sutinen, S., Lohela, P., Pääkkö, P. & Lahti, R. (1982) Accuracy of post-mortem radiography of excised air-inflated human lungs in assessing pulmonary emphysema. *Thorax*, *37*, 906-912

Wagner, C., Pooley, F., Gibbs, A., Sheers, G. & Moncrieff, C. (1986) Inhalation of china clay dust: relationship between the mineralogy of dust retained in the lungs and pathological changes. *Thorax*, *41*, 190-196

# FIBRE TYPE AND BURDEN IN PARENCHYMAL TISSUES OF WORKERS OCCUPATIONALLY EXPOSED TO ASBESTOS IN THE UNITED STATES

A.M. Langer & R.P. Nolan

*Center for Polypeptide and Membrane Research,
The Mount Sinai School of Medicine
of the City University of New York, New York, USA*

*Summary.* Lung tissues obtained from 53 asbestos-exposed workers, and one person exposed in a domestic setting, were studied. Amosite is the most prevalent fibre, occurring in 74% of the specimens. Amosite is always found in the lungs of insulation workers whereas chrysotile is found in only 50% of this population. Crocidolite has been detected in 24% of the lungs examined, but this increases to 40% in workers with shipyard histories. Exposure to chrysotile is widespread; the fibre has been observed in 61% of the tissues studied. Chrysotile occurs as the lone fibre in about 22% of the tissues examined, but tremolite is present in one-third of these. Fibre consumption data cannot be used as indices of exposure in the workplace; amphibole exposure appears to be product- and job-category-related. The assessment of risk to asbestos disease in the general population of the United States, exposed to chrysotile, should be based on appropriate chrysotile-exposed cohorts.

## *Study rationale*

The estimate of risk of asbestos-induced neoplastic diseases among the general population of the United States has been extrapolated from data obtained in many and varied occupational settings. The major assumptions made which are thought to validate this extrapolation include the following: chrysotile exposure constitutes the major fibre exposure of workers in the United States, inasmuch as 95% of the fibre used commercially is of this type; most non-occupational exposures, especially in ambient and building air, are predominantly chrysotile; the exposure to chrysotile in the general population renders the extrapolation biologically relevant, neglecting factors such as fibre length and physicochemical properties (Langer et al., 1971).

Significant objections have, however, been raised concerning the fibre type assumption. Some major cohorts on which the time and 'dose-response' components of the extrapolation are based appear to be primarily, or at least in part, amphibole-exposed. For example, workers in a factory in Paterson, New Jersey, were exposed to amosite in the manufacture of insulation for the US Navy (Seidman et al., 1979), and insulation workers in the United States were exposed to mixtures of amphiboles and chrysotile (Selikoff et al., 1979). The data for these cohorts have provided the basis for the extrapolation of, e.g., mesothelioma risk. Review of cohort data around the world conveys the impression that the amphibole asbestos varieties, crocidolite, amosite, and asbestiform tremolite, may be more important as agents of mesothelioma induction in workers than chrysotile (Doll & Peto, 1985; Langer & Nolan, 1985). The issue then is whether or not cohort data derived from US workers are based on amphibole or chrysotile exposure. Can these data be used to extrapolate asbestos risk, especially for mesothelioma occurrence among, e.g., persons reportedly exposed to low levels of chrysotile in indoor air? As an initial step in answering this question, we began a study of the pulmonary tissues of workers to determine fibre types and burden after exposure in specific trades.

## *Study design*

### Selection of cases

The 54 cases evaluated in this study came to our laboratory as referrals. The specific concerns which arose in these cases were the result of legal issues involving both compensation and tort litigation. Verification of tumour type (especially mesothelioma) and tissue burden study (fibre type and amount) were the two most frequent requirements. Even though occupational asbestos exposure (and domestic asbestos contact in one case) was established for all the cases studied, this group represents what we consider to be an almost worst-case biological outcome. The study has a very strong built-in bias toward asbestos diseases.

### Primary causes of death

The primary causes of death among the 54 cases underscore the selective bias: mesothelioma, 23/54 (42.6%); lung cancer, 20/54 (37.0%); asbestosis, 9/54 (16.7%); 'other causes', 2/54 (3.7%). Cause of death and tumour type were verified by autopsy and/or pathology review of histological material. No occupational cohort studied has ever shown this mortality experience.

### Occupational categories

The 54 cases were grouped into three major occupational categories on the following basis:

(1) *Insulation worker*: if ever in insulation work; no history of shipyard work; with or without experience in other dusty trades (e.g., construction trades). $N=16$.

(2) *Shipyard worker*: if ever in shipyard environment regardless of trade (e.g., welder, electrician, machinist, insulator, etc.). $N=15$.

(3) *Other trades*: never in insulation work, never in shipyard environment. Includes brake repair, plasterer, railroad worker, rubber worker, construction trades, etc. $N=23$.

The rationale for these categories and their selection criteria is based on the perception that insulation work involves exposure to mixed fibre types and carries great risk of asbestos disease, that shipyard work involves exposure to amosite and may carry the greatest potential for crocidolite fibre exposure; 'other trades' include workplaces where exposure to chrysotile only appears most likely.

**Materials studied**

Over 80% of the cases had formalin-fixed bulk pulmonary tissues available for study. The remaining specimens were available from histological blocks. The recovery of the inorganic dust was carried out on tissues ranging in mass from 20 to 5000 mg by the alkali digestion method (Langer & McCaughey, 1982).

**Analytical protocol**

The recovered residues were suspended in a premeasured volume of distilled water and sonicated for dispersal. A single 10-$\mu$l aliquot was placed on each of several carbon-coated, Formvar-support, 200-mesh nickel locator grids. Fibre counts were obtained on 10 fields scanned at magnifications of 10 000–20 000, fibre dimensions were measured directly from photographic plates, and fibre identification was based on morphology, chemistry (determined by energy-dispersive X-ray spectrometry), and structural character (determined by selected area electron diffraction). Results are expressed as fibres per gram (f/g) of dry lung tissue. Only data for fibres greater than 1 $\mu$m in length are shown.

## Results

Amosite, chrysotile and crocidolite, the three major fibres used in commerce in the USA, are found in the pulmonary tissues of the asbestos workers studied (Table 1). Amosite appears to be the most prevalent fibre type in lungs, occurring in 74% of the cases studied. Additionally, almost 25% of these workers have been exposed to crocidolite. Tremolite, actinolite and anthophyllite are also frequently found in pulmonary tissues. Amphibole fibres of commercial types have been observed in 42 of 54 ($\simeq$78%) pulmonary specimens.

In terms of commercial fibre occurrence, amosite and chrysotile are observed to occur as the sole fibre with statistically equal frequency. Crocidolite only rarely occurs as the sole fibre type in the lung (Table 1). Tremolite, actinolite and anthophyllite fibres are never found in pulmonary tissues as the sole fibre type. Although the data are not presented here, anthophyllite and actinolite tend to occur with amosite exposure and tremolite with chrysotile exposure.

In terms of total fibre burden (Table 2), the lungs of insulation workers contain statistically greater concentrations of total fibre as compared with those of shipyard workers. Although much amosite is used in shipyard settings, the insulation workers' lungs may contain slightly greater concentrations. Chrysotile fibre concentrations are universally high in all trades. We note the paradox that crocidolite burdens in the various occupational groups show no statistical difference as between shipyard workers and other occupational categories, but rather between *insulation workers* and *other trades*. If insulation work in shipyards is neglected, insulation workers' lungs contain, statistically, the lowest concentrations of crocidolite. We add the caveat that

**Table 1. Concentrations$^a$ of asbestos fibres in lung tissues of US workers (54 cases)**

| Fibre type | Cases in which fibre observed | | Cases in which fibre observed as sole fibre$^b$ | | Fibre concentration$^c$ | | |
|---|---|---|---|---|---|---|---|
| | No. | % | No. | % | Average | SD | Range |
| Amosite | 40 | 74.1 | 14 | 25.9 | 27.2 | 64.9 | 0.2-350.0 |
| Chrysotile | 33 | 61.1 | 12 | 22.2 | 281.8 | 746.5 | 0.1-4420.0 |
| Crocidolite | 13 | 24.1 | 2 | 3.7 | 2.1 | 7.5 | 0.1-50.0 |
| Tremolite-actinolite | 13 | 24.1 | 0 | 0.0 | 3.0 | 10.9 | 0.1-60.0 |
| Anthophyllite | 16 | 29.6 | 0 | 0.0 | 1.6 | 7.3 | 0.1-50.0 |

$^a$Per g of *dried* tissue

$^b$Those cases among the 54 studied in which the fibre type was the *only* major commercial fibre found in tissues.

$^c$Values for average concentration, standard deviation (SD), and range are expressed in millions of fibres per g of dried lung tissue for fibres >1 μm only.

**Table 2. Concentrations of major commercial types of asbestos fibres in lung tissue, by occupational category**

| Occupational category | No. | Concentration$^a$ | | |
|---|---|---|---|---|
| | | Amosite | Chrysotile | Crocidolite |
| Insulation workers | 16 | 47.5±78.2 | 299.3±1101.1 | 0.5±1.7 |
| Shipyard workers | 15 | 11.4±19.9 | 91.4±177.4 | 5.1±13.4 |
| Other trades | 23 | 25.4±75.6 | 394.7±699.2 | 3.1±3.4 |

$^a$Millions of fibres per g of dried lung tissue.

insulators who have ever worked in shipyards may be at some unusually high risk of asbestos-induced malignancies.

As far as prevalence of fibre type is concerned (Table 3), the lungs of insulation workers always contain amosite and may also contain chrysotile and/or crocidolite. No other occupational category shows a major commercial fibre as universally present. We note that the prevalence of crocidolite fibre is greatest in shipyard workers, this type of fibre being found in some 40% of these workers.

## Discussion and conclusions

Although 95% of the fibre used in the United States over the past 5 decades has been chrysotile, the 2 other commercially important fibres, amosite and crocidolite, were commonly found in the lungs of the workers studied. Chrysotile fibre was anticipated, but amosite occurrence is very much greater (by a factor of more than 25) than would be expected from its commercial use. Only about 2% of the fibre used

**Table 3. Asbestos fibre type and prevalence within occupational categories**

| Occupational category | No. | Prevalence | | |
|---|---|---|---|---|
| | | Amosite | Chrysotile | Crocidolite |
| Insulation workers | 16 | 16/16 (100%) | 8/16 (50%) | 3/16 (19%) |
| Shipyard workers | 15 | 10/15 (67%) | 10/15 (67%) | 6/15 (40%) |
| Other trades | 23 | 14/23 (61%) | 15/23 (65%) | 4/23 (17%) |
| Total | 54 | 40/54 (74%) | 33/54 (61%) | 13/54 (24%) |

during this same period was crocidolite, yet almost 25% of the workers studied had crocidolite in their pulmonary tissue. Crocidolite is a much more common fibre in tissues than previously reported. Remarkably, it has been found in 40% of the shipyard workers examined. This is nearly 20 times greater than could be expected from its commercial use in the United States. Shipyard work may provide significant opportunities for crocidolite exposure, and we conclude that an appreciable mesothelioma risk may be associated with work in the shipyard environment. The same general statements may hold for insulation workers whose lungs contain amosite. Amosite, which accounted for only 3% of fibre used in commerce during this period, is universally present (100%) in the lungs of insulation workers. This is more than 30 times what could be expected from the commercial usage of the fibre in the United States.

In some instances, however, fibre burden is consistent with patterns of product composition and use. We note that chrysotile occurrence alone, as the sole major fibre in pulmonary tissues, is associated with specific products and trades (e.g., brake pads and brake repair). We also note that patterns of fibre burden need not be in line with fibre consumption data. Much chrysotile may end up in products which are neither dusty nor produce significant amounts of dust when installed, used, or removed (e.g., vinyl-asbestos floor tiles), so that 95% of the fibre usage need not produce 95% of the dust exposure in workplaces. On the other hand, small percentages of amphibole-containing products may be a source of a disproportionately large amount of dust and fibre exposure to the worker.

In the cases studied, tremolite tends to occur with chrysotile exposure and anthophyllite and actinolite with amosite exposure. More cases are needed to verify this.

The above data, although derived from a small number of selected cases, suggest that the amphibole minerals and mixed fibre exposures may be more important as agents in the etiology of lung cancer and mesothelioma than chrysotile alone.

Current assessment of risk of disease in the general population of the United States exposed to chrysotile fibre in ambient and indoor air should be based on data obtained for chrysotile exposure in the workplace. We would also add the caveat that appropriate cohorts should be used to incorporate the physicochemical characteristics of the fibre as well. The data generated by the present study suggest that many currently used data sets are of limited value for the assessment of risk in the general population.

## Acknowledgements

The authors wish to acknowledge support from the Société Nationale de l'Amiante of Canada. One of us (RPN) wishes to specifically acknowledge support from the Stony Wold-Herbert Fund and assistance from The Life Sciences Foundation.

## References

Doll, R. & Peto, J. (1985) *Asbestos: Effects on Health of Exposure to Asbestos*, London, Her Majesty's Stationery Office

Langer, A.M. & McCaughey, W.E.T. (1982) Mesothelioma in a brake repair worker. *Lancet, ii*, 1101-1103

Langer, A.M. & Nolan, R.P. (1986) Asbestos in potable water supplies and attributable risk of gastrointestinal cancer. *Northeast. Environ. Sci.*, 5, 41-53

Langer, A.M., Selikoff, I.J. & Sastre, A. (1971) Chrysotile asbestos in the lungs of persons in New York. *Arch. Environ. Health*, 22, 348-361

Seidman, H., Selikoff, I.J. & Hammond, E.C. (1979) Short-term asbestos work exposure and long-term observation. *Ann. N.Y. Acad. Sci.*, 330, 61-89

Selikoff, I.J., Hammond, E.C. & Seidman, H. (1979) Mortality experience of insulation workers in the United States and Canada, 1943-1976. *Ann. N.Y. Acad. Sci.*, 330, 91-116

# AIRBORNE MINERAL FIBRE CONCENTRATIONS IN AN URBAN AREA NEAR AN ASBESTOS-CEMENT PLANT

### A. Marconi

*Laboratory of Indoor Hygiene, Higher Institute of Health, Rome, Italy*

### G. Cecchetti & M. Barbieri

*Industrial Hygiene Centre, Catholic University, Rome, Italy*

*Summary.* Ambient air concentrations of asbestos and total mineral fibres were measured during the period June-July 1985 at several locations near a large asbestos-cement factory located in the proximity of a northern Italian town. Measurements of the number and type of fibres were made by means of analytical scanning electron microscopy and energy-dispersive X-ray analysis (SEM-EDXA), essentially according to the RTM2 reference method of the Asbestos International Association.

Total mineral fibre concentrations (longer than 5 $\mu$m) ranged from below detection limit (D.L., i.e. 0.4 fibres per litre (f/l)) to 227 f/l (single value); mean values ranged from 1.3 to 74.0 f/l. However, elemental microanalysis (EDXA) showed that about 65% of fibres were sulfate fibres, 20% were aluminium silicates or other silicates, and only 15% were asbestos fibres (mainly chrysotile and tremolite-group amphiboles). Asbestos concentrations (fibres more than 5 $\mu$m in length) were in the range of <D.L. to 19.1 f/l, mean values ranging from <D.L. to 11.1 f/l. The results obtained showed large differences in day-to-day concentrations, suggesting that they were affected by the rate of production in the plant and by weather conditions. In particular, wind direction and distance from the source appeared to be of major importance since the highest asbestos fibre concentrations were mostly found at points closer to the source and downwind. When the distance from the source was increased, other local factors appeared to be determining. In addition, the large proportion of non-asbestos mineral fibres and elongated sulfate (or sulfur-containing) particles clearly indicates major contributions from other sources, such as local pollution and natural soil erosion.

## Introduction

Asbestos-cement production is the main manufacturing process utilizing asbestos in Italy, accounting for more than 65% of the industrial consumption of asbestos fibre (Marconi, 1987).

In this study, carried out during the period June-July 1985, several sites around a large asbestos-cement plant located in the suburbs of a city in the north of Italy were monitored for ambient air inorganic fibre concentration. In addition, the results obtained have been evaluated with respect to the distance from the source and the prevailing meteorological parameters.

## Sampling and analysis

Ambient air concentrations of inorganic fibres, including asbestos, were measured during the period 19 June to 31 July 1985 at the 11 sites, including those inside the plant (point S), shown in Figure 1. Nine sites were located around the factory (points A to I) at distances varying between 400 and about 6 km. In addition, 2 sites at a distance of more than 20 km (points M: rural village and N: centre of a small city) were included in order to have non-source-influenced reference values.

**Fig. 1. Sampling locations**

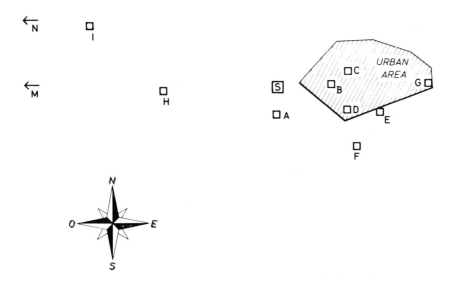

In the sampling and analysis, the AIA RTM2 method has been followed (Asbestos International Association, 1984).

Air samples were collected by means of low-volume, constant-flow pumps (Zambelli and Tecora types), at a flow rate of 8 (±10%) l/min. Sampling times were in the range 4–8 h.

For airborne dust collection, polycarbonate filters, 0.8 $\mu$m pore size, 25 mm diameter, previously coated with a gold layer of thickness of about 40–60 $\mu$m, were used. Under these conditions, a face velocity of about 35 cm/s was achieved. Open-faced field monitor cassettes (Millipore), with conductive plastic cowl, were used as filter holders. They were placed at about 1.5 m from the soil, facing downwards.

Portions of the dust-loaded filters were mounted on aluminium stubs, previously coated with colloidal graphite. For the analysis, a scanning electron microscope (SEM) (Cambridge Stereoscan 180) was used at an accelerating voltage of 20 kV, 12 mm working distance, 2000× magnification, and 45° tilt angle (using correction for fibre length). Fibre counting and sizing were performed directly on the SEM screen after calibration with a test grid (Agar Aids). Counting and sizing criteria were 100 fields or 50 fibres; the fibres counted were those with length $(L)$: diameter $(D)$ ratio $>3:1$, $D<3$ $\mu$m and $L>1$ $\mu$m; fibres with $L>5$ $\mu$m and $<5$ $\mu$m were recorded separately. Fibre identification was effected by energy-dispersive X-ray analysis (EDXA) (Link 860 system), following the fibre characterization scheme proposed in the AIA method (e.g., distinction between inorganic non-asbestos, asbestos and sulfate).

Blank filters (10% of total samples) were included in the analysis; if their fibre counts were greater than 5 fibres per mm$^2$ (f/mm$^2$), the excess was subtracted from the sample counts.

A detection limit of 0.4 f/l was assumed on the basis of 1 fibre in 100 fields, 3 m$^3$ of air and data from blank tests.

During the sampling period, wind speed and direction and relative humidity were recorded at the sampling sites.

In general, concentrations of both total inorganic and asbestos fibres differ markedly from day to day, depending on the rate of production, weather conditions and location. In particular, it appears that, for total inorganic fibres, distance from the source and wind direction (at least for the speeds recorded) are two important influencing factors. Fibre concentrations decrease up to a distance from source of about 1600 m. Higher concentrations are generally associated with winds from the N-NE and with downwind samples, except at site C, located in the centre of the city and therefore subject to the influence of additional local factors. The lowest values are associated with winds from the S-SE, for distances from source greater than 400-500 m. From results obtained at sites at distances of more than 1600 m with variable wind direction, it appears that the general level of total inorganic fibres ($L>5$ $\mu$m) is about 5-6 f/l. A higher background level has been found for sites located in the centre of town (see points C and N). For these fibres, the high percentage of sulfur-bearing and siliceous fibres suggests a contribution from both anthropogenic and natural sources.

For asbestos fibre concentrations ($L>5$ $\mu$m), similar trends were observed (detailed results are given later in Table 3 and Figure 3). Downwind mean values were almost always greater than upwind values for sites at distances less than 1600 m. No downwind values exceeded 8 f/l, except a single value of 19.0 f/l at a distance of 400 m from source.

Upwind values at all sites are less than 6 f/l (point C, town centre) and most are in the range 0.5-2.5 f/l, and often lower than the levels recorded at the reference sites (M and N). For asbestos fibres again, the distance from the source and the wind direction appear to be the dominant factors controlling fibre concentrations, at least for sites located less than 1600 m from the source.

Finally, the presence in the air of calcium-rich amphibole fibres may indicate a contribution from contamination associated with the raw chrysotile used in the asbestos-cement process, but also from wind erosion of soils which, in this northern region of Italy, are often rich in serpentine and amphiboles. A contribution from natural sources to the concentration of asbestos fibres can therefore not be excluded.

## Results and discussion

The results obtained for total inorganic and asbestos fibres are summarized in Tables 1–3 and Figures 2 and 3. Fibre concentrations are shown as a function of distance from the source (S) and of prevailing meteorological parameters. Although values for fibres with $L<5$ µm are reported, the most significant results refer to fibres longer than 5 µm. The low concentrations recorded for fibres with $L<5$ µm probably reflect problems in detecting smaller fibres at the 2000 × magnification used (see also Cherry *et al.*, 1987). Mean total inorganic fibre concentrations lie in the range 1.3–74.0 f/l ($L>5$ µm). The value of 124.0 f/l (due to a single very high value of 227.0 f/l) is probably attributable to construction activities near the sampling site. In fact, the composition of the majority of the fibres detected was consistent with their being calcium or potassium sulfate (probably gypsum).

**Fig. 2. Mean concentrations of total inorganic fibres in relation to distance from source**

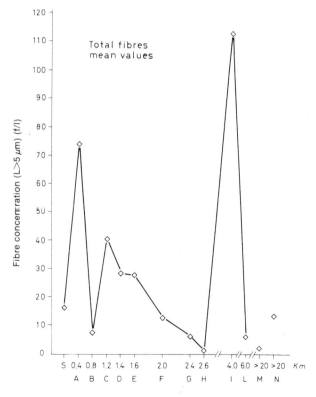

**Table 1. Total inorganic fibre concentrations and meteorological parameters**[a]

| Date | Sampling points | | | | | | | | | | | | | Relative humidity (%) | Wind speed (m/sec) | Wind direction |
|------|----|----|----|----|----|----|----|----|----|----|----|----|----|----|----|----|
| | S | A | B | C | D | E | F | G | H | I | L | M | N | | | |
| 19 June | 9.0 | - | 8.0 D | 40.0 U | 30.0 D | 16.0 D | - | - | 1.0 D | - | - | - | - | - | 6-7 | N |
| 21 June | 2.0 | 20.0 U (2.0) | - | - | 2.0 U | - | - | - | DL D | <DL | - | - | - | 19-78 | 4 | SE |
| 18 July | 63.0 | 128.0 D | - | 68.0 U (1.0) | 78.0 D (16.0) | 87.0 D (14.0) | - | - | 2.0 D (1.0) | 228.0 | - | - | - | 54-88 | 4-5 | N-NE |
| 30 July | 8.0 (1.0) | - | - | 6.5 V (4.0) | 6.0 V | 5.0 V | 13.0 V | 2.0 V | 3.0 V | - | 13.0 (6.0) 4.0 V (1.0) | 4.0 | 16.0 | 57-90 | 3-10 | NW-S-N, variable |
| 31 July | 0.5 (1.0) | - | - | 7.0 V | 26.0 V (1.0) | 3.0 V | - | 11.5 V (1.0) | DL V DL | - | 4.0 V 6.0 V (2.0) | <DL | 10.5 | 20-77 | 3-5 | NW-NE-N, variable |

[a] Fibre concentrations are expressed as f/l for fibres of length >5 μm. Concentrations of fibres of length <5 μm are shown in parentheses. The distance from S (the source) increases in going from A to N. U, upwind; D, downwind; V, variable; DL, detection limit

**Table 2. Asbestos fibre concentrations and meteorological parameters**[a]

| Date | Sampling points | | | | | | | | | | | | | Relative humidity (%) | Wind speed (m/sec) | Wind direction |
|---|---|---|---|---|---|---|---|---|---|---|---|---|---|---|---|---|
| | S | A | B | C | D | E | F | G | H | I | L | M | N | | | |
| 19 June | 3.0 | - | 1.0 D | 6.0 U | 4.5 D | 2.5 D | - | - | <DL | - | - | - | - | - | 6-7 | N |
| 21 June | 0.5 | 3.0 U (DL) | - | - | <DL U | - | - | - | <DL D | <DL D | - | - | - | 19-78 | 4 | SE |
| 18 July | 21.0 | 19.0 D | - | 10.0 U (<DL) | 12.0 D (2.5) | 13.0 D (2.0) | - | - | <DL D (<DL) | 2.5 D | - | - | - | 54-88 | 4.5 | N-NE |
| 30 July | 3.0 (<DL) | - | - | 1.0 V (0.5) | 1.0 V | 1.0 V | 2.0 V | <DL V | 0.5 V | - | 2.0 V  1.0  0.5 V (<DL) | 1.0 | 2.5 | 57-90 | 3-10 | NW-S-N, variable |
| 31 July | <DL (<DL) | - | - | 1.0 V | 4.0 V | DL V | - | 2.0 V (<DL) | <DL V (<DL) | - | 1.0 V <DL V (0.5) | <DL | 1.5 | 20-77 | 3-5 | NW-NE-N, variable |

[a] Fibre concentrations are expressed as f/1 f $\alpha$ fibres of length >5 µm. concentrations of fibres of length <5 µm are shown in parentheses. The distance from S (the source) increases in going from A to N. U, upwind; D, downwind; V, variable; DL, detection limit.

**Fig. 3. Downwind and upwind asbestos fibre concentrations in relation to distance from source**

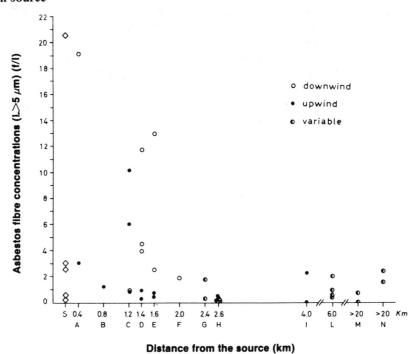

**Table 3. Asbestos fibre concentrations, distance from source and wind direction**

| Site | Distance from source (km) | Mean concentration ($L$>5 µm) (f/l) | | | Overall mean concentration ($L$>5 µm) (f/l) |
|---|---|---|---|---|---|
| | | Downwind | Variable | Upwind | |
| S | 0 | - | - | - | 5.5 |
| A | 0.4 | 19.0 | - | 3.0 | 11.0 |
| B | 0.8 | - | - | 1.0 | 1.0 |
| C[a] | 1.2 | 1.0 | - | 6.0 | 4.5 |
| D | 1.4 | 7.0 | - | 0.5 | 5.5 |
| E | 1.6 | 8.0 | - | 0.5 | 4.0 |
| F | 2.0 | 2.0 | - | - | 2.0 |
| G | 2.4 | - | 1.0 | - | 1.0 |
| H | 2.6 | <DL[b] | - | 0.5 | <DL[b] |
| I | 4.0 | <DL[b] | - | 2.0 | 1.0 |
| L | 6.0 | - | 1.0 | - | 1.0 |
| M | >20 | <DL[b] | - | 1.0 | DL[b] |
| N | >20 | 2.0 | - | 2.5 | 2.0 |

[a]City centre.
[b]Detection limit.

Overall mean asbestos fibre concentrations ($L>5$ µm) range from less than the detection limit to 5.5 f/l, only one value exceeding 10 f/l (Table 3). EDXA analysis showed that, on average, about 65% of the total inorganic fibres contained calcium or potassium and sulfur, suggesting the presence of sulfates, the origin of which is still a matter of discussion (Cherry *et al.*, 1987; Marconi *et al.*, 1987; Marfels & König, 1985; Spurny, 1984) (Figure 4). About 20-25% of total inorganic fibres showed the presence of MgAlSi, often with calcium or potassium and iron (Figure 5). Only 10-15% of fibres had a composition consistent with asbestos. The majority of asbestos fibres can be assigned to chrysotile (Figure 6), but a significant amount appear to belong to the calcium-rich amphibole group, probably tremolite-actinolite (Figures 7 and 8). The proportion of fibres of this type in the asbestos group ranged from 15% to 30%.

**Fig. 4. EDXA spectrum of sulfate fibres**

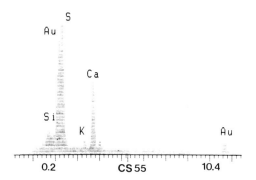

**Fig. 5. EDXA spectrum of non-asbestos mineral fibres**

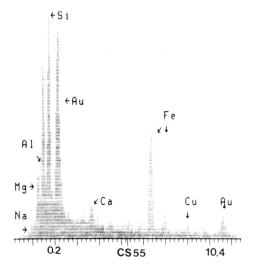

**Fig. 6. EDXA spectrum of chrysotile fibre**

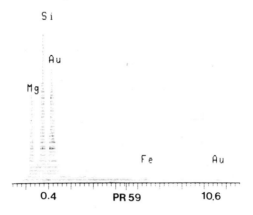

**Figs 7, 8. EDXA spectra of calcium silicate fibres**

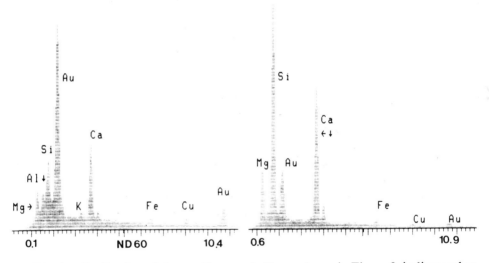

The size distribution of the total inorganic fibres, shown in Figure 9, indicates that fibres longer than 5 $\mu$m account for about 90%, and some 75% of fibres had diameters between 0.4 and 1.6 $\mu$m. These results, however, probably reflect the limitation of the low magnification used. Nevertheless, the majority of fibres are in the range of dimensions assumed to be biologically active.

A comparison between the results obtained in this study and those reported in the literature for source-oriented locations around industrial plants is shown in Table 4 in terms of asbestos concentrations for fibres with $L>5$ $\mu$m. It appears that, in general, good agreement exists betweenis restricted to fibres longer than 5 $\mu$m.

In addition, in these studies, the effects of source distance and wind direction have been generally reported to be similar to those observed in the present survey.

**Fig. 9.** Fibre size distribution (total inorganic fibres)

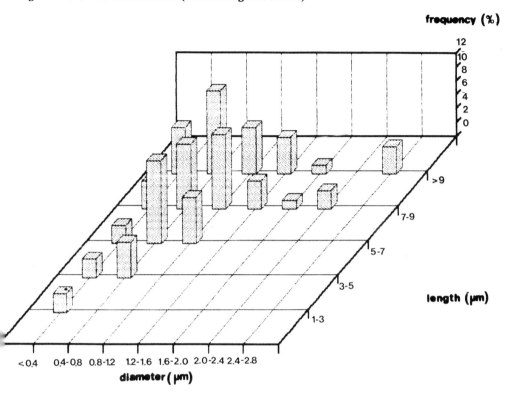

**Table 4. Summary of published data on asbestos fibre ($L>5$ μm) concentrations around industrial plants**

| Type of plant | Method of analysis | Asbestos concentration ($L>5$ μm) (f/l) | | Reference |
|---|---|---|---|---|
| | | Mean | Range | |
| Asbestos cement plant | TEM[a] | 4.0 | - | Lanting and den Boeft, 1983 |
| Asbestos-cement plant | SEM[a] | - | <0.1-1.9 | Teichert, 1980 |
| Industrial plant (unspecified) | SEM | - | 0.6-2.2[b] | Marfels et al., 1984 |
| Chrysotile mine | TEM[c] | 1.8 | 0.04-8.1 | Singh and Thouez, 1985 |
| Amosite board manufacturing plant | TEM[c] | - | 15-60 | Burdett et al., 1984 |
| Textile plant | SEM | - | 0.9-11.5 | Burdett et al., 1984 |
| Chrysotile mine | SEM | 2.4 | 0.6-8.8 | Felbermayer, 1983 |
| Asbestos-cement plant | SEM | 0.5 | 0-2.2 | Felbermayer, 1983 |
| Textile plant | SEM | - | <1.0-6.0 | Cherry et al., 1987 |
| Asbestos-cement plant | SEM | 3.0 | 0.4-11.0 | Marconi et al. (this study) |

[a]TEM, transmission electron microscopy; SEM, scanning electron microscopy.
[b]Peak concentration 15 f/l.
[c]Derived from gravimetric data using the factor 1 ng = 30 fibres ($L>5$ μm).

## References

Asbestos International Association (1984) *Method for the Determination of Airborne Asbestos Fibre and Other Inorganic Fibres by Scanning Electron Microscopy* (*AIA Health and Safety Publication RTM2*)

Burdett, G.J., Le Guen, J.M. & Rood, A.P. (1984) Mass concentrations of airborne asbestos in the non-occupational environment: a preliminary report of U.K. measurements. *Ann. Occup. Hyg.*, *28*, 31-38

Cherry, J.W., Dodgson, J., Groat, S. & Carson, M. (1987) *Comparison of Optical and Electron Microscopy for Evaluating Airborne Asbestos* (*Report TM/87/01*), Edinburgh, Institute of Occupational Medicine

Felbermayer, W. (1983) Ambient air measurements in Austria of fibre levels from asbestos-cement products [in German]. In: Reinisch, D., Schneider, H.W. & Birkner, K.F., eds, *Fibrous Dusts — Measurement, Effects, Prevention* (*VDI-Berichte 475*), Düsseldorf, VDI-Verlag, pp. 143-146

Lanting, R.W. & den Boeft, J. (1983) Ambient air concentrations of mineral fibres in the Netherlands. In: Reinisch, D., Schneider, H.W. & Birkner, K.F., eds, *Fibrous Dusts — Measurement, Effects, Prevention* (*VDI-Berichte 475*), Düsseldorf, VDI-Verlag, pp. 123-128

Marconi, A. (1987) Occupational exposure to asbestos in Italy: considerations on the basis of available data [in Italian]. In: Parolari, G., Gherson, G., Cristofolini, A. & Merler, E., eds, *Occupational and Environmental Cancer Risk from Asbestos*, Verona, Bi & Gi Publishers, pp. 159-170

Marconi, A., Rossi, L., Migliaccio, G., DiGirolamo, P. Scaccia, M., Galoppini, P. & Barbieri, M. (1987) Mineral fibre contamination in building insulated with surfacing fibrous materials: results of a survey in a large office unit [in Italian]. *G. Ig. Ind.*, *12*, 25-63

Marfels, V.H., Spurny, S., Boose, C., Schormann, J., Opiela, H., Althaus, W. & Weiss, G. (1984) Measurements of fibrous dust in ambient air of the Federal Republic of Germany. I. Measurements in the vicinity of an industrial source [in German]. *Staub-Reinhalt. Luft*, *44*, 259-263

Marfels, H. & König, R. (1985) On the 'Gypsum fibre problem' in relation to the measurement of fibrous dusts in ambient air [in German]. *Staub-Reinhalt. Luft*, *45*, 441-444

Singh, B. & Thouez, J.P. (1985) Ambient air concentrations of asbestos fibres near the town of Asbestos, Quebec. *Environ. Res.*, *36*, 144-159

Spurny, K. (1984) On the technical and hygienical evaluation of fiber dust concentration in the atmospheric environment. *Staub-Reinhalt. Luft*, *44*, 456-458

Teichert, U. (1980) Measurements of asbestos fibre concentrations near suspected asbestos emission sources. In: Asbestos International Association and Association Française de l'Amiante, eds, *Proceedings of the Third International Colloquium on Dust Measuring Techniques and Strategy, Cannes*, Paris, AFA Publications, pp. 227-231

# NON-ASBESTOS FIBRE CONTENT OF SELECTED CONSUMER PRODUCTS

### J.C. Méranger & A.B.C. Davey

*Health and Welfare Canada, Environmental Health Directorate, Ottawa, Ontario, Canada*

*Summary.* The use of asbestos in consumer products such as joint compounds and modelling clays has been banned since these products, in normal use, have the potential to produce high levels of fibrous dusts in ambient air. However, we were concerned that the asbestos fibres present in these products would be replaced by other fibrous materials. X-ray diffraction and transmission electron microscopy analyses were performed to identify the crystalline components, including mineral fibres, in a selection of 45 samples. Using quantitative X-ray diffractometry, it was established that the fibrous content of several consumer products was in the range of 1–5%. The widespread use of attapulgite in cat litters, joint and spackling compounds and to a lesser extent in art supplies was confirmed. It was also shown that the attapulgite fibres detected were generally in the same size range as chrysotile, with diameters of 0.03–0.5 $\mu$m and lengths up to 4 $\mu$m.

## Introduction

The use of asbestos in consumer products such as joint compounds and modelling clays has been banned in Canada since it was recognized that these products, in normal use, have the potential to produce high levels of fibrous dusts in ambient air (Government of Canada, 1983). Our efforts to monitor products of these types arose from our concern that the asbestos fibres would be replaced by other fibrous materials which, because of their fibre size distribution, might also have a high toxic potential.

## Materials and methods

### Sampling

Samples of modelling clays, joint and spackling compounds and cat litters representative of those found on the local Ottawa market were purchased from retail outlets. The modelling clays included 10 moist clays, 5 powders, and 2 solid sculpting blocks. Powdered and paste joint and spackling compounds including 6 pastes and 15 powders, and 7 cat litters were also obtained from local retail stores.

### X-Ray diffraction analysis

X-Ray diffraction spectra were obtained at 50 kV, 20 mA, using a molybdenum target. Scans were obtained from 2° to 40° at a scanning speed of 0.5° per min. The spectra were characterized and matched with available International Center for

### Table 1. X-ray diffraction analysis of selected art supplies

| Sample No. | Components identified |
|---|---|
| **Moist clay or clay-like** | |
| 1 | Calcite + calcium hydroxide (trace) |
| 2 | Quartz, kaolinite or serpentine |
| 4 | Not identified |
| 11 | Calcite + kaolinite or serpentine and unknown (trace) |
| 12 | Calcite + unknown (trace) |
| 13 | Quartz, kaolinite or serpentine |
| 14 | Calcite, kaolinite or serpentine |
| 15 | Calcite + kaolinite or serpentine (trace) |
| 16 | Amorphous, sodium chloride |
| 17 | Calcite + kaolinite or serpentine and unknown (trace) |
| **Powders** | |
| 3 | Calcium sulfate + cellulose (trace) |
| 5 | Quartz, kaolinite or serpentine |
| 6 | Kaolinite or serpentine, cellulose |
| 9 | Quartz, kaolinite or serpentine |
| 10 | Calcite, cellulose |
| **Solid blocks** | |
| 7 | Calcite + unknown |
| 8 | Talc + kaolinite or serpentine (trace) |

Diffraction Data (Swarthmore, PA) files. The components identified in these consumer products are shown in Tables 1-3.

### Transmission electron microscopy and identification of fibres

All samples were examined by transmission electron microscopy (TEM) (Figure 1) for the presence of fibres, using the direct carbon-coated polycarbonate filter technique (Anderson & Long, 1980). Mineral fibres were identified by a combination of selected-area electron diffraction and energy-dispersive X-ray analysis (Figures 2 and 3). The results of the analysis are shown in Tables 4-6.

## Discussion

The presence of inorganic fibrous material has been confirmed in several consumer products. In particular, the widespread use of attapulgite in cat litters, joint and spackling compounds and, to a lesser extent, in art supplies is a matter of concern. Using a quantitative X-ray diffraction technique it was established that up to 20% w/w of attapulgite could be added to joint compounds. TEM analysis established that the fibres detected were generally in the same size range as chrysotile, with diameters between 0.03 and 0.5 $\mu$m and lengths up to 4 $\mu$m. Because of their size distribution, these fibres are potentially toxic and will be the subject of further studies.

## Table 2. X-ray diffraction analysis of selected joint and spackling compounds

| Sample No. | Components identified |
|---|---|
| **Powders** | |
| 23 | Calcite + calcium sulfate (trace) |
| 25 | Calcite + calcium sulfate (trace) |
| 26 | Calcite + talc and mica (trace) |
| 27 | Gypsum + kaolinite or serpentine (trace) |
| 29 | Calcite, calcium sulfate + quartz (trace) |
| 32 | Calcite, mica |
| 33 | Calcite, calcium sulfate, mica |
| 36 | Calcium sulfate, dolomite + quartz (trace) |
| 37 | Calcium hydroxide, magnesium oxide + calcite and magnesium hydroxide (trace) |
| 38 | Calcite + calcium hydroxide (trace) |
| 39 | Calcium sulfate |
| 40 | Calcium sulfate |
| 41 | Calcite + mica (trace) |
| 42 | Calcium sulfate, dolomite |
| 43 | Calcium sulfate, dolomite |
| **Pastes** | |
| 24 | Calcite |
| 28 | Calcite, mica + kaolinite or serpentine (trace) |
| 30 | Calcite + mica and talc (trace) |
| 31 | Calcite + mica (trace) |
| 34 | Amorphous |
| 35 | Calcite + talc (trace) |

## Table 3. X-ray diffraction analysis of selected cat litters

| Sample No. | Components identified |
|---|---|
| **Granules** | |
| 51 | Attapulgite, quartz |
| 53 | Attapulgite, quartz |
| 54 | Attapulgite, quartz |
| 55 | Attapulgite, quartz |
| 56 | Attapulgite, quartz |
| 57 | Attapulgite, quartz |
| **Wood shavings** | |
| 52 | Cellulose |

### Table 4. Transmission electron microscopy analysis of selected art supplies

| Sample No. | Fibre type | Other minerals |
|---|---|---|
| **Moist clay or clay-like** | | |
| 1 | None detected | Titanium dioxide, calcium oxide-calcite |
| 4 | Rutile, vermiculite | Talc |
| 11 | None detected | Titanium dioxide, calcium oxide-calcite |
| 12 | Attapulgite | Calcium oxide-calcite, titanium dioxide, aluminium silicate |
| 13 | Rutile, halloysite | Aluminium silicate, kaolinite, dickite, nacrite |
| 14 | Halloysite | Aluminium silicate, kaolinite, dickite, nacrite |
| 15 | Rutile, halloysite, glauconite | Aluminium silicate, glauconite |
| 17 | Talc, calcite | Magnesium alumino-silicate, talc, calcite |
| **Solid blocks** | | |
| 7 | Amosite | Talc |
| 8 | Talc | Platy talc |

### Table 5. Transmission electron microscopy analysis of selected joint and spackling compounds

| Sample No. | Fibre type | Other minerals |
|---|---|---|
| **Powders** | | |
| 26 | Rutile, antigorite, goethite | Platy talc |
| 27 | Halloysite, | Aluminium silicate |
| 32 | Attapulgite, anthophyllite | Mica |
| 36 | Rutile, apatite, glauconite | - |
| 41 | Attapulgite | - |
| **Pastes** | | |
| 28 | Attapulgite, halloysite, mica | Kaolinite, nacrite or dickite |
| 30 | Antigorite, attapulgite | - |
| 31 | Attapulgite | Aluminium silicate |
| 34 | None detected | - |

**Table 6. Transmission electron microscopy analysis of selected cat litters**

| Sample No. | Fibre type | Other minerals |
|---|---|---|
| **Granules** | | |
| 51 | Attapulgite | - |
| 53 | Attapulgite | - |
| 54 | Attapulgite, diatomaceous earth | Magnesium aluminium silicate |
| 55 | Attapulgite, halloysite | Silica, glauconite |
| 56 | Attapulgite | - |
| 57 | Attapulgite | - |
| **Wood shavings** | | |
| 52 | None detected | - |

**Fig. 1. Attapulgite standard; digital STEM image. Field size 15 × 20 $\mu$m**

**Figure 2. Attapulgite standard; SAED pattern**

In this technique, the electrons diffracted by a selected area of the specimen (typically 0.5 μm diameter) form a characteristic pattern

**Figure 3. Attapulgite standard; EDXA spectrum with detector window open**

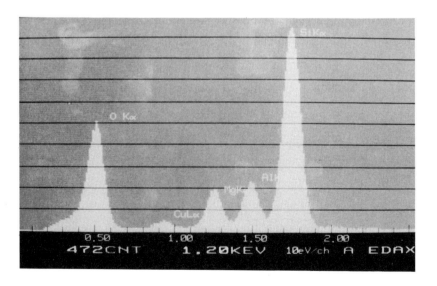

Note oxygen peak

*References*

Anderson, C.H. & Long, J.M. (1980) *Interim Method for Determining Asbestos in Water (EPA-600/4-80-005)*, Athens, GA, US Environmental Protection Agency

Government of Canada (1983) *Hazardous Products Act*, Part 1, Section 26

# MINERAL FIBRES AND DUSTS IN LUNGS OF SUBJECTS LIVING IN AN URBAN ENVIRONMENT

L. Paoletti[1], L. Eibenschutz[1], A.M. Cassano[1], M. Falchi[1],
D. Batisti[1], C. Ciallella[2] & G. Donelli[1]

[1] *Department of Ultrastructures, Istituto Superiore di Sanità Rome, Italy*
[2] *Institute of Forensic Medicine, La Sapienza University, Rome, Italy*

*Summary.* We have undertaken a study on 200 autopsy lung samples collected from subjects who lived in the Rome urban area and were not occupationally exposed to mineral dusts. The samples belonged to subjects who died aged between 15 and 65, both male and female. Subjects suffering from diseases and drug addicts were excluded. The purpose of this investigation was to determine whether any correlation existed between subjects' life-style, in particular smoking habits, and the presence of mineral fibres and dusts in their lungs. The data obtained were compared with those on airborne mineral dusts in the environment of the subjects themselves, particularly data on the concentration and types of mineral fibres present in that environment.

## Introduction

It is known that toxic and carcinogenic substances are present in the airborne particulates of urban environments. Indeed, the higher incidence of cancer and chronic lung diseases in urban as compared with rural areas is generally taken to be a consequence of such environmental conditions. Several authors have recently highlighted the need to identify more precisely the role of environmental pollution in the etiology of lung cancer, in particular with respect to other factors, such as smoking (Vena, 1982).

For this purpose, we studied a population sample of subjects who lived in Rome and were not occupationally exposed to mineral dusts. We looked for possible correlations between the subjects' life-style, particularly smoking habits, and the amount and type of mineral fibres and dusts present in their lungs (Paoletti *et al.*, 1987). The data obtained were compared with those on airborne mineral dusts present in the subjects' environment.

## Materials and methods

The population sample consisted of 200 subjects, 100 males and 100 females, selected from persons aged 15–65 and resident in the Rome urban area, who had died in accidents. We excluded from the sample subjects who were occupationally exposed to mineral dusts, subjects with serious or chronic diseases and drug addicts.

A total of 62 subjects have been studied so far. Autopsy samples of lung tissue were all obtained from the upper lobe of the right lung. Samples were ashed in atomic oxygen plasma. The inorganic component was collected, transferred on to carbon-coated copper grids for transmission electron microscopy (Paoletti *et al.*, 1987), and analysed using a

Philips EM 430 electron microscope equipped with an energy-dispersive spectrometer for X-ray microanalysis.

## Results

### Concentration of mineral particulates in the parenchyma

The concentration of mineral particulates in the subjects' lung parenchyma ranged from 0.1 to $2.8 \times 10^5$ particles per mg dry tissue. Table 1 shows the frequency with which these concentrations were observed. Although there was no direct correlation between particulate concentration and age, it appears to increase with age (Figure 1).

**Table 1. Mineral particle concentrations in lung parenchyma**

| Observed concentrations (particles per mg×100 000) | % of cases |
|---|---|
| 0.00 - 0.25 | 15 |
| 0.25 - 0.50 | 48 |
| 0.50 - 0.75 | 16 |
| 0.75 - 1.00 | 5 |
| 1.00 - 1.25 | 4 |
| 1.25 - 1.50 | 4 |
| 1.50 - 1.75 | 2 |
| 1.75 - 2.00 | 0 |
| 2.00 - 2.25 | 2 |
| 2.25 - 2.50 | 0 |
| 2.50 - 2.75 | 0 |
| 2.75 - 3.00 | 4 |

**Fig. 1. Mineral particle concentrations (particles per mg) in autopsy samples**

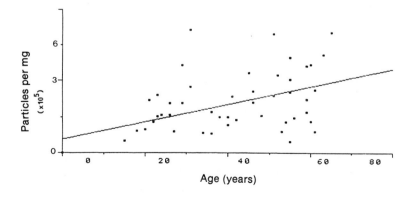

Asbestos fibres were present in 23% of the subjects, with concentrations ranging from 200 to 3000 fibres per mg (f/mg) dry tissue. However, in all cases fibres accounted for 1-2% of the corresponding total inorganic particulate concentrations.

### Types of particle

A total of 16 different types of mineral — mainly silicates — and 22 metals in the form of oxides or sulfates were identified in the inorganic particles observed in the samples. On average, oxides and sulfates accounted for approximately 46% and silicates for approximately 54% of the total. Most of the silicates had a lamellar crystal structure (clays, micas, talc, etc).

Figure 2 shows the frequency with which the various metals were observed in the subjects.

**Fig. 2. Metals (as oxides and sulfates) in the lung tissue samples, with the percentages of samples in which they were observed**

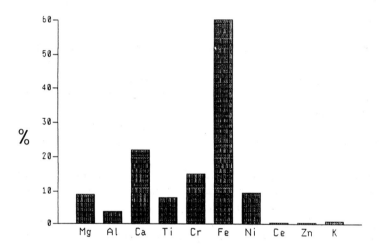

The majority of fibrous particles consisted of chrysotile. Tremolite and crocidolite fibres, as well as talc and rutile (titanium oxide) fibres, were also observed.

No artificial mineral fibres, such as glass wool and rock wool, were found.

### Anthracosis

The incidence of anthracosis in the subjects was evaluated on sections of lung tissue prepared for light microscope examination.

All samples were assigned to one of 4 degrees of anthracosis, from 0=absent to 3=maximum. Table 2, which summarizes the results obtained, shows a good correlation between anthracosis and age.

**Table 2. Frequency of anthracosis as a function of age**

| Age (years) | Frequency[a] Degree of anthracosis | | | |
|---|---|---|---|---|
| | 0 | 1 | 2 | 3 |
| 65 - 55 | - | - | 5 | 5 |
| 55 - 45 | - | - | 6 | 7 |
| 45 - 35 | - | 1 | 6 | 2 |
| 35 - 25 | 1 | 3 | 5 | - |
| <25 | 1 | 5 | 3 | 1 |

[a]Number of subjects per age-group and degree of anthracosis.

## Discussion

### Correlations between particulates in the subjects' environment and in their lung parenchyma

Knowledge of the nature of the airborne particulates present in the subjects' environment can help to show how the amount and type of mineral particulates in the lung parenchyma are correlated with the environmental situation.

The urban environment where the subjects resided is characterized by heavy motor vehicle traffic, while it is relatively distant from large industrial plants and electric power stations.

In such an environment, the airborne particulates consist of silicates and metals (as oxides and sulfates). Figure 3 shows the frequency of the two groups of particulates, as observed in the subjects' lung parenchyma and environment. Figure 4 shows the metals present in the airborne particulates of the urban area in question. Several elements, such as lead and vanadium, frequently found in the urban environment, were not found in the lung parenchyma. This is probably because these elements are released into the urban environment in the form of highly soluble compounds.

Airborne asbestos fibres are present in urban areas with concentrations ranging from 50 to 700 f/m$^3$. When related to the total concentration of airborne particulates (100–300 $\mu$g/m$^3$), it can be estimated that these fibres account for only $1/10^4$–$1/10^5$ of it. This means that they are present at a dilution that is 2 or 3 orders of magnitude greater than that in the lung parenchyma particulates.

Just as in the lung parenchyma, most of the silicates in the environment are lamellar in character so that, because of their high surface/volume ratio, they can effectively adsorb and carry the constituents of environmental aerosols. In particular, approximately 5% of these particles have been observed to contain metals such as vanadium, lead, chromium and nickel that are not present in the mineral forming the particles.

**Fig. 3.** Percentages of silicates and metals in the subjects' lung tissue and in the urban environment

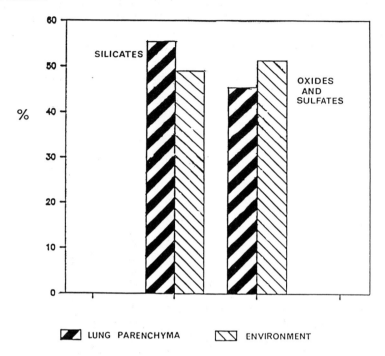

**Fig. 4.** Frequencies of occurrence of metals in the environment

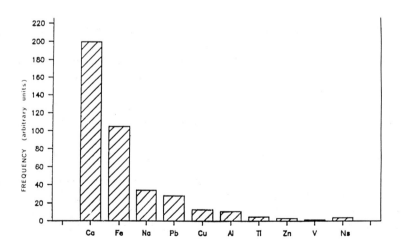

## Correlation between smoking and particulates in the lung parenchyma

The data obtained indicate that, on average, smoking is associated — for the same age-groups — with a greater concentration of mineral particulates in the lung parenchyma. Figure 5 shows the average particulate concentrations in the parenchyma of smokers and non-smokers for the age-groups into which the subjects were divided.

**Fig. 5. Particulate concentrations in the lung parenchyma of smokers and non-smokers**

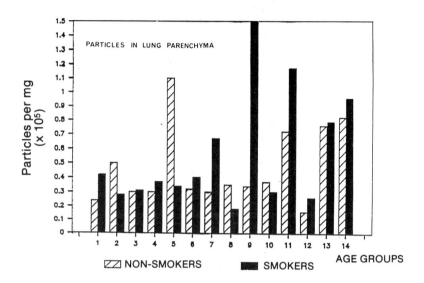

There are indications that smoking is also related to the differences in the composition of the particles present in the lung parenchyma. In particular, in smokers, it is possible to observe that the percentage of oxides and metal sulfates in the particulates increases with age (Figure 6). Some metals (iron, magnesium, nickel and barium) are present in larger amounts in the lungs of smokers.

Fibres were more frequently found in the lungs of smokers.

Finally, the degree of anthracosis correlates well with smoking.

## Conclusions

These preliminary results confirm the high degree of correlation between the mineral particulates present in the lung parenchyma and the environmental situation. In particular, it appears that a population residing in an urban area is highly exposed to toxic and carcinogenic substances released by motor vehicles (metal oxides, asbestos fibres).

**Fig. 6. Percentages of metals in the mineral particles from the lung tissue of smokers function of subjects' age**

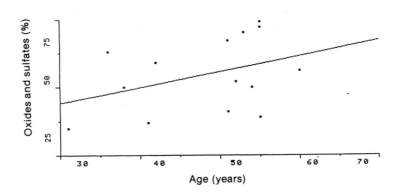

Finally, our results confirm the effect of smoking on the quantity and natu particulates present in the lungs.

## References

Paoletti, L., Batisti, D., Caiazza, S., Petrelli, M.G., Taggi, F., De Zorzi, L., Dina, M.A. & Done
(1987) Mineral particles in the lungs of subjects resident in the Rome area and not occupati
exposed to mineral dust. *Environ. Res.*, *44*, 18-28

Vena, J.E. (1982) Air pollution as a risk factor in lung cancer. *Am. J. Epidemiol.*, *116*, 42-56

# MEASUREMENT OF INORGANIC FIBROUS PARTICULATES IN AMBIENT AIR AND INDOORS WITH THE SCANNING ELECTRON MICROSCOPE

K. Rödelsperger[1], U. Teichert[2], H. Marfels[3], K. Spurny[3],
R. Arhelger[1] & H.-J. Woitowitz[1]

[1]*Institute and Polyclinic for Occupational and Social Medicine,
Justus Liebig University, Giessen, Federal Republic of Germany*
[2]*Society for Dust Measurement Techniques and Occupational Health and Safety,
Neuss, Federal Republic of Germany*
[3]*Fraunhofer Institute for Environmental Chemistry and Ecotoxicology,
Schmallenberg-Grafschaft, Federal Republic of Germany*

*Summary.* In the Federal Republic of Germany, the concentration of inorganic fibrous particles in the ambient air is measured by scanning electron microscopy (SEM) in accordance with draft guideline VDI 3492. Fibres of length $\geqslant 2.5$ µm are counted at a magnification of ×2000.

The concentrations and fibre dimensions of asbestos fibres and other mineral fibres were determined for 231 ambient air samples, 219 indoor air samples taken during asbestos removal and 21 taken at workplaces. In the ambient air measurements in rural and urban areas, the average concentration of other inorganic fibres was $>1000$ fibres of length $\geqslant 5$ µm per m³. On the average, only about 100 or fewer asbestos fibres of length $\geqslant 5$ µm per m³ were observed. These 'asbestos fibres' from ambient air had low aspect ratios. They differed substantially from asbestos fibres measured in the ambient air close to sources of asbestos emissions, indoors during asbestos removal or at workplaces. Furthermore, the size distribution of these 'asbestos fibres' was similar to that of 'other inorganic fibres'. We suggest, therefore, that a substantial portion of the 'asbestos fibres' observed by SEM analysis in ambient air samples from rural and urban areas should be classified instead as 'other mineral fibres'.

## *Introduction*

In the Federal Republic of Germany, concentrations of inorganic fibres in ambient air have been monitored for a number of years by scanning electron microscopy (SEM) in accordance with draft VDI guideline 3492 (König *et al.*, 1985). In order to evaluate these results, ambient air measurements were compared with indoor measurements in buildings with sprayed asbestos insulation and workplaces. The evaluation was aimed at answering the following questions:

(1) How high is the concentration of asbestos and other fibres in the environment?

(2) What size distribution is found in ambient air measurements as compared with indoor and workplace measurements?

(3) What proportion of the fibres with a length:diameter ratio ($L:D$) of $L:D \geqslant 3:1$ have a ratio in the range $3:1 < L:D < 5:1$?

## Materials and methods

According to draft VDI guideline 3492, inorganic fibres $\geqslant 2.5$ μm in length should be counted by SEM at a magnification of ×2000. Fibres are sampled on coated Nuclepore filters which allow plasma ashing of organic particles on the filter. The elementary composition of the fibres is determined by means of energy-dispersive X-ray analysis and fibres are divided into the following groups:

— asbestos fibres;

— calcium sulfate fibres;

— other mineral fibres.

Apart from the calcium sulfate fibres, the length and diameter of each fibre found by SEM is documented. Initially, the original lists were evaluated by the participating institutes by means of special questionnaires covering the concentrations of fibres having certain dimensions and the size distributions for groups of measurements. These amount, in total, to 22 groups and 471 individual measurements, including 231 ambient air measurements made up as follows:

— in the environment ($n=155$); and

— near sources of asbestos emissions ($n=76$).

These are compared with 219 indoor measurements made before, during and after asbestos removal and 21 workplace measurements made during asbestos-cement processing, brake repair work and glass-blowing.

## Results

### Fibre concentrations

The concentrations of fibres $\geqslant 5$ μm in length found in 155 ambient air measurements are shown in Table 1. Only 113 asbestos fibres with $L \geqslant 2.5$ μm and 48 asbestos fibres with $L \geqslant 5$ μm were found. Thus, an average of less than 1 asbestos fibre with $L \geqslant 5$ μm, but up to 12 other mineral fibres are observed per filter. Mean concentrations of asbestos fibres with $L \geqslant 5$ μm are between 13 and 110 fibres per m³ (f/m³). The numbers of fibres found in all 471 measurements are shown in Table 2. Fibres $2.5$ μm $< L < 5$ μm and $L \geqslant 5$ μm are given separately and the percentage of fibres with $L:D < 5:1$ is shown. As expected, the numbers of asbestos fibres obtained during asbestos removal and at the workplace are greater than those for the ambient air. In contrast, in the ambient air measurements the numbers of other mineral fibres are substantially greater than those for asbestos fibres.

### Table 1. Sensitivity and mean fibre concentration for 6 groups totalling 155 ambient air measurements made in non-urban areas and cities by participating institutes

| Environmental measurements | | | Sensitivity[a] | Mean fibre concentration ($L \geqslant 5$ μm) (f/m³) | |
|---|---|---|---|---|---|
| Institute | Area | No. | | Asbestos | Other mineral fibres |
| FhG[b] | Non-urban | 23 | 155±97 | 110 | 1200 |
|  | Large | 2 | 144±23 | - | 650 |
|  | cities | 46 | 199±118 | 110 | 1600 |
| GSA[c] | Large | 57 | 130±21 | 14 | 1500 |
|  | cities | 23 | 304±111 | 13 | 2600 |
| IPAS[d] | Small city | 4 | 708±49 | - | 920 |

[a]Calculated average concentration corresponding to one fibre (arithmetic mean ± standard deviation).
[b]Fraunhofer Institute.
[c]Society for Dust Measurement Techniques and Occupational Health and Safety.
[d]Institute and Polyclinic for Occupational and Social Medicine.

### Table 2. Amounts of asbestos and other mineral fibres in the ambient air, indoors, and at the workplaces

| Source of sample | No. of measurements | No. of fibres with $L:D \geqslant 3:1$[a] | | | |
|---|---|---|---|---|---|
| | | Asbestos | | Other mineral fibres | |
| | | $L<5$ μm | $L \geqslant 5$ μm | $L<5$ μm | $L \geqslant 5$ μm |
| **Ambient air** | | | | | |
| Environment | 155 | 65 (64%) | 48 (33%) | 2234 (55%) | 1441 (45%) |
| Near sources of asbestos emission | 76 | 54 (26%) | 62 (6%) | 991 (57%) | 692 (44%) |
| **Indoor air** | | | | | |
| Asbestos removal (crocidolite, chrysotile) | 219 | 524 (11%) | 1974 (3%) | 3013 (75%) | 2967 (52%) |
| **Workplace** | | | | | |
| Chrysotile | 21 | 308 (7%) | 353 (1%) | 77 (27%) | 24 (25%) |

[a]Percentage of fibres with $L:D<5:1$ in parentheses.

A higher proportion of fibres with $L:D<5:1$ is found among fibres with lengths 2.5 µm $<L<5$ µm than among fibres with $L\geqslant 5$ µm for both asbestos and other mineral fibres.

In the measurements of asbestos fibres in the environment, an astonishingly large proportion (64% and 33%) of fibres with $L:D<5:1$ is found. The proportion of asbestos fibres with $L:D<5:1$ in ambient air measurements near sources of asbestos emission tends to be lower. Asbestos fibres obtained during or before indoor asbestos removal or at the workplace contain only a small number of fibres with $L:D<5:1$. Other mineral fibres, however, contain a constant proportion of fibres with $L:D<5:1$ in all measurements (ambient air, indoor and workplace).

**Distribution of fibre length and diameter**

Asbestos fibres at the workplace generally have mean $L:D$ ratios $>10:1$ (Walton, 1982; Woitowitz & Rödelsperger, 1980). Our measurements, made indoors during asbestos removal as well as at the workplace, are shown in Figure 1. In both cases, the fibres are mostly asbestos with median $L:D$ ratios $\geqslant 10:1$. However, asbestos fibres found in environmental situations where there are no obvious asbestos emitters have $L:D$ ratios $\leqslant 6$. The median values for the asbestos fibres with $L\geqslant 2.5$ µm are:

— fibre lengths between 3.8 and 12.6 µm; and

— fibre diameters between 0.1 and 1.4 µm.

Environmental ambient air measurements, however, produce median lengths $\leqslant 5$ µm and diameters $\geqslant 0.8$ µm. They correspond more or less to the median values for other mineral fibres and their $L:D$ ratios. Markedly higher $L:D$ ratios are observed in ambient air measurements near sources of asbestos emission.

## Discussion

Measurements of concentrations of asbestos fibres with $L\geqslant 5$ µm in the environment with electron microscopy generally give results based on only a few fibres (Bundesamt für Umweltschutz, 1986; Chatfield, 1984; König et al., 1985; Lanting & den Boeft, 1983; Marfels et al., 1984; Teichert, 1983) but concentrations near the detection limit of the method ($\leqslant 400$ f/m³) are very unreliable. As shown in Table 1, surprisingly low concentrations of asbestos fibres with $L\geqslant 5$ µm (between 13 and 110 f/m³) are found in the combined evaluation of 4 groups of 155 ambient air measurements. Here, differences between the results obtained by 2 institutes seem to appear. In all cases, however, the results are based on average fibre concentrations far below the detection limit of single measurements. In addition, the remarkably low $L:D$ ratios of the asbestos fibres detected cast doubt on the identity of these fibres. Fibres were identified merely by means of elementary analysis with the SEM. Even at these low concentrations, therefore, the possibility exists that silicate fibres other than asbestos fibres have been found (Chatfield, 1984). On the other hand, in environmental measurements, numbers of other mineral fibres 30 times greater than those of asbestos fibres are observed. Since asbestos fibres with diameters $<0.2$ µm give, in

**Fig. 1.** Median lengths ($L$) compared with median diameters ($D$) of asbestos and other mineral fibres with $L \geq 2.5$ μm.

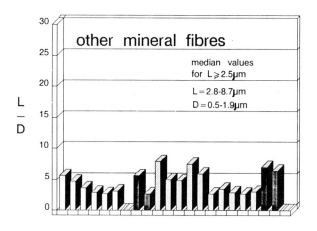

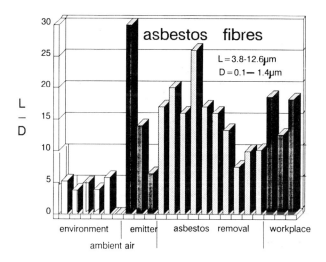

Results of 471 measurements divided into 22 groups of ambient air, asbestos removal and workplace measurements.

part, inadequate elementary spectra, they may be wrongly identified as other mineral fibres. If only 10% of these false identifications are made, an asbestos fibre concentration between 100 and 300 f/m³ would result. Analysis of the three-dimensional length and diameter distribution shows, however, that fibres with lengths ≥5 μm and <0.2 μm in diameter only account for 1% of the other mineral fibres with length ≥5 μm.

Apart from the analytical difficulties, the possibility of recognizing thin fibres in the SEM must be questioned. Comparative measurements with TEM show, however, that especially for fibres $\geqslant 5$ µm in length, the factor for converting from SEM to TEM with high resolution should not exceed 2 (König *et al.*, 1985). All in all, any evaluation of these low results obtained in ambient air with the SEM should be checked by means of comparative measurements with the TEM. On the other hand, there can be no doubt that SEM measurements are suitable for the sensitive determination of asbestos fibres with length $\geqslant 5$ µm close to indoor or outdoor asbestos emitters.

## Acknowledgements

This work was carried out with the financial help of the Umweltbundesamt, Berlin (Project No. 1040 2166).

## References

Bundesamt für Umweltschutz (1986) *Luftbelastung durch Asbestfasern in der Schweiz.* (*Schriftenreihe für Umweltschutz No. 49*), Bern

Chatfield, E.J. (1984) Measurement and interpretation of asbestos fibre concentrations in ambient air. In: Baunach, F., ed., *Proceedings of the Fifth Colloquium on Dust Measuring Technique and Strategy*, Johannesburg, South African Asbestos Producers Advisory Committee, pp. 269-296

König, R., Marfels, H. & Spurny, K.R. (1985) *Felderprobung und Standardisierung von Verfahren zur Messung faserförmiger Stäube in der Aussenluft. Untersuchungen zu standardisierten Messverfahren fur VDI- und ISO-Richtlinien*, (*Forschungsbericht 104 02 256/01-02 im Auftrag des Umweltbundesamtes*), Berlin, Umweltbundesamt

Lanting, R.W. & den Boeft, J. (1983) Ambient air concentrations of mineral fibres in the Netherlands. In: Reinisch, D., Schneider, H.W. & Birkner, K.-F., eds, *Fibrous Dusts — Measurements, Effects, Prevention (VDI-Berichte 475)*, Düsseldorf, VDI-Verlag, pp. 123-128

Marfels, H., Spurny, K.R., Boose, C., Schormann, J., Opiela, H., Althaus, W. & Weiss, G. (1984) Ambient air measurements of fibrous dusts in the Federal Republic of Germany. I. Measurements close to industrial sources. II. Measurements at traffic junction inside a major city [in German]. *Staub-Reinhalt. Luft*, 44, 259-263; 410-414

Teichert, U. (1983) Results of source-related outdoor ambient air measurements of fibrous particles [in German]. In: Reinisch, D., Schneider, H.W. & Birkner, K.F., eds. *Fibrous Dusts — Measurement, Effects, Prevention (VDI-Berichte 475)*, Düsseldorf, VDI-Verlag, pp. 117-122

Walton, W.H. (1982) The nature, hazard and assessment of occupational exposure to airborne asbestos dust: a review. *Ann. Occup. Hyg.*, 25, 117-247

Woitowitz, H.-J. & Rödelsperger, K. (1980) Tumour epidemiology [in German]. In: *Luftqualitätskriterien. Umweltbelastung durch Asbest und andere faserige Feinstäube.* (*Berichte 7/80, Umweltbundesamt*), Berlin, E. Schmidt Verlag, pp. 203-266

# ASBESTOS FIBRE RELEASE BY CORRODED AND WEATHERED ASBESTOS-CEMENT PRODUCTS

### K.R. Spurny

*Division of Aerosol Chemistry,*
*Fraunhofer Institute for Environmental Chemistry and Ecotoxicology,*
*Schmallenberg-Grafschaft, Federal Republic of Germany*

*Summary.* A description is given of portable equipment and a method of sampling and measuring asbestos fibre emissions from solid plane surfaces of asbestos-cement products (roofs and facades).

Asbestos-cement products, e.g., roof tiles, contain as much as 11-12% of chrysotile asbestos. As a result of continuing exposure to the weather and to acid rain, the surface of asbestos-cement products becomes corroded and weathered. Cement particles, asbestos fibres and agglomerates of particles and fibres are therefore released from the surface and dispersed in air and water.

The method described has been used to measure asbestos fibre emissions and ambient air concentrations in the Federal Republic of Germany over the period 1984-1986.

## *Introduction*

Few reliable data on the distribution of asbestos sources or on asbestos emission factors are currently available, although the world consumption of asbestos amounts to more than 6 million tons per year.

Asbestos-cement products, e.g., roof tiles, contain as much as 11-12% of asbestos. As a result of continuing exposure to meteorological influences, such as acid rain, sunshine, wind and frost, as well as to atmospheric pollutants, the surface of asbestos-cement products becomes corroded and weathered. Cement particles, asbestos fibres and agglomerates of particles and fibres are therefore released from the surface and dispersed in the air and water.

In our first investigation (Spurny *et al.*, 1979), higher asbestos ambient air concentrations were found in the vicinity of buildings containing asbestos-cement products. In a subsequent investigation (Spurny *et al.*, 1986), asbestos fibre emissions and ambient air asbestos fibre concentrations were measured in the Federal Republic of Germany during 1984-1986. More than $1 \times 10^9$ m$^2$ of such asbestos-cement surfaces were believed to exist and to be emission sources in the country.

## *Measurement and analyses*

A special device and procedure were developed for sampling purposes (Figure 1). The fibrous emissions are passed through Nuclepore or membrane filters and the numbers of fibres, their size distribution and identities are then evaluated by means of electron microscopy.

**Fig. 1. Diagram of sampling system for the measurement of released fibrous aerosols**

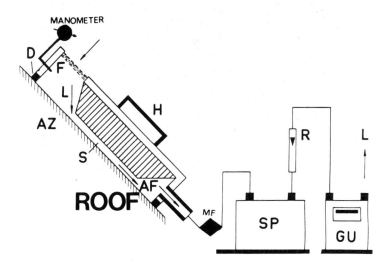

D, polymer ribbons; F, filter; L, clean air; S, slot; AZ, asbestos-cement sheet; H, handle; AF, asbestos fibres; MF, membrane filter; R, flow meter; GU SP, vacuum pump

The sampling chamber, 2–4 litres in volume, is placed on the surface (e.g., on asbestos roofing tiles). Close contact with the surface is achieved by means of polymer ribbons. Ambient air is sucked into the chamber through the filter. The clean air flows through the small slot over the corroded and weathered asbestos-cement sheet. The flow velocities of the simulated wind in the slot used in the measurements were between 1 and 5 m/sec. The sampling chamber can be kept in position on the surface as long as necessary by means of the handle, but in most cases the negative pressure produced is sufficient for this purpose.

Asbestos fibres released from the corroded surface are then sampled on a membrane filter and the fibre concentrations and emissions evaluated by electron microscopy. Other parts of the equipment are used to measure the flow rate and the total flow volume (GU, SP is a vacuum pump with a capacity of about 180 l/min).

The sampling time was 2 h, and each filter sample (Figure 2) was then evaluated quantitatively by scanning electron microscopy. The individual fibres were identified by means of energy-dispersive X-ray analysis. Bulk analysis was effected by X-ray fluorescence spectroscopy. Crystallographic changes in weathered asbestos fibres were evaluated by electron microdiffraction analysis of single fibres (Spurny, 1986).

**Fig. 2. Scanning electron micrograph of asbestos fibres sampled on a Nuclepore filter**

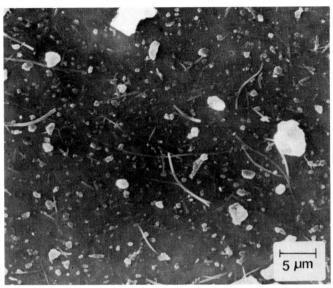

## Results and discussion

The investigations of asbestos-cement products (asbestos sheets from roofs and facades) conducted during 1984 and 1986 in the Federal Republic of Germany, have shown that:

(1) asbestos-cement surfaces corrode and weather as a result of aggressive atmospheric pollution (gases such as sulfur dioxide, aerosols and acid rain);
(2) the surface cement matrix of the material is destroyed, and a thin (approx 0.1–0.3 mm) layer of free deposited asbestos fibres is built up;
(3) the corrosion rate depends strongly on the acidity of the rain and on the concentration of air pollutants;
(4) the wind disperses fibres into the ambient air. Fibre emission factors in the range $10^6$–$10^9$ asbestos fibres per m² per h (f/m² h) have been measured. As already pointed out, the corrosion time, pollution intensity and weather conditions also affect fibre release (Figures 3 and 4);
(5) preliminary evaluations have shown that about 20% of free asbestos fibres are dispersed into the ambient air and 80% are washed out by rain water;
(6) analysis of bulk samples as well as of individual fibres showed chemical and crystallographic changes in the corroded chrysotile fibres. Particulate air pollutants (metal compounds and organic substances) were present and were deposited on free asbestos fibres in the corroded layer of asbestos-cement products;

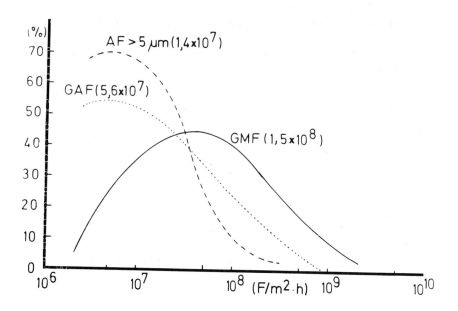

**Fig. 3.** Measured fibre emissions: total mineral fibres (GMF), total asbestos fibres (GAF) and asbestos fibres longer than 5 μm (AF)

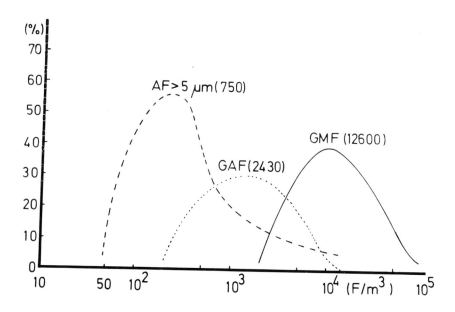

**Fig. 4.** Measured fibre concentrations in the vicinity of buildings with corroded asbestos-cement products

(7) animal experiments (intraperitoneal injection of the fibres released from the corroded asbestos-cement products) have shown that their carcinogenic potency is approximately as high as that of standardized UICC chrysotile fibres (Spurny et al., 1986);

(8) measurements of asbestos fibre concentrations in the vicinity of buildings containing corroded and weathered asbestos-cement products gave asbestos fibre concentrations (for fibres longer than 5 $\mu$m) in the range 200-1200 fibres per m$^3$ air.

## Acknowledgements

I am grateful to the Umweltbundesamt, Berlin, for their interest in these investigations and for financial support.

## References

Spurny, K.R., ed. (1986) *Physical and Chemical Characterization of Individual Airborne Particles*. Chichester, John Wiley

Spurny, K., Weiss, G. & Opiela, H. (1979) On the emission of asbestos fibres from weathered asbestos-cement sheets [in German]. *Staub-Reinhalt. Luft, 39*, 422-427

Spurny, K.R., Marfels, H., Pott, F. & Muhle, H. (1986) *Investigations on Corrosion and Weathering of Asbestos Cement Products as well as on the Carcinogenic Effect of the Weathering Products (Report UBA 104 08 314, 1-172)*, Berlin, Umweltbundesamt

# IV. EPIDEMIOLOGICAL DATA

# EFFECTS ON HEALTH OF NON-OCCUPATIONAL EXPOSURE TO AIRBORNE MINERAL FIBRES

### M.J. Gardner

*Medical Research Council Environmental Epidemiology Unit (University of Southampton), Southampton General Hospital, Southampton, UK*

### R. Saracci

*Unit of Analytical Epidemiology, International Agency for Research on Cancer, Lyon, France*

*Summary.* The most prominent potential marker of disease-related non-occupational exposure to mineral fibres is mesothelioma. Although many cases of mesothelioma have resulted from occupational exposure to asbestos, some have been associated with para-occupational domestic and/or neighbourhood exposure and have been reported in case series, case-control studies and a cohort study among non-occupationally exposed subjects. However, little information is available on mesothelioma as a direct consequence of general environmental asbestos exposure.

Such cases of mesothelioma related to non-occupational exposure to asbestos as have occurred to date are likely to have resulted from past exposures much higher than those prevailing at the present time (in the developed countries); numbers will therefore probably decrease in the future.

Very high rates of mesothelioma have been reported as a result of exposure to erionite.

No studies are available on the effects of non-occupational exposure to man-made mineral fibres but, among occupationally exposed workers, a risk of mesothelioma is not apparent.

There are suggestions of raised lung cancer rates among household contacts of asbestos workers and among individuals exposed to erionite.

Non-malignant parenchymal and pleural abnormalities have been observed in subjects exposed non-occupationally to asbestos and erionite, but these are not necessarily associated with malignant lesions.

Quantitative risk estimates of adverse effects on health have not been derived from these studies, essentially because of the absence of fibre exposure measurements.

## *Introduction*

Among the asbestos-related diseases, the most commonly used marker for any effect of non-occupational exposure to asbestos fibres is mesothelioma. Various reasons make the other asbestos-related conditions largely, though not completely inappropriate. Thus, asbestosis requires heavy exposures which would rarely have occurred outside the occupational environment. In the general population of Western countries, where cigarette smoking is highly prevalent, lung cancer is common and its non-specificity to asbestos makes implication of non-occupational asbestos exposure as a cause difficult. The same applies to other suggested asbestos-related cancers, such as cancers of the gastrointestinal and upper respiratory tracts, where the contribution of other causes would be overwhelming — these tumours, unlike cancer of the lung, are in any case less certainly attributable to occupational exposure to asbestos (Acheson & Gardner, 1983; Doll & Peto, 1985) and are not considered further in this review. Asymptomatic radiological changes and abnormalities of lung function have less often been studied.

Mesothelioma is therefore considered first in this paper, followed by lung cancer and pleural changes. Various general comments are then made on the study methodologies, and the results are critically presented so as to indicate the extent and limitations of the conclusions that can be drawn from them. Because of the extensive literature on the subject, this review is necessarily limited to a selection of representative studies having a bearing on the assessment of the health impact of non-occupational exposure to airborne mineral fibres — including asbestos, zeolites and man-made mineral fibres.

## Mesothelioma

Although mesothelioma became of more widespread concern following its link to asbestos exposure, the disease was known before asbestos was commercially exploited. Estimates of the proportions of mesothelioma cases which can be attributed to asbestos exposure differ widely (for example, Peterson *et al.* (1984) quoted a range of 13–100%, and similarly large differences in estimates have been found for occupational asbestos exposure (see also Tables 1 and 2). Recent comprehensive reviews have been provided by McDonald (1985) and McDonald and McDonald (1987). The latter article discusses other potential causes of mesothelioma, as do Pelnar (1983, 1985) and Peterson *et al.* (1984).

Studies of mesothelioma incidence in the population non-occupationally exposed to mineral fibres have broadly looked at three aspects — domestic exposure, neighbourhood exposure and a wider general environmental exposure. The earliest definitive report will be considered first, after which some of the other reports (there being too many to consider all of them) on these three areas of study will be described in sequence. The time-trends in mesothelioma incidence and the data on lung fibre burden at autopsy in mesothelioma cases and controls provide further information, and these have been considered by McDonald in this volume (pp. 420–427).

A summary of the findings in the case-control studies reviewed below is given in Table 1. It is clear that relatively more cases than controls in each study have reported asbestos exposure — both domestic/neighbourhood as well as occupational — although the proportions vary from study to study. Table 2 summarizes the findings in the various case series. Again, all the studies include cases with reported domestic/neighbourhood asbestos exposure in widely varying proportions, but no comparison data are available for individuals without mesothelioma. The other types of study are too few in number for tabular summary.

**Table 1. Case-control studies of mesothelioma associated with reported domestic/neighbourhood exposure to asbestos**[a]

| Reference | Groups | Reported asbestos exposure[b] | | |
|---|---|---|---|---|
| | | Occupational | Domestic/neighbourhood | None |
| Newhouse & Thompson (1965) | Cases | 31 (41) | 20 (26) | 25 (33) |
| | Controls | 8 (11) | 6 (8) | 62 (82) |
| Vianna & Polan (1978) | Cases | 6 (12) | 9 (17) | 37 (71) |
| | Controls | 2 (4) | 1 (2) | 49 (94) |
| McDonald & McDonald (1980) | Cases | 190 (34) | 14 (3) | 353 (63) |
| | Controls | 78 (14) | 8 (1) | 471 (85) |

[a]Two other case-control studies (Hain et al., 1974; Teta et al., 1983) are mentioned in the text but do not provide data suitable for tabulation.
[b]Numbers of cases or controls with percentages in parentheses.

**Table 2. Case series of mesothelioma associated with reported domestic/neighbourhood exposure to asbestos**

| Reference | Reported asbestos exposure[a] | | |
|---|---|---|---|
| | Occupational | Domestic/neighbourhood | None |
| Wagner et al. (1960) | 18 (55) | 14 (42) | 1 (3) |
| Lieben & Pistawka (1967) | 10 (24) | 11 (26) | 21 (50) |
| Milne (1972) | 26 (81) | 1 (3) | 5 (16) |
| Webster (1973) | 102 (44) | 76 (33) | 54 (23) |
| Vianna et al. (1981) | 17 (55) | 7 (23) | 7 (23) |
| Bianchi et al. (1982) | 59 (84) | 1 (1) | 10 (14) |
| Armstrong et al. (1984) | 101 (73) | 5 (4) | 32 (23) |
| Bianchi et al. (1987) | 35 (88) | 4 (10) | 1 (3) |

[a]Numbers of cases with percentages in parentheses.

### The earliest report

Wagner et al. (1960) first reported on the definite relationship between asbestos exposure and pleural mesothelioma. It is important to note that the observation was made simultaneously among workers in, and people living in the neighbourhood of the crocidolite asbestos mines in the north-western Cape Province of South Africa. Of the initial series of 33 cases (22 male and 11 females) during the period 1956-1960, all but one had a probable history of exposure to blue asbestos — 18 from employment, predominantly in the mines, and the other 14 only from living in the mining area.

In the case of one female mesothelioma patient, her relatives denied that she had ever visited the asbestos mines or had otherwise been exposed. A footnote to the paper mentions 14 more cases of mesothelioma (13 pleural, 1 peritoneal), but gives no information on potential neighbourhood exposure, although details on these and further cases are available (Wagner, personal communication).

**Domestic exposure (asbestos)**

Domestic exposure to asbestos has resulted mainly from para-occupational exposure to asbestos brought home by asbestos workers in the family. This would have occurred particularly during periods when employees came home in their work-clothes, carrying asbestos and other dust with them, and would have decreased markedly when changes of clothing and showering before leaving the workplace became common. Quite high levels of exposure are reported to have occurred from domestic exposure of this type (Nicholson, 1983).

A number of reports of mesothelioma related to such domestic (para-occupational) exposure have appeared in the literature based on approaches of differing epidemiological rigour and are discussed below.

*Case-control studies*

Newhouse and Thompson (1965) reported a case-control study of 76 mesothelioma (both pleural and peritoneal) patients in London, England, each matched by sex and date of birth to within 5 years with hospital in-patients in the same hospital. Examination of the domestic exposure histories showed 9 (7 female, 2 male) of 76 cases, as compared with 76 controls, reporting potential asbestos exposure through this route. The most usual history was that the wife had washed her husband's work-clothes, but the two male cases had sisters who had worked in an asbestos factory when they were children (one of the sisters is known to have died from asbestosis).

Vianna and Polan (1978) reported a case-control study of 52 female mesothelioma (both pleural and peritoneal) patients from New York each matched by sex, race, county of residence, marital status, age and year at death to a control dying from a cause other than cancer. Occupational histories of the cases, controls and members of their family were ascertained by questionnaires, and additional information was obtained from medical records and industrial reports. Of the 46 mesothelioma cases where occupational exposure to asbestos was not reported, 8 were classified as having experienced domestic exposure from a relative's occupation (in 7 cases the husband and in 1 case the father) compared with 1 of the 46 matched controls. In all instances these 8 cases were said to have routinely hand-laundered their husband's or father's clothing.

McDonald and McDonald (1980) reported the largest case-control study undertaken of 557 mesothelioma (both pleural and peritoneal) patients from Canada and the USA, each matched by hospital, sex, age and year of death with a control in whom pulmonary metastases were present from a non-pulmonary malignant tumour. Occupational histories were obtained by interview with relatives. For 8 (6 females, 2 males) of the 557 mesothelioma cases, a history of home exposure to asbestos was reported compared with 2 of the 557 controls; none of these exposures were within the same matched pairs. Of the 8 cases, 5 had been exposed to asbestos in childhood, but neither of the 2 controls had been so exposed. Information was also obtained for the US cases about asbestos exposure from working in the home or hobbies. Among 156 of these mesothelioma cases not reporting occupational asbestos exposure, 5 gave affirmative answers to the question about such exposure compared with 2 of their 156 matched controls.

## Cohort studies

Anderson (1983) reported the preliminary results of a cohort follow-up study, continued until 1980, among 2218 household contacts of amosite workers who were employed between 1941 and 1945. The 663 deaths observed were compared with the numbers expected by cause, using age- and sex-specific rates for the State of New Jersey. There were 3 mesothelioma deaths (sites unspecified), all in the period 20+ years after first household contact with an asbestos worker and all among children (2 female and 1 male) of asbestos workers. No figure is given for the expected number of mesothelioma deaths, but it must be close to zero. The available published information does not make it clear, for example, whether those household members could subsequently have been occupationally exposed to asbestos themselves.

## Case series

Lieben and Pistawka (1967) reported a descriptive survey of 42 mesothelioma cases diagnosed in south-east Pennsylvania. Occupational histories were obtained from the patients themselves, if alive, otherwise from family members or employers. Among the 32 patients for whom occupational exposures to asbestos were not reported, 3 female cases (2 peritoneal and 1 pleural) were classified as exposed to asbestos through family contacts. Of these 3 women, 2 were daughters and 1 the mother of asbestos workers. No controls or comparison group of subjects were studied.

Milne (1972) reported a descriptive survey of 32 mesothelioma (29 pleural and 3 peritoneal) cases diagnosed between 1962 and 1972 in Victoria, Australia. Occupational histories were obtained from the patients themselves, if possible, otherwise from relatives, friends, acquaintances and medical records. Among the 6 patients for whom possible occupational exposures to asbestos were not reported, 1 female (pleural) was classified as possibly exposed to asbestos through a family contact — her father had worked for a short time in an asbestos-cement factory. (Two of the 3 cases of peritoneal mesothelioma were siblings, and there was no definitive evidence of exposure to asbestos, occupational of otherwise, in either.) No control subjects were studied.

Vianna et al. (1981) reported a descriptive survey of mesothelioma cases diagnosed during the period 1973-1978 in New York State (excluding New York City). In

6 counties with high incidence rates, occupational histories and other information on potential sources of asbestos exposure for 31 cases (22 male, 9 female) were obtained from the patient, if alive, or otherwise from a first-degree relative. Among the 14 patients for whom occupational exposures to asbestos were not reported, 7 (6 female) were classified as exposed to asbestos through family contacts. Of the 6 husbands of women with mesothelioma, 5 were farmers and the sixth a fireman. Farmers are not generally regarded as occupationally at risk of asbestos exposure, but the authors cite construction and maintenance work and the natural asbestos content of soils in certain areas as relevant. No control subjects were studied.

Bianchi et al. (1982) reported a descriptive survey of 70 cases of mesothelioma diagnosed during the period 1967-1980 in the shipbuilding area of Trieste in north-east Italy. Lifetime occupational histories were obtained from the patients' relatives by interview. Among the 11 cases for whom occupational exposures to asbestos were not reported, 1 (female) case was classified as related to possible domestic exposure to asbestos. No control subjects were studied.

Bianchi et al. (1987a) later reported a descriptive survey of 40 cases of mesothelioma diagnosed in 1979-1987 in the neighbouring shipbuilding area of Monfalcone. Lifetime work histories were obtained from either the patients or relatives by interview. Among the 5 cases for whom occupational exposures to asbestos were not reported, 4 (all female) cases were classified as related to possible domestic exposure to asbestos. Again, no control subjects were studied.

*Other studies*

Anderson et al. (1979) updated an earlier report (Anderson et al., 1976) on mesothelioma cases among household contacts of asbestos workers. The study group were the household members of workers in an amosite asbestos insulation materials factory who themselves had not reported personal occupational exposure to asbestos. Among the family contacts of the 1664 workers, 5 pleural mesothelioma deaths were identified. Dust brought home on shoes, hair and work-clothes was postulated as the source of contamination — changing rooms and laundered work-clothes were said not to be provided by the company. No control subjects were studied. (The cohort study by Anderson (1983) involves a subset of these household contacts.)

## Neighbourhood exposure (asbestos)

Neighbourhood exposure to asbestos has usually resulted from residence near to asbestos mines, mills, factories or dumps. It would be the category used to indicate potential exposure to asbestos if there was no report of either occupational or para-occupational (mainly domestic — see above) exposure.

A number of reports of mesothelioma related to such neighbourhood exposure have appeared in the literature based on approaches of differing epidemiological rigour and are discussed below.

*Case-control studies*

Newhouse and Thompson (1965) reported a case-control study of 76 mesothelioma (both pleural and peritoneal) patients in London, England, each matched by sex and date of birth to within 5 years with hospital in-patients in the same hospital. Examination of the neighbourhood exposure histories showed 11 (10 female, 1 male) of 76 cases as compared with 5 of 76 controls, reporting potential asbestos exposure through this route. Neighbourhood potential exposure was defined as having lived within a half-mile radius of an asbestos factory.

Hain et al. (1974) reported a case-control study of 150 cases of mesothelioma in the Hamburg area compared with 150 control subjects of similar year of birth and sex. Of the 65 cases without occupational exposure to asbestos, 20 were reported to have lived for at least 5 years within 1 km of an asbestos factory, but unfortunately comparable residential information was not sought from the controls.

Vianna and Polan (1978) reported a case-control study of 52 female mesothelioma (both pleural and peritoneal) patients from New York, each matched by sex, race, county of residence, marital status, age and year of death with a control dying from a cause other than cancer. Occupational and residential histories were obtained by questionnaire, and also from medical records and industrial reports. Of the 46 mesothelioma cases for whom occupational exposure to asbestos was not reported, 1 was classified as having experienced potential neighbourhood exposure (defined as reported to have lived within 8 km of an asbestos factory), as compared with none of the matched controls.

McDonald and McDonald (1980) reported a case-control study of 557 mesothelioma (both pleural and peritoneal) patients from Canada and the USA, each matched by hospital, sex, age and year of death with a control in whom pulmonary metastases were present from a non-pulmonary malignant tumour. Occupational and residential histories were obtained by interview with relatives. For 1 (sex not given) of the 557 mesothelioma cases with no reported occupational or domestic exposure to asbestos, a history of potential neighbourhood exposure was reported, as compared with 4 of the 557 controls. Neighbourhood exposure was defined as having lived within 20 miles of a chrysotile asbestos mine in either Canada or California.

Teta et al. (1983) reported a case-control study of 201 cases of mesothelioma (and 19 cases of other primary malignant pleural tumours) in Connecticut compared with 604 control subjects taken as a random sample from all deaths in the State during the period (1955 1975) in which the cases were diagnosed. Occupational and residential histories were obtained from city directories. Excluding subjects with defined occupational exposure to asbestos, 9.2% of cases and 9.5% of controls were reported to have lived in the vicinity of an asbestos friction material production plant, so that there was no apparent difference in possible residential exposure to asbestos.

*Cohort studies*

Hammond et al. (1979) reported a cohort study of men who were residents of 2 areas in New Jersey. One area was defined as all dwellings located within half a mile of an amosite asbestos factory; the other was a control area several miles away but similar in respect of socioeconomic status, race, national origin and type of dwellings. Residents during 1942–1954 were identified from a population directory and followed up to obtain information on mortality during 1962–1976. Any men in either area who worked in the asbestos factory were excluded. One mesothelioma (in an electrician) among 780 deaths in the 'target' area was reported, as compared with none among 1735 deaths in the 'control' area.

*Geographical studies*

Botha *et al.* (1986) reported a geographical study of mortality rates during 1968-1980 in South African crocidolite mining districts as compared with contiguous control districts. Increased death rates from mesothelioma and/or asbestosis were found for both men and women in the mining districts and the increases were similar within each sex by race. The authors comment that environmental, rather than occupational, exposure may have played a major part for the following two reasons: (1) women had not worked in the mines to any extent; (2) some deaths occurred at early ages.

*Case series*

Lieben and Pistawka (1967) reported a descriptive survey of 42 mesothelioma cases diagnosed in south-east Pennsylvania. Occupational and residential histories were obtained from the patients, if alive, but otherwise from family members or employers. Among the 32 patients for whom occupational exposures to asbestos were not reported, 8 were classified as either having lived in the immediate neighbourhood of an asbestos plant or been employed next to an asbestos plant. No controls or comparison group of subjects were studied.

Webster (1973) reported a descriptive survey of 232 cases of pleural mesothelioma diagnosed in South Africa between 1956 and 1970. Occupational and environmental histories were obtained (source not reported) on all the cases. Of the 130 cases for whom occupational exposures to asbestos (mainly in the blue asbestos mines in Cape Province) were not reported, 76 were classified as having some environmental exposure to asbestos, based mainly on the areas in which they had lived. No control subjects were studied. There is probably some overlap between the early cases in this study and those described by Wagner *et al.* (1960).

Armstrong *et al.* (1984) reported a descriptive survey of 138 mesothelioma cases diagnosed in Western Australia between 1960 and 1982. Occupational histories were taken from all available sources, including hospitals and occupational health services, together with those of employment in the Wittenoom crocidolite mine and mill and of residence nearby. Among the 37 patients for whom occupational exposures to asbestos were not reported, 4 were classified as living near the mine and mill and 1 other had recorded playing as a 10-year-old child among heaps of 'whitish-grey' material in an asbestos tile factory. Between 1983 and 1986 another 2 cases of mesothelioma have been diagnosed in Wittenoom residents without reported occupational exposure (Musk, personal communication). No control subjects were studied.

**General environmental exposure (asbestos)**

*Industrial areas*

Little, if any, direct evidence is available on the risk of mesothelioma (or of other asbestos-related diseases) as a consequence of exposure to asbestos in the general environment in countries where asbestos has been used industrially for many years. Although asbestos fibres are widely found in the ambient environment, the concentrations are very low, as compared with most present, and certainly most past, occupational levels of exposure. Formal studies of groups exposed at these low levels have not been carried out and, if they were attempted, the statistical power to detect an asbestos-related risk would be small, given our knowledge of risks at higher concentrations. Thus children exposed at school to levels typically found in school buildings and members of the general population exposed to levels usually found in public buildings, houses and the outside air would be expected to be at a very low risk. However, it should be noted that surveys of asbestos levels in schools and other buildings have generally been made in situations where asbestos has been known to be present for one reason or another, and more representative samples of such buildings therefore need to be surveyed (Omenn et al., 1986).

The occurrence of mesothelioma in children, however, is well documented. For example, Wasserman et al. (1980) listed 30 cases from the literature and suggested that, since some of these cases were reported to be associated with known asbestos exposure, transplacental exposure might have occurred. However, there is little evidence to support this suggestion.

Although direct evidence of mesothelioma as a consequence of environmental exposure to asbestos, based on studies of individuals known to be exposed to low general environmental concentrations of asbestos fibres, has not been produced, a number of studies have presented indirect evidence to this effect. These will be discussed briefly and are partly based on geographical analyses making use of maps.

McDonald and McDonald (1977) reported mesothelioma rates in areas of Canada between 1966 and 1972 and in areas of the USA during 1972. There were fairly large variations in the rates around an overall figure of 1 per million per year. The rate for the province of Quebec was the highest among the reported figures for Canada, but the excess was reduced after crude adjustment for review of diagnoses by a pathology panel (Liddell, 1983). A study by Theriault and Grand-Bois (1978) examined mesothelioma rates within Quebec, and found that the incidence was much lower in rural regions away from the asbestos-producing areas.

In their study mentioned earlier, McDonald and McDonald (1980) reported results for 146 female cases of mesothelioma in Canada and the USA not related to occupational or domestic exposure to asbestos. The distribution of area of residence for the period 20–40 years before death of the case was similar in both cases and controls with respect to long-term urban/rural residence — 82 cases, as compared with 79 controls and 24 cases, as compared with 31 controls, respectively.

Gardner et al. (1982) and Gardner et al. (1985) reported studies of the geographical distribution of pleural mesothelioma between 1968 and 1978 and peritoneal mesothelioma between 1967 and 1982 respectively in 1366 small local authority areas of England and Wales. There was no real suggestion in either case of the occurrence of tumours in areas away from the main asbestos-using industries. On the maps of mortality from pleural mesothelioma, for example, more than three-quarters of the

11 years, even among men, in whom the rates were higher than among women by a factor of more than 3. It was therefore concluded that occupational and occupationally-related exposure had been most important (Gardner, 1983).

Enterline and Henderson (1987) reported a study of the geographical distribution of mortality from pleural mesothelioma between 1968 and 1981 in the USA at state level and also in more detail by counties. A pattern of potential occupational asbestos exposure was found among the counties with elevated death rates. However, the authors were unable to explain all such elevated rates in terms of industrial use of asbestos, and commented that the impression was that there were other causes less well identified and which needed to be explored. (See the discussion of zeolites below.)

*Non-industrial areas*

The patterns of mesothelioma occurrence described up to this point largely reflect, through domestic and neighbourhood exposure, the presence of asbestos fibres in the workplace. The occurrence of mesothelioma resulting from asbestos exposure has, however, been documented in circumstances unrelated to its industrial production or use.

Most of the evidence of this nature comes from Turkey, where 122 diagnosed cases of mesothelioma (82 males aged 15-71 years and 40 females aged 12-69 years) were reported by Baris and his colleagues (1980, 1987). The cases, of whom only 5 had reported possible occupational exposure to asbestos, occurred in several different villages in the rural part of central Turkey where mineralogical investigations of samples of stucco, soil and airborne dust revealed the presence of tremolite and chrysotile asbestos.

Following an initial retrospective study (Yazicioglu et al., 1978), cases were investigated in greater detail in relation to their source population by Yazicioglu et al. (1980). They compared mesothelioma incidence rates in 5 districts (4 in the south-east province of Diyarbakir and one in the adjacent province of Elazig) where asbestos deposits exist with the incidence in 7 nearby districts in the province of Diyarbakir without asbestos deposits. During the study years of 1977 and 1978, 23 cases of pleural mesothelioma were diagnosed in the 12 districts (12 males aged 41 years and over, and 11 females aged 30-70 years). Relevant diagnostic procedures included chest radiography in all subjects, cytological examinations of pleural effusion in 23 cases, needle pleural biopsy in 10 cases, pleural biopsy at thoracoscopy in 13 cases and thoracotomy in 2 cases. Based on these figures, Table 3 shows the crude incidence rate of pleural mesothelioma in the 5 districts with asbestos deposits as 5 per 100 000 person-years, as compared with a rate of 0.2 in the 7 districts without asbestos deposits (1 case observed in a total population of 217 962). The latter rate is of the same order as mesothelioma (death) rates among the general population in Western countries (McDonald & McDonald, 1977).

The three districts with higher rates in Table 3, namely, Cermik, Cungus and Ergani, are those in which asbestos deposits were reported to be more common and more commonly quarried (by men) for sale elsewhere or for local use as the basic ingredient of whitewash. This is prepared and applied (for women) once a year to

Table 3. Pleural mesothelioma during 1977–1978 in 5 districts with asbestos deposits in south-eastern Turkey[a]

| District | Population (1970 | Cases | Rate per 100 000 person-years |
|---|---|---|---|
| Cermik | 34 297 | 5 | 7 |
| Ergani | 50 766 | 9 | 9 |
| Cungus | 15 738 | 3 | 10 |
| Maden | 36 594 | 2 | 3 |
| Siverek | 90 027 | 3 | 2 |
| Total | 227 422 | 22 | 5 |

[a]Based on Yazicioglu et al. (1980).

house walls and floors. Exposure is thus both occupational and environmental (from house walls). Samples of the whitewash analysed by electron microscopy and by probe analysis of individual fibres revealed the presence of tremolite asbestos. In the lung biopsy specimen of a single individual with lung cancer, most of the fibres were tremolite, only a small fraction being chrysotile.

Tremolite fibres were recently identified as the basic material of a specimen of the whitewash used in Metsovo village in north-western Greece, which is located in an area where, according to a preliminary report (Langer et al., 1987), 6 deaths have occurred among 7 patients affected by pleural mesothelioma (3 male and 4 female, aged 45–57 years). The total number of expected deaths in this population during the period of observation was about 600.

Preliminary information is also available (Boutin et al., this volume, pp. 406–410) on the occurrence of 5 mesothelioma deaths in subjects non-occupationally exposed to dust containing chrysotile and tremolite fibres in north-east Corsica. In the Diyarbakir area of Turkey, and possibly in the Metsovo area of Greece and in north-east Corsica, mesothelioma rates are greater by roughly one order of magnitude than the general population rates in Western countries.

## Non-asbestos mineral fibres

### Zeolite (erionite)

A much greater increase in mesothelioma, by some 3 orders of magnitude (1000-fold), was first reported by Baris et al. (1975, 1978) from another area of Turkey. This represents the most striking instance of mesothelioma, and indeed of cancer, related to the general environment. The latest report (Baris et al., 1987) presents the results of 4- and 5-year investigations of mortality and the environment in 4 villages located in the Cappadocia region of central Turkey (province of Nevsehir) about 250 km south-east of Ankara. Table 4 shows that a total of 141 deaths was

**Table 4. Deaths by sex and cause, including mesothelioma, in 4 villages during various study periods in central Turkey**[a]

| Population and cause of death | Karain 1979-1983 | | Karlik 1979-1983 | | Sarihidir 1980-1983 | | Tuzköy 1980-1983 | |
|---|---|---|---|---|---|---|---|---|
| | M | F | M | F | M | F | M | F |
| Population (aged 20+) | 1491 | 178 | 107 | 121 | 154 | 183 | 1461 | 1458 |
| Deaths: | | | | | | | | |
| All causes | 25 | 25 | 6 | 10 | 14 | 10 | 31 | 20 |
| Pleural mesothelioma | 12 | 9 | 0 | 0 | 2 | 1 | 0 | 5 |
| Peritoneal mesothelioma | 0 | 0 | 0 | 0 | 0 | 0 | 0 | 4 |
| Lung cancer | 2 | 0 | 0 | 0 | 5 | 1 | 9 | 0 |

[a]From Baris et al. (1987).

recorded during the study periods in the 4 villages, of which 29 were from pleural mesothelioma and 4 from peritoneal mesothelioma. The diagnosis of pleural mesothelioma was based on autopsy in 1 case and biopsy (during thoracotomy and thoracoscopy) in 12 cases. The remaining 16 cases, as well as the 4 peritoneal mesothelioma cases, were all diagnosed in hospitals but the research team was not able to retrieve and examine the pertinent records.

Table 5 presents some of the key findings on mortality, together with the results of the environmental survey. In the 3 villages with highest all-causes mortality, namely Karain, Sarihidir and Tuzköy, substantial proportions of the deaths were from mesothelioma. No mesotheliomas were reported from Karlik, the other village. Fibre concentrations in samples of airborne street dust are uniformly low, but the highest values (as reflected in the concentration ranges) were observed in the 3 villages affected by mesothelioma. Similarly, and more clearly, the composition of the airborne fibres indicates a substantially higher proportion of zeolite fibres in the samples from the 3 affected villages, as compared with the unaffected village of Karlik.

The zeolite fibres had previously been identified as having an elemental composition close to that of erionite, a crystalline fibrous form of zeolite (Pooley, 1979). Although the presence of chrysotile, crocidolite and tremolite was detected in some rock samples from the area (Rohl et al., 1982) as well as in biological samples (lungs) from humans and sheep, the pattern of their distribution matches the distribution of the disease less well than that of zeolite (erionite), as shown in Table 6 for the sheep sample measurements. Similarly, the disease distribution is consistent with the

Table 5. Mortality from all causes, pleural and peritoneal mesothelioma, fibre concentrations and proportion of zeolite in street samples in 4 villages in central Turkey[a]

| Item | Karain 1979-1983 | | Karlik 1979-1983 | | Sarihidir 1980-1983 | | Tuzköy 1980-1983 | |
|---|---|---|---|---|---|---|---|---|
| | M | F | M | F | M | F | M | F |
| Standardized death rate from all causes (per 1000 person-years) | 35 | 28 | 7 | 13 | 27 | 21 | 17 | 11 |
| Deaths (%) due to: | | | | | | | | |
| Pleural mesothelioma | 48 | 36 | 0 | 0 | 14 | 10 | 0 | 25 |
| Peritoneal mesothelioma | 0 | 0 | 0 | 0 | 0 | 0 | 0 | 20 |
| Range of fibre concentration (f/l) | 2-10 | | 2-6 | | 1-29 | | 5-25 | |
| No. of street samples | 36 | | 21 | | 24 | | 18 | |
| Proportion of zeolite in street sample material (%) | 80 | | 20 | | 60 | | 85 | |

[a]From Baris et al. (1987).

Table 6. Fibre content of sheep lungs[a]

| Fibre type | Two affected villages | Five unaffected villages | Difference between affected and unaffected villages (%) | Statistical significance ($p=0.05$) |
|---|---|---|---|---|
| Chrysotile | 4.27 | 3.21 | + 33 | Not significant |
| Crocidolite | 0.03 | 0.13 | − 77 | Not significant |
| Zeolite | 0.13 | 0.01 | + 1200 | Not significant |

[a]In $f/g \times 10^6$. Figures are averages for specimens collected in 2 villages in which mesothelioma was reported (affected villages: Sarihidir and Tuzköy) and 5 villages in which no mesothelioma was reported (unaffected villages: Boyali, Bozca, Karlik, Kizilköy, Yesiloz). Source: Baris et al. (1987).

presence or absence of ferruginous bodies identified as zeolite bodies (Sébastien et al., 1981, 1984) in the sputum of subjects from the affected and unaffected villages.

These epidemiological data receive strong support from experimental evidence showing that erionite, as well as having genotoxic activity (Poole et al., 1983), is carcinogenic both by inhalation and by the intrapleural route (Maltoni et al., 1982; Suzuki, 1982; Suzuki & Kohyama, 1984; Wagner, 1982; Wagner et al., 1985) to a much higher degree than any other mineral fibre yet tested (Wagner et al., 1985).

Quantitatively, the risk of mesothelioma in the affected villages is of a magnitude never before reported from any general population; it also exceeds that reported for most populations exposed in the past to high levels of asbestos dust. Thus, in the village of Karain, the most severely affected, 50 cases of mesothelioma were reported, of which 19 were histologically confirmed, in the period 1970-1978, giving a crude rate of 1000 per 100 000 (or a rate of 380 per 100 000 for those cases histologically confirmed). If the age-group 20-69 years is considered, the overall death rate is 1420 per 100 000.

Table 7 shows the age- and sex-specific rates. These indicate a steeply increasing trend with age, well fitted on a log-log scale by a straight line with an exponent of 2.4 for both sexes combined (Saracci *et al.*, 1982). This power relationship of mortality with time is closely paralleled by that observed in some industrial cohorts exposed to asbestos, provided time is taken as time from first exposure or, in some general populations with no recognized exposure to asbestos, provided time is taken as time since birth, i.e., age (Peto *et al.*, 1982). The data thus appear compatible with lifetime exposure starting at birth, but the curve observed in Karain is displaced upwards to levels of mesothelioma incidence considerably higher than in other contexts, despite the apparent low environmental concentrations of zeolite fibres (Table 5). It is still unclear whether the sole explanation capable of reconciling these two findings, namely, very high incidence and low measured environmental levels of fibres, is the high degree of carcinogenicity of erionite already alluded to (Wagner *et al.*, 1985).

**Table 7. Mesothelioma death rates per 1000 person-years by age and sex during 1970-1978 in Karain**

| Age-group (years) | Males | | Females | | Both sexes | |
|---|---|---|---|---|---|---|
| | No. of deaths | Rate | No. of deaths | Rate | No. of deaths | Rate |
| 20-29 | 2 | 5 | 0 | 0 | 2 | 2 |
| 30-39 | 2 | 9 | 3 | 9 | 5 | 9 |
| 40-49 | 10 | 29 | 6 | 12 | 16 | 19 |
| 50-59 | 7 | 19 | 6 | 20 | 13 | 19 |
| 60-69 | 6 | 37 | 21 | 32 | 27 | 34 |

*a*Source: Saracci *et al.* (1982).

In the previously mentioned case-control study of 557 mesothelioma subjects by McDonald and McDonald (1980), 17 cases and 12 controls were reported to have lived for 20-40 years within 20 miles of zeolite deposits in the western USA. When occupational exposure to asbestos was taken into account, the association was less marked.

In the geographical study of Enterline and Henderson (1987), high mesothelioma death rates were found in the Rocky Mountain area. The authors suggest that this might be related to the natural presence there of zeolite, although no direct evidence for this is reported.

*Man-made mineral fibres*

In the case-control study of 557 mesothelioma subjects of McDonald and McDonald (1980) mentioned earlier, 30 cases and 15 controls were reported to have been occupationally exposed to man-made mineral fibres. When occupational exposure to asbestos was taken into account, the association was less marked.

On the other hand, no risk of mesothelioma has been found in two very large studies of man-made mineral fibre production workers in Europe and the USA (Enterline et al., 1987; Simonato et al., 1986). In each study, involving workers in rock, slag and glass wool and continuous filament production, the small number of mesothelioma deaths observed (1 in the European and 3 in the US study) is in line with the number expected based on national rates.

Thus, from these findings, a detectable excess among the general population — again, as for asbestos, exposed at much lower fibre concentrations than occupational groups — is not to be expected. However, no relevant studies have been reported to date.

## Lung cancer

In the occupational environment, exposure to asbestos has led to a larger number of excess deaths from lung cancer than from mesothelioma. However, because of the dominant effect of cigarette smoking and the resulting high rates of lung cancer, any such attributable deaths at low asbestos exposure levels would be difficult to detect. A few studies have, however, produced relevant information.

### Domestic exposure (asbestos)

*Cohort studies*

Anderson (1983) reported the preliminary results of a cohort follow-up study continued until 1980 among 2218 household members of amosite asbestos workers who were employed between 1941 and 1945. The results of this study in terms of mesothelioma have already been described. For respiratory cancer, 25 deaths were observed, as compared with 16.4 expected (SMR=152) using age- and sex-specific death rates for the State of New Jersey, where the majority of the cohort lived for most of their lives. The excess was larger, with 20 deaths observed compared with 10.8 expected (SMR=185) in the period 20+ years after first household contact with an asbestos worker. (No other cancers or other causes of death were in excess, except for accidents, suicides and violence as a group among the male contacts.)

The available published information does not indicate whether or not these household members might subsequently have been occupationally exposed to asbestos themselves, nor is there any information on their cigarette smoking habits.

## Neighbourhood exposure (asbestos)

*Cohort studies*

Hammond *et al.* (1979) reported on a cohort follow-up study of men who were non-asbestos workers and residents in two areas of New Jersey, one located within half a mile of an amosite asbestos factory and the other a control area. The age distributions of the two groups were described as 'very close indeed', and mortality data were therefore presented in terms of proportions of deaths by cause. In the 'target' area, 41 (2.3%) of the 1779 men died from lung cancer during the follow-up period, as compared with 98 (2.6%) of the 3771 men in the control area. There was thus no suggestion of an association between lung cancer and neighbourhood asbestos exposure. No information was presented on cigarette smoking habits in the two groups.

*Geographical studies*

Siemiatycki (1983) updated the results of a study previously published by Pampalon *et al.* (1982). The groups of municipalities surrounding the Quebec chrysotile mining areas of Thetford Mines and Asbestos, where both occupational and non-occupational (domestic and neighbourhood) exposure to asbestos were known to have occurred, were the basic study units. Mortality over the period 1966–1977 by cause and sex in the combined areas was compared with that for Quebec province as a whole. Among men, some 75% of whom were reported as having worked in the mines or mills, 176 lung cancer deaths were observed as compared with 118 expected (SMR=149). By contrast, 23 lung cancer deaths were observed among women, which was close to the 21.5 expected (SMR=107).

The authors provisionally concluded that the male rates were consistent with the results of the occupational cohort follow-up study of chrysotile miners and millers (McDonald *et al.*, 1980), and that the female rates were compatible with the hypothesis that there was no excess lung cancer risk from non-occupational exposure to chrysotile asbestos (no data on the employment of women in the mines and mills were given, but it was assumed to be low). The interpretation is unchanged if lung cancer rates for urban areas of Quebec only are used for comparison (Siemiatycki, personal communication). (No figures for mesothelioma were reported.)

Botha *et al.* (1986) reported a geographical study of mortality rates between 1968 and 1980 in the South African crocidolite mining districts as compared with contiguous control districts. The results of this study in terms of mesothelioma have been described earlier. Increased death rates from lung cancer were found for both men and women in the mining districts.

## Non-asbestos mineral fibres

*Zeolite (erionite)*

In the study of 4 villages in the Cappadocia region of central Turkey, mentioned earlier, an excess of lung cancer was also reported from the 3 villages which had raised levels of mesothelioma (Baris *et al.*, 1987). The results are shown in Table 4 and, as for mesothelioma, the villages with high rates are those with higher proportions of zeolite fibres in samples of street dust (see Table 5). The authors comment that: 'misclassification of malignant pleural mesothelioma as lung cancer (and *vice versa*) may have occurred, which could modify the risk estimates, but not explain away the excess of pleural and parenchymal tumours' (Baris et al., 1987).

*Man-made mineral fibres*

The two large occupational cohort studies in Europe and the USA (Enterline et al., 1987; Simonato et al., 1986) show an excess of lung cancer among rock/slag wool production workers, but not among glass wool or continuous filament production workers. However, no studies have been reported on lung cancer and non-occupational exposure to man-made mineral fibres.

## Other effects

These include parenchymal (pulmonary) and pleural lesions, other than those deriving from lung cancer and mesothelioma, detectable during life or at autopsy. Anderson et al. (1976, 1979) reported asbestos-associated radiographic abnormalities in household contacts of amosite asbestos workers more frequently than in a control group of similar age and sex distribution taken from urban New Jersey residents attending the same medical facility for routine chest X-rays. The authors reported that, among the household contacts, a higher prevalence of abnormalities was associated with an increasing duration of exposure.

Bianchi et al. (1987b), in Monfalcone (north-east Italy), found hyalin plaques present in 75 out of 343 (21.6%) women at autopsy. In 59 of these cases, household exposure to asbestos was reported by relatives, in 2 cases occupational exposure, and in 9 cases both types of exposure.

In the non-industrial areas of Turkey in which an excess of mesothelioma has been found, radiological parenchymal abnormalities have also been reported ('pulmonary fibrosis' by Yazicioglu et al., 1980; small round opacities by Baris et al., 1987). From the Metsovo area in Greece a restrictive lung function decrement has been reported, accompanied by extensive bilateral calcified plaques in some older people (Constantopoulos et al., 1985; Langer et al., 1987), but the reality of this association has been questioned (Bazas, 1987) on the basis of a previous epidemiological investigation carried out in the same area (Bazas et al., 1985).

The presence of non-malignant pleural abnormalities of one or more types (thickening of visceral pleura, parietal pleural plaques, uncomplicated pleural exudates, progressive pleural fibrosis) has been reported from every non-industrial area in which mesothelioma has occurred (Baris et al., 1978, 1981, 1987; Boutin et al., this volume, pp. 406-410; Steinbauer et al., 1987; Yazicioglu et al., 1980). Crude prevalence rates ranging from a few per cent to more than 50% have been reported, increasing with age and, in some studies, with higher rates in males than in females.

The association within the same population of mesothelioma and pleural abnormalities, notably plaques, appears, however, not to be obligatory. Prevalence rates comparable with those just mentioned have been reported, first from Finland

(Kiviluoto, 1960), and and then from areas in Czechoslovakia (Marsova, 1964), Bulgaria (Burilkov & Babadjov, 1970), the USSR (Ginzburg et al., 1973) and Austria (Neuberger et al., 1984) in which no mesotheliomas have been recorded. Nevertheless, it remains completely unclear at present how far this may be due to missed diagnosis of mesothelioma. For instance, the early investigations in the Metsovo area of Greece (Bazas et al., 1985) and north-east Corsica (Steinbauer et al., 1987) did not report any mesotheliomas, whereas more recent, though still preliminary, studies (Boutin et al., this volume, pp. 406-410; Langer et al., 1987) indicate that mesotheliomas may indeed have occurred.

## Study methodologies and risk estimation

The original publication linking pleural mesothelioma with asbestos exposure by Wagner et al. (1960) was a report of 33 cases. This gave information for each case on the association with asbestos, except for one case with no reported history of such association, together with demographic and diagnostic details. The tumour was then generally regarded as uncommon, and was rarely seen elsewhere in South Africa. The link with asbestos exposure was strongly suggested by the evidence and widely accepted.

Since then, a number of other case series in specific locations have been published, dealing mainly with occupational exposures but also some with potential non-occupational exposures. However, the absence in this type of study of a comparison group, among whom the frequency of such potential non-occupational exposure in the particular locality could be evaluated, makes the attribution of cause not necessarily straightforward and the estimation of (relative) risk impossible.

Another problem is the difficulty of obtaining detailed and precise histories of potential non-occupational exposure (as indicated, for example, by the work of Gibbs et al. (this volume, pp. 219-228), which often involves recollection over many years. This is complicated further by the fact that, in many studies, some of the cases will have died, and information then has to be gathered by proxy from relatives or friends, or from records, thus increasing the possibility that the data will be inaccurate.

In case-control studies, the comparison problem mentioned above is theoretically overcome, but it is often difficult to ensure that identical methods and approaches to exposure ascertainment are employed for both cases and controls. Thus, for example, most cases may have died and concurrent live controls are sometimes used, or it proves impossible to 'blind' the interviewer to the case/control status of the patient and/or of other informants, which may influence, even if unintentionally, both the questioning and reporting.

Most case-control studies in the literature are based on individual matching for various relevant factors, but the findings in relation to potential non-occupational exposures are not always reported in that way. Reports have usually concentrated on the results of occupational histories. As a consequence, estimates of risk from non-occupational exposure are rarely available, and have not been reported in this review.

Even where relative risks from, for example, domestic exposure are reported and available, no quantitative data on asbestos exposure levels are available. Thus, there are no studies of asbestos-related disease and non-occupational exposure which make it possible to derive any quantitative dose-response risk estimates. However, some relationships of this kind may become available from the studies of mesothelioma and erionite.

In addition, there is the problem of the diagnosis and certification of mesothelioma. In the study in Quebec, it was found that the number of reported mesotheliomas was reduced after review by a pathology panel, and over-diagnosis was suggested as a possible factor in the apparent higher mesothelioma rate in Quebec (McDonald & McDonald, 1977). Similar results have been found in relation to death certificate mesothelioma cases (Greenberg & Lloyd Davies, 1974). Under-diagnosis may be a possibility, however, in areas remote from asbestos-using industries (Greenberg & Lloyd-Davies, 1974), and cases of mesothelioma have been reported among asbestos workers without the condition being mentioned on their death certificates (Newhouse & Wagner, 1969).

Thus, although it seems incontrovertible that cases of mesothelioma have occurred from non-occupational exposure to airborne mineral fibres, and possibly cases of lung cancer also, the available evidence does not provide any reliable or quantitative estimates of the risks involved either at past or current exposure levels.

## Conclusions

A few main conclusions emerge from the present review:

(a) Mesothelioma and lung cancer have occurred from non-occupational exposure to asbestos either in homes or in the neighbourhood of asbestos mines or asbestos-using industries. These past exposures have generally been higher than present exposures in buildings (for example, offices and schools) in industrialized countries.

(b) No direct epidemiological evidence that mesothelioma and lung cancer have resulted from general environmental exposure to asbestos in industrialized countries is available.

(c) Mesothelioma and lung cancer have occurred as a result of general environmental exposure to erionite and tremolite (plus chrysotile).

(d) No epidemiological data are available on non-occupational exposure to man-made mineral fibres.

(e) No exposure-response relationships are available from studies of non-occupational exposure to asbestos in industrialized countries.

The last of the above conclusions implies that no quantitative risk estimates are derivable from the direct observations made in populations non-occupationally exposed to asbestos in industrialized countries.

## References

Acheson, E.D. & Gardner, M.J. (1983) *Asbestos: the Control Limit for Asbestos*, London, Her Majesty's Stationery Office

Anderson, H.A. (1983) Family contact exposure. In: *Proceedings of the World Symposium on Asbestos*, Montreal, Canadian Asbestos Information Center, pp. 349-362

Anderson, H.A., Lilis, R., Daum, S.M., Fischbein, A.S. & Selikoff, I.J. (1976) Household contact asbestos neoplastic risk. *Ann. N.Y. Acad. Sci., 271*, 311-323

Anderson, H.A., Lilis, R., Daum, S.M. & Selikoff, I.J. (1979) Asbestosis among household contacts of asbestos factory workers. *Ann. N.Y. Acad. Sci., 330*, 387-399

Armstrong, B.K., Musk, A.W., Baker, J.E., Hunt, J.M., Newall, C.C., Henzell, H.R., Blundson, B.S., Clarke-Hundley, M.D., Woodward, S.D. & Hobbs, M.S.T. (1984) Epidemiology of malignant mesothelioma in Western Australia. *Med. J. Aust., 141*, 86-88

Baris, Y. (1980) The clinical and radiological aspects of 185 cases of malignant pleural mesothelioma. In: Wagner, J.C., ed., *Biological Effects of Mineral Fibres (IARC Scientific Publications No. 30)*, Lyon, International Agency for Research on Cancer, pp. 937-947

Baris, Y.I., Özesmi, M., Kerse, I., Özen, E., Sahin, A., Kolaçan, B. & Ogankulu, M. (1975) An outbreak of pleural mesothelioma in the village of Karain/Urgup in Anatolia. *Kanser, 5*, 1-4

Baris, Y.I., Sahin, A.A., Ozesmi, M., Kerse, I., Ozen, E., Kolaçan, B., Altinörs, M. & Göktepeli, A. (1978) An outbreak of pleural mesothelioma and chronic fibrosing pleurisy in the village of Karain/Urgup in Anatolia. *Thorax, 33*, 181-192

Baris, Y.I., Saracci, R., Simonato, L., Skidmore, J.W. & Artvinli, M. (1981) Malignant mesothelioma and radiological chest abnormalities in two villages in central Turkey. *Lancet, i*, 984-987

Baris, I., Simonato, L., Artvinli, M., Pooley, F., Saracci, R., Skidmore, J. & Wagner, J.C. (1987) Epidemiological and environmental evidence of the health effects of exposure to erionite fibres: a four-year study in the Cappadocian region of Turkey. *Int. J. Cancer, 39*, 10-17

Bazas, T. (1987) Pleural effects of tremolites in Northwest Greece. *Lancet, i*, 1490-1491

Bazas, T., Oakes, D., Gilson, J.C., Bazas, B. & McDonald, J.C. (1985) Pleural calcification in North West Greece. *Environ. Res., 38*, 239-247

Bianchi, C., Giarelli, L., Di Bonito, L., Grandi, G., Brollo, A. & Bittesini, L. (1982) Asbestos-related pleural mesothelioma in the Trieste area. *Adv. Pathol., 2*, 545-548

Bianchi, C., Brollo, A., Bittesini, L. & Ramani, L. (1987a) Asbestos-related mesothelioma of the pleura: what occupations at risk? In: *Proceedings, XIth World Congress on the Prevention of Occupational Accidents and Disease*

Bianchi, C., Brollo, A., Bittesini, L. & Ramani, L. (1987b) Pleural hyaline plaques and domestic exposure to asbestos [in Italian]. *Med. Lav., 78*, 44-49

Botha, J.L., Irwig, L.M. & Strebel, P.M. (1986) Excess mortality from stomach cancer, lung cancer, and asbestosis and/or mesothelioma in crocidolite mining districts in South Africa. *Am. J. Epidemiol., 123*, 30-40

Burilkov, T. & Babadjov, L. (1970) A contribution on the endemic occurrence of bilateral pleural calcification [in German]. *Prax. Pneumol., 24*, 433-438

Constantopoulos, S.H., Goudevenos, J.A., Saratzis, N., Langer, A.M., Selikoff, I.J. & Montsopoulos, H.M. (1985) Metsovo lung: pleural calcification and restrictive lung function in North-western Greece. Environmental exposure to mineral fiber as etiology. *Environ. Res., 38*, 319-331

Doll, R. & Peto, J. (1985) *Effects on Health of Exposure to Asbestos*, London, Her Majesty's Stationery Office

Enterline, P.E. & Henderson, V.L. (1987) Geographic patterns for pleural mesothelioma deaths in the United States, 1968-81. *J. Natl Cancer Inst., 79*, 31-37

Enterline, P.E., Marsh, G.M., Henderson, V. & Callahan, C. (1987) Mortality update of a cohort of U.S. man-made mineral fibre workers. *Ann. Occup. Hyg.*, *31*, 625-656

Gardner, M.J. (1983) Tumour incidence after asbestos exposure in Great Britain — with special reference to the cancer risk of the non-occupational population. In: Reinisch, D., Schneider, H.W. & Birkner, K.-F., eds, *Fibrous Dusts — Measurement, Effects, Prevention (VDI-Berichte, 475)*, Düsseldorf, VDI-Verlag, pp. 185-190

Gardner, M.J., Acheson, E.D. & Winter, P.D. (1982) Mortality from mesothelioma of the pleura during 1968-78 in England and Wales. *Br. J. Cancer*, *46*, 81-88

Gardner, M.J., Jones, R.D., Pippard, E.C. & Saitoh, N. (1985) Mesothelioma of the peritoneum during 1967-82 in England and Wales. *Br. J. Cancer*, *51*, 121-126

Ginzburg, E.A., Silova, E.V., Sergejev, A.M. *et al.* (1973) X-ray imaging of non-occupational pleural asbestosis [in German]. *Radiol. Diagn.*, *14*, 307-312

Greenberg, M. & Lloyd Davies, T.A. (1974) Mesothelioma register 1967-68. *Br. J. Ind. Med.*, *31*, 91-104

Hain, E., Dalquen, P., Bohlig, H., Dabbert, A. & Hinz, I. (1974) Catamnestic investigations of the origin of mesothelioma [in German]. *Int. Arch. Arbeitsmed.*, *33*, 15-37

Hammond, E.C., Garfinkel, L., Selikoff, I.J. & Nicholson, W.J. (1979) Mortality experience of residents in the neighbourhood of an asbestos factory. *Ann. N.Y. Acad. Sci.*, *330*, 417-422

Kiviluoto, R., (1960) Pleural calcification as a roentgenologic sign of non-occupational endemic anthophyllite-asbestosis. *Acta Radiol.*, Suppl., *194*, 1-67

Langer, A.M., Nolan, R.P., Constantopoulos, S.H. & Moutsopoulos, H.M. (1987) Association of Metsovo lung and pleural mesothelioma with exposure to tremolite-containing whitewash. *Lancet*, *i*, 965-967

Liddell, F.D.K. (1983) Tumour incidence after asbestos exposure in the general population of Canada. In: Reinisch, D., Schneider, H.W. & Birkner, K.-F., eds, *Fibrous Dusts - Measurement, Effects, Prevention (VDI-Berichte, 475)*, Düsseldorf, VDI-Verlag, pp. 179-183

Lieben, J. & Pistawka, H. (1967) Mesothelioma and asbestos exposure. *Arch. Environ. Health*, *14*, 559-563

Maltoni, C., Minardi, F. & Morisi, L. (1982) Pleural mesotheliomas in Sprague-Dawley rats by erionite: first experimental evidence. *Environ. Res.*, *29*, 238-244

Marsova, D. (1964) A contribution on the etiology of pleural calcification [in German]. *Z. Tuberk.*, *121*, 329-334

McDonald, A.D. & McDonald, J.C. (1980) Malignant mesothelioma in North America. *Cancer*, *46*, 1650-1656

McDonald, A.D. & McDonald, J.C. (1987) Epidemiology of malignant mesothelioma. In: Antman, K. & Aisner, J., eds, *Asbestos-related Malignancy*, Orlando, Grune & Stratton, pp. 31-55

McDonald, J.C. (1985) Health implications of environmental exposure to asbestos. *Environ. Health Perspect.*, *62*, 319-328

McDonald, J.C. & McDonald, A.D. (1977) Epidemiology of mesothelioma from estimated incidence. *Prev. Med.*, *6*, 426-446

McDonald, J.C., Liddell, F.D.K., Gibbs, G.W., Eyssen, G.E. & McDonald, A.D. (1980) Dust exposure and mortality in chrysotile mining, 1910-75. *Br. J. Ind. Med.*, *37*, 11-24

Milne, J.E.H. (1972) Thirty-two cases of mesothelioma in Victoria, Australia: a retrospective survey related to occupational asbestos exposure. *Br. J. Ind. Med.*, *33*, 115-122

Neuberger, M., Kundi, M. & Friedl, H.P. (1984) Environmental asbestos exposure and cancer mortality. *Arch. Environ. Health*, *39*, 261-265

Newhouse, M.L. & Thompson, H. (1965) Mesothelioma of pleura and peritoneum following exposure to asbestos in the London area. *Br. J. Ind. Med.*, *22*, 261-269

Newhouse, M.L. & Wagner, J.C. (1969) Validation of death certificates in asbestos workers. *Br. J. Ind. Med.*, *26*, 302-307

Nicholson, W.J. (1983) Tumour incidence after asbestos exposure in the USA: cancer risk of the non-occupational population. In: Reinisch, D., Schneider, H.W. & Birkner, K.-F., eds, *Fibrous Dusts — Measurement, Effects, Prevention (VDI-Berichte, 475)*, Düsseldorf, VDI-Verlag, pp. 161-177

Omenn, G.S., Merchant, J., Boatman, E., Dement, J.M., Kuschner, M., Nicholson, W., Peto, J. & Rosenstock, L. (1986) Contribution of environmental fibers to respiratory cancer. *Environ. Health Perspect.*, 70, 51-56

Pampalon, R., Siemiatycki, J. & Blanchet, M. (1982) Environmental asbestos pollution and public health in Quebec. *Union Med. Can.*, 111, 475-489

Pelnar, P.V. (1983) *Non-asbestos Related Malignant Mesothelioma: A Review of the Scientific and Medical Literature*, Montreal, Canadian Asbestos Information Center

Pelnar, P.V. (1985) *Further Evidence of Non-asbestos Mesothelioma*, Montreal, Canadian Asbestos Information Center

Peterson, J.T., Greenberg, S.D. & Buffler, P.A. (1984) Non-asbestos-related malignant mesothelioma: a review. *Cancer*, 54, 951-960

Peto, J., Seidman, H. & Selikoff, I.J. (1982) Mesothelioma mortality in asbestos workers: implications for models of carcinogenesis and risk assessment. *Br. J. Cancer*, 45, 124-135

Poole, A., Brown, R.C., Turver, C.J., Skidmore, J.W. & Griffiths, D.M. (1983) In vitro genotoxic activities of fibrous erionite. *Br. J. Cancer*, 47, 697-705

Pooley, F.D. (1979) Evaluation of fibre samples taken from the vicinity of two villages in Turkey. In: Lemen, R. & Dement, J.H., eds, *Dust and Disease*, Park Forest South, IL, Pathotox Publishers, pp. 41-44

Rohl, A.N., Langer, A.M., Moncure, G., Selikoff, I.J. & Fischbein, A. (1982) Endemic pleural disease associated with exposure to mixed fibrous dust in Turkey. *Science*, 216, 518-520

Saracci, R., Simonato, L., Baris, Y., Artvinli, M. & Skidmore, J. (1982) The age-mortality curve of endemic pleural mesothelioma in Karain, central Turkey. *Br. J. Cancer*, 45, 147-149

Sébastien, P., Gaudichet, A., Bignon, J. & Baris, Y.I. (1981) Zeolite bodies in human lungs from Turkey. *Lab. Invest.*, 44, 420-425

Sébastien, P., Bignon, J., Baris, Y.I., Awad, L. & Petit, G. (1984) Ferruginous bodies in sputum as an indication of exposure to airborne mineral fibers in the mesothelioma villages of Cappadocia. *Arch. Environ. Health*, 39, 18-23

Siemiatycki, J. (1983) Health effects on the general population: mortality in the general population in asbestos mining areas. In: *Proceedings of the World Symposium on Asbestos*, Montreal, Canadian Asbestos Information Center, pp. 337-348

Simonato, L., Fletcher, A.C., Cherrie, J., Andersen, A., Bertazzi, P.A., Charnay, N., Claude, J., Dodgson, J., Esteve, J., Frentzel-Beyme, R., Gardner, M.J., Jensen, O., Olsen, J., Saracci, R., Teppo, L., Westerholm, P., Winkelmann, R., Winter, P.D. & Zocchetti, C. (1986) The man-made mineral fiber European historical cohort study: extension of the follow-up. *Scand. J. Work. Environ. Health*, 12 (Suppl. 1), 34-47

Steinbauer, J., Boutin, C., Viallat, J.R., Dufour, G., Gaudichet, A., Massey, D.G., Charpin, D. & Mouries, J.C. (1987) Pleural plaques and environmental asbestos exposure in northern Corsica [in French]. *Rev. Mal. Resp.*, 4, 23-27

Suzuki, Y. (1982) Carcinogenic and fibrogenic effects of zeolites: preliminary observations. *Environ. Res.*, 27, 433-445

Suzuki, Y. & Kohyama, N. (1984) Malignant mesothelioma induced by asbestos and zeolite in the mouse peritoneal cavity. *Environ. Res.*, 35, 277-292

Teta, M.J., Lewinsohn, H.C., Meigs, J.W., Vidone, R.A., Mowad, L.Z. & Flannery, J.T. (1983) Mesothelioma in Connecticut, 1955-1977. *J. Occup. Med.*, 25, 749-756

Theriault, G.P. & Grand-Bois, L. (1978) Mesothelioma and asbestos in the Province of Quebec, 1969-1972. *Arch. Environ. Health*, 33, 15-19

Vianna, N.J. & Polan, A.K. (1978) Non-occupational exposure to asbestos and malignant mesothelioma in females. *Lancet*, i, 1061-1063

Vianna, N.J., Maslowsky, J., Roberts, S., Spellman, G. & Patton, R.B. (1981) Malignant mesothelioma: epidemiologic patterns in New York State. *N.Y. State J. Med.*, 5, 735-738

Wagner, J.C. (1982) Health hazards of substitutes. In: *Asbestos, Health and Safety*, Montreal, Canadian Asbestos Information Center, p. 244

Wagner, J.C., Sleggs, C.A. & Marchand, P. (1960) Diffuse pleural mesothelioma and asbestos exposure in the North Western Cape Province. *Br. J. Ind. Med., 17,* 260-271

Wagner, J.C., Skidmore, J.W., Hill, R.C. & Griffiths D.M. (1985) Erionite exposure and mesotheliomas in rats. *Br. J. Cancer, 51,* 727-730

Wasserman, M., Wasserman, D., Steinitz, R., Katz, L. & Lemesch, C. (1980) Mesothelioma in children. In: Wagner, J.C., ed., *Biological Effects of Mineral Fibres (IARC Scientific Publications No. 30),* Lyon, International Agency for Research on Cancer, pp. 253-257

Webster, I. (1973) Asbestos and malignancy. *South Afr. Med. J., 47,* 165-171

Yazicioglu, S., Oktem, K., Ilcayto, R., Balci, K. & Sayli, B.S. (1978) Association between malignant tumors of the lungs and pleurae and asbestosis: a retrospective survey. *Chest, 73,* 52-56

Yazicioglu, S., Ilcayto, R., Balci, K. & Sayli, B.S. & Yorulmaz, B. (1980) Pleural calcification, pleural mesotheliomas, and bronchial cancers caused by tremolite dust. *Thorax, 35,* 564-569

# RELATION OF ENVIRONMENTAL EXPOSURE TO ERIONITE FIBRES TO RISK OF RESPIRATORY CANCER

L. Simonato[1], R. Baris[2], R. Saracci[1], J. Skidmore[3,4] & R. Winkelmann[1]

[1] Unit of Analytical Epidemiology, International Agency for Research on Cancer, Lyon, France
[2] Department of Chest Diseases, Hacettepe University, Ankara, Turkey
[3] Pneumoconiosis Unit, Medical Research Council, Penarth, UK
[4] Present address: Tegeris Laboratories Inc., 9705 N. Washington Blvd, Laurel, MD, USA

*Summary.* During the period 1979-1983, IARC, in collaboration with the Department of Chest Diseases of Hacettepe University in Ankara and the MRC Pneumoconiosis Unit in Penarth, conducted an epidemiological and environmental survey in 4 villages in central Turkey affected by a high incidence of mesothelial tumours. Recent data point to erionite, a zeolite fibre, as the most plausible etiological agent. From animal experiments, erionite appears to be the most powerful carcinogenic fibre so far known. During the study period, 17 pleural mesotheliomas and 7 lung cancer cases have been reported among the villagers. These cancer cases are analysed in relation to exposure to fibres. We assume exposure to occur from birth onwards and therefore consider duration of exposure equal to age. On this basis, the incidence of mesothelial and lung tumours is analysed in relation to age and cumulative exposure to fibres computed using the airborne fibre levels measured during the survey.

## *Introduction*

Following initial observations of cases of malignant pleural mesothelioma in a small village of central Anatolia (Baris *et al.*, 1975, 1978), a team of investigators from the Department of Chest Diseases of Hacettepe University, Ankara, the MRC Pneumoconiosis Unit, Penarth, the Department of Mineral Exploitation, University College, Cardiff, and the Unit of Analytical Epidemiology, IARC, Lyon, carried out an environmental and epidemiological survey in 4 villages in the same area during the period 1979-1983. The results of the survey have been published in the scientific literature (Baris *et al.*, 1981, 1987; Saracci *et al.*, 1982). Erionite, a fibre belonging to the zeolite family, was found both in the environment and in lung tissue of mesothelioma patients. These fibres, when tested in experimental systems, show a high carcinogenic potential after either intrapleural injection or inhalation (Maltoni *et al.*, 1982; Suzuki, 1982; Suzuki & Kohyama, 1984; Wagner *et al.*, 1985). According to the IARC, there is sufficient evidence to show that erionite fibres are carcinogenic both in humans and in animals (IARC, 1987). We present here a reanalysis of part of the data collected in 1979 and in 1980 in the villages of Karain and Sarihidir, with the aim of estimating the relationship between the mortality rates for pleural mesothelioma and lung cancer and the cumulative exposure to fibres present in the environment.

## Materials and methods

The villages of Karain and Sarihidir were studied because of the reported high rates of mesothelioma in the former and, in the latter, the presence of erionite fibres in bedrock and road dust samples. The study consisted of interviews of people aged 20 or over, full-size postero-anterior chest X-ray and brief physical examination. Deaths occurring between June 1979 and July 1983 in Karain and between July 1980 and July 1983 in Sarihidir were recorded using a variety of sources (health centre records, clinical files at Hacettepe University, key informants). A third village in the same province, Tuzköy, already surveyed by researchers from Hacettepe University (Artvinli & Baris, 1979), was also included for mortality recording. The village of Karlik in the same area, where no case of mesothelioma or lung cancer has been reported, was chosen as the control population.

Taking advantage of the fact that the population of Karain and Sarihidir had been identified at individual level, the deaths occurring up to 30 June 1983 in these 2 villages were linked to the files of residents there who were interviewed in the 1979 and 1980 survey. Mortality rates per $10^6$ person-years of observation were computed. It was not possible to ascertain how many people left the villages during the study period, and this may lead to some underestimation of the rates. Given the magnitude of the risk, however, this should not present a major problem.

Airborne fibre levels were monitored on several occasions during the study period. The cumulative fibre exposure was estimated by taking the mean exposure level in fibres per ml (f/ml) in each village and multiplying it by the years since first exposure of each individual enrolled in the survey in the specific village, assuming first exposure to occur at birth. The categories of cumulative fibre exposure were chosen according to the frequency distribution of the subjects. A linear relationship between estimated cumulative fibre exposure and the mortality rates in the 2 villages combined was assumed and linear regressions were computed using a maximum likelihood procedure (Baker & Nelder, 1978).

## Results

The geography of the area where the villages under study are located is shown in Figure 1. As already mentioned, a high incidence of mesothelial and respiratory tumours has been reported from Tuzköy, Sarihidir and Karain, while no case has ever been reported in the control village, Karlik. Only the populations of Sarihidir and Karain are included in the present analysis.

Table 1 shows the concentrations of airborne fibres found in the villages of Karain and Sarihidir; total fibre levels are higher in Sarihidir, but the content of zeolites is lower than in Karain. Exposures to zeolite fibres in the two villages can thus be regarded as similar enough to justify the pooling of the two populations.

**Fig. 1. Area of Central Anatolia under study**

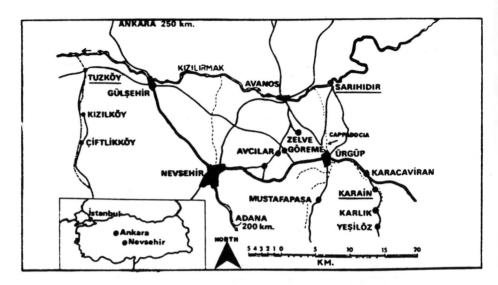

**Table 1. Concentrations of airborne fibres in Karain and Sarihidir**

| Village | Range (f/ml) | No. of samples | Mean (f/ml) | Identity |
|---|---|---|---|---|
| Karain | 0.002-0.010 | 36 | 0.006 | ~80% zeolite + calcium oxide and sulfate |
| Sarihidir | 0.001-0.029 | 24 | 0.009 | ~60% zeolite + calcite, quartz, glass and tremolite |

The results of an analysis of all causes of death, by village, are reported in Table 2. The numbers of observed deaths and the mortality rates differ from those given in a recent publication (Baris et al., 1987) because of the different criteria adopted for inclusion of the observed deaths and the different method used for computing rates. Mortality appears to be higher in Karain than in Sarihidir. This difference is more likely to be due to the less efficient registration of the causes of deaths other than cancer in Sarihidir than to a real difference in mortality. In fact, after exclusion of mesothelioma and lung cancer, the mortality rate is only 600 per 100 000 in Sarihidir, as compared with 1075 in Karain.

Table 3 gives the results of the analysis of mortality rates by time since first exposure for pleural mesothelioma and lung cancer, separately and combined, for the entire population on follow-up The very large excess reported in this table is increasing with time since first exposure and is more evident when the two sites are combined.

**Table 2. Age-specific mortality rates per 100 000 person-years by village and by time for both sexes**[a]

| Time since first exposure (years) | Karain | | | Sarihidir | | |
|---|---|---|---|---|---|---|
| | Person-years | Observed | Rate per 100 000 person-years | Person-years | Observed | Rate per 100 000 person-years |
| 20-29 | 242 | 3 | 1242 | 330 | 0 | - |
| 30-39 | 247 | 4 | 1620 | 184 | 0 | - |
| 40-49 | 228 | 4 | 1756 | 172 | 1 | 581 |
| 50+ | 588 | 19 | 3234 | 315 | 13 | 4126 |
| Total | 1304 | 30 | 2301 | 1001 | 14 | 1399 |

[a]It is assumed that first exposure occurred at birth.

**Table 3. Age-specific mortality rates per 100 000 person-years for pleural mesothelioma and lung cancer in Karain and Sarihidir by time since first exposure for both sexes combined**[a]

| Time since first exposure (years) | Pleural mesothelioma[b] | | Lung cancer | | Both[b] | |
|---|---|---|---|---|---|---|
| 20-29 | (1) | 175 | (1) | 175 | (2) | 350 |
| 30-39 | (2) | 464 | (0) | 0 | (2) | 464 |
| 40-49 | (4) | 1000 | (0) | 0 | (4) | 1000 |
| 50+ | (10) | 1108 | (6) | 665 | (16) | 1773 |
| Total | (17) | 738 | (7) | 304 | (24) | 1041 |

[a]It is assumed that first exposure occurred at birth.
[b]Numbers of deaths in parentheses.

The relationship between the mortality rates for pleural mesothelioma and lung cancer is analysed in Table 4 and shown graphically in Figure 2 for pleural mesothelioma and in Figure 3 for both sites combined. Both pleural mesothelioma and lung cancer mortality rates increase with increasing cumulative fibre dose. The regression equations are:

For pleural mesothelioma: rate/$10^5$ = $-245.3 + 1241$ fibres.year/ml
For both sites combined: rate/$10^5$ = $-38.6 + 3029$ fibres.year/ml

Combining the two sites may be warranted by the limitations of the diagnostic criteria (in approximately 50% of the cases, no revision of the histology was undertaken), which increase the chance of misclassification. The mortality rates appear to increase somewhat more rapidly with increasing cumulative dose when the two sites are combined.

**Fig. 2. Relation between mortality from pleural mesothelioma and cumulative fibre exposure in Karain and Sarihidir, 1979-1983, both sexes combined**

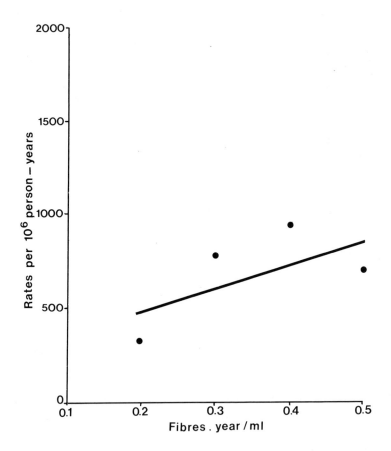

## Discussion and conclusions

The elevated mortality from pleural mesothelioma and lung cancer among residents in the villages of Karain and Sarihidir is highly correlated with cumulative fibre dose.

The dose-response analysis given here is based on three assumptions: (1) exposure starts at date of birth; (2) the exposure has been constant over time; (3) average

**Fig. 3. Relation between combined mortality from pleural mesothelioma and lung cancer combined and cumulative fibre exposure in Karain and Sarihidir, 1979–1983, both sexes combined**

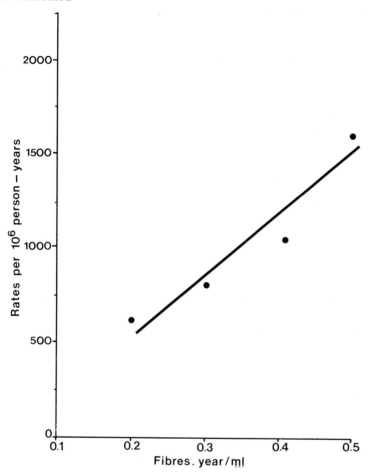

**Table 4. Age-specific mortality rates per 100 000 person-years for pleural mesothelioma and lung cancer in Karain and Sarihidir by cumulative exposure for both sexes combined[a]**

| Cumulative exposure (fibres, year/ml) | Pleural mesothelioma | | Lung cancer[b] | | Both[b] | |
|---|---|---|---|---|---|---|
| ≤0.2  | (1)  | 336 | (1) | 336 | (2)  | 671  |
| ≤0.3  | (6)  | 773 | (0) | 0   | (6)  | 773  |
| ≤0.4  | (6)  | 905 | (1) | 151 | (7)  | 1056 |
| >0.4  | (4)  | 705 | (5) | 881 | (9)  | 1587 |
| Total | (17) | 738 | (7) | 304 | (24) | 1041 |

[a] It is assumed that first exposure occurred at birth.
[b] Numbers of deaths in parentheses.

exposure applies to all individuals in the villages. While there should be little doubt about the first assumption, we cannot be sure that the levels of airborne fibres measured during the survey correspond to past exposure levels nor that they apply equally to all individuals in the survey.

Community life in the villages has not changed much, for at least several decades, and no confounding from occupational exposure to fibres could be observed. However, since the beginning of the 1970s, an increasing number of people have tended to abandon the old houses built of stone from the nearby quarries and to build new, modern houses with bricks. This decrease in the use of stone from the quarries, some of which is contaminated with veins of zeolite fibres, might have decreased the airborne levels and the cumulative dose might therefore have been overestimated. However, as it is hard to believe that major changes have occurred, the dose capable of inducing such a large excess rate appears to be extremely low (all mesothelioma cases are virtually in excess, the number of cases expected in the study population for this rare tumour being approximately 0.005).

A cumulative dose of 1 fibre.year/ml, in fact, appears capable of inducing, for pleural mesothelioma, a rate of 996 per 100 000 person-years in the exposed population. The magnitude of the risk at such a low exposure is not comparable with that for any other (occupational) exposure to asbestos fibres. These findings receive support from the experimental evidence showing that the carcinogenic potency of erionite fibres appears greater than that of any other mineral fibre so far studied (Wagner et al., 1985).

However, it should be emphasized that there are many uncertainties associated with the assumption that daily ambient average airborne fibre levels measured during a short period (1-3 weeks) are indicative of the cumulative exposure of a population exposed every day since birth. There must, in fact, be variations over time in relation to daily activity, both indoor and outdoor, seasonal and climatic changes and secular changes. It is possible that the airborne fibre levels used in the present analysis represent only the 'background' exposure for the populations concerned.

While the size characteristics of erionite do not seem markedly different from those of other carcinogenic mineral fibres, the relatively high content of fibres in the lung tissue of mesothelioma cases suggests that they possess high durability. Erionite fibres tend to accumulate in the lungs in doses much higher than those of other fibres present in the environment at higher levels (Baris et al., 1987).

One important limitation of this study is the difficulty in assessing the causes of death. Only half the cases have, in fact, been histologically confirmed, mainly through biopsy, so that there is some uncertainty about the diagnosis of mesothelioma. We cannot, therefore, exclude a certain degree of misclassification as between mesothelial and parenchymal malignancies. According to available mortality statistics (Firat, 1980), the lung cancer rates in Turkey are about 35 deaths per 100 000, which gives approximately 0.8 expected cases in the population under study. Misclassification would certainly affect the risk ratio between mesothelioma and lung cancer, and to quantify the risk for the two sites separately presents serious difficulties that have still to be overcome, although it is greatly elevated for both.

Smaller numbers of deaths are included in the present analysis, as compared with those in our previous publication (Baris *et al.*, 1987). This appears to be due to the fact that people who were seriously ill in 1979 and 1980 could not be included in the survey and, as a consequence, were neither followed up until June 1983 nor included in the present analysis. Actual mortality rates, particularly for causes of death other than cancer, may therefore have been underestimated.

In conclusion, this supplementary analysis, the results of which agree with those of our previous publications (Baris *et al.*, 1981, 1987; Saracci *et al.*, 1982), represents an attempt to assess the relationship between cumulative fibre dose, as measured on the basis of current airborne fibre levels, and the mortality rates for pleural mesothelioma and for mesothelioma and lung cancer combined. The results indicate that, albeit low, increasing cumulative fibre doses are associated with an increase in the mortality rates for the sites of interest.

## References

Artvinli, M. & Baris, Y.I. (1979) Malignant mesotheliomas in a small village in the Anatolian region of Turkey: an epidemiologic study. *J. Natl Cancer Inst.*, 63, 17-22

Baker, R.J. & Nelder, J.A. (1978) *Generalised Linear Interactive Modelling*. Oxford, Numerical Algorithms Group

Baris, Y.I., Özesmi, M., Kerse, I., Özen, E., Sahin, A., Kolaçan, B. & Ogankulu, M. (1975) An outbreak of pleural mesothelioma in the village of Karain/Urgup, Anatolia. *Kanser*, 5, 1-4

Baris, Y.I., Sahin, A.A., Özesmi, M., Kerse, I., Özen, E., Kolaçan, B., Altinörs, M. & Göktepeli, A. (1978) An outbreak of pleural mesothelioma and chronic fibrosing pleurisy in the village of Karain/Urgup in Anatolia. *Thorax*, 33, 181-192

Baris, Y.I., Saracci, R., Simonato, L., Skidmore, J.W. & Artvinli, M. (1981) Malignant mesothelioma and radiological chest abnormalities in two villages in central Turkey. *Lancet*, ii, 984-987

Baris, Y.I., Simonato, L., Artvinli, M., Pooley, F., Saracci, R., Skidmore, J. & Wagner, J.C. (1987) Epidemiological and environmental evidence of the health effects of exposure to erionite fibres: a four-year study in the Cappadocian region of Turkey. *Int. J. Cancer*, 39, 10-17

Firat, D. (1983) *Cancer Mortality in Turkey and in the World, 1980-1981*, Ankara, Turkish Association for Cancer Research and Control

IARC (1987) *IARC Monographs on the Evaluation of the Carcinogenic Risk of Chemicals to Humans Vol. 42, Silica and Some Silicates*, Lyon, International Agency for Research on Cancer

Maltoni, C., Minardi, F. & Morisi, L. (1982) Pleural mesotheliomas in Sprague-Dawley rats by erionite: first experimental evidence. *Environ. Res.*, 29, 238-244

Saracci, R., Simonato, L., Baris, Y., Artvinli, M. & Skidmore, J. (1982) The age-mortality curve of endemic pleural mesothelioma in Karain, central Turkey. *Br. J. Cancer*, 45, 147-149

Suzuki, Y. (1982) Carcinogenic and fibrogenic effects of zeolites: preliminary observations. *Environ. Res.*, 27, 433-445

Suzuki, Y. & Kohyama, N. (1984) Malignant mesothelioma induced by asbestos and zeolite in the mouse peritoneal cavity. *Environ. Res.*, 35, 277-292

Wagner, J.C., Skidmore, J.W., Hill, R.C. & Griffiths D.M. (1985) Erionite exposure and mesotheliomas in rats. *Br. J. Cancer*, 51, 727-730

# BILATERAL PLEURAL PLAQUES IN CORSICA: A MARKER OF NON-OCCUPATIONAL ASBESTOS EXPOSURE

### G. Boutin, J.R. Viallat & J. Steinbauer

*Department of Pneumology, Conception Hospital, Marseilles, France*

### G. Dufour & A. Gaudichet

*Laboratoire d'Etude des Particules Inhalées, Paris, France*

*Summary.* In north-east Corsica, asbestos outcrops are a source of environmental pollution, as assessed by airborne concentrations of chrysotile and tremolite that are significantly higher in the north-east than the north-west. For this reason we compared the frequencies of patients with asbestos pleural plaques in these 2 regions by checking the X-rays of 1721 patients born in north Corsica who had not undergone any occupational exposure to asbestos. The percentage of plaques was respectively 3.7% in the north-east, and 1.1% in the north-west ($p<0.05$). The relative risk of 3.3 demonstrates a strong relationship between environmental asbestos and the pleural plaques found in north-east Corsica.

## Introduction

Since the initial reports of Lynch and Canon in 1948, and of Cartier in 1949, the asbestos etiology of bilateral pleural plaques has been well documented. Calcified plaques and hyaline non-calcified plaques both indicate asbestos exposure, usually occupational. However, Kiviluoto (1960) described an outbreak of environmental calcified pleural plaques in subjects living around Finnish anthophyllite mines but without occupational exposure. Subsequently the existence of a pleural asbestos pathology of environmental origin has been confirmed in Bulgaria (Burilkov & Babadjov, 1970), Czechoslovakia (Navratil *et al.*, 1975), Austria (Neuberger *et al.*, 1978), Turkey (Baris *et al.*, 1979; Yazicioglu, 1976), Greece, the USSR and India.

In a survey of ex-miners from the Canari asbestos mine in Corsica we found that the percentage of bilateral pleural plaques was 27.7% (Viallat & Boutin, 1980; Viallat *et al.*, 1983). In addition, some miners exhibited pleural plaques only a few years after the beginning of occupational exposure, instead of the usual 30 years after first exposure. In the unexposed control group chosen for the survey, the percentage of bilateral plaques was also unexpectedly high (3.8%). The geology of north-east Corsica, where serpentine, chrysotile, amphiboles and outcrops of glossy schists are present, could be the explanation for this high frequency of plaques.

The procedure adopted in the present study was to identify chest radiographs of patients admitted to Tattone Hospital (Corsica) during the previous decade, and to classify patients with bilateral pleural plaques according to their place of birth. Our hypothesis was that the frequency of bilateral pleural plaques would be significantly higher among patients born in the north-east of Corsica, as a result of an environmental exposure to asbestos outcrops, dating back to childhood.

## Methods

### Selection of patients and radiographs

The study was restricted to patients born in north Corsica. The pathology of patients admitted to Tattone Hospital was mainly related to pneumology, internal medicine or gerontology. All patients having occupational asbestos exposure were excluded from the study, especially those who had worked at the Canari asbestos mine. The study finally included 1721 patients, of whom 1457 were born in north-east and 264 in north-west Corsica.

### Radiographic interpretation

Anteroposterior chest radiographs had the standard dimensions of 43 × 35.5 cm and were taken at high kV. All chest radiographs were read by two experts who were not aware of the geographical origin of the subjects. The reading was performed according to the 1980 ILO Classification (International Labour Office, 1980). In case of disagreement, the radiograph was read by a third expert and a consensus reached.

Pleural plaques, whether fibrohyaline or calcified, had to be bilateral in order to be considered as related to asbestos exposure.

### Metrology of airborne sampling

In order to assess airborne asbestos exposure, air samples for further analysis by the Laboratoire d'Etude des Particules Inhalées (Paris) were collected in 4 villages in north-east Corsica and 4 villages in the rest of the island, 3 samples being collected from open-air sites, 1 from indoors (usually in town halls). Briefly, airborne particles were collected for 24 h by means of an air pump (output 16 1/min) on 0.45-$\mu$m Millipore filters. After incineration in a low-temperature oxygen plasma oven, mineral particles were transferred on to electron microscope grids for further analysis with a transmission electron microscope and an energy-dispersive X-ray spectrometer (Billon-Galland *et al.*, 1988).

## Results

### Prevalence of pleural plaques

Of the 1721 chest radiographs available from patients born in north Corsica, 67 showed bilateral pleural plaques. Among these patients, 8 had been admitted because of an 'anomalous thoracic radiograph'; this reason for hospitalization might have biased our study, and these patients were therefore excluded. Inquiries at patients' homes revealed that 3 of those with bilateral plaques had worked temporarily in the Canari asbestos mine; they were also excluded from the study. A total of 56 patients were thus studied (Table 1), of whom 53 came originally from north-east

Corsica, where there are deposits of asbestos, and 3 from the north-west. Among the latter, 2 subjects had had a temporary occupational exposure to asbestos as shown by a detailed inquiry; they were retained in the study, however, in order to balance any risk of error which might occur in the other group. The difference in frequency of pleural plaques as between north-east (3.7%) and north-west Corsica (1.1%) is significant: $X^2 = 4.45$ ($p<0.05$).

Table 1. Distribution of 56 bilateral pleural plaques by place of birth[a]

| Place of birth | Subjects with | | % |
|---|---|---|---|
| | No plaques | Bilateral plaques | |
| North-east | 1393 | 53 | 3.66 |
| North-west | 261 | 3 | 1.14 |

[a] $X^2 = 4.45$; $p<0.05$; relative risk, 3.3.

### Characteristics of patients with bilateral plaques born in the north-east

A total of 43 patients had bilateral calcified pleural plaques, and 10 exhibited either a bilateral thickening or a pleural hyaline thickening in conjunction with contralateral calcified plaques.

The average age of patients with calcified plaques was 70.9±1.5 years, as compared with 68.8±2.3 years for the patients having bilateral plaques of which at least one was not calcified. The mean age of the whole group was 70.5±1.3 years. The percentage of women was 26.3%, and that of smokers 62.3%.

### Airborne concentrations of asbestos

A significant difference in concentrations of tremolite was observed as between the western and eastern air samples (Table 2). In western villages, airborne asbestos concentrations never exceeded 2 ng/m³, while in eastern villages concentrations ranged from 0.1 to 128 ng/m³.

Table 2. Concentrations of airborne asbestos (ng/m³) in Corsican villages

| Type of village | Location of sampling site | Chrysotile | Tremolite |
|---|---|---|---|
| Exposed[a] | Indoors | 14.3 ± 15.7 | 59.8 ± 48 |
| | Open air | 15.5 ± 10.9 | 12.0 ± 6.6 |
| Unexposed | Indoors | 0.6 ± 0.5 | ND[b] |
| | Open air | 0.4 ± 0.2 | 0.25 ± 0.25 |

[a] Rutali, Murato, Campile and Moita.
[b] ND, not determined.

## Discussion

The exceptional geological situation of Corsica and the numerous asbestos deposits present, which are strictly limited to the north-east of the island, are associated with an unexpectedly high frequency of pleural asbestos plaques of non-occupational origin: 3.7% as compared with 1.1% in north-west Corsica. The relative risk is thus 3.3.

The mean age of the subjects is high, and the environmental exposure is, on the whole, moderate, which may explain the late radiological emergence of pleural plaques.

The percentage of Corsican women having bilateral pleural plaques (26.3%) is similar to the percentage of women admitted to the Tattone Hospital (28%). This is in line with a non-occupational origin of these plaques; indeed, if exposure had been occupational, we would have found women to be under-represented, because few women of this age have ever been in employment. In addition, metrology of air samples confirmed the existence of significant pollution by airborne tremolite fibres in villages in north-east Corsica, as compared with villages in the north-west (Billon-Galland et al., 1988).

The environmental contact with asbestos of patients having bilateral plaques can be traced back to their childhood; for instance, in a village built on a deposit of asbestos, one of the children's diversions was to search for the longest fibres on the ground in order to make lighter wicks. For this reason, we associated the existence or non-existence of plaques with the birth-place instead of the domicile at the time of hospitalization. There was, however, little difference in the results when this was done, since the Corsican population is exceedingly stable.

As far as we know, this is the first description of pleural plaques of environmental origin in Western Europe. Their frequency (3.7% in north-east Corsica) lies in the same range as those reported in the literature from different countries, such as Austria, Finland, Turkey, Greece or Bulgaria. In Turkey (Baris et al., 1981; Yazicioglu et al., 1978), this excess of pleural plaques is connected with a significant excess of mesothelioma and cancer of the bronchi. Whether this is the case in Corsica cannot be ascertained from the available data; up to the present, however, we have recorded 5 cases of pleural mesothelioma in patients born in north-east Corsica, without occupational asbestos exposure but with a heavy pulmonary asbestos burden.

## References

Baris, Y.I., Artvinli, M. & Sahin, A.A. (1979) Environmental mesothelioma in Turkey. *Ann. N.Y. Acad. Sci.*, 330, 423-432

Baris, Y.I., Saracci, R., Simonato, L., Skidmore, J.W. & Artvinli, M. (1981) Malignant mesothelioma and radiological chest abnormalities in two villages in central Turkey. *Lancet*, i, 984-987

Billon-Galland, M.A., Dufour, G., Gaudichet, A., Boutin, C. & Viallat, J.R. (1988) Environmental airborne asbestos pollution and pleural plaques in Corsica. In: *Inhaled Particles VI*, Proceedings of the 6th International Symposium on Inhaled Particles, Cambridge, 2-6 September 1985 (in press)

Burilkov, T. & Babadjov, L. (1970) A contribution on the endemic occurrence of bilateral pleural calfication [in German]. *Prax. Pneumol.*, *24*, 433-438

International Labour Office (1980) *Classification of Radiographs for Pneumoconiosis*, Geneva

Kiviluoto, R. (1960) Pleural calcification as a roentgenologic sign of non-occupational endemic anthophyllite asbestosis. *Acta. Radiol., Suppl.*, *194*, 1-67

Navratil, M., Moravkova, K. & Trippe, F. (1975) Follow-up study of pleural hyalinosis in individuals not exposed to asbestos dust. *Environ. Res.*, *114*, 1433-1438

Neuberger, M., Grundorfer, W., Haider, M., Konigshofer, R., Muller, H.W., Raber, A., Riedmuller, G. & Schwaighofer, B. (1978) Endemic pleural plaques and environmental factors. *Zentralbl. Bakteriol.*, *167*, 391-404

Viallat, J.R. & Boutin, C. (1980) Radiographic changes in chrysotile mine and mill ex-workers in Corsica. (A survey 14 years after cessation of exposure). *Lung*, *157*, 155-163

Viallat, J.R., Boutin, C., Pietri, J.F. & Fondarai, J. (1983) Late progression of radiographic changes in Canari Chrysotile Mine and Mill ex-workers. *Arch. Environ. Health*, *38*, 54-58

Yazicioglu, S. (1976) Pleural calcification associated with exposure to chrysotile asbestos in southeast Turkey. *Chest*, *70*, 43-47

Yazicioglu, S., Oktem, K., Ilcayto, R., Balci, K. & Sayli, B.S. (1978) Association between malignant tumors of the lungs and pleurae and asbestosis; a retrospective survey. *Chest*, *73*, 52-56

# MESOTHELIOMA IN CYPRUS

## K. McConnochie

*Department of Tuberculosis and Chest Diseases, University of Wales College of Medicine, Llandough Hospital, Penarth, South Glamorgan, UK*

## L. Simonato

*Unit of Analytical Epidemiology, Division of Epidemiology and Biostatistics, International Agency for Research on Cancer, Lyon, France*

## P. Mavrides & P. Christofides

*The Chest Clinic, Nicosia General Hospital, Nicosia, Cyprus*

## R. Mitha

*Institute of Materials, University College, Cardiff, UK*

## D.M. Griffiths & J.C. Wagner

*External Scientific Staff, MRC Team on Occupational Lung Diseases, Llandough Hospital, Penarth, South Glamorgan, UK*

*Summary.* For many years, the main source of asbestos in Cyprus was thought to be the chrysotile mine in the central mountains. When a woman, who had no connection with the mine, developed mesothelioma, it was surprising to discover tremolite asbestos bodies within her lung. However, further studies have shown that tremolite occurs as a contaminant within the chrysotile ore body. In this study we have shown that both chrysotile and tremolite can be found in domestic and environmental samples throughout the mountain region; in particular, numerous fine fibres of both materials are present in stucco. Preliminary radiological studies have shown pleural disease in the village population and 5 out of 13 known cases of mesothelioma have arisen in persons unconnected with the mine. This suggests an environmental contribution to asbestos-related disease on the island.

## *Introduction*

In 1964, the Geographical Pathology Committee of the International Union against Cancer made several recommendations on the future direction of investigation into asbestos-related diseases (UICC, 1965). Amongst these, it was suggested that epidemiologists should focus attention on areas where there had been exposure of populations to single fibre types. Workers in industries which use asbestos have invariably been exposed to both amphibole and serpentine asbestos. However, in many instances, mine workers have had pure exposure. There was grave concern that chrysotile might behave in a similar manner to crocidolite (Wagner et al., 1960).

Cyprus has a chrysotile mine in the central mountains but no other appreciable source of industrial asbestos and was therefore considered as one of the possible places for further studies. In 1965, contact was made between the MRC Pneumoconiosis Unit in Penarth and chest physicians in Nicosia. Three years later, lung tissue from a woman with mesothelioma became available for further analysis. This woman had no connection with the mine, though she lived in a village nearby. It was therefore with surprise that amphibole asbestos bodies were identified in a fragment of lung adhering to her tumour. Further investigation showed that this amphibole was tremolite and that there were tremolite inclusions within the ore body of the mine. However, this still did not satisfactorily explain her disease. Unfortunately, 1974 saw hostilities on the island which disrupted communications. Investigations were resumed in 1977 and in the subsequent 10 years a further 13 cases of mesothelioma have been positively identified. Five of these cases have occurred in chrysotile mine workers, 3 cases are miners' wives, and the remainder have had no occupational exposure to asbestos (McConnochie *et al.*, 1987). The incidence of asbestos-related disease in the community is unknown but we report some preliminary information. However, the main purpose of this study is to identify the type and source of asbestos to which these people have been exposed.

## Materials and methods

### Fibre analysis

Fibres analysed have been obtained from 4 sources. Human lung tissue suitable for study has been obtained from a mine worker with mesothelioma, another with asbestosis, a farmer with lung cancer and a city-dwelling housewife. Six samples of sheep lung have been obtained from animals which grazed for several years around villages in the vicinity of the mine. These specimens were coded and mixed with other material to disguise their origin from the examining microscopist. All lung tissue samples were digested by the usual method (Pooley & Clark, 1979). Wet specimens were examined, after standard preparation, using an analytical transmission electron microscope. Numbers of fibres per g and fibre dimensions were noted. Environmental specimens were collected from several sources. Dust was obtained from rafters of the roof eaves of houses in villages close to the mine. Samples of residue were obtained from the beds of the Kouris and Limnatis rivers, which dry up during the summer months (Figure 1). The Kouris rises in the Troodos Mountains and runs through the mine, while the Limnatis flows several miles to the east. Samples were also taken from the estuary where both rivers run into Episkopi Bay. Fibre analysis has also been performed on stucco which is used to whitewash local houses.

### X-ray reading

The opportunity arose during a visit to the island for one of the authors to view radiographs from two different sources. The total workforce of the mine is now around 350 and each worker has an annual X-ray. Of these films, 118 from the current workforce were examined. This was not a random sample, though there was no reason

### Fig. 1. Villages and river beds where samples were collected

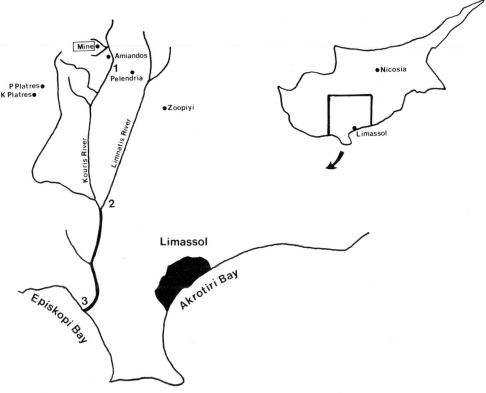

Source: McConnochie et al. (1987)

to suspect any particular bias. In addition, 43 recall films from a recently completed mass miniature radiography (MMR) survey of 966 villagers from the area around the mine were examined. Any MMR film thought to be abnormal resulted in a recall and full-size chest X-ray.

## Results

### Fibre analysis

Human lung tissue contained chrysotile and tremolite (Table 1). Sheep lung also contained a small amount of crocidolite (Table 2, previously reported, see McConnochie et al., 1987). Environmental samples all contained asbestos (Table 3). Antigorite, a non-fibrous mineral, was widespread. Specimens of stucco contained numerous chrysotile and tremolite fibres. In many instances the chrysotile had formed very fine fibrils, even though preparation had not entailed crushing. Tremolite found was generally in fine fibrous form (Figures 2-5).

## Table 1. Asbestos content of human lung

| Diagnosis | Dry weight (mg) | Asbestos content ($f/g \times 10^6$) | | |
|---|---|---|---|---|
| | | Chrysotile | Tremolite | Amosite |
| Mesothelioma | 20 | 115 | 220 | - |
| Asbestosis | 25 | 175 | 38 | 0.9 |
| Lung cancer | 6 | 1212 | 72 | - |
| Control | 5 | 46 | - | 5 |

## Table 2. Asbestos content of sheep lung[a]

| Dry weight (mg) | Asbestos content ($f/g \times 10^6$) | | |
|---|---|---|---|
| | Chrysotile | Tremolite | Crocidolite |
| 115 | 39.7 | 0.02 | 0.02 |
| 169 | 7.8 | 0.56 | 0.07 |
| 143 | 15.7 | 0.33 | 0.49 |
| 91 | 78.5 | 7.00 | 0.52 |
| 102 | 26.6 | 1.39 | 0.69 |
| 125 | 25.2 | 0.57 | 0.19 |

[a]Source: McConnochie et al. (1987).

## Table 3. Asbestos content of environmental samples[a]

| Source | Chrysotile | Antigorite | Tremolite |
|---|---|---|---|
| All areas of mine, including general office dust, 1983 | Yes | Yes | No |
| Dust from roof eaves, 1983: | | | |
|   Kato Amiandos | Yes | Yes | Yes |
|   Pelendria | Yes | Yes | Yes |
| River-beds: | | | |
|   Site No. 1 | Yes | Yes | No |
|   Site No. 2 | Yes | No | Yes |
|   Site No. 3 | Yes | Yes | Yes |

[a]See Figure 1 for location of river-bed sampling sites.

**Fig. 2. Electron micrograph of stucco from a village house 2 miles from the mine**

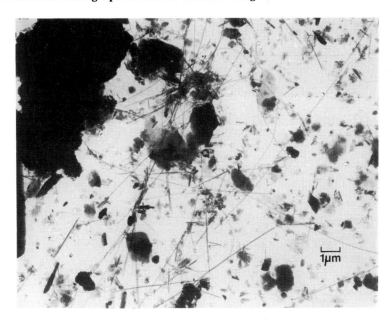

The stucco contains chrysotile and tremolite fibres. ×10 000

**Fig. 3. Electron micrograph of stucco from a village house one mile from the mine**

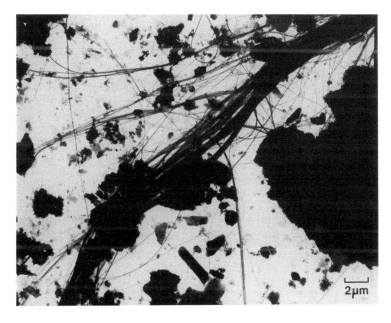

The stucco contains long tremolite fibres. ×15 000

**Fig. 4. Electron micrograph of stucco from a village house 4 miles from the mine**

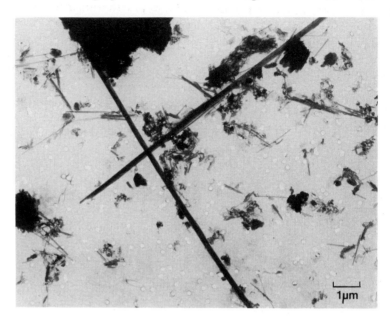

The stucco contains chrysotile and tremolite fibres. ×6500

**Fig. 5. Electron micrograph of stucco from a village house 2 miles from the mine**

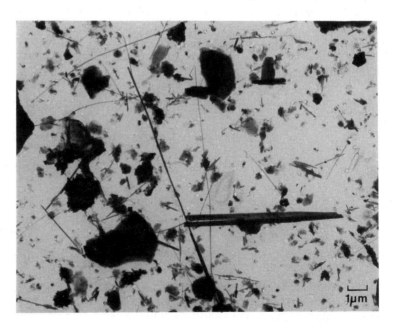

The stucco contains chrysotile and tremolite fibres. ×10 000.

**X-ray reading**

No ILO standard films were available, and X-rays seen were therefore examined for the presence or absence of pleural and interstitial disease. All the MMR recall films were abnormal; 32 belonged to former asbestos mine workers and 11 to villagers with no connection with the mine. The villagers all had pleural disease; none was considered to have interstitial disease. Conversely, 21 of the former workers were thought to have evidence of interstitial disease, with 13 of these exhibiting pleural disease as well. The remaining 10 had pleural disease alone. The 118 films belonging to the current workforce were examined in the same way; 14 films showed evidence of both pleural and interstitial disease. This information is of little worth in its own right but it does suggest that proper evaluation may be fruitful.

## Discussion

The Troodos mountains within which the mine is situated were originally part of the sea bed. During their development, conditions were ideal for the formation of serpentine asbestos and for the amphibolization of other rock. This process produced chrysotile, picrolite, tremolite and nephrite throughout a wide area. In a comprehensive geological survey carried out in the 1950s, Wilson and Ingham (1959) reported veins of tremolite which were up to three feet thick. These veins could be found close to main roads in the mountain region.

Chrysotile was not commercially exploited at the mine until 1904. As the fibres are generally short, they could not be used for textiles and only found application when asbestos was added to cement. The asbestos is obtained from an open pit which is on several levels. The seams are not very thick, on average less than 1 cm. Thus large volumes of rock must be crushed to make the operation commercially viable. Mining has to stop during the winter months because of adverse weather conditions but the mill continues to operate using stored asbestos. The workforce is very stable. In the 1930s there were around 4000 employees, but this had dropped to around 1000 by the late 1950s. In the mid 1980s, the number of employees had dwindled to around 350. Part of this reduction in the workforce was the result of mechanization. Many workers and their families lived in a specially constructed village on top of the mine, but this is no longer used as the required number of workers can be drawn from the local village labour force.

In 1969, Constantinides et al. (1972) carried out a survey of 645 current and former employees of the mine. This showed that pleural thickening was related mainly to age rather than dust exposure and that ill-defined cardiac outline was also strongly age-related. The prevalence of pleural calcification rose sharply with age, again with no relation to dust exposure, as did that of irregular opacities. Of the population studied, 2.3% had radiological disease which reached category 2 or more on the scale used in the ILO classification of radiographs for pneumoconiosis (1980). There was some question about the accuracy of the dust measurements used in the analysis, but the overall finding was that the mine produced little disease. The opposite view, in the light of current knowledge, is that environmental exposure may have given rise to some of these results.

By the mid 1970s it had become clear that several forms of asbestos occurred naturally, but, unfortunately, further studies had to be abandoned due to the political situation on the island.

Tremolite is found throughout the world. It can occur as coarse flake through to fine fibre, but in all forms it is a very brittle material with little commercial value. It has been found as a contaminant in other chrysotile deposits. Pooley (1976) found tremolite fibre in 11 out of 20 cases of asbestosis from the Canadian chrysotile mining industry. In 7 cases, amphibole was the dominant fibre, suggesting preferential accumulation in the lung after exposure to mixed dust clouds.

Churg et al. (1984) studied 6 Canadian chrysotile workers with mesothelioma and found more tremolite-group amphiboles in their lung tissue than chrysotile. They therefore suggested that tremolite-group amphiboles may be important in the pathogenesis of mesothelioma.

Wagner et al. (1982) have shown that tremolite is carcinogenic in the animal model. More recently, Langer et al. (1987) have reported the association of pleural mesothelioma with exposure to tremolite-containing whitewash in north-western Greece.

In Cyprus, we found tremolite and chrysotile in all the domestic dust samples analysed and both forms of asbestos were found in dry river beds. Because of the difficulties in obtaining human lung for analysis, we examined sheep lungs from animals that had lived within 5 miles of the mine. This technique was used to good effect in Turkey by Baris et al. (1987), who showed that zeolite fibres found within sheep lung correlated with the geographical distribution of disease. We found tremolite and chrysotile fibres in all sheep lung samples analysed. There was also a low concentration of crocidolite fibre, which may be the result of recent industrial operations. A similar concentration was found in the Turkish sheep. The stucco examined came from 3 villages within a 10-mile radius of the mine. This material was heavily contaminated with fine fibrils of chrysotile and long, thin tremolite fibres.

To date, 13 cases of mesothelioma have been confirmed on the island. In a recent workshop on the biological effects of chrysotile, Berry (1986) estimated that, at most, only 10 cases of mesothelioma reported in the literature could be attributed without doubt to commercial-grade chrysotile. Five of our cases had no known contact with the mine, though 2 lived in the vicinity. Either the chrysotile on Cyprus is behaving differently to that found elsewhere or some other factor is at work. We suspect the amphibole, tremolite, is playing a significant role in the etiology of these cases.

Studies in Cyprus have so far not allowed us to come to any definite conclusions but evidence is accumulating which suggests that naturally occurring tremolite may be causing disease.

## *Acknowledgements*

We should like to thank the International Agency for Research on Cancer for financial support.

# References

Baris, Y., Simonato, L., Artvinli, M., Pooley, F., Saracci, R., Skidmore, J. & Wagner, J.C. (1987) Epidemiological and environmental evidence of the health effects of exposure to erionite fibres: a four-year study in the Cappadocian region of Turkey. *Int. J. Cancer, 39*, 10-17

Berry, G. (1986) Conclusions of the epidemiology and statistics group. In: Wagner, J.C., ed., *Accomplishments in Oncology, The Biological Effects of Chrysotile*, Philadelphia, Lippincott, pp. 158-160

Churg, A., Wiggs, B., Depaoli, L., Kampe, B. & Stevens, B. (1984) Lung asbestos content in chrysotile workers with mesothelioma. *Am. Rev. Respir. Dis., 130*, 1042-1045

Constantinides, M.G., Kronides, G. & Gilson, J.C. (1972) Asbestos health hazards in Cyprus. In: *Proceedings of the East Mediterranean Medical Congress and 15th Annual Meeting of the British Medical Association, Cyprus, 11-15 April 1972*

International Labour Office (1980) *Classification of Radiographs for Pneumoconiosis*, Geneva

Langer, A.M., Nolan, R.P., Constantopoulos, S.H. & Moutsopoulos, H.M. (1987) Association of Metsovo lung and pleural mesothelioma with exposure to tremolite-containing whitewash. *Lancet, i*, 965-967

McConnochie, K., Simonato, L., Mavrides, P., Christofides, P., Pooley, F.D. & Wagner, J.C. (1987) Mesothelioma in Cyprus — the role of tremolite. *Thorax, 42*, 342-347

Pooley, F.D. (1976) An examination of the fibrous mineral content of asbestos lung tissue from the Canadian chrysotile mining industry. *Environ. Res., 12*, 281-298

Pooley, F.D. & Clark, N.J. (1979) Quantitative assessment of inorganic fibrous particulates in dust samples with an analytical transmission electron microscope. *Ann. Occup. Hyg., 22*, 253-271

UICC (1965) Report and recommendations of the working group on asbestos and cancer. Convened under the auspices of the Geographical Pathology Committee of the International Union against Cancer (UICC). *Br. J. Ind. Med., 22*, 165-171

Wagner, J.C., Sleggs, C.A. & Marchand, P. (1960) Diffuse pleural mesothelioma in the North Western Cape Province. *Br. J. Ind. Med., 17*, 260-271

Wagner, J.C., Chamberlain, M., Brown, R.C., Berry, G., Pooley, F.D., Davies, R. & Griffiths, D.M. (1982) Biological effects of tremolite. *Br. J. Cancer, 45*, 352-360

Wilson, R.A.M. & Ingham, F.T. (1959) *The Geology of the Xeros-Troodos Area with an Account of the Mineral Resources* (Geological Survey Department, Cyprus Memoir No. 1), Margate, Eyre and Spottiswood

# EPIDEMIOLOGICAL OBSERVATIONS ON MESOTHELIOMA AND THEIR IMPLICATIONS FOR NON-OCCUPATIONAL EXPOSURE

J.C. McDonald, P. Sébastien, A.D. McDonald & B. Case

*Dust Disease Research Unit, School of Occupational Health, McGill University, Montreal, Canada*

Primary malignant mesothelial tumours, because of their highly specific causal relationship to mineral fibre exposure and distinctive diagnostic features, are particularly valuable indicators of environmental impact. Systematic ascertainment of these tumours through pathologists has been suggested as a cost-effective method of monitoring the asbestos hazard, both occupational and non-occupational (McDonald, 1979). This approach was first adopted by our group in Canada in 1966 and has been maintained, somewhat intermittently, ever since. The initial survey covered a period beginning in 1960 and the most recent, still under analysis, has brought the period of observation up to 1984. For one year, 1972, the survey was extended to cover the whole of North America — Canada and the USA.

At first, the objective was to test hypotheses concerning the relation of mesothelioma to asbestos, certain other occupational exposures and cigarette smoking (McDonald *et al.*, 1970). A second phase focused on the increasing incidence of the disease in Canada in its relation to occupation (McDonald & McDonald, 1973). Extensive case-referent enquiries based on the North American survey of 1972 showed a very high relative risk associated with insulation work, some evidence of indirect occupational risk in households, but none from the general environment (McDonald & McDonald, 1980). These data had also been used to demonstrate the concentration of cases in shipyard cities of Europe and North America (McDonald & McDonald, 1977). Because of strong epidemiological indications from these and other studies that amphiboles, not chrysotile, constituted the greater hazard, subsequent surveys concentrated on the use of electron-microscopic analysis of lung tissue for mineral fibre content. These investigations (McDonald, 1980), in parallel with studies in the United Kingdom (Jones *et al.*, 1980), provided further evidence of the importance of differences of fibre type and demonstrated the potential of this approach, which our most recent survey, begun in 1982, was designed to exploit.

In the present paper, data of two kinds, derived from surveys of mesothelioma in general populations, will be reviewed. The first set of data relate to incidence trends in Canada and certain other countries; the second are based on the preliminary findings from our most recent survey on lung dust analysis. The implications of both sets of data will then be discussed.

## Incidence trends

McDonald (1985) drew attention to the importance of sex differences in mesothelioma mortality. It has been evident for at least 25 years that these tumours are more common in men than women; our surveys in Canada had shown that this reflected the more frequent occupational exposure of men to asbestos. Etiologically, cases fall into 4 main categories — 3 related to asbestos exposure (occupational, household and environmental) and a background group due to unknown causes. Whereas occupational asbestos exposure is the dominant factor for males, the other three categories are relatively more important for females.

The commercial exploitation of asbestos began in Canada, Russia and South Africa in the 1890s. Consumption rose very steeply during the first half of this century, flattened out and, from about 1970, declined sharply. All things being equal, we would expect the incidence of mesothelioma to approximately parallel the trend in consumption, after a latency period of 30–40 years. However, this simple correlation might be affected by the pattern of use of chrysotile and the amphibole fibre types. Only now might we begin to see the effect of dust-control measures introduced after the end of the Second World War. In the USA, cases resulting from the sharp increase in consumption in the First World War would have become manifest in the 1950s. This was the period when, in fact, cases were first noted in the mining region of South Africa (Wagner et al., 1960), the Cape Asbestos plants in London (Newhouse & Thompson, 1965) and Hamburg (Bohlig et al., 1970), and the shipyards of Belfast (Elmes et al., 1965) and Tyneside (Ashcroft, 1973).

With these facts in mind, we proposed a simple predictive model, illustrated in Figure 1, with which past and future incidence data could be compared. This was based on the assumption that, prior to the impact of exposure to asbestos directly or indirectly related to its industrial use, there existed a background incidence of the tumour, perhaps equal in the two sexes. As men were much more frequently exposed occupationally than women, the trends would be expected to separate from about 1950 onwards, that for males rising more steeply. However, as some women worked with asbestos and many others were exposed in the home to asbestos dust brought back by husbands and other, usually male, household members, some upward trend in female incidence would also be expected. There then remained the possibility of cases in both males and females attributable to general urban pollution with asbestos fibre, whether from mines and factories, construction sites, automobile brakes, or deteriorating insulation materials in public buildings. The statistical picture would also be affected by changing diagnostic and coding practices; before the 1960s, malignant mesothelioma was scarcely recognized as a cause of death by physicians or pathologists. This may be the reason why mortality in the USA from pleural mesothelioma showed no increase or difference in incidence between the sexes until 1968, whereas these changes must surely have begun at least 10 years earlier. Archer and Rom (1983), who were probably the first to recognize the full significance of these trends, noted that the divergence between the sexes was only evident above age 45.

**Fig. 1. Conceptual model for mesothelioma mortality, assuming complete ascertainment (from McDonald, 1985)**

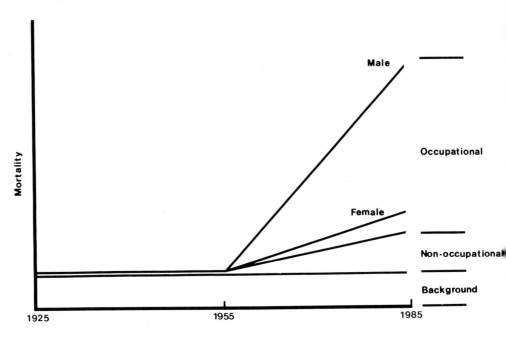

The best set of data for testing the predicted pattern shown in Figure 1 are those available since 1974 from the Surveillance, Epidemiology and End Results (SEER) Program of population-based cancer registries in five states and five city areas in the USA (Biometry Branch, National Cancer Institute, unpublished data), supplemented by figures from the Third National Cancer Survey for 1970–1972. Figure 2, taken from the report of McDonald (1985), shows male and female mortality trends for the period 1970–1980. Since then, a report by Connelly et al. (1987) indicates that the male rate for 1980 was out of line with the prevailing trend; this fell back in 1981 and 1982 to the 1979 level and only regained the 1980 level in 1984. The incidence in females, however, remained much the same for the whole 15-year period. Data from the Connecticut Tumor Registry showed essentially the same pattern, though with a steadily increasing incidence, mainly due to the diverging upward trend in males from about 1950 onwards (Lewinsohn et al., 1980).

Other countries in which national statistics have been published include England and Wales (Gardner et al., 1982) and Finland (Nurminen, 1975), but only for pleural mesothelioma. In the data for England and Wales, mortality in females remained unchanged at an average of 41 deaths per annum over the period 1968–1978 inclusive. In the same period, male deaths increased from 96 in 1968 to 199 in 1978. Elmes & Simpson (1976), using cases of mesothelioma reported to a national panel of pathologists in the United Kingdom, showed an increased frequency of notifications

**Fig. 2.** Mesothelioma incidence in the USA, 1970–1980, age-adjusted to the national population (from McDonald, 1985)

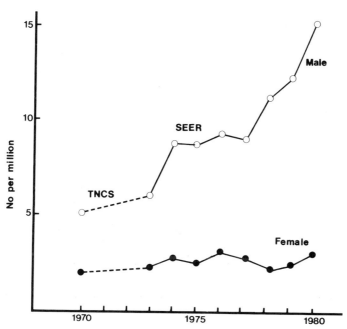

from the mid-1960s, with a coincident elevation of the male trend. During the 10-year period 1960–1969, the annual number of female cases remained much the same. Cases of pleural mesothelioma reported to the Finnish Cancer Registry over the period 1953–1969 showed a similar number of male and female cases until 1966, when the male and female trends began to separate.

There remain the results of our own efforts at systematic ascertainment of mesothelioma through Canadian pathologists. The method used was essentially the same over the 25-year period 1960–1984. In each survey, all pathologists were contacted first by mail or, if necessary, by telephone and asked whether they had seen any fatal cases of primary malignant mesothelial tumour, diagnosed at autopsy or biopsy.

A reply was obtained from all but a few of them. Those who had seen a case were visited by a physician, the details were recorded and one or more referents selected from the same register. Cases were accepted initially if, 'on balance', the pathologist considered the diagnosis more likely than not. The first enquiry was sent in 1966 and it is likely that there was under-reporting for the years 1960–1964. The last enquiry was made in 1984 and the results are therefore incomplete for that year; moreover, British Columbia was omitted because a separate survey was being conducted there by Dr Andrew Churg. The annual numbers of reported cases by sex are shown in Figure 3; the figures for 1982–1983 include a 12.5% correction (based on population) to allow for the omission of British Columbia. It can be seen that male incidence has continued to follow a steep upward trend whereas female incidence has shown no appreciable change.

**Fig. 3. Fatal cases of malignant mesothelioma reported by pathologists in Canada, 1961–1983**

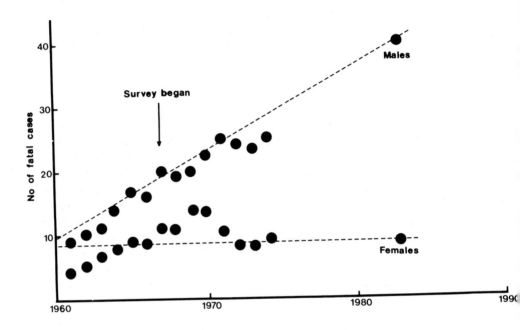

Figures for 1961–1975 are 3-year moving averages, that for 1982–1983 is the average for those years (based on McDonald et al., unpublished data)

## Lung dust analysis

With rare exceptions, e.g., in the immediate vicinity of the mines and mills of Thetford Mines, Quebec, only chrysotile is found in air samples from the general external environment. At Thetford, trace amounts of fibrous tremolite (up to 1%) have been found (Sébastien et al., 1986). Apart from buildings in which amphiboles have been used, the same applies to indoor air. The special circumstances of exposure in the homes of asbestos workers will not be considered here; such exposures were frequently high, often at levels no longer permitted in the workplace (Nicholson et al., 1980). In assessing the hazard of mesothelioma as a consequence of general environmental exposure, account must therefore be taken of fibre type and, in particular, of the degree of risk associated with uncontaminated chrysotile. As clinical epidemiological studies lack the required precision and, in any event, have not so far evaluated the quantitative relationship between mesothelioma and exposure to any type of asbestos fibre, we have chosen to study mineral fibre concentrations in lungs at autopsy of mesothelioma cases and referents (McDonald et al., unpublished data).

Using the fatal cases of mesothelioma in Canada (less British Columbia) ascertained through pathologists, lung tissue samples were obtained from autopsied cases over the period 1980-1984. The referents selected from the same autopsy register had died from causes other than respiratory or malignant disease, and were matched for sex, date of death, type of tissue and, as far as possible, date of birth, with the mesothelioma cases. The paired samples were examined blind and in parallel by both optical microscopy and analytical transmission electron microscopy. The tumours were also classified by Dr W.T.E. McCaughey at the Canadian Tumor Reference Centre. Multivariate analyses indicated that two-thirds of the mesothelioma cases could be attributed to long (>8 $\mu$m) amphibole fibres, including tremolite, but that chrysotile itself made no significant contribution.

## Discussion

During the past 20-30 years, the pattern of mesothelioma incidence or mortality has been remarkably similar in the United States, Canada, the United Kingdom, and perhaps also in Finland. The data suggest that, until the 1950s, there were about 2 deaths from this cause per million population in both males and females. Since then, male mortality has risen steeply, roughly in parallel with the industrial consumption of asbestos 30-40 years earlier. It thus seems likely that this trend will continue upward, though hopefully with a diminishing gradient, for another 20-30 years. Female mortality, on the other hand, has shown little or no increase, in spite of the fact that: (*a*) there have certainly been cases in women as a result of both direct and indirect (household) occupational exposure; and (*b*) physicians and pathologists are more ready to recognize and diagnose mesothelioma now than in the past. Thus there would appear to be no room left for cases attributable to general non-occupational environmental exposure. Various estimates, ranging from about 4 to 50 deaths per million, have been made for the lifetime risk of mesothelioma associated with average urban asbestos fibre concentrations in the USA (about 5 ng/m$^2$) by extrapolation from assumed linear exposure-response models (see McDonald, 1985). However, the deaths occurring now are of persons who can only have been exposed for about half a lifetime (i.e., since, say, 1945) and represent a cohort only about half as numerous as the current population. From these very approximate estimates, we would therefore expect about 1-12 deaths per million annually, based on the current population. The present Canadian female population of 10 million might therefore expect to experience 10-120 *excess* mesothelioma deaths per annum, whereas, in 1982-1983, we ascertained an average of 8 cases per annum in all (95% confidence limits, 3.2-14.4), but with no evidence of an excess. The American SEER data (Figure 2), based on more complete ascertainment than our Canadian surveys, show an average annual incidence in females of about 2.5 per million since 1980, with little evidence of an increase since 1970. Risk estimates based on linear extrapolation would have forecast an annual *excess* of 1-12 deaths. Thus the numbers of cases observed do not support even the lowest risks forecast.

This conclusion, derived by deduction from observed incidence data, is supported inductively by our findings on retained asbestos fibres in lung tissue. The relative risk of mesothelioma associated with long chrysotile fibres, although slightly raised, does not differ significantly from unity; long amphibole fibres, on the other hand, could explain two-thirds of the cases. Bearing in mind how few asbestos fibres in the general environment reach a length of 8 $\mu$m, and that they are seldom other than chrysotile, forecasting methods which take no account of fibre type and little of fibre length may well overestimate the risk.

## References

Archer, J.E. & Rom, W.N. (1983) Trends in mortality of diffuse malignant mesothelioma of pleura. *Lancet, ii*, 112-113

Ashcroft, T. (1973) Epidemiological and quantitative relationships between mesothelioma and asbestos on Tyneside. *J. Clin. Pathol.*, 26, 832-840

Bohlig, H., Dabbert, A.F., Dalquen, P., Hain, E. & Hinz, I. (1970) Epidemiology of malignant mesothelioma in Hamburg — Preliminary report. *Environ. Res.*, 3, 365-372

Connelly, R.R., Spirtas, R., Myers, M.H., Percy, C.L. & Fraumeni, J.F. (1987) Demographic patterns for mesothelioma in the United States. *J. Natl Cancer Inst.*, 78, 1053-1059

Elmes, P.C., McCaughey, W.T.E. & Wade, O.L. (1965) Diffuse mesothelioma of the pleura and asbestos. *Br. Med. J.*, 1, 350-353

Elmes, P.C. & Simpson, M.J.C. (1976) The clinical aspects of mesothelioma. *Q. J. Med., 179*, 427-449

Gardner, M.J., Acheson, E.D. & Winter, P.D. (1982) Mortality from mesothelioma of the pleura during 1968-78 in England and Wales. *Br. J. Cancer*, 46, 81-88

Jones, J.S.P., Roberts, G.H., Pooley, F.D., Clark, N.J., Smith, P.G., Owen, W.G., Wagner, J.C., Berry, G. & Pollock, D.J. (1980) The pathology and mineral content of lungs in cases of mesothelioma in the United Kingdom in 1976. In: Wagner, J.C., ed., *Biological Effects of Mineral Fibres (IARC Scientific Publications No. 30)*, Lyon, International Agency for Research on Cancer, pp. 187-199

Lewinsohn, H.C., Meigs, J.W., Teta, M.J. & Flannery, J.T. (1980) The influence of occupational and environmental asbestos exposure on the incidence of malignant mesothelioma in Connecticut. In: Wagner, J.C., ed., *Biological Effects of Mineral Fibres (IARC Scientific Publications No. 30)*, Lyon, International Agency for Research on Cancer, pp. 665-660.

McDonald, A.D. (1979) Mesothelioma registries in identifying asbestos hazards. *Ann. N.Y. Acad. Sci*, 330, 441-454

McDonald, A.D. & McDonald, J.C. (1973) Epidemiologic surveillance of mesothelioma in Canada. *Can. Med. Ass.*, 109, 359-362

McDonald, A.D. & McDonald, J.C. (1980) Malignant mesothelioma in North America. *Cancer*, 46, 147-154

McDonald, A.D., Harper, A., El Attar, O.A. & McDonald, J.C. (1970) Epidemiology of primary malignant mesothelial tumours in Canada. *Cancer*, 26, 914-919

McDonald, J.C. (1980) Asbestos-related disease: an epidemiological review. In: Wagner, J.C., ed., *Biological Effects of Mineral Fibres (IARC Scientific Publications No. 30)*, Lyon, International Agency for Research on Cancer, pp. 587-601

McDonald, J.C. (1985) Health implications of environmental exposure to asbestos. *Environ. Health Perspect.*, 62, 319-328

McDonald, J.C. & McDonald, A.D. (1977) Epidemiology of mesothelioma from estimated incidence. *Prev. Med.*, 6, 426-446

Newhouse, M.L. & Thompson, H. (1965) Mesothelioma of pleura and peritoneum following exposure to asbestos in the London area. *Br. J. Ind. Med.*, 22, 261-269

Nicholson, W.J., Rohl, A.N., Weisman, I. & Selikoff, I.J. (1980) Environmental asbestos concentrations in the United States. In: Wagner, J.C., ed., *Biological Effects of Mineral Fibres (IARC Scientific Publications No. 30)*, Lyon, International Agency for Research on Cancer, pp. 823-827

Nurminen, M. (1975) The epidemiologic relationship between pleural mesothelioma and asbestos exposure. *Scand. J. Work Environ. Health*, 1, 128-137

Sébastien, P., Plourde, M., Robb, R., Ross, M., Nadon, B. & Wypnuk (1986) *Ambient Air Asbestos Survey in Quebec Mining Towns, Part 2, Main Study Report (Environment Canada Report No. EPS 5/AP/RQ-2E)*, Montreal, Supply and Services Canada

Wagner, J.C., Sleggs, C.A. & Marchand, P. (1960) Diffuse pleural mesothelioma and asbestos exposure in the North Western Cape Province. *Br. J. Ind. Med.*, 17, 260-271

# EPIDEMIOLOGICAL STUDIES ON INGESTED MINERAL FIBRES: GASTRIC AND OTHER CANCERS

## M.S. Kanarek

*Department of Preventive Medicine, Medical School
and the Institute for Environmental Studies,
University of Wisconsin-Madison, Madison, WI, USA*

*Summary.* The epidemiological studies of ingested asbestos fibres conducted world-wide are reviewed and evaluated. Most of the studies have been done in the United States and Canada and have involved community exposures via natural contamination of drinking-water supplies. One or more studies found associations between asbestos fibres in drinking-water supplies and cancer incidence or mortality associated with many body sites, including oesophagus, stomach, small intestine, colon, rectum, gall-bladder, lungs, pancreas, peritoneum, pleura, prostate, kidneys, brain and thyroid. Each study has methodological limitations or weaknesses that limit the ability to assess risk from ingested asbestos. There is no agreement between the results of the various studies, but an association between ingested asbestos fibres and cancer of the stomach and pancreas has been found with some degree of consistency.

## *Introduction*

There have been a number of recent reviews of the studies on the non-occupational ingestion of asbestos fibres. New research in this area has slowed down in the last 5 years, due to a decline in research funding for this subject, but there have been a number of summaries. Especially noteworthy are the excellent contributions by the Working Group for the DHSS Committee (1987), Levine (1985), Toft *et al.* (1984), National Research Council (1983), Erdreich (1983), and Marsh (1983).

## *Review of studies*

The following 11 major epidemiological studies of the possible association between asbestos fibres in drinking-water supplies and cancer incidence or mortality are reviewed here from 5 areas of the United States and Canada: Mason *et al.* (1974), Levy *et al.* (1976), Wigle (1977), Harrington *et al.* (1978), Meigs *et al.* (1980), Sigurdson *et al.* (1981), Toft *et al.* (1981), Kanarek *et al.* (1980), Conforti *et al.* (1981), Polissar *et al.* (1982), Sigurdson (1983) and Polissar *et al.* (1984). Other potentially relevant studies were studied, but not included, because of major defects. Table 1 summarizes the exposure details for each of the 5 areas, including fibre type, counts, population and duration. Table 2 summarizes the findings for gastrointestinal (GI) cancer sites and Table 3 for non-GI cancer sites from these studies.

## Table 1. Asbestos exposures in drinking-water in various studies[a]

|  | Duluth | Connecticut | Quebec | San Francisco Bay Area | Puget Sound |
|---|---|---|---|---|---|
| Type of asbestos | Amphibole | Chrysotile | Chrysotile | Chrysotile | Chrysotile |
| Concentration (f/l) | $1\text{-}30 \times 10^6$ | $\text{BDL-}0.7 \times 10^6$ | $1.1\text{-}1300 \times 10^6$ | $0.025\text{-}36 \times 10^6$ | $7.3\text{-}206.5 \times 10^6$ |
| Population exposed | 100 000 | 576 800 | 420 000 | 3 000 000 | 200 000 |
| Maximum duration of exposure (years) | 15-20 | 23-44 | >50 | >40 | >40 |

[a]BLD, below detectable limit.

## Duluth, Minnesota

In 1974, Cook *et al.* and Nicholson reported the presence of 1–30 million fibres per litre (f/l) of amphibole-like asbestos in the Duluth, Minnesota, drinking-water supply. The contamination was a result of the dumping by the Reserve Mining Company of taconite iron ore tailings into Lake Superior at Silver Bay, Minnesota, 50 miles north-east of Duluth, from 1955 onwards.

Mason *et al.* (1974) compared age-adjusted 1950–1969 cancer death rates for Duluth with those for the state of Minnesota and those for Minneapolis. Significantly higher rates were found for stomach and rectal sites in both sexes and for pancreas in females. Obvious defects in this study include the fact that less than 15 years of exposure were covered and that there was lack of control for major potential confounding factors, such as ethnicity, occupation and others.

Levy *et al.* (1976) compared 1969–1971 Third National Cancer Survey (TNCS) GI cancer incidence rates for Minneapolis and St Paul with the rates found by these investigators for Duluth. In general, the Duluth rates were lower, but increases were observed in Duluth for stomach in males, and for pancreas and 'peritoneum, retroperitoneum and abdomen not otherwise specified' in both sexes. Since the contamination of the water supply did not start until 1955, the cancer data in this study were related only to 14–16 years of exposure.

Sigurdson *et al.* (1981, 1983) extended the study of Duluth 1969–1971 cancer incidence rates to include rates for 1974–1976 and 1979–1980, and added lung cancer. There were no gastrointestinal or other abdominal sites consistently significantly higher in Duluth than in the control areas. The male stomach cancer rate increased with time in Duluth (18.9 in 1969–1971; 22.0 in 1972–1974) and was higher than for either Minneapolis (17.2) or St Paul (14.1).

## Connecticut

The studies in Connecticut involved exposures from asbestos-cement drinking-water pipe. Harrington *et al.* (1978) compared tumour incidence from the Connecticut tumour registry (1935–1973) for townships using asbestos-cement drinking-water

Table 2. Studies of asbestos fibres in drinking-water and gastrointestinal (GI) cancer risk[a]

| Gastrointestinal cancer site | ICD 7th Rev. Codes | Duluth | | | Connecticut | | Quebec | | San Francisco Bay area | | Puget Sound | |
|---|---|---|---|---|---|---|---|---|---|---|---|---|
| | | Mason et al. | Levy et al. | Sigurdson et al. | Harrington et al. | Meigs et al. | Wigle | Toft et al. | Kanarek et al. | Conforti et al. | Polissar et al. (1982) | Polissar et al. (1984) |
| All GI | 150-159 | ++ | -- | 00 | NS | NS | 00 | +0 | ++ | ++ | NS | 00 |
| Oesophagus | 150 | +- | 00 | 00 | NS | NS | 00 | 00 | 0+ | ++ | 00 | NS |
| Stomach | 151 | ++ | +0 | 00 | 00 | 00 | +0 | +0 | ++ | ++ | 00 | +- |
| Small intestine | 152 | NS | 00 | 00 | NS | NS | NS | NS | 00 | 00 | ++ | NS |
| Colon | 153 | 00 | 00 | 00 | 00 | 00 | 00 | 00 | 00 | +0 | 00 | 00 |
| Rectum | 154 | ++ | -- | 00 | 00 | 00 | 00 | 00 | 00 | 00 | 00 | 00 |
| Biliary/liver | 155-156A | 00 | 00 | 00 | NS | NS | NS | NS | 00 | 00 | 00 | NS |
| Gall-bladder | 155.1 | NS | 00 | 00 | NS | NS | NS | NS | 0+ | 00 | 00 | 00 |
| Pancreas | 157 | 0+ | ++ | 0+ | NS | +0 | 0+ | 00 | 0+ | ++ | 00 | 00 |
| Peritoneum | 158 | NS | 00 | 00 | NS | NS | NS | NS | ++ | 0+ | 00 | NS |

[a]Male results are listed on the left, female results on the right in each column; +, positive association; 0, no association; NS, not studied.

Table 3. Studies of asbestos fibres in drinking-water and non-gastrointestinal (GI) cancer risk[a]

| Non-gastro-intestinal cancer site | ICD 7th Rev. Codes | Duluth Mason et al. | Levy et al. | Sigurdson et al. | Connecticut Harrington et al. | Meigs et al. | Quebec Wigle | Toft et al. | San Francisco Bay area Kanarek et al. | Conforti et al. | Puget Sound Polissar et al. (1982) | Polissar et al. (1984) |
|---|---|---|---|---|---|---|---|---|---|---|---|---|
| Buccal cavity, pharynx | 140-148 | NS | NS | NS | NS | NS | 00 | 00 | NS | NS | 00 | +- |
| Bronchus, trachea, lungs | 162-163 | +0 | NS | 00 | NS | 00 | +0 | +0 | +0 | 00 | 00 | 00 |
| Pleura | 162.2 | NS | NS | NS | NS | NS | NS | NS | 0+ | 0+ | NS | NS |
| Prostate | 177 | NS | NS | NS | NS | NS | 0 | 0 | 0 | + | + | NS |
| Kidneys | 180 | NS | NS | NS | NS | 00 | 00 | 00 | 0+ | 00 | 00 | 00 |
| Bladder | 181 | NS | NS | NS | NS | 00 | 00 | 00 | 00 | 00 | 00 | 00 |
| Brain/CNS | 193 | 00 | NS | NS | NS | NS | 00 | 00 | 00 | 00 | +- | NS |
| Thyroid | 194 | NS | NS | NS | NS | NS | NS | NS | 00 | 00 | ++ | NS |
| Leukaemia | 204 | 00 | NS | NS | NS | NS | 00 | 00 | 00 | 00 | + | NS |

[a]Male results are listed on the left, female results on the right in each column; +, positive association; -, negative association; 0, no association; NS, not studied.

supply pipe as compared with townships which used other types of pipe. The authors found no associations between tumour incidence and use of asbestos-cement pipe. Meigs et al. (1980) extended the examination of township asbestos-cement pipe use by including actual measurements of asbestos fibres and a much more detailed epidemiological analysis, including control for socioeconomic status. Few associations were found, but exposures to fibres in drinking-water in Connecticut appeared to have been extremely low. A suggestive association was seen for male pancreatic cancer.

### Quebec

Studies have been conducted in the chrysotile asbestos mining areas of Canada. Wigle (1977) studied cancer mortality in 22 communities in Quebec after grouping them by assumed exposure to asbestos fibres in drinking-water based on location and geology. Significant associations were found for stomach and lung in males and pancreas in females. The author dismissed this finding because of lack of consistency between the sexes and the possible contributions from occupational exposures associated with the asbestos mines.

Toft et al. (1981) studied Canadian measurements of chrysotile fibres in drinking-water samples and compared them with cancer mortality rates in 71 communities. They found statistically significant associations for stomach and lung in males for two of the asbestos mining areas. The authors attributed these findings to occupational (mining) exposures.

### San Francisco Bay Area, California

There is wide variation of exposure to naturally occurring chrysotile fibres in the different drinking-water supplies in the San Francisco Bay Area. Kanarek et al. (1980) reported comparisons of asbestos measurements in census tracts (geographical subdivisions containing some 4000-10 000 residents) and 1961-1971 TNCS cancer incidence rates for the San Francisco-Oakland Standard Metropolitan Statistical Area. Significant associations, after controlling for age, race, marital status, socioeconomic status, occupation and ethnicity, were found for peritoneal and stomach in both sexes; gall-bladder, pancreas, oesophagus, pleura and kidney in females; and lung in males.

Conforti et al. (1981) reported a re-analysis of the above study using 'super tracts', i.e., groups of census tracts, so that the geographical boundaries were the same in both the 1970 and 1980 censuses and it was thus possible to extend the cancer data to 1974. The same variables were controlled as in the study by Kanarek et al. (1980). Positive associations were found for oesophagus, stomach and pancreas in both sexes, colon and prostate in males; and pleura and peritoneal sites in females.

Tarter et al. (1983) suggested that population density might be an important potential confounder in the studies by Kanarek et al. (1980) and Conforti et al. (1981). Conforti (1983) re-analysed the data from the 1981 study, adding population density as a variable in the regression analysis. After controlling for population density, however, the same sites were still significantly associated with asbestos fibre counts. It appeared that population density was distributed in such a way across the San Francisco Bay Area that it had little effect on the associations previously observed in the studies by Kanarek et al. (1980) and Conforti et al. (1981).

**Puget Sound, Washington**

As in the San Francisco Bay Area, there are high exposures to naturally occurring chrysotile fibres in certain drinking-water supplies in the Puget Sound (which includes Seattle) Area. Polissar et al. (1982) reported the results of an analysis of cancer incidence and asbestos fibre measurements in that area. No statistically significant associations were found after controlling for age, socioeconomic status, occupation and population density. A case-control study (Polissar et al., 1984) based on 382 cancer cases and 462 controls showed a significantly increased odds ratio for stomach and pharynx cancer in males. The authors dismiss these findings as possibly the result of chance alone due to the large number of analyses conducted.

## Discussion

**Occupational studies**

There are well proven relationships between occupational exposures to asbestos fibres and excess pleural mesothelioma, peritoneal mesothelioma, and cancers of the lung, oesophagus, stomach, colon-rectum, pharynx and buccal cavity, larynx, kidney (Selikoff et al., 1979) and pancreas (Selikoff & Seidman, 1981). Exposures are primarily through inhalation and the fibres subsequently migrate throughout the body.

The National Research Council (1983) published an extrapolation of the risks of GI cancer, well established in occupational cohorts exposed to asbestos fibres primarily via inhalation, to the hypothetical risks of GI cancer from swallowed asbestos fibres in drinking-water. A linear dose-effect model was assumed and inhaled numbers of fibres in the occupational groups were converted into an equivalent number of fibres swallowed in drinking-water. The resulting equations showed a relative risk of 1.011 per $15 \times 10^6$ fibres swallowed over a 20-year period. It was estimated for men that '... drinking water containing $(1/9.1060) \times 10^6 = 0.11 \times 10^6$ fibres/litre may lead to one GI cancer case per $10^5$ persons exposed over a 70-year lifespan' (National Research Council, 1983, p. 134). The equivalent estimate for females was $0.17 \times 10^6$ f/l.

These risk estimates were then compared with the results from the completed studies in Duluth, Connecticut, San Francisco Bay Area, and Puget Sound Area. They concluded that, given a 1973 average of $8 \times 10^6$ f/l in the Duluth drinking-water supply since 1955, the risk equations would predict a relative risk of 1.004 and '... relative risks of this order of magnitude are far too small for any epidemiological study to detect, let alone differentiate from possible confounding variables' (National Research Council, 1983, p. 135). They label the study in Connecticut as '... fundamentally flawed by the lack of data on actual levels of exposure to asbestos' (National Research Council, 1983, p. 136). Using a highest level estimate of 700 000 f/l for 1950-1973, a relative risk of 1.0006 was derived, which would also be undetectable.

The relative risks for GI tract cancers combined, as found in the San Francisco Bay Area study (Kanarek et al., 1980), were 1.11 for males and females at $8\times10^6$ f/l and 1.28 for males and 1.19 for females at $26\times10^6$ f/l, '... results that are not incompatible with predictions made...' by the National Research Council (1983). Likewise, the results in the Puget Sound Area (Polissar et al., 1982) were fairly compatible with those calculated from the prediction equations.

**Fibre size**

The carcinogenic hazard of asbestos (development of pleural mesothelioma) arises in the critical fibre size range of diameter less than 0.25 $\mu$m and length greater than 8 $\mu$m (Stanton et al., 1981). Thus the lack of positive results for the Puget Sound Area, as compared with the San Francisco Bay Area in California, has been attributed to the shorter fibre lengths in the state of Washington as compared with those in California (Kanarek, 1983).

**Power considerations**

Erdreich (1983) compared the completed epidemiological studies from the point of view of their statistical ability to detect a low cancer risk. She estimated that the San Francisco study was capable of detecting GI relative risk of 1.1 and the Puget Sound study one of 1.5. The other studies would be less powerful because of their smaller sample sizes, and thus might not be able to detect the low ($<1.5$) significant relative risks, such as those detected in the Kanarek et al. (1980) study. Thus, population size and sufficient statistical power may be the simple reason for the findings of a more positive association in the studies by Kanarek et al. (1980) and Conforti et al. (1981) as compared with those in other studies.

Nicholson (1983) also considered the power of the studies completed to date to detect a possible association between asbestos fibres in drinking-water and GI cancer rates. His basic conclusion was that, in the studies that he reviewed, exposures were too low and population numbers too small to allow a possible association to be detected.

Marsh (1983) systematically reviewed the completed epidemiological studies. His basic conclusion was that the number of observations of significant associations of oesophagus, stomach, pancreas and prostate cancer were greater than what would be expected by chance ($p<0.05$). He emphasized the lack of agreement between the results for males and females.

**Confounders**

The following points should be noted: (1) There was a slight increase in the risk of gastric cancer in males in the case-control study in western Washington (conducted by Polissar et al. (1984)), but the finding was confounded by increased cigarette and coffee consumption in the cases as compared with the controls. (2) The mobility of the populations of the USA and Canada reduces the stability of the linkage of exposure with effect, especially in diseases with a long latent period. (3) It has been assumed, of necessity, in all studies that numbers of asbestos fibres in the water sources studied remain reasonably constant, but counts may vary due to unstudied variables, such as turbulence, changes in flow rate, etc. (4) The ubiquity of asbestos in both the occupational and non-occupational environments (asbestos in the ambient air, in indoor air in old school buildings, etc.) poses questions about the exposure status of control groups.

All of the studies, except the case-control study of Polissar et al. (1984), are ecological in character, geographically defined area-wide cancer rates being compared with asbestos fibre counts in water supplies; they are thus indirect epidemiological studies (Yerushalmy, 1966). The usefulness and limitations of such studies in environmental epidemiology have been described by Morgenstern (1982), Kanarek (1981) and Langbein and Lichtman (1978).

**Future studies**

Consistent results were obtained in all the studies for pancreatic and stomach cancers. Table 4 lists the significant findings for these sites and for peritoneum. Peritoneal cancer was not included in most of the studies, but must be strongly considered in any future studies on non-inhalation-associated asbestos exposures. Future incidence-based case-control epidemiological studies on these sites in appropriate exposure areas should include variables relating to drinking-water. Data should be gathered on residential history and drinking-water habits. Mortality studies, where surviving relatives are questioned, may also be relevant. The general methodology for case-control studies is well described in Schlesselman (1982) and the particular techniques required for drinking-water studies in Packham (1986). The addition of a small section on drinking-water (asbestos or other contaminants, such as volatile organics) to planned case-control epidemiological studies would be a means of obtaining additional information without a large increase in research costs.

**Table 4. Possible associations between asbestos fibres in drinking-water and gastrointestinal cancer risk**

| Cancer site | Duluth | Connecticut | Quebec | San Francisco Bay Area | Puget Sound |
|---|---|---|---|---|---|
| Stomach | Males | None | Males | Females and males | Males |
| Pancreas | Females | Males | Females | Females and males | None |
| Peritoneum | None | Not studied | Not studied | Females and males | None |

*References*

Conforti, P.M. (1983) Effect of population density on the results of the study of water supplies in five California counties. *Environ. Health Perspect.*, 53, 69-78

Conforti, P.M., Kanarek, M.S., Jackson, L.A., Cooper, R.C. & Murchio, J.C. (1981) Asbestos in drinking water and cancer in the San Francisco Bay Area: 1969-1974 incidence. *J. Chron. Dis.*, 34, 211-224

Cook, P.M., Glass, G.E. & Tucker, J.G. (1974) Asbestiform amphibole mineral: detection and measurement of high concentration in municipal water supplies. *Science, 185*, 853-855

Erdreich, L.S. (1983) Comparing epidemiologic studies of ingested asbestos for use in risk assessment. *Environ. Health Perspect., 53*, 99-104

Harrington, J.M., Craun, G., Meigs, J.W., Landrigan, P.J., Flannery, J.T. & Woodhall, R.S. (1978) An investigation of the use of asbestos cement pipe for public water supply and the incidence of gastrointestinal cancer in Connecticut, 1935-1973. *Am. J. Epidemiol., 107*, 96-103

Kanarek, M.S. (1981) Indirect epidemiology studies in environmental hazard risk assessment. In: Richmond, C.R., Walsh, P.J. & Copenhaver, E.D., eds, *Health Risk Analysis*, Philadelphia, Franklin Institute Press, pp. 87-96

Kanarek, M.S. (1983) The San Francisco Bay epidemiology studies on asbestos in drinking water and cancer incidence: relationship to studies in other locations and pointers for further research. *Environ. Health Perspect., 53*, 105-106

Kanarek, M.S., Conforti, P.M., Jackson, L.A., Cooper, R.C. & Murchio, J.C. (1980) Asbestos in drinking water and cancer incidence in the San Francisco Bay Area. *Am. J. Epidemiol., 112*, 54-72

Langbein, L.I. & Lichtman, A.J. (1978) *Ecological Interference (Series on Quantitative Applications in the Social Science, No. 07-010)*, Beverly Hills, CA, Sage Publishers

Levine, D.S. (1985) Does asbestos exposure cause gastrointestinal cancer? *Dig. Dis. Sci., 30*, 1189-1198

Levy, B.S., Sigurdson, E., Mandel, J., Laudon, E. & Pearson, J. (1976) Investigating possible effects of asbestos in city water: surveillance of gastrointestinal cancer incidence in Duluth, Minnesota. *Am. J. Epidemiol., 104*, 523-526

Marsh, G.M. (1983) Critical review of epidemiologic studies related to ingested asbestos. *Environ. Health Perspect., 53*, 49-56

Mason, T.J., McKay, F.W. & Miller, R.W. (1984) Asbestos-like fibers in Duluth water supply. *J. Am. Med. Assoc., 228*, 1019-1020

Meigs, J.W., Walter, S., Hestbon, J., Millette, J.R., Craun, G.G., Woodhull, R.S. & Flannery, J.T. (1980) Asbestos-cement pipe and cancer in Connecticut, 1955-1974. *Environ. Res., 42*, 187-197

Morgenstern, H. (1982) Uses of ecologic analysis in epidemiologic research. *Am. J. Public Health, 72*, 1336-1344

National Research Council (1983) *Drinking Water and Health*, Vol. 5, *A Report of the Safe Drinking Water Committee, Commission on Life Sciences*, Washington DC, National Academy Press, pp. 123-144

Nicholson, W.J. (1974) Analysis of amphibole asbestiform fibers in municipal water supplies. *Environ. Health Perspect., 9*, 165-172

Nicholson, W.J. (1983) Human cancer risk from ingested asbestos: a problem of uncertainty. *Environ. Health Perspect., 53*, 111-113

Packham, R.F., ed. (1986) *Evaluation of Methods for Assessing Human Health Hazards from Drinking Water (Technical Report No. 86/001)*, Lyon, International Agency for Research on Cancer

Polissar, L., Severson, R.K., Boatman, E.S. & Thomas, D.B. (1982) Cancer incidence in relation to asbestos in drinking water in the Puget Sound Region. *Am. J. Epidemiol., 116*, 314-328

Polissar, L., Severson, R.K. & Boatman, E.S. (1984) A case-control study of asbestos in drinking water and cancer risk. *Am. J. Epidemiol., 119*, 456-471

Schlesselman, J.J. (1982) *Case-control Studies*, New York, Oxford University Press

Selikoff, I.J. & Seidman, H. (1981) Cancer of the pancreas among asbestos insulation workers. *Cancer, 47*, 1470-1473

Selikoff, I.J., Hammond, E.C. & Seidman, H. (1979) Mortality experience of insulation workers in the United States and Canada 1943-1976. *Ann. N.Y. Acad. Sci., 330*, 91-116

Sigurdson, E.E. (1983) Observations of cancer incidence in Duluth, Minnesota. *Environ. Health Perspect., 53*, 61-67

Sigurdson, E.E., Levy, B.S., Mandel, J., McHugh, R., Michienzi, L.J., Jagger, H. & Pearson, J. (1981) Cancer morbidity investigations: Lessons from the Duluth study of possible effects of asbestos in drinking water. *Environ. Res., 25*, 50-61

Stanton, M.F., Layard, M., Tegeris, A., Miller, E., May, M., Morgan, E. & Smith, A. (1981) Relation of particle dimension to carcinogenicity in amphibole asbestos and other fibrous minerals. *J. Natl Cancer Inst., 67*, 965-975

Tarter, M.E., Cooper, R.C. & Freeman, W.R. (1983) A graphical analysis of the interrelationships among waterborne asbestos, digestive system cancer and population density. *Environ. Health Perspect., 53*, 79-89

Toft, P., Wigle, D.T., Méranger, J.C. & Mao, Y. (1981) Asbestos and drinking water in Canada. *Sci. Total Environ., 18*, 77-89

Toft, P., Meek, M.E., Wigle, D.T. & Méranger, J.C. (1984) Asbestos in drinking water. *CRC Crit. Rev. Environ. Control, 14*, 151-197

Wigle, D.T. (1977) Cancer mortality in relation to asbestos in municipal water supplies. *Arch. Environ. Health, 32*, 185-190

Working Group for the DHSS Committee to Coordinate Environmental and Related Programs, Subcommittee on Risk Assessment (1987) Report on cancer risks associated with the ingestion of asbestos. *Environ. Health Perspect., 72*, 253-265

Yerushalmy, J. (1966) On inferring causality from observed associations. In: Ingelfinger, F.J., Relman, A.S. & Finland, M., eds, *Controversy in Internal Medicine*, Philadelphia, W.B. Saunders, pp. 659-668

# ASBESTOS FIBRE CONTENT OF LUNGS WITH MESOTHELIOMAS IN OSAKA, JAPAN: A PRELIMINARY REPORT

K. Morinaga[1], N. Kohyama[2], K. Yokoyama[3], Y. Yasui[4],
I. Hara[4], M. Sasaki[5], Y. Suzuki[6] & Y. Sera[3]

[1] Department of Field Research, Center for Adult Diseases, Osaka, Japan
[2] Ministry of Labor, Institute of Industrial Health, Kawasaki, Japan
[3] National Kinki-Chuo Hospital for Chest Disease, Sakai, Japan
[4] Department of Public Health, Kansai Medical University, Moriguchi, Japan
[5] Department of Pathology, Osaka Red Cross Hospital, Osaka, Japan
[6] Environmental Sciences Laboratory, Mount Sinai School of Medicine, New York, NY, USA

*Summary.* That crocidolite and amosite are both associated with the development of mesothelioma is now well established, but earlier studies have failed to find an excess of chrysotile in lungs with mesotheliomas as compared with the amounts in lungs of unaffected controls. In an attempt to clarify the importance of fibre type in tissue, an examination of a series of mesotheliomas is being undertaken in Osaka, Japan. A total of 23 mesotheliomas and 5 rejected cases reviewed by the Osaka Mesothelioma Panel were examined for the types of asbestos and semiquantitative fibre content by means of a transmission electron microscope equipped with energy-dispersive X-ray analyser. Asbestos fibres were detected in 19 of the 23 mesotheliomas (21 pleura, 1 pericardium, 1 peritoneum; 19 males, 4 females). Amphibole fibres were found in 13 cases. Five pleural and one peritoneal mesothelioma were found to have only chrysotile fibres. One female pleural mesothelioma with neighbourhood exposure had short chrysotile fibres. Among the 5 rejected cases, only one case with occupational exposure had both chrysotile and amosite fibres. A group of 17 controls were also examined and asbestos fibres were found in 5. Our data, while not definitive, suggest that mesotheliomas can be induced in humans, not only by crocidolite and amosite, but also by chrysotile, though possibly to a lesser extent.

## *Introduction*

In Japan, only one chrysotile mine is in operation and it produces about 1500 tons of low-grade asbestos annually. Almost all asbestos used in Japan therefore comes from foreign countries (Table 1). The import of raw asbestos to Japan has increased since the 1950s and reached about 300 000 tons in the 1970s, which exceeded US total consumption in 1983 (Figure 1).

Osaka, which is the second largest commercial and industrial urban area in the country, has the largest number of asbestos factories and produces about one-third of all asbestos in Japan. Judging by the number of death certificates and cancer registry

Table 1. Asbestos imports in Japan in 1965 and 1985

| Country | Imports (tons) | |
|---|---|---|
| | 1965 | 1985 |
| Canada | 71 767 | 102 979 |
| South Africa | 35 731 | 57 955 |
| USSR | 15 032 | 45 493 |
| USA | 10 031 | 10 780 |
| Australia | 890 | – |
| Zimbabwe | – | 29 253 |
| Italy | – | 7 414 |
| Greece | – | 6 152 |
| Others | 363 | 1 622 |
| **Total** | 146 294 | 261 648 |

Fig. 1. Annual US consumption of asbestos from 1890 to 1986 and annual Japanese asbestos imports from 1913 to 1986

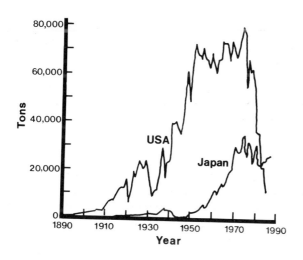

ards notified as mesothelioma, the mortality from this condition has been increasing together with the number of death certificates mentioning asbestosis (Table 2). To ascertain incidence rates of mesotheliomas and to assess the relationship of this tumour to asbestos, the Mesothelioma Register and Panel was started in 1980. This is the first report from Japan on the types of asbestos in lungs with mesotheliomas.

Table 2. Death certificates and cancer registry cards mentioning mesothelioma and asbestosis, Osaka, 1967-1985

| Year of death | Death certificates mentioning mesothelioma | Registry cards mentioning mesothelioma | All certificates mentioning asbestosis |
|---|---|---|---|
| 1967 | 1 | 4 | 1 |
| 1968 | 0 | 4 | 2 |
| 1969 | 1 | 2 | 1 |
| 1970 | 0 | 3 | 4 |
| 1971 | 4 | 6 | 1 |
| 1972 | 1 | 4 | 6 |
| 1973 | 2 | 4 | 1 |
| 1974 | 3 | 5 | 4 |
| 1975 | 1 | 3 | 5 |
| 1976 | 6 | 8 | 5 |
| 1977 | 4 | 6 | 3 |
| 1978 | 4 | 8 | 6 |
| 1979 | 9 | 12 | 7 |
| 1980 | 4 | 9 | 3 |
| 1981 | 7 | 11 | 8 |
| 1982 | 13 | 20 | 5 |
| 1983 | 9 | 16 | 7 |
| 1984 | 10 | 15 | 9 |
| 1985 | 11 | 13 | 5 |

## Materials and methods

From June 1981 to October 1986, 49 autopsied cases were reviewed by the Osaka Mesothelioma Panel in the same way as described by Kannerstein et al. (1979). A total of 33 cases were classified as definite or probable mesotheliomas, 2 possible, and 14 unsubstantiated. Blocks of tissues were requested to be more than 2×2×2 cm in size and were collected from 23 mesotheliomas and 5 rejected cases. In each case, 2 controls matched for hospital, age and year of death were obtained from a case of lung cancer and one of a non-respiratory disease. For mesotheliomas and rejected cases, lung tissue was obtained from the middle basal area of the unaffected lower lobe, and for controls, tissues from the same area of the same-side lower lobe were used. They were embedded in paraffin and cut into several serial sections of thickness 20 $\mu$m. After deparaffinization in xylene, those specimens were ashed at low temperature and transferred on to a nickel grid by a carbon extraction technique. The specimens were examined by one of the authors (N.K.) using a Hitachi H-500 transmission electron microscope fitted with a Kevex model 5100 energy-dispersive X-ray analyser. Asbestos types were identified by their characteristic morphologies, chemical composition and electron diffraction patterns. The amounts of asbestos fibres observed were graded as follows: (+++) dozens of fibres observed in almost all areas arbitrarily selected under low magnification (5000–10 000) in the microscope; (++) 10 or so fibres on average detected in every 2 or 3 areas; (+) a few fibres found in the entire specimen; (-) no fibres detected. A fibre was defined as a particle having an aspect ratio of 3:1 or greater.

## Results

The results obtained are shown in Table 3. All the mesotheliomas, 17 controls, and the 5 rejected cases were examined to determine the mineral fibre content (see Table 4). Asbestos fibres were frequently found in mesotheliomas (19/23), but infrequently in the control series (5/17). Amounts of asbestos fibres graded as (++) or (+++) occurred only in the mesothelioma series except for one rejected case with occupational exposure, which was correctly diagnosed as peripheral lung carcinoma. Five cases were graded (+) or (+∼++) in the control series; of these, 3 were lung cancer patients, whose occupations had been farmer, housewife and part-time office worker. The other 2 controls were a farmer who had died of myocardial infarction and a man with renal cancer who had worked for a delivery service company. Table 5 shows the data on types of asbestos in the study series. Amphiboles were detected in 13 mesotheliomas, of whom 3 had been shipbuilders, 3 had been employed in a plant which manufactured asbestos products, 2 in a factory which made railway carriages, and a chemical engineer who had been intermittently exposed to asbestos while replacing boiler insulation. The remaining 4 were 3 clerks and one worker employed in a plant which manufactured light bulbs; it is not known whether or not they had a history of asbestos exposure. Four controls and one rejected case also had amphiboles, while 6 mesotheliomas were found to have chrysotile only (Table 6). One plastic factory worker with occupational exposure was graded (+++) and had long, thin chrysotile fibres, while a housewife living near an asbestos factory was graded (+∼++) and had short chrysotile fibres. The latent period in her case from initial neighbourhood exposure to detection of the disease was 35 years. We could not ascertain whether the design engineer and the surgeon had had any contact with asbestos.

Table 3. Number, sex and mean age of cases examined

| Series | Men | Women | Total | Mean age (years) |
|---|---|---|---|---|
| Mesotheliomas | 19 | 4 | 23 | 62±10 |
| Controls 1[a] | 10 | 3 | 13 | 63±6 |
| Controls 2[b] | 11 | 3 | 14 | 62±11 |
| Rejected cases | 4 | 1 | 5 | 61±13 |

[a]Lung cancer.
[b]Non-respiratory disease.

### Table 4. Asbestos fibre content in mesotheliomas, controls and rejected cases

| Series | Fibre content | | | | | |
|---|---|---|---|---|---|---|
| | − | + | +~++ | ++ | +++ | Total |
| Mesotheliomas | 4 | 3 | 2 | 4 | 10 | 23 |
| Controls 1[a] | 5 | 2 | 1 | 0 | 0 | 8 |
| Controls 2[b] | 7 | 2 | 0 | 0 | 0 | 9 |
| Rejected cases | 3 | 1 | 0 | 1 | 0 | 5 |

[a]Lung cancer.

[b]Non-respiratory disease.

### Table 5. Types of asbestos in the study series

| Series | Number examined | Serpentines | Amphiboles[a] | | |
|---|---|---|---|---|---|
| | | Chrysotile | Amo | Cro | Ac-Tr |
| Mesotheliomas | 23 | 12 | 13 | 3 | 3 |
| Controls 1[b] | 8 | 0 | 1 | 1 | 1 |
| Controls 2[c] | 9 | 1 | 0 | 0 | 1 |
| Rejected cases | 5 | 1 | 1 | 0 | 1 |

[a]Amo, amosite; Cro, crocidolite; Ac-Tr, actinolite-tremolite.

[b]Lung cancer.

[c]Non-respiratory disease.

### Table 6. Clinical, occupational and semiquantitative fibre content data on 6 mesotheliomas having only chrysotile fibres[a]

| No. | Age (years) | Sex | Site | Occupation | Fibre content |
|---|---|---|---|---|---|
| 2 | 62 | M | Pl | Welder | + |
| 8 | 76 | M | Pl | Plater | ++ |
| 22 | 69 | F | Pl | Housewife | +~+++ |
| 26 | 54 | M | Per | Design engineer | + |
| 32 | 69 | M | Pl | Plastic factory worker | +++ |
| 38 | 70 | M | Pl | Surgeon | +~++ |

[a]M, male; F, female; Pl, pleura; Per, peritoneum.

## Discussion

An association between malignant mesothelial tumours and exposure to commercial amphiboles is well established. With regard to chrysotile, however, it has been reported (Jones et al., 1980; McDonald et al., 1982; Wagner et al., 1982) that chrysotile fibres were not found to a greater extent in the lungs of mesotheliomas than in those of controls. The deposition and clearance of chrysotile from the lungs are different from those of amphiboles, and chrysotile may disappear more rapidly than amphiboles in the course of time. In addition, the digestion methods commonly used for the preparation of lung tissue specimens may produce a more distorted estimation of the fibre burden of the lung for chrysotile than for amphiboles. Our method of using the ashing technique alone may be useful for differentiating between occupational and non-occupational exposures. Our data show that, if amphibole and/or an amount of chrysotile graded as more than (++) are found, mesotheliomas may be occupationally induced. This is only a preliminary report, but it nevertheless indicates that 32% (6/19) of mesotheliomas found to contain asbestos fibres contained only chrysotile fibres. Of 3 mesotheliomas found to contain actinolite-tremolite, all also had an amount of amosite graded as more than (++). Our study suggests, therefore, that chrysotile can also induce mesotheliomas in humans, but possibly to a lesser extent than crocidolite and amosite. Thus chrysotile may have a larger role in causing mesothelioma than previously found.

## Acknowledgements

This study was partly supported by a Grant in Aid for Scientific Research from the Ministry of Education, Culture and Science, Japan. We thank Dr I. Fujimoto, A. Hanai (Osaka Cancer Registry) and the members of the Mesothelioma Study Group for their co-operation in data collection.

## References

Jones, J.S.P., Roberts, G.H., Pooley, F.D., Clark, N.J., Smith, P.G., Owen, W.G., Wagner, J.C. & Berry, G. (1980) The pathology and mineral content of lungs in cases of mesothelioma in the United Kingdom in 1976. In: Wagner, J.C., ed., *Biological Effects of Mineral Fibres (IARC Scientific Publications No. 30)*, Lyon, International Agency for Research on Cancer, pp. 187-200

Kannerstein, M., Churg, J. & McCaughey, W.T.E. (1979) Functions of mesothelioma panel. *Ann. N.Y. Acad. Sci.*, 330, 433-439

McDonald, A.D., McDonald, J.C. & Pooley, F.D. (1982) Mineral fibre content of lungs in mesothelial tumours in North America. *Ann. Occup. Hyg.*, 26, 417-422

Wagner, J.C., Pooley, F.D., Berry, G., Seal, R.M.E., Munday, D.E., Morgan, J. & Clark, N.J. (1982) A pathological and mineralogical study of asbestos-related deaths in the United Kingdom. *Ann. Occup. Hyg.*, 26, 423-431

# CORRELATION BETWEEN LUNG FIBRE CONTENT AND DISEASE IN EAST LONDON ASBESTOS FACTORY WORKERS

J.C. Wagner[1], M.L. Newhouse[2], B. Corrin[3],
C.E. Rossiter[2] & D.M. Griffiths[1]

[1] *MRC Pneumoconiosis Research Unit, Llandough Hospital, Penarth, South Glamorgan, UK*
[2] *TUC Centenary Institute of Occupational Health, London School of Hygiene and Tropical Medicine, London, UK*
[3] *The London Chest Hospital, London, UK*

*Summary.* The lungs from 36 former workers at an East London asbestos factory dying of asbestos-related disease were compared with lung tissue from 56 matched control patients operated on in East London for carcinoma of the lung. The severity of asbestosis and the presence of pulmonary carcinoma or mesothelioma of the pleura or peritoneum were correlated with an asbestos exposure index and with the type and amount of mineral fibre of the lungs. Asbestosis was associated with far heavier fibre burdens than mesothelioma. Moderate or severe asbestosis was more common among those with carcinoma of the lung than in those with mesothelial tumours. Crocidolite and amosite asbestos were strongly associated with asbestosis, carcinoma of the lung and mesothelial tumours, whereas no such correlation was evident with chrysotile or mullite. It is suggested that greater emphasis should be placed on the biological differences between amphibole and serpentine asbestos fibre.

## *Introduction*

Wagner *et al.* (1986), in a study of the asbestos fibre content of the lungs of former workers at the Royal Naval Dockyard, showed that a good correlation existed between the total lung fibre content and the severity of asbestosis. Furthermore, mesotheliomas occurred most commonly in those subjects with minimal or slight asbestosis, in contrast to pulmonary carcinomas, which were commoner in those with the severer grades of asbestosis. We have now conducted a similar study on a group of asbestos workers with a different type of asbestos exposure.

## *Materials and methods*

The study group comprised 36 former workers at a factory in East London where asbestos textiles and other asbestos products were manufactured before the introduction of the 1969 Asbestos Regulations. Crocidolite, chrysotile and amosite were all used extensively. The mortality of workers at this factory has been described previously (Newhouse *et al.*, 1985).

The subjects in the present study had died between 1976 and 1984 and after post mortem their lungs had been submitted to the London Boarding Centre for Respiratory Diseases. Three standard pieces of lung tissue, two from the base and one from the lateral border of a lower lobe, were selected. After a block had been taken for light microscopy, the remainder was digested in potassium hydroxide and the mineral fibre content assessed quantitatively and qualitatively by analytical electron microscopy (Wagner et al., 1980). Asbestosis was graded as minimal, slight, moderate or severe.

The subjects were identified in the records held by one of us (MLN) of former workers at this factory. Exposure was graded on the scale 1–6 according to the degree of dust exposure involved in a particular job. An exposure index was calculated by multiplying the exposure grade by the number of months an individual had worked at the factory.

A control group comprised 56 patients operated on at the London Chest Hospital in 1983–1984 for carcinoma of the lung. These patients had never been occupationally exposed to asbestos and generally lived in East London. Normal lung tissue was taken from the surgical specimens from these patients and examined in the same way as that from the study group.

## Results

The control group consisted of 44 males and 12 females, whereas the study group comprised 25 males and 11 females. The mean age of the controls at the time of their operation and the mean age of the study cases at death were 61 and 62 years, respectively.

In the factory group, there were 19 deaths from mesothelioma (9 pleural and 10 peritoneal), 14 from carcinoma of the lung and 3 from asbestosis. Subjects with mesothelial tumours often showed minimal or slight asbestosis, whereas those dying of carcinoma of the lung usually showed moderate or severe asbestosis (Table 1).

The mean total fibre counts for the controls and for each diagnostic category are shown in Table 2. The heaviest fibre content was found in those dying of certified asbestosis and lung cancer. Counts of each fibre type were made on each lung examined, and the proportion of each fibre calculated; the means of these proportions

**Table 1. Percentages of cases in each diagnostic category by degree of asbestosis**[a]

| Degree of asbestosis | Carcinoma of the lung | Pleural mesothelioma | Peritoneal mesothelioma | Asbestosis |
|---|---|---|---|---|
| Minimal | 0 | 22.2 | 11.1 | 0 |
| Slight | 21.4 | 33.3 | 55.6 | 0 |
| Moderate | 42.9 | 33.3 | 22.2 | 33.3 |
| Severe | 35.7 | 11.1 | 11.1 | 66.7 |

[a]There were 14 cases of carcinoma of the lung, 9 of pleural mesothelioma, 10 of peritoneal mesothelioma (the degree of asbestosis was not recorded for one of these cases) and 3 of asbestosis.

are shown in Table 3. The contrast between the proportion of amphibole asbestos in the controls and in the other categories is striking — less than 6% in the former but ranging from 60% in the pleural mesothelioma to 77% in the asbestosis lungs. Except in the lungs of the pleural mesothelioma cases, there was a higher proportion of crocidolite asbestos than amosite. The highest proportion of chrysotile fibres was found in the control lungs, and here only approximately 30% of the total fibre was asbestos.

### Table 2. Mean fibre content and diagnostic category

| Diagnostic category | No. | Mean total fibre count$\times 10^6$/g |
|---|---|---|
| Controls | 56 | 35.8 |
| Carcinoma of the lung | 14 | 1141.7 |
| Pleural mesothelioma | 9 | 262.9 |
| Peritoneal mesothelioma | 10 | 565.7 |
| Asbestosis | 3 | 1720.7 |

### Table 3. Mean percentage of fibre type by diagnostic category

| Fibre type | Controls | Carcinoma of the lung | Pleural mesothelioma | Peritoneal mesothelioma | Asbestosis |
|---|---|---|---|---|---|
| Amosite | 2.6 | 30.4 | 39.3 | 17.8 | 17.7 |
| Crocidolite | 2.8 | 56.7 | 20.2 | 53.9 | 60.1 |
| Chrysotile | 25.9 | 6.9 | 17.0 | 13.3 | 4.0 |
| Mullite | 64.4 | 5.2 | 20.7 | 14.6 | 17.9 |
| Other fibres | 4.4 | 0.9 | 2.8 | 0.4 | 0.3 |

The proportions of crocidolite and amosite fibres increased with the severity of asbestosis whereas the proportions of chrysotile and non-asbestos fibres decreased (Table 4).

### Table 4. Mean percentage of types of asbestos by grades of asbestosis

| Grade of asbestosis | Total fibres ($\times 10^6$/g) | Amosite (%) | Crocidolite (%) | Chrysotile (%) | Mullite (%) | Other fibres (%) |
|---|---|---|---|---|---|---|
| Controls ($N=56$) | 35.8 | 2.6 | 2.8 | 25.9 | 64.4 | 4.4 |
| Minimal ($N=3$) | 77.7 | 12.3 | 32.2 | 35.5 | 18.7 | 1.2 |
| Slight ($N=11$) | 373.4 | 22.4 | 37.6 | 18.2 | 21.1 | 0.8 |
| Moderate ($N=12$) | 753.8 | 36.8 | 50.9 | 3.7 | 7.0 | 1.6 |
| Severe ($N=9$) | 1735.6 | 25.8 | 62.5 | 4.0 | 6.7 | 1.1 |

A constant relationship between each fibre type and the exposure index does not exist (Table 5), but the proportions of amosite and crocidolite tend to rise in those with the higher indices, and for both it is highest in the highest exposure index group. Conversely, the proportion of chrysotile decreases, and it accounts for less than 1% of the total fibres in the lungs of the 10 workers with the highest exposure index. The proportion of mullite also decreases as the index rises.

### Table 5. Exposure index and mean percentage of types of fibres

| Exposure index | Total fibres ($\times 10^6$/g) | Amosite (%) | Crocidolite (%) | Chrysotile (%) | Mullite (%) | Other fibres (%) |
|---|---|---|---|---|---|---|
| 0 (controls) ($N = 56$) | 35.8 | 2.6 | 2.8 | 25.9 | 64.4 | 4.4 |
| <100 ($N = 8$) | 136.7 | 8.4 | 45.3 | 20.4 | 25.1 | 0.8 |
| 100-500 ($N = 12$) | 697.7 | 32.2 | 42.1 | 10.0 | 14.1 | 1.6 |
| 501-1000 ($N = 6$) | 733.4 | 23.0 | 46.8 | 17.1 | 12.3 | 0.8 |
| 1001-2560 ($N = 10$) | 1530.3 | 41.9 | 54.6 | 0.8 | 1.5 | 1.2 |

## Discussion

In this study, the lung asbestos burdens found were in accordance with present-day concepts of the epidemiology of asbestos-related diseases, namely that asbestosis is associated with very heavy fibre concentrations and mesothelioma with far lower ones. As in the Naval Dockyard Study (Wagner et al., 1986), asbestosis tended to be more severe in the subjects with carcinoma of the lung than in those with mesothelial tumours.

A further confirmatory finding in this series is that crocidolite and amosite are strongly associated with asbestosis, carcinoma of the lung (among those exposed to asbestos) and mesothelial tumours of both sites, whereas differences between the chrysotile contents of the lungs of the controls and the subjects exposed to asbestos were less marked than for amphibole fibres. The proportion of chrysotile fibres is lower in exposed subjects than in the controls. This becomes even more significant when allowance is made for the fact that each amosite fibre is at least 100 times heavier than its chrysotile equivalent. As has been shown by Timbrell (1973) and Pooley (1976), poorer pulmonary penetration of the curly chrysotile fibres and their physicochemical dissolution in the lung may be responsible for this. It is possible that the damage has been caused by chrysotile fibres that have subsequently disappeared, but a more likely explanation is that chrysotile plays only a minor role in the causation of the diseases associated with exposure to asbestos dust. In all the diseases, the increases in amphibole asbestos were very much greater than that of chrysotile. This finding is similar to those reported by McDonald et al. (1982) and Churg (1982). We believe, therefore, that chrysotile is the least harmful form of asbestos in every respect and that greater emphasis should be placed on the different biological effects of the various amphibole fibres.

## References

Churg, A. (1982) Asbestos fibres and pleural plaques in a general autopsy population. *Am. J. Pathol., 109*, 88-96

McDonald, A.D., McDonald, J.C. & Pooley, F.D. (1982) Mineral fibre content of lung in mesothelial tumours in North America. *Ann. Occup. Hyg., 26*, 417-422

Newhouse, M.L., Berry, G. & Wagner, J.C. (1985) Mortality of factory workers in East London. *Br. J. Ind. Med., 42*, 4-11

Pooley, F.D. (1976) An examination of the fibrous mineral content of asbestos lung tissue from the Canadian chrysotile mining industry. *Environ. Res., 12*, 281-298

Timbrell, V. (1973) Physical factors as etiological mechanisms. In: Bogovski, P., Gilson, J.C., Timbrell, V. & Wagner, J.C., eds, *Biological Effects of Asbestos (IARC Scientific Publications No. 8)*, Lyon, International Agency for Research on Cancer, pp. 295-303

Wagner, J.C., Moncrieff, C.B., Coles, R., Griffiths, D.M. & Munday, D.E. (1986) Correlation between fibre content of the lungs and disease in naval dockyard workers. *Br. J. Ind. Med., 43*, 391-395

# EFFECT ON HEALTH OF MAN-MADE MINERAL FIBRES IN KINDERGARTEN CEILINGS

### A. Rindel & C. Hugod

*Department of General Hygiene, The National Board of Health, Copenhagen, Denmark*

### E. Bach

*The Danish Institute for Clinical Epidemiology, Copenhagen, Denmark*

### N.O. Breum

*The Danish National Institute of Occupational Health, Copenhagen, Denmark*

*Summary.* The relation between the presence of readily visible man-made mineral fibre (MMMF) products in the ceilings and the presence/frequency of symptoms and diseases, and the correlation between the presence/frequency of symptoms and diseases and the concentration of MMMF in the indoor environment was investigated in 24 kindergartens. A combination of traditional epidemiological techniques and a technical analysis of a number of indoor air parameters did not support the hypothesis that release of MMMF from readily visible MMMF products in the ceilings was mainly responsible for the occurrence of symptoms or diseases related to indoor exposure in kindergartens.

## Introduction

In the industrial environment, man-made mineral fibres (MMMF) of diameter $>3$ $\mu$m may cause discomfort in the form of irritation of the upper respiratory passages, skin and mucosa of the eye. The consequences to health of contamination by MMMF in the home environment, where the concentrations are much lower, have not yet been fully elucidated, although one study (Alsbirk et al., 1983) has indicated that MMMF from indoor exposure may cause eye irritation. Other symptoms included in the sick building syndrome (irritation of the eye, nose, pharynx and throat, dry cough, tendency to running or stuffed nose, itching of the skin and middle ear problems) have also been described (Jørgensen-Birch et al., 1986; Petersen et al., 1985). As a result of public suspicion that readily visible MMMF products in the ceilings of kindergartens were responsible both for the sick building syndrome and for infections of the respiratory tract, the Danish National Board of Health initiated the present study (Rindel et al., 1985) with the purpose of investigating the relation between readily visible MMMF products in ceilings, MMMF in the indoor environment and the presence/frequency of symptoms and diseases.

## Materials and methods

The study was performed in 24 kindergartens in Denmark (Frederiksborg county) and included approximately 900 children and 200 adults. The institutions were divided into 3 groups: A = kindergartens with ceilings covered with MMMF products containing water-soluble binder; B = kindergartens with ceilings covered with MMMF products containing resin binder; and C = control group (ceilings without readily visible MMMF products).

The health symptoms (irritation of the eye, upper respiratory tract and skin, headache and tiredness) and diseases (infections of the eye, respiratory tract and middle ear) were evaluated using a questionnaire given daily to parents and employees in the kindergarten over a 12-week period in the autumn of 1984 and by a clinical investigation performed by a physician who visited the kindergarten once during the period.

To assess pollution with MMMF, 12 kindergartens were visited 3 times during the 12-week period, when samples of both airborne and settled fibres were collected. Airborne fibres were collected on membrane filters for approximately 1 h. The samples were then analysed by phase-contrast light microscopy. Settled MMMF were collected by means of finger-print lifters and analysed by phase-contrast light microscopy. Finger-print lifters are special sticky foils used for forensic purposes. These foils can follow slightly irregular surfaces and have good sticking properties, and excellent optical properties. The foils have previously been used for sampling particulate matter (Schneider et al., 1987).

Since the sick building syndrome may be related to a large number of confounding factors, the study included health and socioeconomic parameters, information on the buildings and their use and technical investigations of the concentrations of airborne non-MMMF, airborne dust, carbon dioxide, volatile organic components and formaldehyde, and the thermal environment, the air change rate and cleaning standards.

All the risk, outcome and confounding variables are categorical. Therefore control for confounder influence was performed by log linear analysis for contingency tables (Anderson et al., 1980).

The proportions of symptoms/diseases in kindergartens in groups A and B were compared with those in control institutions with respect to the cluster sampling technique, the clusters being the kindergartens (Cochran, 1977). The relationship between the technical measurements and the proportion of persons with symptoms/diseases was calculated by non-parametric correlation analysis (Spearman).

## Results

Some of the results of the health investigations are given in Table 1. Some symptoms and diseases were reported more frequently from institutions with MMMF ceilings than from control institutions: eye irritation in adults, irritation of the respiratory tract in adults (only from the questionnaire), common cold and sore throat in children and adults, infection of the middle ear in children, irritation of the skin

## Table 1. Frequency of symptoms/diseases in 3 groups of kindergartens[a]

| Symptoms/ disease | | Children | | | Adults | | |
|---|---|---|---|---|---|---|---|
| | | A | B | C | A | B | C |
| Eye irritation | F(%) | 4 | 5 | 25 | 21 | 8 | - |
| | p | 0.59 | 0.73 | - | 0.01 | 0.10 | - |
| Irritation of the nose | F(%) | 18 | 13 | 14 | 26 | 18 | 0 |
| | p | 0.24 | 0.75 | - | 0.000 | 0.02 | - |
| Irritation of the skin | F(%) | 16 | 17 | 11 | 24 | 36 | 10 |
| | p | 0.05 | 0.02 | - | 0.02 | 0.001 | - |
| Common cold | F(%) | 59 | 57 | 44 | 49 | 43 | 23 |
| | p | 0.04 | 0.03 | - | 0.002 | 0.09 | - |
| Sore throat | F(%) | 16 | 10 | 7 | 36 | 32 | 9 |
| | p | 0.02 | 0.39 | - | 0.000 | 0.01 | - |
| Otitis media | F(%) | 8 | 10 | 6 | - | - | - |
| | p | 0.47 | 0.003 | - | - | - | - |

[a]Data extracted from the questionnaires. Frequencies in kindergartens in groups A and B are compared with those in kindergartens in Group C. F, frequency; p, level of statistical significance.

(only from the questionnaire and medical examination). Other symptoms/diseases were reported uniformly: inflammatory eye disease and irritation, irritation of the throat and respiratory tract in children, tiredness and headache in adults.

Some of the results of the technical investigations of MMMF are given in Table 2. No significant difference was found between the mean concentrations of respirable MMMF in the 3 groups of kindergartens. The mean concentrations of airborne non-respirable MMMF were the same in kindergartens in groups A and B. In the control institutions, the level was often lower than the analytical detection limit (40–80 f/m$^3$). No significant differences between the 3 groups of kindergartens were found with regard to settled fibres on regularly or occasionally cleaned surfaces, or for the other technical parameters.

Coefficients for correlation between MMMF measurements and the reported symptoms/diseases were calculated. For each kindergarten the mean values of the measurements and the frequency of persons reporting a symptom/disease were used. Some of the results are given in Table 3.

Among children, no correlation was found between symptoms/diseases and the results of the measurements on the MMMF. Among adults, the concentrations of airborne respirable/non-respirable MMMF were positively correlated with eye irritation ($p=0.03/p=0.004$). The presence of settled non-respirable MMMF on surfaces occasionally cleaned was positively correlated with skin irritation among adults ($p=0.005$).

**Table 2. Concentrations of airborne man-made mineral fibres (MMMF) (f/m³) and of settled non-respirable man-made mineral fibres (f/cm²) in 3 groups of kindergartens[a]**

| Type of MMMF | Groups of kindergartens | | |
|---|---|---|---|
| | A | B | C |
| Airborne non-respirable | 23 (8;38) | 40 (11;69) | 0-77 |
| Airborne respirable | 110 (60;160) | 97 (43;150) | 41 (10;70) |
| Settled ('low')[b] | 0-0.7 | 0-0.4 | 0-0.4 |
| Settled ('high')[c] | 0-11 | 0-120 | 0-5 |

[a] Figures in parentheses are 95% confidence limits.
[b] Surfaces cleaned regularly.
[c] Surfaces cleaned occasionally.

**Table 3. Correlation between symptoms in adults and mean values of measurements on man-made mineral fibres (MMMF)[a]**

| Symptom | | Airborne respirable MMMF | Airborne non-respirable MMMF | Settled MMMF ('low') | Settled MMMF ('high') |
|---|---|---|---|---|---|
| Eye irritation | $\rho$ | +0.64 | +0.77 | +0.19 | +0.48 |
| | $p$ | 0.03 | 0.004 | 0.55 | 0.12 |
| Irritation of the nose | $\rho$ | +0.17 | +0.48 | −0.26 | +0.73 |
| | $p$ | 0.60 | 0.11 | 0.42 | 0.007 |
| Irritation of the skin | $\rho$ | +0.25 | +0.38 | +0.01 | +0.75 |
| | $p$ | 0.43 | 0.22 | 0.97 | 0.005 |

[a] Based on replies to questionnaires; $\rho$, Spearman's correlation coefficient; $p$, level of statistical significance. For significance of 'low' and 'high', see Table 2.

## Discussion

In evaluating the results of the present study, the potential bias introduced into the epidemiological part of the study must be taken into account. For practical reasons, a double-blind technique could not be used in the study, because ceiling products were readily visible.

The elevated frequency of eye irritation and irritation of the upper respiratory tract and skin among adults in kindergartens containing MMMF compared with controls is in accordance with earlier investigations (Alsbirk et al., 1983; Jørgensen-Birch et al., 1986; Petersen et al., 1985).

The measured concentrations of MMMF were more than 10 000 times lower than the suggested Danish threshold limit values concerning the occupational environment ($10^6$ f/m³) and almost ten times lower than the exposure level in kindergartens from which severe sick building syndrome has been reported. The presence of MMMF in control kindergartens must be ascribed to the ubiquitous occurrence of MMMF in outdoor air and to indoor contamination with MMMF products used as insulation.

In evaluating the correlations between symptoms/diseases and technical measurements, it must be remembered that, with the large number of possible combinations, random events may on their own give rise to significant correlations. Statistically significant correlations associated with results indicating the same tendency — although not significant — should be considered as carrying greater weight than where the results do not do so.

## Conclusions

Eye and skin symptoms were reported more frequently in kindergartens in groups A and B than in control kindergartens. The frequencies of symptoms and diseases were uniformly distributed in the former. On their own, the health investigations may indicate that certain symptoms are related to the presence of readily visible MMMF products in the ceilings.

The measured level of MMMF pollution (average <200 f/m³) and the contribution to the pollution for sources other than the ceilings indicate that MMMF release from ceilings can hardly be responsible for the relationship.

For children, no correlation was found between the average concentration of MMMF in each institution and symptoms or diseases. For adults, however, the average concentrations of airborne MMMF were positively correlated with eye symptoms, and the presence of settled MMMF on surfaces occasionally cleaned was positively correlated with skin irritation.

## References

Alsbirk, K.E., Johansen, M. & Petersen, R. (1983) Ocular symptom and exposure to mineral fibres in boards for sound-insulation of ceilings [in Danish, English summary]. *Ugeskr. Laeg.*, 145, 43-47

Anderson, S., Auquier, A., Hauck, W.W., Oakes, D., Vandaele, W. & Weisberg, H. (1980) *Statistical Methods for Comparative Studies*, Chichester, New York, John Wiley

Cochran, W.G. (1977) *Sampling Techniques*, Chichester, New York, John Wiley

Jørgensen-Birch, L., Elbrønd, O., Kristiansen, L. & Nielsen, A.L. (1986) The conditions in the middle ears of pre-school children in day institutions with and without mineral wool ceilings [in Danish, English summary]. *Ugeskr. Laeg.*, 148, 1426-1429

Petersen, R., Rønne, H. & Sabroe, S. (1985) Health problems in housing with mineral wool ceilings [in Danish, English summary]. *Ugeskr. Laeg.*, 147, 3190-3195

Rindel, A., Bach, E., Breum, N.O., Hugod, C., Nielsen, A. & Schneider, T. (1985) *Mineral Wool Ceilings in Kindergartens* [in Danish, English summary], Copenhagen, The Working Environment Fund

Schneider, T. *et al.* (1987) Easy method for measuring the quality of cleaning. *Proceedings of the 4th International Conference on Indoor Air Quality and Climate*, Vol. 1, West Berlin, 17-21 August 1987

# V. RISK EVALUATION

# FIBRE CARCINOGENESIS AND ENVIRONMENTAL HAZARDS

## J. Peto

*Section of Epidemiology, Institute of Cancer Research,
Sutton, Surrey, UK*

The main aim of this paper is to re-examine the epidemiological assumptions and exposure measurements that underlie asbestos risk assessment. Extrapolation from observations on asbestos workers to predict the cancer risk caused by exposure in other industrial or non-occupational situations involves three steps: measuring dose, measuring the resulting excess cancer risk, and establishing a formula for the relationship between exposure and risk. There is little difficulty in measuring the risk, provided that the excess is substantial and suitable comparison rates are available for the local population; but measuring exposure and choosing an appropriate dose-response model are very much more difficult.

Three other aspects of fibre carcinogenesis are also briefly discussed here: (1) the widespread belief that one thin fibre is more dangerous than one thick fibre appears to be the opposite of the truth; (2) the suggestive epidemiological evidence that some synthetic fibres may be more potent bronchial carcinogens than chrysotile asbestos at the same fibre level is therefore not surprising; and (3) the evidence that asbestos can cause gastrointestinal cancer is much weaker than is generally supposed.

Much of the material referred to is described in more detail in a review prepared for the UK Health & Safety Commission (Doll & Peto, 1985). A formal description of the models for lung cancer and mesothelioma incidence can be found in Peto *et al*. (1985).

## *Dose measurement*

Prolonged heavy exposure has in the past caused relative risks for lung cancer of between about 2 and 10 in mining, manufacture and use of the major types of asbestos. The only sectors of the industry in which such high risks have not been observed are those in which asbestos, particularly chrysotile, is bonded into other material, as in the manufacture of chrysotile asbestos-cement and friction products. Estimates of the dust levels that caused these high risks have varied widely, however, resulting in much wider variation in risk estimates than in observed cancer rates. This is illustrated by the comparison of the chrysotile textile workers and chrysotile miners and millers studied by McDonald and his colleagues (McDonald *et al*., 1980, 1983). The most heavily exposed workers in these cohorts suffered relative risks for lung cancer of about 10 for textiles and about 3 for mining. Historical exposure estimates indicated that the highest exposures were more than ten times greater in the mines than in the textile factory, however, resulting in a 50-fold difference in risk estimates (Figure 1).

**Fig. 1. Standardized mortality rates for respiratory cancer in relation to dust exposure accumulated to age 45 in chrysotile mining and milling, and in chrysotile textile manufacture**

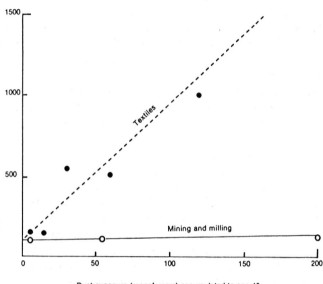

*Source:* McDonald et al. (1983)

Historical exposure estimates are thus crucial to asbestos risk assessment. There are only three cohort studies of asbestos workers in which substantial excess cancer risks have been observed, and extensive particle counts taken in the early 1950s or before, together with the basis for their conversion to fibre counts, have been described in reasonable detail. These are the 2 studies described above, and a cohort of English chrysotile textile workers. The dose-response estimates derived from these three studies are often cited, but few reviewers have examined the reliability of the dust measurements on which they are based, perhaps because the original data have not been readily accessible. The exposure data on the English textile factory were not adequately described until 1985 (Peto et al., 1985), although they had been used in several earlier dose-response analyses (e.g., Peto, 1978), and detailed exposure data on the Quebec miners and millers (McDonald et al., 1980) and the South Carolina textile workers studied independently by Dement et al. (1982) and McDonald et al. (1983) were published separately (Dagbert, 1976; Dement, 1980) and only briefly summarized in the main reports. There are 4 major limitations of historical exposure measurements, each of which casts serious doubt on the validity of any quantitative risk assessment for asbestos:

1. No simple or consistent relationship exists between particle counts taken in the past with different instruments, or between particles and fibres. Comparisons of measurements with different instruments have revealed inconsistencies in the ratio of particle to fibre counts between instruments, between personal and static samples, between different sectors of the industry, and even between different areas within a single factory. In view of these inconsistencies, the ratio of particle to fibre counts observed in parallel studies provides a very unreliable basis for converting earlier (and often much higher) particle counts to fibre counts.
2. In each instance (English and South Carolina textile factories, and Quebec chrysotile miners), the dose-response relationship deduced was strongly influenced by the high mortality experienced by workers who were heavily exposed in periods and areas where little or no routine dust measurement was done. These exposures could only be guessed.
3. The conclusions drawn have emerged from the handful of workplaces in which extensive historical sampling was carried out. No exposure data are available from which useful risk estimates can be calculated for any type of crocidolite or amosite exposure or for many sectors of the chrysotile industry.
4. Even modern fibre counts by electron microscopy are unlikely to provide a reliable measure of carcinogenic effect. Carcinogenicity and transforming potency per fibre vary with fibre dimension and fibre type in animal studies and *in vitro*, but these relationships are still poorly understood. Furthermore, effects in man may be quite different from those in animals, and there are no useful data on the effects of fibre type or dimensions on human carcinogenicity.

After reviewing the evidence underlying these assessments, we concluded that:

'When so much work has been done in collecting and analysing measures of ambient pollution, we hesitate to suggest that the results are insufficiently reliable to justify making any quantitative extrapolation from past experience to the effects of current exposures. Nevertheless, this may, in fact, be the case and we may have to be satisfied with qualitative conclusions based on knowledge of the direction in which progress has been made and epidemiological observations of the effects of qualitatively different types of exposure.' (Doll & Peto, 1985)

## Dose-response relationships for lung cancer and mesothelioma

The majority of published dose-specific estimates of the cancer risk caused by asbestos exposure have been based on the following statistical models:

1. The increase in relative risk for lung cancer is proportional to cumulative asbestos exposure, and the effects of asbestos and cigarette smoking multiply each other.
2. The increase in absolute mesothelioma incidence caused by each brief period of asbestos exposure is proportional to the amount of that exposure (exposure level × duration) and to a power of the time since it occurred,

independent of age or smoking history. The power of time appears to be about 3 (Peto et al., 1982).

The qualitative predictions of these models for a mixed population including smokers and non-smokers are illustrated in Table 1. The lung cancer risk would be approximately 50% greater in smokers, and only about 10% of these rates in non-smokers, among whom the mesothelioma risk exceeds the lung cancer risk. In relation to environmental exposure, an important prediction is that a higher risk for mesothelioma, but not lung cancer, is caused by early exposure.

**Table 1. Predicted excess numbers of deaths before age 80 per 1000 men due to lung cancer or mesothelioma caused by asbestos exposure during working hours at 1.0 fibre/ml**

| Age at first exposure (years) | | Excess deaths per 1000[a] | | | |
|---|---|---|---|---|---|
| | Years exposure: | 5 | 10 | 20 | 40 |
| 0 | Lung cancer | 3.7 | 7.3 | 14.4 | 28.4 |
| | Mesothelioma | 7.5 | 13.1 | 20.1 | 24.7 |
| 20 | Lung cancer | 3.7 | 7.4 | 14.7 | 27.4 |
| | Mesothelioma | 2.1 | 3.5 | 5.0 | 5.6 |
| 40 | Lung cancer | 3.7 | 7.3 | 13.2 | 16.8 |
| | Mesothelioma | 0.3 | 0.5 | 0.6 | 0.7 |

[a]Calculated by extrapolation from observations on English chrysotile textile workers (Peto et al., 1985).

The widespread adoption of these or similar models as a basis for asbestos risk assessment has created the false impression that they are reliably established. They were originally proposed on the grounds that they were qualitatively consistent with available data and corresponded to plausible models of carcinogenesis, but both observation and common sense suggest that they are at best useful approximations, and cannot provide reliable predictions far beyond the observed range of dose, age or duration of exposure.

## Predictions of the lung cancer model

The model for lung cancer implies that the excess relative risk is: (a) proportional to average duration of exposure for a given level of exposure (fibre/ml); and (b) proportional to average level of exposure at a given duration of exposure. The first of these predictions may be approximately true, although anomalously high risks have sometimes been observed in short-service workers. The second prediction has rarely been examined. Various studies have shown a roughly linear relationship between relative risk and cumulative dose, which increases with duration of exposure and (during employment) with time since exposure. Such a relationship does not, however, necessarily imply that exposures were consistently estimated. Indeed, a cohort in which all exposure levels were equal, or even one in which different exposure levels were assigned to individual workers at random, would exhibit an apparently consistent increase in risk with increasing cumulative dose by virtue of the correlation between risk and duration of exposure.

The model also implies that the relative risk remains constant after exposure has ceased, but there is evidence from several cohorts that the relative risk eventually falls (Doll & Peto, 1985; Walker, 1984). This eventual decline appears to be more marked for chrysotile than for the amphiboles.

## Predictions of the mesothelioma model

The model for mesothelioma was proposed to provide a simple explanation of the observation that mesothelioma incidence continues to increase as the third or fourth power of time since first exposure to asbestos even after exposure has ceased, and is independent of age at exposure (Peto et al., 1982). In spite of its widespread use in risk assessment, however, it has only once been formally fitted in a cohort with individual exposure histories (Peto et al., 1985). This analysis was based on only 10 cases, but it suggested that the risk is lower for shorter exposures and higher for longer exposures than the model predicts.

## Relationship to conventional models of carcinogenesis

In terms of conventional notions of multi-stage carcinogenesis, the lung cancer model implies that the rate of the final transition in a multi-stage process is increased in proportion to cumulative dose. Asbestos is thus assumed to act exclusively as a 'promoter' affecting only a late stage in bronchial carcinogenesis. In contrast, the mesothelioma model can be correct only if asbestos acts primarily as an initiator of mesothelial carcinogenesis. There is mounting evidence that many, and perhaps all, 'initiating' carcinogens also act at a later stage in carcinogenesis. A more plausible model may therefore be that the strongest effect happens to be at an early stage for mesothelioma and at a late stage for lung cancer, but that there are both early and late stage effects at both sites. For such a two-stage effect, increases in both exposure level and duration of exposure should cause disproportionate increases in incidence. This cannot be tested adequately, as no accurate data exist on the separate effects of uniform exposure for different durations or equal durations at different measured exposures. The evidence that prolonged exposure causes a disproportionate increase in mesothelioma risk lends some support to a two-stage model, at least for mesothelioma. Cancer incidence increases as a power of duration for various animal carcinogens, and for lung cancer in cigarette smokers.

A further difficulty in formulating any simple model of asbestos carcinogenesis is that inhaled or ingested fibres remain in the body for many years, although they may be chemically and physically altered. They are also progressively dispersed, inactivated or removed by migration, dissolution, encapsulation or phagocytosis, and the extent of these processes varies with asbestos type and fibre size. The relationship between inhaled and effective dose is therefore likely to be intermediate between that of a chemical carcinogen which is rapidly eliminated and the persistent chronic effect of an increasing cumulative dose. The suggestion that the relative risk for lung cancer eventually falls, and that the reduction is greater for chrysotile than for the amphiboles, is consistent with a late-stage effect modified by progressive elimination of residual fibres. Both human and animal studies suggest that chrysotile is eliminated or inactivated more rapidly than the amphiboles.

In the absence of better evidence, such armchair speculations are of little practical value in formulating quantitative risk assessment models. They may, however, indicate directions for further analysis of existing data. In particular, more detailed analysis of the relationship between mesothelioma risk and duration of exposure in cohorts such as the Quebec chrysotile miners and millers and the Australian crocidolite miners would be informative (Hobbs et al., 1980). Better data on changes over time in lung cancer rates, subdividing observations beyond 20 years after first exposure into 5- or 10-year intervals, could also be obtained from a number of studies.

## Implications for industrial control

Many of the uncertainties in these specific models for lung cancer and mesothelioma induction are relatively unimportant in relation to industrial control. Most cohort studies span a wide range of ages at first exposure and durations of exposure, and include workers who have been followed for many years. The only assumption that need be made in relation to these temporal variables is therefore that the patterns observed in the past will persist in the future, and these models provide a convenient device for formalizing this assumption.

The important assumption is that of a linear dose-response relationship, which means that if the exposure level is halved in a particular sector of the industry, the resulting risk will also be halved. Past exposure measurements are so unreliable that this is difficult to test, but most human and animal data on other carcinogens suggest that reductions in exposure level usually produce reductions in risk that are at least proportional, and sometimes greater.

The serious limitation of this common-sense approach to industrial regulation is the absence of any satisfactory measure of exposure level. There may be an approximately linear relationship between average particle and fibre counts within particular sectors of the industry, although this has certainly not been demonstrated; and it may be reasonable to assume that the highest observed dose-specific risk, that based on studies of chrysotile textile workers, will tend to overestimate the risk caused by other uses of chrysotile. No such assumption can be justified for crocidolite or amosite, however, for which there are no cohorts with measured historical exposures from which dose-specific risks can be calculated. The evidence that the amphiboles, and particularly crocidolite, are more dangerous than chrysotile under similar working conditions may reflect higher fibre levels, a higher risk per fibre, or both.

## Environmental exposure

The reliability of these models in predicting the effect of environmental exposure to asbestos is very much more dubious, for two reasons. First, the assumption of linear dose-response must be assumed to hold over more than four orders of magnitude of exposure level, from industrial studies of workers exposed at levels of 10 fibre/ml or more to environmental situations in which average asbestos levels rarely exceed 0.001 fibre/ml. This difficulty is compounded by differences in the distribution of fibre sizes encountered in environmental and occupational settings. Second, the effects of childhood exposure cannot be predicted. The models described above imply a roughly fourfold increase in risk for mesothelioma, but not for lung cancer, when exposure begins soon after birth rather than at age 20, reflecting the cubic residence time assumption (Table 1). Such an age-related effect would be expected for any carcinogen which initiates the induction of a multi-stage carcinogenic process; but this prediction takes no account of the possibility that children are particularly susceptible to carcinogenesis by virtue of factors such as stem cell expansion during growth and development. The risks caused by exposure in childhood may therefore be substantially greater than those predicted for both mesothelioma and lung cancer.

## *Experimental evidence on the effects of fibre length and diameter*

The evidence that different types of fibre differ in solubility, biological activity and transforming or carcinogenic potential has been discussed in other contributions to this volume. Unfortunately, these experimental differences are unlikely to correlate consistently with the carcinogenic effects of different fibre types in man. Human carcinogenesis occurs over decades rather than days or months, and differences *in vivo* in migration, encapsulation and dissolution over such long periods are difficult to study and impossible to translate into predictions of risk. It may nonetheless be possible to establish a simple relationship between fibre dimension and transforming potency *in vitro* or carcinogenic potency *in vivo* following pleural implantation. Different fibres of similar shape and size may have different potencies, but it is reasonable to assume, at least provisionally, that the risk will vary with length and diameter in much the same way for all fibre types. The dependence of biological activity on length and diameter is still not well established, but the following hypothesis, which was proposed by Stanton *et al.* (1977), seems consistent with the available data:

(1) Fibres less than 8 $\mu$m in length may have little or no transforming or carcinogenic activity, perhaps because they are engulfed by phagocytes following pleural implantation (Stanton *et al.*, 1977). Mesothelioma rates following pleural implantation in rats suggest a threshold between about 4 $\mu$m and 8 $\mu$m (Stanton *et al.*, 1977, 1981), and inhalation studies in rats (Davis *et al.*, 1986) suggest that fibres shorter than 5 $\mu$m rarely or never cause pulmonary tumours or fibrosis.

(2) Above this length, the carcinogenic and/or transforming potency per fibre (not per unit mass of fibres) increases both with increasing length (see e.g., Stanton *et al.*, 1981) and with increasing diameter (Hesterberg & Barrett, 1984). This last point is not generally appreciated, as the effects of different fibres are usually compared at similar total masses rather than at similar fibre

concentrations. Thus, for example, Hesterberg & Barrett (1984) observed that the ratio of the transforming potencies of thin versus thick glass fibres was 20:1 for equal masses, but 1:40 (800 times less) for equal numbers of fibres.

Stanton et al. (1977) suggested that the carcinogenic potency of fibres may be proportional to the total surface area that escapes phagocytosis, so that the risk per fibre is very low for fibres less than about 8 $\mu$m long, but for long fibres the effect per fibre is proportional to the product of length and diameter. This simple hypothesis has yet to be tested adequately, but it seems at least qualitatively consistent with experiment. In particular, there is no evidence either of an upper limit for length or a lower limit for diameter beyond which such a relationship ceases to hold. This is exactly the opposite of the almost universal belief that a thin fibre is more dangerous than a thick one. The basis of this belief seems to be simply that thin fibres are more dangerous than thick ones per unit mass, and most experiments have been conducted with equal masses of fibres. A moment's reflection reveals the fallacy of assuming the same relationship with diameter for equal fibre concentrations.

## Carcinogenicity of man-made mineral fibres

Analysis of effects per fibre rather than per unit mass is also relevant to the comparison of asbestos and man-made mineral fibres (MMMF) in the induction of pulmonary tumours by inhalation. The results shown in Table 2, for example, indicate that a similar mass of chrysotile produced a greater tumour incidence than MMMF but that the risk per fibre may have been similar for chrysotile and glass wool, and possibly highest for rock wool, although the numbers are small (Wagner et al., 1984). All groups were exposed to approximately 10 mg/m$^3$, but the fibre count was more than 10 times higher for chrysotile than for rock wool or glass wool. A separate experiment, again at 10 mg/m$^3$, in which chrysotile, crocidolite and amosite were compared gave similar results to those shown in the table for chrysotile (8 carcinomas and 7 adenomas in 40 rats), but only 1 mesothelioma and 3 adenomas in 37 rats inhaling crocidolite or amosite (Davis et al., 1978). These differences disappeared when the results were expressed as risks per fibre/ml over 20 $\mu$m in length, however, as the fibre count was much higher for chrysotile than for the amphiboles. The low incidence of tumours produced by inhalation for both MMMF and amphiboles may thus be entirely due to the difficulty in achieving high concentrations of long respirable fibres rather than to a lower risk per fibre.

Most MMMFs are thicker than airborne asbestos fibres, and the belief that thick fibres are less carcinogenic than thin ones has made the observation of excess lung cancer among certain MMMF workers exposed to quite low fibre levels seem anomalous (Doll, 1987). If thick fibres are indeed more dangerous than thin ones, however, this anomaly disappears. There is thus a striking and disturbing similarity between the experimental and epidemiological results for MMMF and amphiboles. Both appear less dangerous than chrysotile in experimental carcinogenesis, although these differences may be accounted for by differences in fibre counts; and both are

**Table 2. Lung tumour induction in rats by chronic inhalation**

| Material (No. of rats) | Adeno-carcinoma | All neoplasms | Dose (fibre/ml hours) at 12 months | Tumours per 100 fibre/ml hours |
|---|---|---|---|---|
| UICC chrysotile (48) | 11 | 12 | 656 | 1.8 |
| Glass microfibre (48) | 1 | 1 | 223 | 0.4 |
| Rock wool (48) | 0 | 2 | 39 | 5.1 |
| Glass wool with resin (48) | 1 | 1 | 55 | 1.8 |
| Glass wool resin-free (47) | 0 | 1 | 41 | 2.4 |
| Controls (48) | 0 | 0 | - | - |

Source: Wagner et al. (1984).

suspected to be more liable than chrysotile to cause lung cancer in humans at the same fibre level, although epidemiological data on their dose-specific effects are still inadequate. This is rather a superficial analogy, as the amphiboles are less liable to dissolution in the lung than chrysotile, while the converse is probably true of many MMMFs. Certain MMMFs may, however, be more carcinogenic than chrysotile in the bronchus, possibly because they are longer and thicker. There is no epidemiological evidence that inhaled MMMF causes an appreciable risk of mesothelioma. Only very fine inhaled asbestos fibres are likely to reach the pleura (Harington, 1981), and most MMMF may be too thick to migrate extensively.

## *Ingestion of asbestos and gastrointestinal cancer*

The possibility that asbestos may cause gastrointestinal cancers was first suggested by Selikoff *et al.* (1964), who observed a highly significant excess of stomach and colorectal cancer in a heavily exposed cohort of asbestos workers. Smaller but still statistically significant excesses have been observed in a few subsequent studies, and it has become widely accepted that the effect is real, and is caused by ingestion of asbestos. Some doubt has remained, however, as asbestos has consistently failed to cause any cancer in animal feeding experiments and the excesses observed in humans have been small, with relative risks rarely exceeding 2.

An alternative explanation is that the excess is due largely, or even entirely, to occasional misdiagnosis of mesothelioma and lung cancer, both of which were common in the cohorts showing marked excesses of gastrointestinal cancer (Doll & Peto, 1985). Plots of the relative risks for gastrointestinal cancer (Figure 2a) and all other cancers except lung (Figure 2b) against the relative risk for lung cancer in various studies show marked and similar correlations, the only exception being the hypothesis-generating study (Selikoff *et al.*, 1964), in which there was a large excess of gastrointestinal cancer and no excess at other non-respiratory sites. Subsequent studies thus suggest that the relative risks for cancers of virtually all sites except lung cancer and mesothelioma are increased by more or less the same amount, and that the excess is proportional to, although much less than, the excess of lung cancer. The evidence that gastrointestinal cancers are caused by asbestos is thus neither stronger nor weaker than the evidence that almost all forms of cancer are. The conclusion, albeit tentative, that any excess of gastrointestinal cancer is due largely, and perhaps entirely, to misdiagnosis seems preferable to the only reasonable alternative, that asbestos causes cancer of almost every site. This is an important environmental issue, as many water supplies are contaminated with asbestos released from asbestos-cement pipes or from natural sources, and it has been suggested that this might cause an unacceptable risk of gastrointestinal cancer in the general population supplied with water from these sources.

## Discussion

This review has concentrated on the uncertainties inherent in asbestos risk assessment. Certain widespread beliefs, notably that thin fibres are more dangerous than thick ones, and that gastrointestinal cancers are more likely to be caused by asbestos than other non-respiratory cancers, are unsupported or even contradicted by the evidence. The dose-response models generally adopted for lung cancer and mesothelioma are only weakly supported, and certain observations suggest that they are at best only useful approximations. Their predictions should certainly be critically examined in the available cohorts. The virtual absence of historical exposure measurements for many sectors of the industry is generally recognized; but the weaknesses and internal inconsistencies of the exposure data on the few workplaces in which measurements were taken in the 1940s and 1950s are less well known, if only because most of the data have not appeared in widely accessible publications. Reviews of the evidence have often presented the results of the few studies for which measured exposures are available together with those from the larger number for which exposure levels were guessed on the basis of a handful of historical (or even recent) measurements, conveying the spurious impression of a substantial and varied set of dose-response estimates spanning a range that is likely to include the 'true' value.

In spite of these uncertainties, one practical question has to be addressed. Society must decide when, if ever, the risk from environmental exposure to asbestos or other fibres is high enough to justify the cost, hazard and inconvenience of removing its source. The conventional answer is that the risk assessment calculations shown in Table 1 predict a negligible hazard. Average fibre counts in contaminated buildings, at least in the UK, are usually less than 0.001 fibre/ml, and the corresponding predicted lifelong risk is of the order of 1 in 100 000 for 10 years' occupancy. The number of people exposed at this sort of level is not known, but some idea of what a risk of 1 in 100 000 means may be conveyed by assuming that a risk of this magnitude is suffered

**Figure 2. Relative risks for (a) gastrointestinal cancers and (b) cancers other than lung and gastrointestinal plotted against the corresponding relative risks for lung cancer in various cohorts of male asbestos workers**

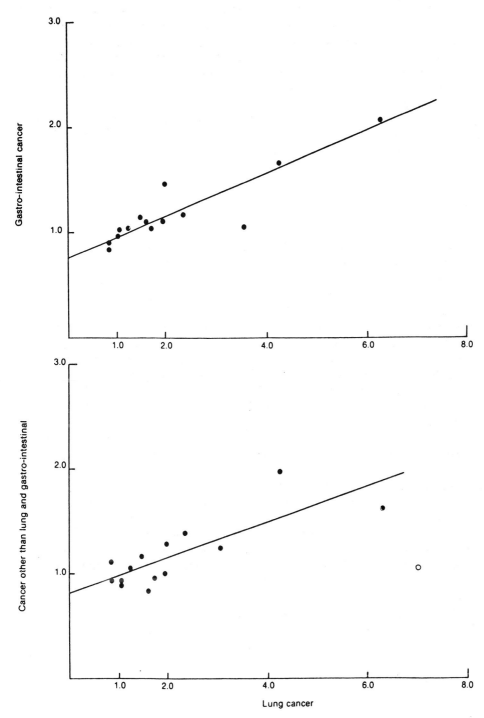

Open circles denote the results reported by Selikoff et al. (1964).
*Source:* Doll & Peto (1985)

by 1 in 5 of the population. This would imply about one death per year in the UK, or 5 per year in the USA, and corresponds to an average loss of expected life of about 15 minutes. This prediction cannot be tested directly. As McDonald has observed (this volume, pp. 420-427), however, data from various countries suggest that the incidence of mesothelioma in those who have no history of any asbestos exposure and do not have unusually high fibre concentrations in their lungs is of the order of 1 per million per year, corresponding to a lifelong risk of about 1 in 10 000. This rate has not increased much, and perhaps not at all, over the last 10 or 20 years, a period in which the incidence due to occupational exposure has risen sharply. The average lifelong risk to the population due to environmental exposure is thus unlikely to be much more than 1 in 10 000, and is probably substantially less.

We thus have an inevitably unreliable risk assessment for chrysotile that suggests a negligible effect of contaminated buildings and a higher, but still very low, upper limit for the overall mesothelioma incidence due to environmental exposure. On this evidence the possibility remains that certain asbestos fibres, and particularly the amphiboles, cause substantially higher individual risks at fibre concentrations of the order of 0.001 fibre/ml. There is no direct evidence on the dose-specific effects of the amphiboles, but the very high mesothelioma rates that have occurred among Turkish villagers apparently exposed to average erionite levels of the order of 0.01 fibre/ml (Simonato et al., this volume, pp. 398-405) suggest that erionite is over 1000 times more potent than chrysotile asbestos. Further studies of these Turkish populations, and particularly case-control studies of fibre concentrations in lung tissue, are needed to exclude the possibility that heavy occasional exposures rather than ambient levels caused these mesotheliomas; but these results, together with evidence that erionite is an extraordinarily potent carcinogen in animals (Wagner et al., 1985), suggest that certain fibre types are so dangerous that no measurable exposure can be regarded as safe. Better understanding of the reasons underlying this remarkable potency is needed to ensure that the dimensions, durability and surface properties of newly introduced synthetic fibres do not produce similar effects.

The pooled observations on MMMF workers suggest an upper limit for the dose-specific effect of certain traditional MMMFs on lung cancer incidence that may well be higher than that for chrysotile; but they also show that most parts of the industry have never suffered the high lung cancer risks that were commonplace in asbestos workers, and mesothelioma has rarely or never been caused (Doll, 1987). MMMFs do not in most applications produce high fibre concentrations, and sensible working practices, combined with a regulated maximum concentration similar to that for chrysotile, should provide acceptable control in the industry. (The maximum chrysotile concentration in the UK is 0.5 fibre/ml. Since the introduction of this limit, average concentrations rarely exceed 0.25 fibre/ml and in many sectors are much less.) There is, however, an urgent need for much larger inhalation experiments of the type shown in Table 2. The numbers studied by Wagner et al. (1984) were so small that the tumour rates did not differ significantly between the MMMF-exposed groups and the controls; but the results, expressed as risks per unit dose, suggest that some MMMFs may have been at least as carcinogenic as chrysotile asbestos.

In relation to environmental control, my personal view is that the low levels of chrysotile commonly found in most buildings are unlikely to produce lifelong risks much exceeding 1 in 100 000, but that amphiboles and newly introduced synthetic fibres must be regarded more cautiously until the mechanisms of fibre carcinogenicity are better understood. Environmental levels of traditional MMMFs are also usually very low. As for asbestos, however, remedial action may be indicated for friable MMMF on accessible surfaces, which can give rise to relatively high fibre releases (Gaudichet et al., this volume, pp. 291-298). Those who demand the removal and substitution of all asbestos, irrespective of fibre type or level of contamination, should note that removal can actually increase cumulative doses to both workers and occupants (Burdett et al., this volume, pp. 277-290), and that substitutes for asbestos may be less innocuous than has generally been assumed. The campaign to eliminate all asbestos on the grounds that 'one fibre can kill', besides being a cost-benefit absurdity, may thus actually increase the risk.

## References

Dagbert, M. (1976) *Etudes de Correlation de Mesures d'Empoussiérage dans l'Industrie de l'Amiante. Document 5 (Beaudry Report)*, Montreal, Quebec, Comité d'Etude sur la Salubrité dans l'Industrie de l'Amiante

Davis, J.M.G., Beckett, S.T., Bolton, R.E., Collings, P. & Middleton, A.P, (1978) Mass and number of fibres in the pathogenesis of asbestos-related lung disease in rats. *Br J. Cancer*, 37, 673-688

Davis, J.M.G., Addison, J., Bolton, R.E., Donaldson, K., Jones, A.D. & Smith, T. (1986) The pathogenicity of long versus short fibre samples of amosite asbestos administered to rats by inhalation and intraperitoneal injection. *Br. J. Exp. Path.*, 67, 415-430

Dement, J.M. (1980) Estimation of dose and evaluation of dose-response in a retrospective cohort mortality study of chrysotile asbestos textile workers. (Doctoral dissertation). University of North Carolina

Dement, J.M., Harris, R.L., Symons, M.J. & Shy, C. (1982) Estimates of dose-response for respiratory cancer among chrysotile asbestos workers. *Ann. Occup. Hyg.*, 26, 869-887

Doll, R. (1987) Symposium on Man-made Mineral Fibres, Copenhagen, October 1986. Overview and conclusions. *Ann. Occup. Hyg.*, 31, 805-820

Doll, R. & Peto, J. (1985) *Effects on Health of Exposure to Asbestos*. A report to the Health and Safety Commission. London, Her Majesty's Stationery Office

Harington, J.S. (1981) Fiber carcinogenesis: epidemiologic observations and the Stanton hypothesis. *J. Natl Cancer Inst.*, 67, 977-989

Hesterberg, T.W. & Barrett, J.C. (1984) Dependence of asbestos and mineral dust-induced transformation of mammalian cells in culture on fiber dimension. *Cancer Res.*, 44, 2170-2180

Hobbs, M.S.T., Woodward, S.D., Murphy, B., Musk, A.W. & Elder, J.E. (1980) The incidence of pneumoconiosis, mesothelioma and other respiratory cancer in men engaged in mining and milling crocidolite in Western Australia. In: Wagner, J.C., ed., *Biological Effects of Mineral Fibres*, Vol. 2 (*IARC Scientific Publications No. 30*), Lyon, International Agency for Research on Cancer, pp. 615-625

McDonald, A.D., Fry, J.S., Woolley, A.J. & McDonald, J.C. (1983) Dust exposure and mortality in an American chrysotile textile plant. *Br. J. Ind. Med.*, 40, 361-367

McDonald, J.C., Liddell, F.D.K., Gibbs, G.W., Eyssen, G.E. & McDonald, A.D. (1980) Dust exposure and mortality in chrysotile mining, 1910-75. *Br. J. Ind. Med.*, 37, 11-24

Peto, J. (1978) The hygiene standard for chrysotile asbestos. *Lancet*, i, 484-489

Peto, J., Seidman, H. & Selikoff, I.J. (1982) Mesothelioma mortality in asbestos workers: implications for models of carcinogenesis and risk assessment. *Br. J. Cancer*, 45, 124-135

Peto, J., Doll, R., Hermon, C., Binns, W., Clayton, R. & Goffe, T. (1985) Relationship of mortality to measures of environmental asbestos pollution in an asbestos textile factory. *Ann. Occup. Hyg.*, *29*, 305-355

Selikoff, I.J., Churg, J. & Hammond, E.C. (1964) Asbestos exposure and neoplasia. *J. Am. Med. Ass.*, *188*, 22-26

Stanton, M.F., Layard, M., Tegeris, A., Miller, E., May, M. & Kent, E. (1977) Carcinogenicity of fibrous glass: pleural response in the rat in relation to fiber dimension. *J. Natl Cancer Inst.*, *58*, 587-603

Stanton, M.F., Layard, M., Tegeris, A., Miller, E., May, M., Morgan, E. & Smith, A. (1981) Relation of particle dimension to carcinogenicity in amphibole asbestos and other fibrous minerals. *J. Natl Cancer Inst.*, *67*, 965-975

Wagner, J.C., Berry, G.B., Hill, R.J., Munday, D.E. & Skidmore, J.W. (1984) Animal experiments with MMM(V)F — effects of inhalation and intrapleural inoculation in rats. In: *Biological Effects of Man-made Fibres, Vol. 2,* Copenhagen, World Health Organization, Regional Office for Europe, pp. 209-233

Wagner, J.C., Skidmore, J.W., Hill, R.C. & Griffiths, D.M. (1985) Erionite exposure and mesothelioma in rats. *Br. J. Cancer*, *51*, 727-730

Walker, A.M. (1984) Declining relative risks for lung cancer after cessation of asbestos exposure. *J. Occup. Med.*, *26*, 422-425

# DEVELOPMENT AND USE OF ASBESTOS RISK ESTIMATES

### J.M. Hughes & H. Weill

*Tulane University School of Medicine, Department of Medicine, Pulmonary Diseases Section, New Orleans, LA, USA*

*Summary.* Several groups of researchers, at the request of government agencies in various countries, have reviewed the literature on the health effects of asbestos exposure and derived quantitative estimates of the risks from such exposures. Because of differences in the area of responsibility of the different agencies, these estimates have inevitably been based on different assumptions concerning exposure (concentration, number of years of exposure, initial age, and number of hours per week). The lifetime cancer risk estimates derived in 6 reports are compared here, after adjustment to a common set of exposure assumptions appropriate for students attending schools containing asbestos products. Estimates are found to be reasonably similar, remaining differences being primarily due to differences in assumptions concerning parameter values of the models and the background risk of lung cancer.

The various administrative processes used by the US agencies are also compared, and recommendations made for US government agencies involved in the derivation of risk estimates for exposures of general public concern.

Government agencies in various countries have arranged for scientists to review the literature on the health effects of asbestos exposure and to make quantitative estimates of the risks from such exposures. The lifetime cancer risk estimates derived in 6 reports are considered here (Consumer Product Safety Commission, 1983; Health and Safety Executive, 1983; National Research Council, Committee on Nonoccupational Health Risks, 1984; Ontario Royal Commission, 1984; Environmental Protection Agency, 1985; US Department of Labor, Occupational Safety and Health Administration, 1986), as well as the varying administrative approaches used in the development of the US reports. Based on the US experience, recommendations are made for US government agencies involved in the derivation of risk estimates for exposures of general public concern.

Because of differences in the area of responsibility of the different agencies, the various groups of researchers involved in deriving estimates of cancer risk from asbestos exposure have inevitably made different assumptions concerning exposure (concentration, number of years of exposure, initial age, and number of hours per week). In order to compare these estimates, adjusted to common exposure assumptions, the risks to students attending schools containing friable asbestos products will be considered.

The Ontario Royal Commission, after a complete review of the data available on asbestos concentration levels inside buildings, concluded that the best estimate of the average asbestos concentration inside buildings containing asbestos products was 0.001 fibres per millilitre of air (f/ml), with peaks of 0.01 f/ml occurring occasionally (Ontario Royal Commission, 1984). Our review of the available buildings exposure data indicates that these estimates reasonably reflect average prevailing conditions. For purposes of comparison of the risk estimates, an average concentration of 0.001 f/ml will be assumed. Students will be assumed to begin exposure in schools at age 10, and continue this exposure for 5 school years (35 h per week, 36 weeks per year). Generally, a mixed fibre exposure will be assumed.

On these assumptions, the adjusted risk estimates based on 5 reports are given in Table 1. Estimates were generally reported separately for men and women, and, in some reports, separately for smokers and non-smokers. In adjusting these estimates, it was assumed that the population would be half male and half female, and that 50% of both males and females would be smokers. To the extent that smoking rates may be lower in the future (currently, 28% of US adults smoke), this assumption will lead to risk overestimation. The estimates in the US Occupational Safety and Health Administration (OSHA) report (US Department of Labor, Occupational Safety and Health Administration, 1986) were very similar to those in the Environmental Protection Agency (EPA) report, and were provided by the same consultant; they are

**Table 1. Estimated lifetime risks from asbestos exposure in schools[a]**

| Report | Lung cancer | | Mesothelioma estimate | Total |
|---|---|---|---|---|
| | Slope[b] | Estimate | | |
| EPA (1986) | 1.0 | 1.5 | 5.8 | 7.3 |
| CPSC (1983) | 0.3-3.0 | 0.3-2.8 (0.93)[c] | 1.6-16.0 (5.3)[c] | 1.9-18.8 (6.2)[c] |
| NRC (1984) | 2.0 | 2.7 | 10.0 | 12.7 |
| ORC (1984)[d] | 1.01 | 2.5 | 8.0 | 12.5 |
| HSE (1983)[e] | 1.0 | 2.4 | <1[f] | <3.4[f] |

[a]Total deaths over a lifetime attributable to asbestos exposure per one million students exposed to 0.001 f/ml of mixed fibres for 5 school years beginning at age 10. Based on Environmental Protection Agency (1986); Consumer Product Safety Commission (1983); National Research Council, Committee on Nonoccupational Health Risks (1984); Ontario Royal Commission (1984); Health and Safety Executive (1983).
[b]Slope of the dose-response line; increase in standardized mortality ratio per unit increase in cumulative asbestos exposure (f/ml-years).
[c]Estimates based on a slope of 1.0 for lung cancer and a value of $1.0 \times 10^{-8}$ for the mesothelioma model parameter.
[d]77% males.
[e]Males only.
[f]Chrysotile exposure.

therefore not included in the table. In their report, the National Research Council (NRC) failed to use lifetable methods for estimating mesothelioma risk, or to adjust lung cancer risk to continuous rather than work-place exposure. The appropriate corrections have been made in calculating the estimates given in the table.

There is reasonably good agreement between the various estimates, as would be expected, since the same general methodologies and risk models were used. The estimate in the Health and Safety Executive (HSE) report was for males only, which at least partially accounts for the somewhat higher value.

As in all mathematical modelling, the choice of the parameter values for the lung cancer and mesothelioma models is critically important. For lung cancer risk, a slope (the increase in the Standardized Mortality Ratio per unit increase in cumulative asbestos exposure in f/ml-years) of 1.0 was used in the reports by the EPA and the HSE, whereas the NRC used a value of 2.0, and the Consumer Product Safety Commission (CPSC) employed a range of slopes. Although the NRC report stated that the median slope from the studies considered was 1.1, this value was rounded upwards to 2.0 in calculating the lung cancer risk. The epidemiological literature suggests that a value of 1.0 is likely to be appropriate in many situations, although the data from mining and the manufacture of friction products suggests a lower slope for these exposures. As in the lung cancer model, the NRC selected a relatively high parameter value for the mesothelioma model, resulting in a higher estimate of risk than in the other reports.

Even when the same slope (1.0) and US rates are used, differences in the lung cancer risk estimates remain. For example, the EPA risk estimate is approximately 50% higher than that of the CPSC. These differences are primarily due to differences in the assumptions concerning the background risk of lung cancer in the absence of asbestos exposure. Although, for this application, the difference in absolute numbers is small (approximately one case as compared with 1.5 cases, over the lifetime of the one million students), a 50% increase in the number of attributable deaths could be substantial in some applications, and especially in occupational exposures. The assumed background risk of lung cancer is therefore an important factor in the risk estimate; more detailed information in the reports concerning this background risk would assist in comparing the methodologies used in them.

The HSE mesothelioma estimate is for chrysotile exposure only. A review of the available data (Hughes & Weill, 1986) suggests that the mesothelioma risk from chrysotile is approximately one-fifth that from a mixed fibre exposure in general. If this adjustment is made to the other estimates, they are then in reasonably close agreement with the HSE estimate for chrysotile exposure.

Although the final risk estimates are similar, the administrative processes used by the various US groups differed substantially, and will be reviewed briefly.

In the USA, EPA was the first government agency to begin compilation of a report concerned with asbestos risks, including quantitative risk estimates. In considering health risks from school asbestos exposures, EPA used a sequential approach, releasing a series of widely circulated draft reports (Environmental Protection Agency, 1980, 1981, 1982, 1984, 1985, 1986) and reacting to subsequent feedback.

There were considerable changes in risk estimates in the early drafts. For example, the 1980 draft estimated '100 to 8000 premature deaths' attributable to school exposures, while this became 'a total of 40 to 400 deaths' in the 1981 draft, without any changes in the exposure assumptions. The 1982 draft contained no quantitative estimates. The 1984 and 1985 drafts, and the final (1986) report (Environmental Protection Agency, 1986), contained estimates consistent with those in Table 1, although the 1984 estimates were for exposures lasting 40 h per week, while the later reports were for continuous exposures (24 h per day, 7 days per week).

The changing risk estimates in the early reports unfortunately only added to public confusion on the issue of risks from school asbestos exposures. For example, in 1984, State of Louisiana environmental and education officials convened a task force to address the problem of asbestos in the schools, based on the 1980 draft report; at that time, these officials were unaware that risk estimates had been substantially lowered in subsequent reports. It is likely that this was not an isolated occurrence, and that there was both considerable confusion and undue concern regarding the magnitude of the potential risk to students.

In 1982, during the period when EPA was releasing its series of draft reports, the CPSC appointed a panel of 7 scientists, including 3 experts in quantitative methods, to prepare a report advising the Commission on its approach to rulemaking for consumer products containing asbestos. It was Commission policy that no preliminary drafts be released to anyone outside the panel and its staff. The final draft report, which was released for public comment, and the subsequent final report (1983) therefore represented a wide range of views, on some of which it had been possible to reach a compromise; the issues on which this had not been possible were also indicated.

The NRC, under contract to EPA to provide a report concerning health risks from non-occupational exposures to asbestiform fibres, also convened a panel of experts (in 1983), which included 13 members from various fields. However, only one expert in quantitative methods was included. Possibly as a result, several errors in calculations made it necessary to withdraw the original published report, and other errors remained undetected in the revised version. Given the complexity of the necessary calculations and the limited time usually available to experts (in the case of both the CPSC and the NRC, they were acting as consultants), the inclusion in any such panel of several members experienced in risk assessment is advisable.

The experience in the USA enables a number of recommendations to be made to US agencies concerned with developing quantitative risk assessments for potentially toxic exposures. It is clear that the release of early draft reports, without a wide variety of scientific input and debate, does not serve the public well and can lead to considerable confusion (and either unwarranted alarm or complacency) on an issue of great public concern. Thus, input from a wide range of experts before even draft reports are disseminated is essential.

Additionally, when quantitative risk estimates are involved, it is probably wise to include input from more than one expert in quantitative methods. It would seem that possible errors are better avoided in this way, while at the same time a range of views on the appropriate underlying assumptions, such as background lung cancer risk and choice of model parameters, is obtained.

There are two areas in which greater detail would probably be helpful in quantitative risk assessment reports. As previously pointed out, the assumed background lifetime lung cancer risk should be stated. Additionally, it would be useful to many readers if instructions were included for applying the reported risk estimates to other situations, e.g., to continuous, workplace or school exposures.

Finally, because it is often difficult to place quantitative risk estimates in perspective, it is useful to compare such estimates with other commonplace risks. The inclusion of risk estimates for a variety of potential hazards, both voluntary (e.g., cycling) and involuntary (e.g., lightning), would therefore assist the reader in understanding the level of risk under consideration and aid government officials in setting public policy, where the problem of allocating limited resources often has to be faced (Weill & Hughes, 1986).

## References

Consumer Product Safety Commission (1983) *Report of the Chronic Hazard Advisory Panel on Asbestos*, Washington DC

Environmental Protection Agency (1980) *Support Document for Final Rule on Friable Asbestos-containing Materials in School Buildings. Health Effects and Magnitude of Exposure*, Washington DC (draft report)

Environmental Protection Agency (1981) *Support Document for Final Rule on Friable Asbestos-containing Materials in School Buildings. Health Effects and Magnitude of Exposure*, Washington DC (draft report)

Environmental Protection Agency (1982) *Support Document for Final Rule on Friable Asbestos-containing Materials in School Buildings. Health Effects and Magnitude of Exposure*, Washington DC (draft report)

Environmental Protection Agency (1984) *Asbestos Health Assessment Update*, Washington DC (draft report)

Environmental Protection Agency (1985) *Airborne Asbestos Health Assessment Update*, Washington DC

Environmental Protection Agency (1986) *Airborne Asbestos Health Assessment Update*, Washington DC

Health and Safety Executive (1983) *Asbestos: The Control Limit for Asbestos*, London, Her Majesty's Stationery Office

Hughes, J.M. & Weill, H. (1986) Asbestos exposure — quantitative assessment of risk. *Am. Rev. Respir. Dis.*, *133*, 5-13

National Research Council, Committee on Nonoccupational Health Risks (1984) *Asbestiform Fibers — Nonoccupational Health Risks*, Washington DC, National Academy Press

Ontario Royal Commission (1984) *Report on Matters of Health and Safety Arising From the Use of Asbestos in Ontario*, Toronto, Ontario, Ministry of Government Services

US Department of Labor, Occupational Safety and Health Administration (1986) Occupational exposure to asbestos, tremolite, anthophyllite, and actinolite. Final rules. *Federal Register*, *51* (119), 22644

Weill, H. & Hughes, J.M. (1986) Asbestos as a public health risk: disease and policy. *Ann. Rev. Public Health*, *7*, 171-192

# ESTIMATIONS OF RISK FROM ENVIRONMENTAL ASBESTOS IN PERSPECTIVE

### B.T. Commins

*Maidenhead, UK*

*Summary.* Because very recent data on asbestos in the environment are available, some fairly firm conclusions can now be drawn regarding current exposure to asbestos fibres. If account is taken of occupational exposure data (where past exposure levels were very high indeed, leading to a very significant risk to workers at the time), it is possible to make some reasoned estimates of the risk from ambient air. In the past, there was considerable confusion regarding the degree of risk for both occupational and environmental conditions. In estimating the risk, account needs to be taken, in particular, of the fact that: (*a*) occupational exposures in the past were frequently higher than reported; (*b*) asbestosis (a disease only associated with very heavy occupational exposures) would seem to be mechanistically involved in the development of lung cancer associated with asbestos exposure; (*c*) chrysotile asbestos is now the commonest form of fibre used unlike in the past, when greater quantities of crocidolite and amosite were used, the latter types being much more closely associated with mesothelioma than chrysotile; (*d*) overall levels of asbestos in environmental ambient air are lower than they used to be; (*e*) ingested asbestos seems to be associated with a negligible degree of risk as indicated by animal and human studies. The estimated values of risk provided here are smaller than those published some years ago but are similar to those given in very recent key publications. The level of environmental lifetime risk from exposure to airborne asbestos would appear to be about 1 in 100 000 or even lower. Such a level of risk is exceedingly low, and bearing in mind the criteria of both WHO and the Royal Society of London, it would appear to represent an acceptable 'rare-event' extremely low-level risk, like the cancer risk from the cosmic radiation adsorbed when flying across the Atlantic or from eating charcoal broiled meat, or the risk of being killed by lightning.

## *Introduction*

Concern regarding ambient levels of asbestos arose from the fact that industrial airborne exposure to asbestos, especially in factories in the past, caused serious health problems, including asbestosis, lung and mesothelial cancers (Ontario Royal Commission, 1984; Selikoff, 1976; UK Advisory Committee, 1979; WHO, 1986), giving rise to understandable anxieties among the general public. In retrospect, however, the great concern regarding the health significance of environmental asbestos has been largely unwarranted because environmental exposures to airborne asbestos for the general public (apart from a few very special situations many years ago) have been found to be relatively low (Commins, 1985; Ontario Royal Commission, 1984; UK Advisory Committee, 1979); normal exposures have been several orders of magnitude lower than those for asbestos dust in the air of factories in the past associated with what is now known to be a high occupational health risk. Only recently, in addition, has it become possible to evaluate with greater confidence the significance of exposure to the very low levels of asbestos which exist in the general environment (Commins, 1983, 1985; Doll & Peto, 1985; Environmental Protection Agency Workshop, 1983; US National Research Council, 1984; WHO, 1987). Such an evaluation has come about because of improvements in the measurement of environmental levels and assessment of exposure, the completion and evaluation of several very important toxicological and epidemiological studies, the gathering of various types of information from different parts of the world, and studies of risk and its estimation (Commins, 1983, 1985; Doll & Peto, 1985; EPA Workshop, 1983; Royal Society of London, 1983; US National Research Council, 1984; WHO, 1984, 1986, 1987). Asbestos is now probably one of the most extensively studied single substances in relation to human exposure.

It is hoped that this paper will help to clarify some of the earlier misunderstandings regarding the issue of environmental asbestos; the acquisition of some very recent data referred to here now makes it possible to draw much firmer conclusions than were previously possible.

## Sources of environmental asbestos

In the past, not insignifiant quantities of asbestos were released to the environment near some mining and milling activities, in some cases when asbestos and its products were being transported, and also from other industrial sources, including waste tips, etc. In many parts of the world, the introduction of good control measures in recent years has reduced the degree and extent of such contamination (Bragg, 1986; du Toit, 1987; UK Advisory Committee, 1979). Water sources can become contaminated by contact with asbestos in natural situations (Commins, 1983; Toft, 1984). Drinking-water may also become slightly contaminated as a result of the use of asbestos-cement products, including pipes and tanks, although in fact the quantities contributed are generally smaller than those from natural sources of asbestos (Commins, 1983; Toft, 1984) because the cement in asbestos-cement pipes binds asbestos fibres tightly. This binding of asbestos in various types of asbestos-cement products also accounts for fibre release to the air being minimized when such products are used for various purposes. In addition, contrary to some expressed views, the release of asbestos fibres *per se* from brake linings is very low because, in use, most of the asbestos is physically and chemically degraded, entering the environment as a biologically inert material (Ontario Royal Commission, 1984).

## Environmental exposure to asbestos

Numerous studies have been carried out to determine the levels of asbestos in environmental air. A long-term overall concentration (representative of many years of exposure, taking into account indoor, including schoolroom, situations and outdoor conditions) appears often to be in the range 0.0002 to around 0.001 fibres per ml (f/ml) air (Commins, 1985; US National Research Council, 1984; WHO, 1987). The fibres normally counted are those representative of equivalent microscopy measurement of fibres over 5 $\mu$m long, i.e., those known to be the biologically more dangerous when inhaled in high concentrations (WHO, 1987). Although short-period exposures well below and in some cases considerably higher than the above levels will occur, it is the overall long-term exposure which seems to be related to any possible health effects (Commins, 1985; US National Research Council, 1984). By far the commonest form of fibre currently detected is chrysotile (WHO, 1987), whereas in the past, crocidolite and amosite were used to a much greater extent. The most recent measurements (Bragg, 1986; Ontario Royal Commission, 1984; Toft, 1984; du Toit, 1987; US National Research Council, 1984) suggest that environmental levels in air are declining. A present-day level of 0.0005 f/ml air would seem to be a reasonable estimate (Commins, 1985; WHO, 1987). The World Health Organization (WHO, 1987) has recently evaluated the situation and considers that median values are in the range 0.0004–0.0005 f/ml.

Typical drinking-water seems likely to contain between 0.2 and 2 million fibres of chrysotile asbestos per litre (Commins, 1983). However, some drinking-waters in Canada have been reported to contain several hundreds of millions of fibres per litre (Chatfield & Dillon, 1979).

## Health significance of asbestos in drinking-water

There now seems to be no firm evidence whatsoever that the ingestion of environmental asbestos can cause any disease or in fact any ill effect in man (Commins, 1983; Ontario Royal Commission, 1984; WHO, 1986). Although a few earlier occupational studies (US National Research Council, 1984) did indicate a possible association between occupational asbestos exposure and gastrointestinal cancers, recent evaluations (Commins, 1985; US National Research Council, 1984) suggest that the asbestos may not be the cause of such occupational cancers. In particular, very recent work (Doll & Peto, 1985) suggests that misdiagnosis of gastrointestinal cancers is a likely explanation for some earlier suggestive associations with occupational asbestos exposure. In addition, a recently reported evaluation of various large-scale, high-dose animal studies and detailed epidemiological studies (EPA Workshop, 1983; WHO, 1987) failed to support any causal relationship between ingested asbestos (whatever the source) and disease.

Thus it may concluded with a fair degree of certainty that asbestos ingested in drinking-water (whether present in small amounts from asbestos-cement pipes or at the often higher levels due to contamination by natural sources of asbestos) seems to be of sensibly zero significance (Commins, 1983, 1984), i.e., the health risk (if any exists at all) is at the most, exceedingly low.

## Health significance of present-day levels of asbestos in environmental ambient air

It is appropriate to evaluate the health impact of exposure to present-day levels of asbestos in environmental ambient air since the risk for present and future populations can then be estimated for this level of exposure. Even for the low exposures occurring today, the risk, although it may be very small, is unlikely to be zero. That a risk may exist can be seen from the fact that, in the past, a few people developed mesotheliomas as a result of their very unusual environmental exposure (to predominantly crocidolite asbestos). Future risks will probably be lower still because of improvements in overall control measures, and the continuing decline in asbestos usage in some countries and especially of crocidolite asbestos, which is now being replaced by chrysotile (Ontario Royal Commission, 1984).

It is important, in particular, to distinguish between the manifest adverse health effects associated with very heavy occupational asbestos exposures (Doll & Peto, 1985; Ontario Royal Commission, 1984; Selikoff, 1976; UK Advisory Committee, 1979; US National Research Council, 1983; WHO, 1987), and the lack of demonstrable health effects associated with today's general environmental levels of asbestos (Commins, 1985, 1986, 1987; Doll & Peto, 1985).

Recently, certain important conclusions have been drawn regarding the significance of exposure to asbestos.

Where occupational exposure has been heavy, mesothelioma has seldom followed exposure to chrysotile asbestos alone (WHO, 1986). This is particularly important since most workers these days (and also the general public) will be exposed predominantly to the chrysotile form rather than to the much more dangerous amphibole forms (crocidolite and amosite), widely used in the past (Ontario Royal Commission, 1984). The difference between the health effects of the two types of fibre has been demonstrated in recent epidemiological studies in situations in which workers were exposed to predominantly chrysotile asbestos where no or a possible very low risk was reported for mesothelioma (Gardner & Powell, 1986; Hughes et al., 1987). This is in clear contrast to some other occupational studies, where diseases such as mesothelioma have been very significantly associated with exposure to the other (amphibole) forms of asbestos (WHO, 1987).

In addition, asbestosis has been associated with the very heavy occupational exposures of the past and is relatively speaking of little importance in relation to the much reduced occupational exposures of today; the risk is indeed completely negligible for the minute levels to which the general public is exposed (Ontario Royal Commission, 1984; WHO, 1986).

In relation to asbestosis, some very recent findings are of special importance. It is well known that lung cancer (even for non-smokers) is statistically associated with heavy prolonged occupational exposure to asbestos of all types (Ontario Royal Commission, 1984; WHO, 1987); the data are firm enough to conclude that asbestos is capable of acting as a human lung carcinogen under some occupational exposure conditions. However, recent findings (Browne, 1986a,b,c) suggest that only when exposure levels have been high enough and sufficiently prolonged (i.e., under conditions where asbestosis could develop) was the risk of lung cancer significantly increased. Thus an initial conclusion might be that asbestosis *per se* is a necessary precondition before lung cancer can develop. More correctly, however, it should perhaps be concluded that, associated with the conditions where asbestosis is induced in the human lung, a related or even an indirectly related mechanism leads to the development of cancer. Although several studies of occupationally exposed workers (Browne, 1986a,b,c) indicate a fairly strong statistical association between the development of lung cancer and asbestosis, the precise mechanism whereby asbestosis (or some condition directly or indirectly associated with it) may act so as to induce cancer cannot be properly defined at present. However, it seems probable that fibrosis as such may not be involved, but that, where asbestos exposure is high enough, incomplete macrophage engulfment and digestion of the longer (i.e., the pathogenic fibres) will occur. Activated macrophages are known to produce a fibroblast-stimulating factor (Brody, 1986) and also superoxides (Gee, 1980). Epithelial proliferation is also well known to occur (Davis *et al.*, 1986), perhaps in response to cell damage caused by superoxide release. It has been stressed recently (Slack, 1986) that epithelial metaplasia deserves more attention than it has so far received, and that the occurrence of metaplasia in areas of chronic tissue regeneration (such as occurs with fibrosis, for example) must tend to support models of epithelial organization involving cell division.

The implications of these findings in relation to asbestosis are of particular importance, since it seems that only when exposure has been very high (i.e., to an extent such that asbestosis can develop) does asbestos become clearly carcinogenic. Thus, for the general public exposed to very low levels of asbestos (where asbestosis does not manifest itself) (WHO, 1986), the risk of developing lung cancer (due to asbestos exposure alone) may perhaps be negligible. This may imply that a sort of threshold exists in relation to carcinogenic response. It is, in fact, not surprising that asbestos should behave differently from some other more conventional carcinogens because it is considered not to be a 'complete' carcinogen (Mossman, 1983). Nevertheless, it must be recognized that quite small doses of certain forms of asbestos (e.g., crocidolite) seem to be capable of causing pleural mesotheliomas (with no obvious indication of a threshold), but perhaps this may be explained by differences in tissue type and conditions existing in the pleura as compared with the bronchi, where lung cancer frequently develops.

## *Risk estimates related to exposure to present-day environmental levels of asbestos in ambient air*

It must be stressed that no definite evidence exists of any risk for the general public exposed to present-day levels of asbestos in environmental ambient air (Commins, 1985; Ontario Royal Commission, 1984). This does not mean, however, that in fact no risk exists, although clearly it must be low. In the past there was some evidence of a very localized health risk (some cases of mesothelioma developed associated with fairly large exposures in a few isolated situations close to uncontrolled industrial emissions (Ontario Royal Commission, 1984); such conditions are no longer known to occur. In these situations, it was likely that the mesotheliomas were associated with significant exposure to crocidolite and/or amosite dust, and this would be compatible with the fact that, even where large populations have been exposed (as in Canada) to fairly high levels of chrysotile (considered to be far less pathogenic in comparison with crocidolite or amosite), no health effects whatever were detectable (Churg, 1986).

For mesothelioma and lung cancer, it is appropriate to make predictions of the exposure risk for the general public, although the World Health Organization (WHO, 1986) has recently concluded that 'the risk of mesothelioma and bronchial cancer, attributable to asbestos exposure in the general population, is undetectably low'. In order to attempt to predict risk, however small, it is necessary to assume that:

1. A linear relationship exists between lung cancer and exposure over the whole range from occupational exposures down to environmental levels. This may not be true, however, since it neglects the possibility that a threshold may exist. The World Health Organization (WHO, 1987) considers that such an approach is likely, anyway, to overestimate rather than underestimate the risk at low levels. Very recently, too, the World Health Organization (WHO, 1987) has reported that the extent of metabolism of some carcinogens at high doses where bioassays are conducted is less than that at environmental exposure levels; in consequence, linearized cancer risk models may yield considerable overestimates of the human carcinogenic potential for some environmental carcinogens. (For mesothelioma, various complex power relationships have been considered (WHO, 1986); these are necessary in order to obtain a reasonably satisfactory estimate of risk for this particular disease.)

2. The published industrial exposure data are reasonably reliable in terms of levels of asbestos and types of asbestos fibre used. In many cases, however, this may be open to considerable doubt, especially with regard to the heavy exposures occurring many years ago.

3. The disease incidence data for the occupational studies are reliable both in terms of rates and types of disease. Unfortunately, this is also open to doubt in a number of cases, especially when workers were exposed to various types of chemical carcinogen.

Based on the above assumptions, however, various risk estimate values have been reported; these are reviewed by Peto in this volume (pp. 457–470). In the appraisal of such risk estimates, it should be recognized that there are a number of reasons why quoted risk values are often exaggerated:

1. Estimates of the critical occupational exposures to asbestos in the past were often too low (Robock, 1983), poor sampling and analytical methodology being mainly to blame for this. Eye-witness accounts (Ontario Royal Commission, 1984) would support much higher figures. In addition, little if any ventilation was provided, so that 'pockets' of high pollution will have existed which would probably not have been sampled.

2. Some very high exposures in the past would have affected the lung's natural ability to clear asbestos dust; this would mean that the true effective exposure would be higher than usually indicated by airborne measurements alone.

3. Asbestos workers were exposed not only at work but also at home (through dusty overalls, etc., taken home with them); thus, again, effective exposures may have been higher than those associated with work-place conditions alone.

4. Because the existence of a threshold is not normally taken into account, the predicted risk at low exposures may be overemphasized.

5. Inadequate corrections may be applied to the lung cancer risk associated with non-asbestos factors, especially those related to tobacco smoking and other industrial carcinogens.

6. The effective exposure for industrial workers is not the same as that for the general public because the former breathe more quickly and more deeply; this was especially true in the past, when mechanical aids were less commonly available.

7. It is not always possible to distinguish adequately between the type of fibre to which workers are exposed (e.g., chrysotile may be confused with amphibole, the latter being more pathogenic).

Recent data on environmental risks for exposure to asbestos have been provided (as a range of values) by the World Health Organization (WHO, 1987). For an exposure of 0.0005 f/ml air, the predicted lifetime risk per 100 000 is quoted as 0.1-1 for lung cancer and 1-10 for mesothelioma. With regard to these particular WHO data, it is important to note, firstly, that separate lung cancer risk values for smokers and non-smokers are not provided, a figure for smokers and non-smokers combined being quoted. From other data (US National Research Council, 1984), however, it is clear that the lung cancer risk for non-smokers is substantially lower than that for smokers exposed to environmental asbestos. Thus it is reasonable to consider that the non-smokers' lung cancer risk should be towards the lower end of the range given by WHO. Secondly, the risk data provided by WHO are conservatively cautious in order to protect health and, in the calculations, exposure to chrysotile has been assigned precisely the same risk as the potentially much more dangerous (from the point of view of developing mesothelioma) crocidolite and other amphibole forms. Thus the mesothelioma risk values at the lower end of the range would seem more appropriate in defining the risk associated with present-day levels of asbestos which, in any case, is now predominantly chrysotile (a concentration of 0.0005 f/ml air is quoted by WHO).

If account is taken of various published risk-estimate data and the fact that extrapolation from occupational to environmental levels can readily lead to exaggerated risk values, it is suggested that a total lifetime cancer risk of around 1 in 100 000 applies for non-smokers exposed primarily to chrysotile concentrations of 0.0005 f/ml air. However, if the threshold concept for lung cancer applies (as suggested by recent asbestosis case studies; see above), then the overall lifetime cancer risk could be even lower than the above-mentioned figure of around 1 in 100 000.

## Environmental asbestos risk in perspective

Various published lifetime risk values are shown in Table 1 in relation to common situations and human activities; these can be compared with the lifetime risk of around 1 in 100 000 for environmental asbestos at an exposure of around 0.0005 f/ml (primarily chrysotile). Such a general level of risk has been classified as acceptable by the World Health Organization, and the Royal Society of London considers that a lifetime risk of nearly 1 in 100 000 (0.7 in 100 000) is a level where further controls

**Table 1. Lifetime risk values for selected situations**

| Situation | Lifetime risk per 100 000 |
|---|---|
| *Extra high risk* | |
| Smoking (all causes of death) | 21 900 |
| Smoking (cancer only) | 8 800 |
| *High risk* | |
| Motor vehicle, USA, 1975 (deaths) | 1 600 |
| *Elevated risk* | |
| Frequent airline passenger (deaths) | 730 |
| Cirrhosis of liver, moderate drinker (deaths) | 290 |
| Motor accidents, pedestrians, USA, 1975 (deaths) | 290 |
| Skiing, 40 h per year (deaths) | 220 |
| *Moderate risk* | |
| Light drinker, one beer per day (cancer) | 150 |
| Drowning deaths, all recreational causes | 140 |
| Air pollution, USA, benzo[*a*]pyrene (cancer) | 110 |
| Natural background radiation, sea level (cancer) | 110 |
| Frequent airline passenger, cosmic rays (cancer) | 110 |
| *Low risk* | |
| Home accidents, USA, 1975 (deaths) | 88 |
| Cycling (deaths) | 75 |
| Person sharing room with smoker (cancer) | 75 |
| Diagnostic X-rays, USA (cancer) | 75 |
| Risk level where few would commit their own resources to reduce risk: Royal Society, London (1983) | 70 |
| *Very low risk* | |
| Person living in brick building, additional natural radiation (cancer) | 35 |
| Vaccination for smallpox, per occasion (death) | 22 |
| One transcontinental air flight per year (death) | 22 |
| Saccharin, average USA consumption (cancer) | 15 |
| Consuming Miami or New Orleans drinking-water (cancer) | 7 |
| Risk level where very few would consider action necessary, unless clear causal links with consumer products; Royal Society, London (1983) | 7 |
| *Extremely low 'rare-event' risk* | |
| One transcontinental air flight per year, natural radiation (cancer) | 4 |
| Lightning (deaths) | 3 |
| Hurricane (deaths) | 3 |
| Charcoal-broiled steak, one per week (cancer) | 3 |
| Environmental asbestos risk[a], (1985) (cancer) | 1 |
| 'Acceptable' risk for drinking-water (cancer) (WHO, 1984) | 1 |
| Further control not justified, Royal Society, London (1983) | 0.7 |

[a]Excludes possible effects of smoking.

would certainly not be required. Thus it seems that a level of lifetime risk of around 1 in 100 000 for environmental asbestos may reasonably be classified as acceptable and is anyway within the same range as, or lower than 'rare' event low-level risks, such as those due to cancer from one transcontinental flight per year, or eating one charcoal-broiled steak per week, or being killed by lightning.

## References

Bragg, G.M. (1986) Asbestos in the environment – an industry viewpoint. In: *Proceedings, International Conference on Chemicals in the Environment, Lisbon, 1-3 July 1986*
Brody, A.R. (1986) Pulmonary cell interactions with asbestos fibres *in vivo* and *in vitro*. *Chest* (suppl.), *89*, 155S-159S
Browne, K. (1986a) Is asbestos or asbestosis the cause of the increased risk of lung cancer in asbestos workers? *Br. J. Ind. Med.*, *43*, 145-149
Browne, K. (1986b) A threshold for asbestos-related lung cancer. *Br. J. Ind. Med.*, *43*, 556-558
Browne, K. (1986c) The absence of increased lung cancer risk at low levels of asbestos exposure. In: *Proceedings, Third International Conference on Environmental Lung Disease, American College of Chest Physicians, Montreal, 1986*
Chatfield, E.J. & Dillon, M.J. (1979) *A National Survey for Asbestos Fibre in Canadian Drinking Water Supplies*, Ottawa, National Health and Welfare, Canada (79-EHD-34)
Churg, A. (1986) Lung asbestos content in long-term residents of a chrysotile mining town. *Am. Rev. Respir. Dis.*, *134*, 125-127
Commins, B.T. (1983) *Asbestos Fibres in Drinking Water*, Pippins, Altwood Close, Maidenhead, UK, Commins Associates (STR1)
Commins, B.T. (1984) Asbestos fibres in drinking-water with special reference to asbestos-cement pipe usage. *Pipes Pipelines Int.*, *29*, 7-13
Commins, B.T. (1985) *The Significance of Asbestos and Other Mineral Fibres in Environmental Ambient Air*, Pippins, Altwood Close, Maidenhead, Commins Associates (STR2)
Commins, B.T. (1986) Asbestos industry and environmental asbestos. *Ind. Environ.*, *9*, 24-25
Commins, B.T. (1987) Workplace exposure to asbestos including reference to ambient environmental exposure. *Ind. Environ*, *10*, 19-23
Davis, J.M.G., Bolton, R.E., Brown, D. & Tully, H.E. (1986) Experimental lesions in rats corresponding to advanced human asbestosis. *Exp. mol. Pathol.*, *44*, 207-221
Doll, R. & Peto, J. (1985) *Effects on Health of Exposure to Asbestos*, London, Her Majesty's Stationery Office
Environmental Protection Agency Workshop (1983) Ingested asbestos. *Environ. Health Perspect.*, *53*, 1-204
Gardner, M.J. & Powell, C.A. (1986) Mortality of asbestos cement workers using almost exclusively chrysotile fibre. *J. Soc. Occup. Med.*, *36*, 124-126
Gee, J.B.L. (1980) Cellular mechanisms in occupational lung disease. *Chest*, *78*, 384-387
Hughes, J.M., Weill, H. & Hammad, Y.Y. (1987) Mortality of workers employed in two asbestos-cement manufacturing plants. *Br. J. Ind. Med.*, *44*, 161-174
Mossman, B. (1983) Asbestos: mechanisms of toxicity and carcinogenicity in the respiratory tract. *Ann. Rev. Pharmacol. Toxicol.*, *23*, 595-599
Ontario Royal Commission (1984) *Report of the Royal Commission on Matters of Health and Safety Arising from the Use of Asbestos in Ontario*, Toronto, Ontario Government Bookshop
Robock, K. (1983) Analytical methods for airborne asbestos. In: *Sesto Convegno di Igiene Industriale, Rome, 1983*
Royal Society of London (1983) *Risk Assessment; A Study Group Report*
Selikoff, I.J. (1976) Lung cancer and mesothelioma during prospective surveillance of 1249 asbestos insulation workers. *Ann N.Y. Acad. Sci.*, *271*, 448-456

Slack, J.M.W. (1986) Epithelial metaplasia and the second anatomy. *Lancet, ii*, 268-270
du Toit, R.S.J. (1987) Asbestos levels in South African mining towns drop dramatically (quoted in *The Asbestos Report*, No. 8, February 1987)
Toft, P. (1984) Asbestos in drinking water, *CRC Crit. Rev. Environ. Control, 14*, 151-197
UK Advisory Committee on Asbestos (1979) *Final Report*, London, Her Majesty's Stationery Office
US National Research Council (1984) *Asbestiform Fibres: Non-occupational Health Risks*, Washington, DC, National Academy Press
WHO (1984) *Guidelines for Drinking Water Quality*, Geneva, World Health Organization
WHO (1986) *Asbestos and Other Natural Mineral Fibres (Environmental Health Criteria 53)* Geneva, World Health Organization
WHO (1987) *Air Quality Guidelines*, Copenhagen, World Health Organization, Regional Office for Europe

# MESOTHELIOMAS — ASBESTOS EXPOSURE AND LUNG BURDEN

### G. Berry

*Department of Public Health, University of Sydney, Australia*

### A.J. Rogers

*National Occupational Health and Safety Commission, University of Sydney, Australia*

### F.D. Pooley

*Department of Mining, Geological and Minerals Engineering, University College, Cardiff, Wales, UK*

*Summary.* The assessment of asbestos fibres in the lungs at post mortem in groups of mesotheliomas, groups occupationally exposed to asbestos, and controls has shown that all these groups contain significant levels of asbestos as a lung burden. The amounts in each group are dependent on the degree of past exposure, being highest in those cases with a known or extrapolated occupational exposure, less in those cases with recorded neighbourhood or environmental exposure, and less again in those cases with no evidence of exposure to asbestos and in controls. Relative risk estimates and the use of models developed for occupational situations do not provide good estimates of the relevance of environmental fibres in producing mesotheliomas in the general population. This may be the result of differences between the groups in their time periods of exposure and long-term elimination of asbestos from the lungs. The number of mesotheliomas that might be due to low-level environmental exposure to asbestos cannot be determined from lung contents alone, but an assessment based on detailed occupational histories from the Australian Mesothelioma Surveillance Program show that the problem is not one of great importance when compared with other public health issues.

## *Introduction*

Since the work of Wagner *et al.* (1960), the link between asbestos exposure and mesothelioma has become well established. Most of the evidence has come from studies of groups with known occupational exposure to asbestos, although it has also been shown that para-occupational exposure, such as household contact with exposed workers, or residence near to an asbestos source can result in mesothelioma (Newhouse & Thompson, 1965; Anderson *et al.*, 1976). Studies on asbestos-exposed groups have indicated that the risk of mesothelioma induction is dependent on the type of asbestos as well as the severity and duration of exposure.

National or regional studies indicate that mesotheliomas also occur in those for whom no evidence of either occupational or environmental exposure was obtained from a detailed history. Greenberg and Davies (1974) gave data on 234 mesotheliomas occurring in England, Wales or Scotland in 1967–1968, for whom it had been possible to obtain a history of possible asbestos exposure. Of these, 183 (78%) had definite or possible occupational exposure, 13 (6%) had neighbourhood, domestic or hobby exposure, whilst for 38 (16%) careful enquiry failed to elicit any exposure. Ferguson et al. (1987) reported on 690 mesotheliomas occurring in Australia in 1979-1985 for whom an adequate history was available. Definite, probable or possible occupational exposure was found for 456 (66%), neighbourhood, domestic or hobby exposure for 43 (6%), but there remained 191 (28%) with no history of exposure. The annual mesothelioma rate in adults with no history of asbestos exposure is about 1.5 per million (McDonald & McDonald, 1977; Peto, 1984). The etiology of these cases is unknown although some local isolated clusters have similar exposures to environmental agents (Peterson et al., 1984). Low-level environmental exposure to asbestos and asbestiform minerals has been postulated as a factor in these cases (Omenn et al., 1986).

One problem in assessing the effect of general environmental asbestos exposure is that of obtaining a measure of the amount of such exposure and relating this to the exposure experienced by those occupationally exposed. Since exposure to asbestos leads to the inhalation and retention of fibres in the lungs, the amount of asbestos in the lungs is, to some extent, a measure of exposure during life. The study of women employed in the assembly of gas-masks (Jones et al., 1980b) showed that there was considerable crocidolite in the lungs (median 86 million fibres per g) 25-30 years after a short period of exposure.

In this paper, an attempt is made to estimate differences in exposure by comparing the lung contents of those with mesothelioma and controls in Australia with similar data from previous series. Attention is restricted to the two commercial amphibole fibres, amosite and crocidolite, because of their known strong association with the induction of mesothelioma.

## Materials and methods

The sources of lung tissue obtained post mortem have been described previously and are only briefly summarized here.

### Australian series

This consists of 189 cases of mesothelioma obtained during 1979-1985 by the Australian Mesothelioma Surveillance Program (Ferguson et al., 1987). Based on the examination of a detailed occupational history, these cases were divided into 3 groups: those with occupational exposure to asbestos, those with identified environmental exposure to asbestos, and those with no identified exposure to asbestos. A control series of 50 cases was also included, consisting of male urban dwellers aged 60-79 from a series of consecutive necropsies (Rogers, 1984).

### United Kingdom series (1976)

A series of 86 mesothelioma cases was obtained from pathologists (Jones et al., 1980a). The pathologists also supplied 56 age-matched controls who had died either of bronchial carcinoma or cerebrovascular disease.

### United Kingdom series (1977)

This series contained three groups (Wagner et al., 1982), the first being a group of 145 cases from the Pneumoconiosis Medical Panels (PMP) in which exposure to asbestos had been considered a factor in the cause of death. About 30% of this group had a mesothelioma and another 30% lung cancer. The second group consisted of 25 mesotheliomas. Some of these cases were known to have had occupational exposure to asbestos. The third group was made up of 94 controls from consecutive necropsies in adults from 6 hospitals, chosen to represent areas of severe, moderate and low industrial pollution.

### North American series

This series consisted of 99 mesotheliomas collected from pathologists in the USA and Canada in 1976, and from 100 controls matched for age and sex (McDonald et al., 1982).

### Tissue analysis

Preparation of the specimens for electron microscopy was described in the earlier reports (Jones et al., 1980a; McDonald et al., 1982; Rogers, 1984; Wagner et al., 1982). Mineral fibre analysis was carried out on the United Kingdom and North American series in Cardiff using transmission electron microscopy and an energy-dispersive X-ray analysis (EDXA) system (Pooley & Clark, 1979, 1980). Fibres of all sizes that were resolved by the electron microscope were included in the counts, provided that they had an aspect ratio greater than 3 to 1. The Australian series was analysed in Sydney using a transmission electron microscope fitted with the EDXA system. Because of problems of minor contamination from short fibres on the collection filter, only fibres longer than 2 $\mu$m were reported (Rogers, 1984).

### Presentation of results

Fibre counts are expressed in units of millions per gram of dry tissue. Results are presented for amosite, crocidolite, and amosite plus crocidolite combined. The distribution of counts is given as relative frequencies (%) in the categories less than 1, 1-9.9, 10-99.9, and 100 million fibres or more per gram. The median fibre counts and interquartile range (25 and 75 percentiles) are also given. In some cases, the lower quartile (and the median) could not be estimated directly since, for more than 25% (or 50%) of the samples, no fibres of the type being considered were identified. In these cases, the quartile (and median) were estimated from a cumulative probability plot on log-probability paper. Such plots showed that the distributions of counts were approximately lognormal.

## Results

The distributions, medians and interquartile ranges are given in Tables 1, 2 and 3 for amosite, crocidolite, and amosite and crocidolite combined, respectively. For the Australian series, the occupationally exposed mesotheliomas contained the highest levels of crocidolite and amosite. The environmentally exposed mesotheliomas contained slightly more crocidolite than those with no known exposure to asbestos. The mesotheliomas not exposed to asbestos contained a similar quantity of amosite and crocidolite to that of the controls. The median content of amosite and crocidolite combined was 5 times higher for mesotheliomas with occupational exposure than for controls, and there was considerable overlap between the distributions.

### Table 1. Distribution of amosite in lungs

| Series and group | No. of cases | Relative frequency (%) at a concentration (million fibres per g) of | | | | Median | Interquartile range |
|---|---|---|---|---|---|---|---|
| | | <1 | 1-9.9 | 10-99.9 | >100 | | |
| *Australia 1979-1985* | | | | | | | |
| Mesotheliomas: | | | | | | | |
| Occupationally exposed | 140 | 70.7 | 17.9 | 10.7 | 0.7 | 0.55 | 0.16-1.4 |
| Environmentally exposed | 12 | 91.7 | 8.3 | 0.0 | 0.0 | $0.10^a$ | $0.04^b$-0.26 |
| Not exposed | 37 | 100.0 | 0.0 | 0.0 | 0.0 | 0.14 | $0.07^b$-0.27 |
| Controls | 50 | 96.0 | 4.0 | 0.0 | 0.0 | $0.09^a$ | $0.04^b$-0.19 |
| *United Kingdom 1976* | | | | | | | |
| Mesotheliomas | 86 | 51.2 | 30.2 | 11.6 | 7.0 | 0.93 | 0.13-5.5 |
| Controls | 56 | 85.7 | 12.5 | 0.0 | 1.8 | $0.05^a$ | $0.01^b$-0.35 |
| *United Kingdom 1977* | | | | | | | |
| Pneumoconiosis Medical Panel cases | 145 | 32.4 | 29.0 | 26.2 | 12.4 | 4.2 | 0.38-37 |
| Other mesotheliomas | 25 | 60.0 | 40.0 | 0.0 | 0.0 | 0.50 | $0.05^b$-1.2 |
| Controls | 94 | 84.0 | 13.8 | 2.1 | 0.0 | 0.18 | $0.04^b$-0.60 |
| *North America 1972* | | | | | | | |
| Mesotheliomas | 99 | 72.7 | 16.2 | 7.1 | 4.0 | $0.05^a$ | $<0.01^b$-1.7 |
| Controls | 100 | 91.0 | 7.0 | 2.0 | 0.0 | $0.03^a$ | $<0.01^b$-0.20 |

$^a$Estimated from lognormal plot; in more than 50% of samples no amosite fibres were detected.
$^b$Estimated from lognormal plot; in more than 25% of samples no amosite fibres were detected.

In the United Kingdom series, the mesotheliomas and the Pneumoconiosis Medical Panel cases had more amosite and crocidolite than the controls. The differences between the groups were larger than in the Australian series but there was nevertheless a considerable overlap between the distributions.

The North American series differed from the other series by containing more amosite than crocidolite. Also the amounts of amosite and crocidolite were less than for the United Kingdom series for both mesotheliomas and controls.

## Discussion

In an attempt to determine the proportion of mesotheliomas that may be a result of exposure to low levels of asbestos fibres present in the general environment, two approaches were tried.

### Relative risk based on amphibole lung burden

The Australian, United Kingdom 1976 and North American series are essentially case-control studies of mesotheliomas. In a similar study of mesothelioma in Norway, Mowé et al. (1985) calculated an odds ratio (relative risk) of 8.5 (95% confidence interval (CI), 2.3-31) for a lung content of more than 1 million fibres per gram of dry tissue compared with lower lung contents. A similar approach here, using amosite and crocidolite combined, gave a relative risk of 8.0 (95% CI, 3.0-21) for Australia, 7.4 (95% CI, 3.5-16) for the United Kingdom in 1976 and 3.8 (95% CI, 1.8-8.0) for North America.

The usefulness of such an approach depends on the extent to which the amount of asbestos fibre in the lungs at post mortem is a valid measure of the risk of mesothelioma as a result of exposure to asbestos. If it is assumed that the amount of fibre deposited at any time is proportional to the concentration of fibre in the air, and long-term elimination is ignored, then the amount of asbestos in the lungs is proportional to the cumulative exposure. One of the disadvantages of this measure is that it does not take into account the time at which the exposure took place, and it is well established that mesothelioma incidence increases with time since exposure. Occupational exposure to asbestos has often come to an end several years, or even decades, before a mesothelioma occurs. The greater part of the cumulative exposure will have taken place by the end of the occupational exposure. In contrast, background environmental exposure continues throughout life. Thus the ratio of asbestos in the lungs after occupational exposure to that after environmental exposure will be less than the ratio of the mesothelioma incidences.

If there is also long-term elimination of dust from the lungs, then the discrepancy between asbestos lung burden and mesothelioma incidence noted above will be greater, since the lung burden of those who have ceased to be occupationally exposed will decline, whilst the mesothelioma incidence increases. Elimination has been demonstrated in animal experiments to take place at a rate of about 20% per year (Wagner et al., 1974).

Another difference between occupational and environmental exposures is that the fibres are shorter and generally finer in environmental situations. Studies on the relative rates of clearance from the lung of fibres of various sizes indicates that short fibres are more rapidly removed and that a higher proportion of long fibres remain even after extended periods (Morgan et al., 1978).

### Risk based on degree of exposure and lung clearance

Peto (1984) gave a mathematical formula relating mesothelioma incidence, level of exposure and time since exposure. This formula may be applied to the gas-mask workers studied by Jones et al. (1980b), who had a mesothelioma incidence 1500 times

## Table 2. Distribution of crocidolite in lungs

| Series and group | No. of cases | Relative frequency (%) at a concentration (million fibres per g) of | | | | Median | Interquartile range |
|---|---|---|---|---|---|---|---|
| | | <1 | 1-9.9 | 10-99.9 | >100 | | |
| *Australia 1979-1985* | | | | | | | |
| Mesotheliomas: | | | | | | | |
|   Occupationally exposed | 140 | 55.7 | 33.6 | 8.6 | 2.1 | 0.84 | 0.24-2.9 |
|   Environmentally exposed | 12 | 83.3 | 8.3 | 8.3 | 0.0 | 0.25 | 0.09$^a$-0.67 |
|   Not exposed | 37 | 94.6 | 5.4 | 0.0 | 0.0 | 0.15 | 0.05$^a$-0.42 |
| Controls | 50 | 96.0 | 4.0 | 0.0 | 0.0 | 0.22 | 0.11$^a$-0.47 |
| *United Kingdom 1976* | | | | | | | |
| Mesotheliomas | 86 | 40.7 | 27.9 | 18.6 | 12.8 | 1.6 | 0.43-20 |
| Controls | 56 | 75.0 | 21.4 | 0.0 | 3.6 | 0.14 | 0.02$^a$-1.0 |
| *United Kingdom 1977* | | | | | | | |
| Pneumoconiosis Medical Panel cases | 145 | 26.9 | 26.9 | 30.3 | 15.9 | 8.3 | 0.84-49 |
| Other mesotheliomas | 25 | 44.0 | 32.0 | 16.0 | 8.0 | 1.5 | 0.12-6.7 |
| Controls | 94 | 86.2 | 11.7 | 2.1 | 0.0 | 0.07 | 0.01$^a$-0.43 |
| *North America 1972* | | | | | | | |
| Mesotheliomas | 99 | 81.8 | 9.1 | 8.1 | 1.0 | 0.02$^b$ | <0.01$^a$-0.40 |
| Controls | 100 | 94.0 | 6.0 | 0.0 | 0.0 | 0.01$^b$ | <0.01$^a$-0.07 |

$^a$Estimated from lognormal plot; in more than 25% of samples no crocidolite fibres were detected.
$^b$Estimated from lognormal plot; in more than 50% of samples no crocidolite fibres were detected.

that in the unexposed, to give an estimated asbestos level in the factory about 100 000 times that in the general environment, if it is assumed that all mesotheliomas are the result of exposure to asbestos (see Appendix). However, the gas-mask workers had at the most 150 times more amphibole fibre in their lungs than controls. This could occur if the elimination rate was about 15-17.5% a year (see Appendix).

The gas-mask workers are atypical of the occupationally exposed in both their high mesothelioma incidence and short duration of exposure. As a more typical situation consider occupational exposure leading to a 250-fold relative risk of mesothelioma (this corresponds to about 2% of deaths being due to mesothelioma) and suppose exposure from age 25 to age 35 gives this excess at age 60. Then the

### Table 3. Distribution of amosite and crocidolite in lungs

| Series and group | No. of cases | Relative frequency (%) at a concentration (million fibres per g) of | | | | Median | Interquartile range |
|---|---|---|---|---|---|---|---|
| | | <1 | 1-9.9 | 10-99.9 | >100 | | |
| *Australia 1979-1985* | | | | | | | |
| Mesotheliomas: | | | | | | | |
|   Occupationally exposed | 140 | 40.0 | 37.9 | 20.0 | 2.1 | 1.6 | 0.67-7.4 |
|   Environmentally exposed | 12 | 83.3 | 8.3 | 8.3 | 0.0 | 0.52 | 0.13$^a$-0.76 |
|   Not exposed | 37 | 91.9 | 8.1 | 0.0 | 0.0 | 0.31 | 0.12$^a$-0.62 |
| Controls | 50 | 90.0 | 10.0 | 0.0 | 0.0 | 0.30 | 0.11$^a$-0.52 |
| *United Kingdom 1976* | | | | | | | |
| Mesotheliomas | 86 | 22.1 | 41.9 | 19.8 | 16.3 | 5.9 | 1.1-37 |
| Controls | 56 | 67.9 | 26.8 | 1.8 | 3.6 | 0.42 | 0.10-1.5 |
| *United Kingdom 1977* | | | | | | | |
| Pneumoconiosis Medical Panel cases | 145 | 16.6 | 26.2 | 27.6 | 29.7 | 20.0 | 2.5-124 |
| Other mesotheliomas | 25 | 28.0 | 48.0 | 16.0 | 8.0 | 2.4 | 0.76-7.8 |
| Controls | 94 | 73.4 | 24.5 | 2.1 | 0.0 | 0.36 | 0.09$^a$-1.2 |
| *North America 1972* | | | | | | | |
| Mesotheliomas | 99 | 65.7 | 18.2 | 11.1 | 5.1 | 0.20 | 0.01$^a$-3.0 |
| Controls | 100 | 88.0 | 10.0 | 2.0 | 0.0 | 0.04$^b$ | <0.01$^a$-0.30 |

$^a$Estimated from lognormal plot; in more than 25% of samples no amosite or crocidolite fibres were detected.
$^b$Estimated from lognormal plot; in more than 50% of samples no amosite or crocidolite fibres were detected.

occupational exposure level would be 2500 times that in the environment. With an elimination rate of 15% per year, the occupational cases would contain about 50 times more asbestos in their lungs than the unexposed population whilst, if the elimination rate was 17.5% per year, the ratio would only be 25.

These ratios are greater than that found for the Australian cases (Table 3), where the occupationally exposed had median lung contents only 5 times that of controls. This could be either because the category of occupational exposure corresponded to an average relative risk of less than 250, or because the elimination rate was higher. In the former case, a greater proportion of the population would have had to be exposed in order for about two-thirds of the mesotheliomas to occur in the occupationally exposed (for a relative risk of 250, 1 person in 125 would have to be occupationally exposed). In the Australian study, the term occupational exposure did not necessarily imply that asbestos processing or handling was a main activity nor that exposure was necessarily high. For the Pneumoconiosis Medical Panel cases in the United Kingdom in 1977, the median content was 60 times that of controls. To be accepted by the Pneumoconiosis Medical Panel, there would have had to be evidence of considerable exposure.

The above calculations are based on the assumption that all mesotheliomas are due to asbestos exposure. If only a proportion of mesotheliomas are due to asbestos, then a higher level of occupational exposure and a higher elimination rate would be necessary to fit the data.

This analysis does not provide an answer to the question of what proportion of mesotheliomas without occupational or para-occupational exposure to asbestos may be due to background environmental exposure to asbestos. What it does suggest is that there may not be any major discrepancy between the different relative amounts of asbestos in the lungs and the relative risks of mesothelioma. Such differences, as well as the overlap of the distributions of the occupationally exposed and controls, may be explained by different time patterns of exposure and long-term elimination of fibres from the lungs.

The assessment of lung fibre burden at post mortem is important in detecting or confirming heavy exposure, and in identifying the types of asbestos to which there has been exposure. Examination of controls periodically could be useful in monitoring environmental pollution due to asbestos. However, such determinations alone cannot indicate the number of mesotheliomas due to low levels of asbestos in the environment.

Extrapolation of the risk data for air borne exposures, even at relatively high environmental levels, as in asbestos-containing buildings, only accounts for a relatively small proportion of non-occupational mesotheliomas (Doll & Peto, 1985).

An upper limit to the number of mesotheliomas due to background environmental exposure is provided by ascertainment of most cases and detailed occupational histories. By such means (Ferguson *et al.*, 1987), it has been found that about 45 mesotheliomas occur each year in Australia without known asbestos exposure. The importance of this in public health terms is not high when compared with 2900 deaths annually due to motor vehicle accidents, and with 5700 bronchogenic cancers, many attributable to cigarette smoking. The differences between these figures indicate the relative priorities to be assigned to preventive measures aimed at reducing road accidents and smoking, as compared with the removal of asbestos from buildings.

## Acknowledgements

The Australian data come from the Australian Mesothelioma Surveillance Program, and we are grateful to Professor D. Ferguson for permission to use these data.

## References

Anderson, H.A., Lilis, R., Daum, S.M., Fischbein, A.S. & Selikoff, I.J. (1976) Household-contact asbestos neoplastic risk. *Ann. N.Y. Acad. Sci.*, 271, 311-323

Doll, R. & Peto, J. (1985) *Effects on Health of Exposure to Asbestos*, London, Her Majesty's Stationery Office

Ferguson, D., Berry, G., Jelihovsky, T., Andreas, S., Rogers, A., Fung, S.C., Grimwood, A. & Thompson, R. (1987) The Australian Mesothelioma Surveillance Program, 1979-1985. *Med. J. Aust.*, 147, 166-172

Greenberg, M. & Davies, T.A.L. (1974) Mesothelioma Register 1967-68. *Br. J. Ind. Med.*, 31, 91-104

Jones, J.S.P., Pooley, F.D., Clark, N.J., Owen, W.G., Roberts, G.H., Smith, P.G., Wagner, J.C., Berry, G. & Pollock, D.J. (1980a) The pathology and mineral content of lungs in cases of mesothelioma in the United Kingdom in 1976. In: Wagner, J.C., ed., *Biological Effects of Mineral Fibres (IARC Scientific Publications No. 30)*, Lyon, International Agency for Research on Cancer, pp. 187-199

Jones, J.S.P., Smith, P.G., Pooley, F.D., Berry, G., Sawle, G.W., Wignall, B.K., Madeley, R.J. & Aggarwal, A. (1980b) The consequences of exposure to asbestos dust in a wartime gas-mask factory. In: Wagner, J.C., ed., *Biological Effects of Mineral Fibres (IARC Scientific Publications No. 30)*, Lyon, International Agency for Research on Cancer, pp. 637-653

McDonald, J.C. & McDonald, A.D. (1977) Epidemiology of mesothelioma from estimated incidence. *Prev. Med.*, 6, 426-446

McDonald, A.D., McDonald, J.C. & Pooley, F.D. (1982) Mineral fibre content of lung in mesothelial tumours in North America. *Ann. Occup. Hyg.*, 26, 417-422

Morgan, A., Talbot, R.J. & Holmes, A. (1978) Signficance of fibre length in the clearance of asbestos fibres from the lung. *Br. J. Ind. Med.*, 35, 146-153

Mowé, G., Gylseth, B., Harviet, F. & Skaug, V. (1985) Fiber concentration in lung tissue of patients with malignant mesothelioma — a case-control study. *Cancer*, 56, 1089-1093

Newhouse, M.L. & Thompson, H. (1965) Mesothelioma of the pleura and peritoneum following exposure to asbestos in the London area. *Br. J. Ind. Med.*, 22, 261-269

Omenn, G.S., Merchant, J., Boatman, E., Dement, J.M., Kuschner, M., Nicholson, W., Peto, J. & Rosenstock, L. (1986) Contribution of environmental fibers to respiratory cancer. *Environ. Health Perspect.*, 70, 51-56

Peterson, J.T., Greenberg, S.D. & Buffler, P.A. (1984) Non-asbestos related malignant mesothelioma – a review. *Cancer*, 54, 951-960

Peto, J. (1984) Dose and time relationship for lung cancer and mesothelioma in relation to smoking and asbestos exposure. In: Fischer, M. & Meyer, E., eds, *Zur Beurteilung der Krebsgefahr durch Asbest*, München, MMV Medizin Verlag, BGA Schriften, 2/84, pp. 126-132

Pooley, F.D. & Clark, N.J. (1979) Quantitative assessment of inorganic fibrous particulates in dust samples with an analytical transmission electron microscope. *Ann. Occup. Hyg.*, 22, 253-271

Pooley, F.D. & Clark, N.J. (1980) The chemical and physical characteristics of fibrous particles detected in human post-mortem lung tissue. In: Wagner, J.C., ed., *Biological Effects of Mineral Fibres (IARC Scientific Publications No. 30)*, Lyon, International Agency for Research on Cancer, pp. 79-86

Rogers, A.J. (1984) Determination of mineral fibre in human lung tissue by light microscopy and transmission electron microscopy. *Ann. Occup. Hyg.*, 28, 1-12

Wagner, J.C., Berry, G., Skidmore, J.W. & Timbrell, V. (1974) The effects of the inhalation of asbestos in rats. *Br. J. Cancer*, 29, 252-269

Wagner, J.C., Pooley, F.D., Berry, G., Seal, R.M.E., Munday, D.E., Morgan, J. & Clark, N.J. (1982) A pathological and mineralogical study of asbestos-related deaths in the United Kingdom in 1977. *Ann. Occup. Hyg.*, 26, 423-431

Wagner, J.C., Sleggs, C.A. & Marchand, P. (1960) Diffuse pleural mesothelioma and asbestos exposure in the North Western Cape Province. *Br. J. Ind. Med.*, 17, 260-271

## *Appendix*

Following the formulation given by Peto (1984), the mesothelioma incidence $t$ years after the beginning of continuous exposure to a level $d$ is $dt^{3.5}$. If the environmental level is denoted by $d$, then the mesothelioma incidence at age $t$, $m_E$, is given by:

$$m_E = d_E t^{3.5} \tag{1}$$

Now consider occupational exposure to level $d_O$ from ages $t_1$ to $t_2$. Assuming that $d_O \gg d_E$ so that environmental exposure can be ignored, then the mesothelioma risk, $m_O$, at age $t(>t_2)$ is:

$$m_O = d_O[(t-t_1)^{3.5} - (t-t_2)^{3.5}] \tag{2}$$

The corresponding amounts of fibre in the lungs, $A_E$ and $A_O$, in the absence of long-term elimination, are:

$$A_E = k d_E t \tag{3}$$

and

$$A_O = k d_O (t_2 - t_1) \tag{4}$$

where $k$ is a constant representing the retention rate. If elimination takes place exponentially at a rate of $\lambda$ per year, then the amounts would become:

$$A_E' = k d_E \lambda^{-1}(1 - e^{-\lambda t}) \tag{5}$$

and

$$A_O = k d_O \lambda^{-1}[e^{-\lambda(t-t_2)} - e^{-\lambda(t-t_1)}] \tag{6}$$

Applying the above to the gas-mask workers (Jones *et al.*, 1980b), the mesothelioma incidence 30 years after a mean exposure of 1 year was 2200 per million per year, about 1500 times the incidence in the unexposed, that is:

$$m_O/m_E = 1500$$

If the mesotheliomas in the unexposed represent 55 years of environmental exposure, then using equations (1) and (2) we find:

$$d_O/d_E = 100\,000$$

Fibre burdens are available for 14 cases with mean duration of employment 22.5 months. The median lung burden was 120 million amphibole fibres per g, compared with about 0.4 for United Kingdom controls. The ratio of lung contents is 300, but scaling down the lung contents of the mesotheliomas to a mean duration of exposure of 1 year gives approximately:

$$A'_O/A'_E = 150$$

Then from equations (5) and (6), $\lambda = 0.15$.

Although these calculations are based on assumptions that are probably oversimplistic, and lung burdens in the gas-mask workers were not available from a representative sample of the group but mainly from those who had died of mesothelioma, they give some indication of the situation. The relative mesothelioma incidence was 10 times the relative amounts of asbestos in the lungs, and this could have occurred as a result of long-term elimination of fibre from the lungs at a rate of 15% per year. If the lung burden in the group as a whole was less than that of the mesotheliomas, then a higher elimination rate would be estimated, e.g., $A'_O/A'_E = 75$ gives $\lambda = 0.175$.

An elimination rate of 15% per year means that only 22% of fibre in the lungs at the cessation of exposure remains 10 years later; for an elimination rate of 17.5%, only 17% would remain.

# PUBLIC PERCEPTION OF RISK AND ITS CONSEQUENCES: THE CASE OF A NATURAL FIBROUS MINERAL DEPOSIT

### G. Major

*C.M. Consultants Pty Limited, Wahroonga, NSW, Australia*

### G.F. Vardy

*Public Works Department, Eden, NSW, Australia*

*Summary.* A public authority building a breakwater and other harbour facilities at a small seaport (population 3000) had short-term requirements for 261 000 tonnes of rock and ultimately for 1 000 000 tonnes. A suitable quarry was found about 11 km from the port but unfortunately the rock was found to be contaminated to a small extent with a fibrous mineral identified with the analytical transmission electron microscope as a non-commercial type of fine amphibole with many long fibres. Quarrying only was intended and there were no plans to crush the rock, but the projected work soon brought complaints from local residents, who expressed fears concerning risks to health from what soon became known as 'the asbestos mine'. These complaints posed a dilemma for both the construction and health authorities; they were forcefully expressed, and residents were supported by local newspapers, municipal authorities and regional politicians. The Land and Environment Court ordered (by consent) that the construction authority 'take all reasonable measures to ensure that no loose asbestos material and no rock with any asbestos material exposed on the surface (is) removed from the site'. Personal monitoring of quarry workmen by the membrane filter method and ambient air monitoring near residents' homes with analysis by electron microscope showed that only insignificant concentrations of airborne fibres were present. The breakwater was ultimately completed after much delay and extra expense. Other and greater risks to health and safety, such as the transport of liquid chlorine through the centre of the town to the fish processing plant and the storage, distribution and transport of petroleum products from the nearby regional facilities, were not perceived as such by the residents.

## *Introduction*

Rock deposits containing fibrous material in non-commercial quantities and their possible contribution to levels of contamination of the ambient air when disturbed for road building and construction purposes were the subject of discussion and speculation, principally in the United States, as long ago as 1971 (Levine, 1978, p. 51). Although the use of wastes containing commercial asbestos or tailings from asbestos mining and milling for road making is prohibited by the US National Asbestos Air Emission Standard, wastes that may contain non-commercial asbestos as a contaminant have not been regulated (Levine, 1978, p. 57).

This paper is concerned with a civil engineering project where the perception of risk from environmental contamination with fibres by a small, isolated community resulted in the effective prohibition of the use of uncrushed rock contaminated with isolated and sparse fibrous veins of an amphibole which is not a commercial type of asbestos but is still erroneously referred to by the community and others as tremolite. The full implications of the community response to the proposed use of this rock have yet to be realized; it handicapped the engineers responsible for the particular project and added significantly to its cost, and similar reactions could inhibit civil engineers in future work and not only in the particular area concerned.

The work described in this paper is not unique in Australia (a deposit of fibrous serpentine was once encountered and disturbed in an extensive and deep road cutting during a major highway construction programme about 500 km from the area under discussion) and similar incidents no doubt occur in other countries. Fibrous contamination of minerals is not uncommon — about half the area of the United States of America is said to contain asbestiform minerals in the bedrocks (Campbell et al., 1977) — and their disturbance in quarrying and excavation work might not be infrequent. Fibres are, of course, often found in the mining of, for example, iron ore deposits.

In the case reported here, the community response to possible increased contamination of the local environment with air- and water-borne natural mineral fibres arose from its interpretation of available knowledge concerning exposure to asbestos and health and in particular the risks for cancer. Some of the fears expressed by the community were (are) bizarre, others arose from a statement attributed to IARC and published in a local newspaper — 'it (IARC) had been unable to identify any level below which disease would not occur'. Yet others resulted from a (true) statement attributed to the Clean Air Authority that 'levels of asbestos fibres within the community areas cannot be practically monitored'. Some of the experts attending this Symposium might be surprised to know that photocopies of papers written by them were used to good effect by a residents action committee in an area 500 km from the nearest library holding the journal concerned.

## Historical aspects

The work described in this paper was carried out at Eden, NSW (37 03S lat., 149 55E long.), a small port with a population of 3420, rising to about 40 000 during the summer holiday season; this is of some importance, albeit minor, to the argument. The regional centre (population 4800) is 60 km away and the nearest town with a population exceeding 10 000 is 300 km distant. Three tabloid newspapers (with a circulation of 2000–3000 each) are published in the area and they all extensively reported the subject of this paper.

The town is on a peninsula with the Pacific Ocean on one side and an extensive, deep-water bay on the other. The bay could probably hold the entire United States navy and, in any other part of the world, the area might support a population of 300 000, not 3000. First settled in 1846, Eden developed slowly with fishing and off-shore whaling, and served the requirements of the dairy farming and forestry activities carried on in the surrounding area. A wood-chip mill with its associated (private) wharf facilities routinely handling 65 000-tonne ships was established on the bay opposite the town in 1971, doubling its population by the direct and indirect employment opportunities that it provided. This also has some importance to the argument in that it led to disputes concerning environmental protection and flora and fauna conservation, further influenced by the fact that the wood chips were to be exported for the manufacture of paper products and chip-board. These disputes and other environmental factors had some influence on the community response to the threat of an increased fibre contamination of the environment, but it would be churlish to suggest that all the fears entertained were not *bona fide*.

A timber jetty was erected in a cove off the bay in 1860. Extensions were made to this structure in 1911 but, in spite of routine maintenance, it deteriorated beyond the point of economic repair and could no longer meet the needs of an increasingly sophisticated and larger fishing fleet. Consequently, in the 1970s, new port facilities, including reclamation work, shore developments and a multi-purpose jetty, were designed and built in stages until their completion in 1984. At the same time, developments in the tourist industry and a general upgrading of facilities brought to the surrounding area new public works — roads, bridges, car parks, a water supply dam and, in an isolated part of the bay near Eden, a boat-launching ramp protected by a breakwater 125 m long.

Much of this construction work called for rock, some of which was basalt quarried from an area 8 km from Eden (some was sandstone, which has associated silicosis risks for the quarry workers). This basalt was quarried by the normal methods — drilling, blasting and loading of the broken rock into trucks for haulage over the public road system to the construction site. Quarrying and use of basalt was uneventful for at least 5 years until early in 1983, when a road worker noticed some fibre in rocks in a retaining wall for an already completed bridge (parts of the wall are under water at certain tides) and drew the matter to the attention of his labour union and one of the state regulatory agencies concerned with occupational health and safety. (Three such regulatory agencies exist, something that is not unique to New South Wales, and all played some part in the matter which is the subject of this paper — the Division of Occupational Health, the Construction Safety Inspectorate and the Mines Inspection Division. Ultimately the Clean Air Division, the Clean Water Division, the Soil Conservation Service, the Land and Environment Court and, of course, the Local Government Authority were all involved and contributed to some of the problems subsequently experienced by civil engineers.)

It was reported in the local newspapers that the Division of Occupational Health was less than delighted to find fibrous material on the surface of rocks on the retaining wall of a bridge. Further investigations were made at other completed works and at the quarry from which the rock was obtained. A petrological examination reported that the fibrous material was tremolite/actinolite and from then onwards it was referred to as tremolite or tremolite asbestos or merely asbestos. About 2% of all the rocks counted on the breakwater at the boat ramp had exposed fibre on their surfaces (this 2% became significant at a later date) and this was judged to be unacceptable because the breakwater is used by professional fishermen and, for recreational purposes, by other fishermen. Remedial action to remove or cover all exposed fibre here and at all other places where it could be found was recommended and the further use of rock which might contain tremolite or other forms of asbestos was deprecated.

Some rocks with visible fibrous contamination were removed and replaced with clean substitutes but, of course, this could not be done with most rocks in the existing structures. Abrasive blasting, using copper slag brought 400 km from a smelter for the purpose, successfully removed the fibrous layer from some but this practice was abandoned because it generated airborne fibres. Ultimately it was decided to cover most of the rocks to which visible fibrous material adhered with cement mortar, which was applied by a spray technique by a swimming pool contractor (30–40 mm of mortar was applied according to the spray-man's judgement; where appropriate, the rocks were first washed with fresh water at low tide. The work appears to have been successful in that little deterioration is evident after the lapse of 4 years). Mortar was spread on some rocks with a trowel, while others were coated with a tar-epoxy compound to bind the fibrous material to the substrate and prevent its release to the environment. Tar-epoxy compounds are not without risks to health both in their manufacture and use.

## Further developments

Although the multipurpose jetty was completed in April 1984, it was always the Government's intention to extend the breakwater which protected it and the moored fishing fleet (of 36 vessels) by another 150 m, thus doubling the breakwater length. Plans were therefore in hand to extend the breakwater even before the jetty was completed. This work did not require the preparation of a new Environmental Impact Statement but the public authority responsible for the construction published a Review of Environmental Factors involved in the extension and invited public comments by advertisement in the local newspapers in December 1983. At that time it was expected that work would begin by July 1984 and take up to 18 months to complete.

The extension of the breakwater was estimated to require 268 000 tonnes of rock, while other long-term requirements of the construction authority are for about 1 000 000 tonnes. The authority purchased a potential, undeveloped, quarry site in the same area as that which produced the fibre-bearing rock which had produced the adverse reactions, located about 11 km from the breakwater. Previous studies had established that the basalt/dolerite of a local volcanic complex is the only rock type near Eden suitable for the production of large-sized natural breakwater armour and that minor fibrous veins are widespread throughout all occurrences of basalt/dolerite within about 100 km. The alternative to using natural breakwater armour is to use concrete blocks each weighing up to 20 tonnes; concrete can be made from rock drawn from quarries which cannot produce large-sized natural breakwater armour. Quite apart from questions of expense, the use of concrete is associated with other health and safety risks; not the least of these is that of transporting thousands of tonnes of raw materials for concrete, ready-mixed concrete or concrete blocks 30 or more km over the (two-lane) highway linking two state capitals (separated by a distance of 1000 km).

It is interesting and appropriate to break down the quantity of rock required to build the breakwater into a number of different categories. The estimates were:

| | |
|---|---|
| 20-tonne rocks | 14 000 tonnes (i.e., 700 rocks) |
| 15-tonne rocks | 45 000 tonnes (i.e., 3000 rocks) |
| 8–12-tonne rocks | 13 000 tonnes |
| 4–8- tonne rocks | 18 000 tonnes |
| 2–4-tonne rocks | 38 000 tonnes |
| Quarry run | 140 000 tonnes (most larger than 25 kg) |

To win these very large rocks, the operating procedure adopted in a quarry must be the opposite of that used in one producing concrete aggregate, where large rocks, far from being desired, are a liability. Less dust is generated in the blasting operation when large rocks are being sought. The fact that so much of the 'armour' is in the form of discrete rocks which required individual handling made it possible for the engineers to comply with some of the limitations imposed on them. It is also interesting to note that a 20-tonne rock occupies an entire transport lorry and that any fibre attached to it is relatively easy to detect and also unlikely to be released to the environment during transport.

## Residents' response

The engineers expected some opposition to the project. It was estimated that a fleet of 7 20-tonne trucks would make 14 journeys per hour (through the main shopping centre of Eden), 6 days per week, 10 hours per day (between 0700 and 1700) for 26 weeks (assuming good weather), so that complaints concerning truck movements and noise nuisance were to be expected. The main street of the town is the only one strong enough to carry heavy weights. It is also used by road tankers carrying petroleum fuel from the storage facilities in the jetty area; the jetty itself was designed to take 44-tonne trucks, which would need to pass through the main shopping centre of the town independently of the breakwater construction. The engineers were prepared to take reasonable and practical steps to minimize complaints concerning road traffic and noise; the adverse effect of haulage during the main tourist season was recognized and an undertaking was given at an early stage to refrain from this activity during the 2-week Christmas–New Year period. No doubt other compromises would have been negotiated but the events and the fibres of 1983 brought unexpected problems.

The newspaper advertisement concerning the Review of Environmental Factors brought complaints from residents in the vicinity of the proposed quarry and the terms 'asbestos mine' and 'asbestos quarry' came into widespread use. At a public meeting (attended by 28 persons) they 'agreed unanimously' to a statement that the proposal should be opposed 'because of the asbestos in the rock and the health hazard that the asbestos would pose to quarry workers, truckdrivers, workers on the breakwater and residents in the area'. Quarry workers, truckdrivers and workers on the breakwater were ultimately quite unconcerned (their exposure to airborne fibres was below the reliable detection limit) and only the local residents strongly expressed fears concerning the effects on health of the fibres. The total population of the area is about 100 and the population density was less than one person per $km^2$; the nearest house was 1 km from the centre of the proposed quarry and no others were within 2 km. The residents are principally farmers or 'hobby farmers' living in the area for the life-style.

The opposition was forceful and resulted in many local newspaper articles opposing the project. An early letter to the editor of one paper ran across 7 columns of tabloid letterpress (albeit half empty columns) beneath the headline 'Asbestos — Residents oppose quarry'. It contained such eye-catching statements as: 'Tremolite is estimated to occur in 'less than 2 percent' of the rock by the Environmental Impact Study prepared for the Public Works Department. Less than 2 percent implies more than 1 percent, and 1 percent (of 200 000 tonnes) represents in excess of 2000 tonnes of contaminated rock!' (The figure of 2% was obtained by counting the number of rocks with fibre on their surface on an already constructed breakwater.) One of the more bizarre speculations was that 'There is the possibility, also, that some asbestos material will wash out of the breakwater extension, especially during construction, and find its way onto nearby beaches. This will be of concern to Eden tourist promoters as well as residents because, naturally, people will not wish to holiday in an area with contaminated beaches.' ('Contaminated beaches' is an expression in common use in New South Wales; Sydney, the State capital, has beaches contaminated with raw sewage on occasions.)

The Residents' Committee had clearly received information from some person familiar with the asbestos literature — the letter to the editor contained the reference to the IARC 1977 statement quoted above. Photocopies of up-to-date material intended for the expert — 'some of these minerals may present a hazard to man if used indiscriminately' (Wagner, 1980) and 'Tremolite thus proved to be the most dangerous mineral that we have studied' (Davis et al., 1985) – had wide currency amongst the residents, but another statement by Wagner and his colleagues 'Tremolite sample C, would be a human health hazard if present *in sufficient airborne concentrations*' (our emphasis) (Wagner et al., 1982) was apparently withheld from or disregarded by them. Of course, recent views (Peto et al., 1985) concerning short exposure periods and mesothelioma were not known to the residents.

The residents undertook much lobbying of local and central government politicians. One newspaper carried the headline 'Minister's assurance on asbestos rock', together with the comment: 'If there is any danger whatsoever to residents from the proposed quarrying of rock at Nethercote for the Eden breakwater extensions, we will definitely not use the rock'. A report of a local (60 km distant) council meeting under the headline 'Councillors query health hazard' stated that 'A trial blast will be carried out at the Nethercote quarry to determine if asbestos dust can be kept to 0.1 fibres per millilitre of air. This was the major restriction that the Bega Valley Shire Council has placed on hard rock mining at the quarry at its full meeting last week'. Another condition made known at that meeting was that 'the Development Consent (for the quarry) only applies to the provision of rock for the Eden Breakwater construction and, at the end of such work, the quarry becomes redundant and must be rehabilitated in accordance with a Plan approved by Council'. Six weeks later (November 1984) another newspaper reported in large letters 'Trial blast impressive but residents fight on'.

## The legal outcome

In January 1985, the constructing authority was granted approval by the Local Government Authority to 'establish a quarry for the supply of rock for the extension of the existing breakwater at the port of Eden' subject to 39 conditions, many of which arose directly from the fibrous contamination. The residents were dissatisfied and, represented by the solicitor for the Environmental Defenders' Office Ltd, applied to the Land and Environment Court to restrain the development, citing the local government authority and the construction authority as respondents. At the end of May 1985 the Court ordered, by consent, that the application to establish the quarry be granted subject to the conditions originally imposed by the local authority but with the addition of some distinctly more onerous ones, among them the requirement that the constructing authority 'will take all reasonable measures to ensure that no loose asbestos material and no rocks with any asbestos material exposed on the surface are removed from the site'; the 'no rocks are removed...' condition was a major restraint. In effect, an agreement had ultimately been reached between the applicant Residents' Committee and the respondents which, in the words of a local newspaper, contained 'a range of conditions not originally contained in the Council agreement on the mining of rock containing asbestos in the area.' A spokesman for the residents was reported in the same newspaper as saying that he believed that the agreement contained in the court order 'was the most that could be hoped for' and 'that with the dust suppressing measures in force at the site the risk *to residents* (our emphasis) will be nil' — the views of a person inexperienced in dust suppression in mines and quarries.

## The operations

The court order is, of course, a legally binding document and, amongst other things, requires that 'the time weighted average exposure to airborne *tremolite* (our emphasis) fibre over an eight hour working day shall not exceed 0.1 fibres of asbestos per millilitre of air ... using a four hour sampling period'. This limit is that for crocidolite and amosite introduced into the Construction Safety Act and similar New South Wales legislation in 1984. Because the Construction Safety Act prohibits the use of crocidolite and amosite in new work, the residents sought also to prohibit that of tremolite, asserting (in newspapers) that its omission from the legislation was 'an oversight' and that all amphiboles should be subject to the same restrictions.

This paper is concerned with non-occupational exposure to mineral fibres but it is worthy of note that the exposure of quarry workmen, as determined by a consultant to the constructing authority, never reached 0.1 fibres per ml (f/ml) and that the mine safety authorities always reported their measurements as below the detectable level. The Quarry Operations Manual prepared for the constructing authority to enable it to comply with the requirements effectively converted the quarry into an asbestos mine but also required measurements of ambient dust levels by means of deposit gauges, high-volume samplers and the membrane filter method with estimation performed with the scanning electron microscope. The results obtained during the operation of the quarry never exceeded the targets; deposit gauges were never to exceed existing levels by more than 20%, total suspended particulates measured with the high-volume sampler were never to exceed background levels by more than 10%. No membrane filter sample taken at the nearest house exceeded the detection limit with the electron microscope.

The Quarry Operations Manual required work to be performed with the least possible generation of dust. Water was applied to all stockpiles of rock and at all loading and tipping points in the quarry. The load on all trucks leaving the quarry was thoroughly wetted and all vehicles (trucks and cars) were required to pass through a truck wash station (referred to as 'the decontamination system' in the approval document) to ensure that no free fibre taken from the quarry was lost on the 11 km road journey to the breakwater site. A water truck drove over the quarry roads applying water for dust suppression; it passed any given point about once each hour. The water requirements were estimated as 12 593 000 litres per year, which required the construction of a water storage dam to hold 7 megalitres. Of course, water run-off resulting from these operations contained solids not normally conducted to the nearby creek and, although the Clean Air Authority had expressed no interest in the quarry because no crushing or screening was being undertaken, the water to be discharged to the creek was of interest to the Clean Water Authority. This led to the construction of a settlement dam with a wall capable of withstanding a once in 10-year flood, and a condition in the approval document that 'dirty waters shall be collected, treated, and then either discharged to a creek or re-used'. A suitable subterranean water source was required in case the supply dam was exhausted during the working period – months can pass without rain in the area. None of this work, and its consequent delay and expense, would have been required under normal quarrying operations; construction of the dam was the first work undertaken at the quarry.

A dust collector was incorporated into the drilling equipment for the blast holes. It included an approved and tested filter system and a reverse-pulse-jet facility to remove the dust from the filter and direct it into a holding hopper which was emptied at intervals into polyethylene bags. These bags were buried as 'asbestos waste', although it was impossible, at reasonable expense, to identify fibre amongst the drilling dust.

The man attending to the dust bags wore a respirator approved by the health authorities. Whilst we applaud the use of dust-suppression devices with rotary/percussion rock-drilling equipment in quarries, we believe that a silicosis risk to the drilling crew was more credible than one of an asbestos-related disease. Under the conditions of use, neither the elaborate dust collection and disposal system nor the use of respiratory protection would have been warranted had the rock been quartz. Of course, the dust collection and burial system would have been a comfort to the residents, who had the right under the Court Order to enter the site 'for the purposes of inspecting the method by which monitoring of exposure to airborne tremolite fibre is being carried out'.

An important person in the quarrying production area was the 'asbestos spotter' (employed and paid by the contractor operating the quarry), whose task was to mark with paint from a spray can all rocks carrying visible fibre. These rocks were collected by a front-end loader driver and taken to a reject dump. One condition of the approval for the quarry was that 'The site to be used for the stockpiling of tremolite-bearing rocks shall be clearly defined and any subsequent burial of such material shall be within a specifically designated area'. (Rock containing fibre could be found occurring naturally in a nearby creek bed.) The asbestos spotter had no discretion — if any fibrous material was visible the rock was discarded in order to ensure that no 'tremolite' was taken on the roads or deposited at the breakwater site. A representative of the construction authority also had a spray paint can and rejected any rocks which might have escaped the vigilance of the spotter. A further asbestos spotter was at the breakwater to reject any rocks which might have passed through the quarry net. Although the fibre contamination of the rock was sparse, the magnitude of the quarrying operation was such that thousands of tonnes of otherwise valuable material was rejected, ultimately to be covered with over-burden when the area is 'restored'; approval for the quarry was limited to the provision of armouring rock for the Eden breakwater.

At one stage in the operations an insufficient number of 'uncontaminated' 15–20-tonne rocks was available and fibre on some was covered with a tar-epoxy compound and transported to, and used at the breakwater. It was argued that this was within the letter of the approval conditions; these rocks did not have 'any asbestos material exposed on the surface' and the residents did not press their objections. These rocks were placed in water 3 m below low tide level.

The quarry workers were provided with changing rooms of the type found in factories which use raw asbestos and work-places where asbestos is being removed from buildings and plant. The principle of 'clean' and 'dirty' changing rooms was well executed and well controlled. The 'clean' changing room was clearly labelled 'quarantine area; no entry during shift'. This together with the excellent bathroom, ensured that workmen would not take contamination from the quarry to the 'outside' environment. All work clothes were laundered at the work site by a woman especially engaged for this work. Air samples taken in the changing room area (and including those from personal samplers worn by the 'laundry girl') were always satisfactorily low.

The quarry was required to appoint an approved medical officer and the frequency of X-ray examinations was set at once every 3 years (the quarry was supposed to be rehabilitated and revegetated within 18 months, with a possible extension to 2 years). Of the first 16 persons medically examined (including one woman in the site office), one had a history of an extended period of asbestos exposure, 3 others some identifiable occupational asbestos exposure and two-thirds had worked in dusty occupations (they were mostly habitual quarry workers and miners). Nearly all were tobacco smokers but were nevertheless healthy (except for those who bore the stigmata of previous traumatic injuries).

The rehabilitation and revegetation of the quarry was addressed in the development approval, even to noxious weed and pest control: 'It will also be necessary to control rabbits and kangaroos on site to ensure adequate revegetation. Details shall be provided detailing the type and standard of fencing, gating [sic] as well as rabbit control programming to the Soil Conservation Service for its views.' Rehabilitation and revegetation of quarries is not customary; indeed, it is deprecated in some cases.

Work at the breakwater itself was done in such a manner as to ensure that all dust generation was as low as reasonably achievable (the ALARA principle introduced by the health physicists). Although the rock in the trucks was often still wet on arrival at the site, those carrying 'run of mine' or 'core' material were sprayed with sea water before tipping so that visible dust was seen infrequently. After tipping, this material was pushed about by bulldozer or lifted and placed in position with an excavator; the water ensured that this was not a dusty job. A water truck kept the approach road and other areas adequately under control. The large crane and rock grab which were used to place the larger (up to 25-tonne) rocks in position did not generate dust other than the occasional puff when a rock was hit forcefully.

## *Risk: its perception, assessment and acceptability*

It is not the intention of the authors of this paper to be judgemental other than to say that a local amenity was made more expensive and its construction delayed because a small group perceived a risk to its health arising out of the mere presence of a natural mineral fibre in its environment. Some members of the general community actively discounted the fears of the concerned group; one wrote to a local newspaper asserting that the fears were bogus.

Much of the residents' perception of risk was based on what might be termed an assessment of that risk by an official regulating agency. The residents held the views of the agency in high regard: 'The quarry contains a form of asbestos known as Tremolite. The Division of Occupational Health which is the Government Department controlling Occupational Health has said that no rocks or materials containing asbestos should be removed from the quarry'. It can only be assumed that, in expressing its concern about the presence of mineral fibre on rocks in a retaining wall of a bridge and a breakwater, this agency had formed the view that the presence of the fibre was unacceptable, despite the fact that people in the community do not habitually sleep beneath bridges and that both professional and recreational fishermen accept certain risks as normal. The risk from the presence of this fibre in the bridge and breakwater might be viewed in the light of the risk estimates proposed by various authors concerning sprayed asbestos in office buildings and schools (Doll & Peto, 1985; Ontario Royal Commission, 1984). Doll and Peto make some comments concerning crocidolite and amosite but our amphibole is unknown to epidemiologists.

Whilst the small residents group found the risks associated with the mineral fibre unacceptable, the community at large accepted the passage of tankers carrying liquid and gaseous petroleum products through the main street of the town from the storage depot adjacent to the breakwater. Liquid chlorine in 2-tonne containers also passes through the town to the water-treatment plant for the fish-processing facilities.

Risk and its perception, assessment and acceptability deserves, even demands, more than lip service from regulating authorities and other official agencies. It is a specialized discipline, often understood only superficially by those persons with scientific and/or legal experience who are required to make judgements on health issues. Of course, the issue is not unique to the asbestos problem, although the unnecessary and expensive removal of much sprayed asbestos has highlighted it. Workers in the field of radiation and health are only too familiar with the matter and the difficulties in communicating it to the public (Slovic *et al.*, 1981). The issue is far from being a merely scientific one. In requiring a uranium mine to prevent the run-off to nearby creeks of rain-water which is not radioactive and not contaminated by uranium, the Australian Government, against the judgement of its own specialist scientists and through an intelligent and well-educated Attorney-General, has said that 'The social factors so-called, which include community and interest group reaction to water management, have to be considered and were considered by the Government because they are part of the definition of best practicable technology' (Evans, 1987). The dam required to store the rain-water and withstand a once-in-ten-years flood cost the uranium miners $A1.5 million, only slightly more than the extra expense which the fibre added to the Eden breakwater extension.

## References

Campbell, W.J., Blake, R.L., Brown, L.L., Cather, E.E. & Sjoberg, J.J. (1977) *Selected Silicate Materials and Their Asbestiform Varieties*, Washington DC, US Department of the Interior, Bureau of Mines (Information Circular 8751), p. 2

Davis, J.M.G., Addison, J., Bolton, R.E., Donaldson, K., Jones, A.D. & Miller, B.G. (1985) Inhalation studies on the effects of tremolite and brucite dust in rats. *Carcinogenesis*, 6, 667-674

Doll, R. & Peto, J. (1985) *Asbestos, Effects on Health of Exposure to Asbestos*, London, Her Majesty's Stationery Office, p. 47

Evans, G. (1987) *The Sydney Morning Herald*, 31 March 1987, p. 6

Levine, R.J., ed. (1978) *Asbestos, An Information Resource*, Bethesda, MD, US Department of Health, Education, and Welfare, pp. 51 and 57

Ontario Royal Commission (1984) *Report of the Royal Commission on Matters of Health and Safety Arising from the Use of Asbestos in Ontario*, Vol. 2, Toronto, Ontario Government Bookshop, p. 585

Peto, J., Doll, R., Hermon, C., Binns, W., Clayton, R. & Goffe, T. (1985) Relationship of mortality to measures of environmental asbestos pollution in an asbestos textile factory. *Ann. Occup. Hyg.*, 29, 305-355

Slovic, P., Fischoff, B. & Lichtenstein, S. (1981) Informing the public about the risks from ionizing radiation. *Health Phys.*, *41*, 589-598

Wagner, J.C. (1980) Opening discussion – environmental and occupational exposure to natural mineral fibres. In: Wagner, J.C., ed., *Biological Effects of Mineral Fibres (IARC Scientific Publications No. 30)*, Lyon, International Agency for Research on Cancer, pp. 995-997

Wagner, J.C., Chamberlain, M., Brown, R.C., Berry, G., Pooley, F.D., Davies, R. & Griffiths, D.M. (1982) Biological effects of tremolite. *Br. J. Cancer*, *45*, 352-360

# VI. CONCLUDING SESSION

# MINERAL FIBRES IN THE NON-OCCUPATIONAL ENVIRONMENT: CONCLUDING REMARKS

### R. Doll

*Imperial Cancer Research Fund Cancer Epidemiology and
Clinical Trials Unit, Radcliffe Infirmary,
Oxford, UK*

## Introduction

The symposium on which this volume is based came at a particularly opportune time, when people in many countries had been subjected to so much publicity and misinformation about the effects of asbestos, that many believed that exposure to one fibre carries with it a material hazard of developing cancer. Much anxiety has been, in consequence, caused by the realization that the ambient air contains fibres of asbestos or of other materials that can be, and often are, described as being like asbestos, and action is called for that has social costs out of all proportion to the possible benefits, as is described so vividly by Major & Vardy. There is, therefore, an urgent need for scientists to determine precisely which fibres are carcinogenic, how great the concentration of fibres is likely to be under different conditions, and the size of the risks that exposure to these concentrations is liable to produce.

## Characteristics of carcinogenic fibres

The information that was given in part II together with that reviewed recently by the International Programme on Chemical Safety (World Health Organization, 1986) and the International Agency for Research on Cancer (1987a,b, 1988) go a long way to enabling us to reach a conclusion about the characteristics of those fibres that are liable to cause cancer. These, it is clear, are not limited to the many varieties of asbestos, but pertain to many of the mineral fibres irrespective of whether they occur in nature, such as erionite and attapulgite, or are made by man from slag, rock or glass or from blends of silica, alumina, zirconia, and other materials fused to make ceramics. Two characteristics that all these materials have in common are a capacity to persist in animal tissues for months, if not for years, and a physical form such that their length is often more than 5 $\mu$m while their diameter is less than 2 $\mu$m. Properly speaking, no particle should be described as a fibre unless it is at least 5 $\mu$m long and the diameter is less than one-third of its length. This, however, is commonly overlooked and much shorter particles are described as fibres if they are made of the same material as long ones, and I shall follow the common practice. It is unfortunate, however, that this practice has evolved, as there is increasing evidence that short fibres (properly described as elongated particles) are much less carcinogenic, if they are carcinogenic at all. It is now 15 years since Stanton and Wrench (1972) and Pott and Friedrichs (1972) independently found that the physical dimensions of fibres were a major factor in determining their ability to cause cancer when injected intrapleurally or intraperitoneally; however, the difficulty in obtaining sufficient numbers of fibres of defined sizes made it difficult to be sure that the same was true when fibres were inhaled. The data that Davis reported now make it highly probable that the physical dimensions of the fibres are equally important in these circumstances and, taken in conjunction with the many studies of the effect of fibres on intrapleural, intraperitoneal, or intratracheal injection, they indicate that, to quote Davis, 'fibres <5 $\mu$m in length may be innocuous in lung tissue'.

This conclusion, if borne out, is of profound importance, for the great majority of so-called fibres that are normally found in air in and out of doors are less than 5 $\mu$m in length.

No-one would, I think, now question the conclusion that short so-called fibres are less likely to cause cancer than long ones, but the conclusion that there is a cut-off point somewhere in the region of 5–10 $\mu$m, below which fibres are not carcinogenic, is more dubious. It may be correct, but would be easier to accept if the mechanism were known by which fibres caused cancer and if it were possible to show that the mechanism was triggered only by fibres longer than (say) 5 $\mu$m.

In our present state of incomplete understanding this cannot now be done. It is still not even clear whether asbestos fibres are generally genotoxic. Jaurand's review makes clear that genotoxicity has been demonstrated in hamster embryo cells and in rat mesothelial cells, but, despite many experiments, asbestos has never been shown to be genotoxic in tracheobronchial epithelium. Fibres do, however, lead to the release of superoxide, and this may cause collagen to be formed and fibrosis produced. It may be, therefore, that asbestos is a complete carcinogen only for the mesothelium and that it acts to produce bronchial carcinoma less directly — a conclusion that is supported by the epidemiological data on the differential effects of age at first exposure for the two types of cancer reported by Professor Peto. In these circumstances, the essential difference between long and short fibres would seem to be that fibres less than 10 $\mu$m long are removed more readily by macrophages. We should note also, however, that the longer fibres cause more superoxide to be released and that, in some special types of cell, they have been shown to induce transformation more readily, interfere more with the movement of chromosomes at mitosis, and cause more aneuploidy.

## Concentration of fibres in the environment

It seems, therefore, that for practical purposes the best we can now do is to work on the assumption that all fibres that meet the criteria of the International Agency for Research on Cancer (IARC, 1988) for proven carcinogenicity in animals should be regarded as potentially carcinogenic to humans, but that we should base our estimate of potential risk on both the chemical constitution of the fibres and their size, counting only those fibres that are respirable and more than 5 $\mu$m long. This may lead to an underestimation of risk if short fibres also have some effect and are in the great majority, but it will be less misleading than if we base our estimate on the total fibre count or the total mass. Unfortunately, this means that we must measure ambient pollution by electron microscopy, as optical microscopy, which is fine in an occupational setting where nearly all fibres will be of one specific type, is useless for measuring ambient pollution, as it is unable to distinguish mineral fibres from others that are commonly the predominant type in the general environment, leave alone distinguish one type of mineral fibre from another.

It must be admitted, however, that fibre counting by electron microscopy fails to take cognizance of the clumps of fibres that may occasionally occur, as Professor Nicholson and Dr Burdett pointed out, and that this could be a matter of some concern if any substantial proportion of the clumps penetrate to the lung.

With or without this qualification it is evident that fibres that are potentially carcinogenic are present ubiquitously, due partly to the weathering of geological formations and partly to man's activities. According to the recent report of the International Programme on Chemical Safety (World Health Organization, 1986), more fibres are probably emitted due to the former than the latter. Those emitted from natural sources are, however, of less practical importance as they are dispersed throughout sparsely populated areas, whereas those emitted from the sources listed by Professor Nicholson (the operation of mines and mills, the construction and demolition of buildings, vehicle braking, and the wear and tear of domestic appliances and of the material used in house construction) disperse fibres principally in areas of high population density.

Most of the reports in this volume agree that the concentration of such fibres, even in densely populated areas, is low. Typical data for asbestos fibres are presented by Burdett. These showed that concentrations of fibres more than 5 $\mu$m in length might be as great as 0.012 fibres per ml (f/ml) in a room with a large area of damaged asbestos, but that the air in less than a quarter of the buildings (9/39) which contained sprayed asbestos, asbestos plaster, or warm air heaters containing asbestos gave average concentrations greater than their limits of quantification, taken (for this study) to be a count of 4 fibres. Four sets of samples above the limits of quantification gave on average figures of 0.0009 f/ml for buildings with sprayed or trowelled asbestos insulation or plaster, while 5 for buildings with air heaters containing asbestos gave an average of 0.0008 f/ml. The other 30 similar buildings with sample counts below the limit of quantification must have given a considerably lower average, and an overall figure of 0.0005 f/ml is more likely to be an overestimate than an underestimate. Chrysotile and amosite were each identified in over half the buildings (56%), but crocidolite was identified in only 2 (5%). In one-sixth of the buildings (18%), no type could be identified as no asbestos fibres were seen.

Further information that Burdett and his colleagues provided is, I think, important for purposes of control: namely, that even with the adoption of complex systems for the containment of contamination, the removal of asbestos from parts of a building led to substantial contamination in other parts and that higher fibre counts than had been present previously (sometimes an order of magnitude higher) persisted for many weeks.

Out of doors, counts of asbestos fibres made in the recommended way have generally been less than 0.0005 f/ml. Too few have, however, been reported to allow any representative figure to be given, except for Japan where Kohyama reported a mean of 0.0004 f/ml, based on a massive series of hundreds of counts throughout the country. These, in agreement with other observations, revealed that counts were consistently raised in areas of heavy traffic.

Fewer data still are available for other types of fibre, the carcinogenicity or potential carcinogenicity of which has been appreciated only relatively recently, but it seems clear that the concentrations are generally less. Two sets of figures have been cited, one for 3 German cities, where the concentrations of glass fibres averaged about one-quarter of that for asbestos, and another for Paris, where Dr Gaudichet and her colleagues obtained counts of synthetic fibres out of doors (using a polarized optical microscope) that were nearly 2 orders of magnitude lower, namely $4 \times 10^{-6}$ f/ml. Indoors, within buildings in which synthetic fibres were known to have been applied by spraying, the counts were 40–60 times greater. All these counts were, however, of total fibres and so can give only an upper limit to any postulated cancer hazard.

One surprising finding, in the light of the epidemiological data, is that of the relatively low fibre counts in the Turkish villages, where Dr Simonato and his colleagues confirmed the existence of a high mortality from mesothelioma and lung cancer. The fibre counts that they obtained were certainly higher than in industrialized countries, averaging about 0.006 f/ml, but even in homes and other places where people lived or (in the case of children) played, the counts were seldom greater than 0.1 f/ml, and the highest, recorded in a cave used as a home, was only 0.3 f/ml. There was, moreover, not much difference between the 3 affected villages and the one unaffected. In one village, asbestos contributed up to 10% of the short fibres, but most of the fibres, particularly in the 3 affected villages, were of erionite which, according to Wagner et al. (1985), has a greater carcinogenic potency than any of the other mineral fibres that they have tested.

## Assessment of risk

Assessment of any risks associated with these low counts can, for the most part, be only indirect, as the fibre counts to which people have been exposed are so far below the levels at which it has been possible to detect risks in industry. The one possible exception is that to which I have just referred: namely, the high risks of mesothelioma and lung cancer in the Turkish villages of Karain, Sarihidir, and Tuzköy. Mesotheliomas accounted for 50% of adult deaths in one of these villages in a 4-year period and 16% and possibly more in 2 of the others, giving annual mesothelioma death rates varying from 2.5 to 14.1 per 1000, as compared with typical non-occupational rates of under 0.002 per 1000 in Europe and North America and a maximum rate in asbestos insulation workers of 4 per 1000. Lung cancer death rates of approximately 6 per 1000 in men were, in contrast, only about 3 to 4 times greater than in Britain and perhaps 17 times greater than in the whole of Turkey. Simonato suggests that exposures in the past, when houses tended to have been built of local stone, may have been higher than they are now, when houses tend to be built of brick, and this is supported by Professor Sébastien's report that the mean number of coated fibres in the sputum of people living in affected villages was 200 times that of people living elsewhere. Otherwise it seems that erionite must be accorded a bizarrely high risk, even compared with crocidolite, if local environmental pollution is accepted, as I think it must be, as the principal cause of the disease.

In assessing the hazards from other types of environmental exposure, we must distinguish between the risks of mesothelioma, lung cancer, and other types of cancer, between man-made mineral fibres and asbestos, and between the various types of asbestos.

So far as man-made mineral fibres are concerned, we have heard very little about them, doubtless for the reason that they have not been unequivocally demonstrated to cause cancer in humans, despite many thousands of men and women having been exposed occupationally for several decades to substantially greater concentrations than are likely to be met in the general environment. The observations that have been made on occupationally exposed men and women led the International Agency for Research on Cancer (IARC, 1988) to conclude that there was limited evidence for the carcinogenicity to man of rock or slag wool fibres, but no clear evidence of any such effect from glass wool fibres and no data at all that would help to determine the carcinogenicity of ceramic fibres. This balanced opinion is perhaps a little conservative (Doll, 1987), but it is certainly the case that there is no reason to suppose that any man-made mineral fibres have caused mesotheliomas (perhaps because most of them are insufficiently durable in human lungs) while any risk of lung cancer can be only small and, at present, unquantifiable. It is obviously wise to keep an eye on the trends in environmental exposure and to avoid unnecessary exposure to such fibres, in so far as this can be achieved without great social cost, but the environmental risk may well be so small that an informed society would wish to ignore it.

Asbestos, however, presents a far more difficult problem. Firstly, there is the difference between the effects of chrysotile and amphiboles, which is so great in relation to mesothelioma that it is possible to argue that chrysotile does not cause mesothelioma at all and that the relatively few cases that have occurred in men occupationally exposed to chrysotile have been due to the presence of an unintended contamination with minute amounts of tremolite. Secondly, there is the possibility that the dose-response relationships differ quantitatively for mesothelioma and lung cancer, and thirdly there is the question of whether the ingestion of asbestos causes cancer in sites other than the lung and the epithelial lining of the pleura.

That pure chrysotile does not cause mesothelioma is strongly suggested by the low incidence of the disease in groups of men and women occupationally exposed only to chrysotile (with or without some contamination by tremolite) and by the results of tissue analyses which have repeatedly shown that the lungs of people who have died of mesothelioma contain very little (if any) more chrysotile than the lungs of those who have died of non-asbestos-related diseases, once the amount of any associated amphibole is taken into account. This was illustrated beautifully by the new data that

Dr Gibbs and his colleagues and Professor McDonald and his colleagues presented. That chrysotile should produce less pleural disease than amphibole asbestos is understandable, in view of the greater speed with which it is removed from the lung, but I hesitate to give it a completely clean bill of health. The data from Japan, presented by Professor Morinaga, provide perhaps the strongest evidence that mesotheliomas may be produced by pure chrysotile. In his series of cases graded qualitatively (+, ++, +++, ...) for chrysotile presence, the lungs of 4 out of 6 mesothelioma patients which did not appear to contain any amphibole asbestos (including any tremolite) were graded ++ or higher for chrysotile, whereas only one out of 17 control patients dying of other diseases was so categorized. As a working hypothesis I would suggest that, for similar amounts of exposure, chrysotile carries a risk of producing mesothelioma that is no more than 5% of that associated with an average mix of amphiboles and may even be less.

In contrast to these findings, there is no firm evidence to suggest that chrysotile carries a smaller risk of lung cancer than amphibole asbestos (though the risk has been notably low in miners and friction product and cement workers) and for the time being we must, I think, regard all types of asbestos as carrying an equal risk of cancer of the lung.

How great a risk of either disease people are likely to have incurred in non-occupational settings cannot be estimated directly, as no measurements have been made of the exposures that have caused cancer in the homes of asbestos workers or in the neighbourhood of asbestos mines, factories or dumps, and the concentrations to which the mass of the population are exposed in and out of doors are so low that it is impossible to design studies that will measure the difference between the different levels of risk involved. Mesotheliomas are, however, normally so rare in the absence of exposure to asbestos that the case-control and geographical studies reviewed by Professor Gardner have demonstrated, without possibility of cavill, that this disease has been produced by exposure to asbestos in the home or in the neighbourhood of sources of environmental pollution. It seems probable, therefore, that lung cancer may have been produced in the same way, but it has not been possible to demonstrate it epidemiologically. This may be because the background incidence of the disease is relatively so high — even in non-smoking women. Alternatively, it may be because this disease is not produced by low levels of exposure at all.

At present we can estimate the risks associated with current levels of non-occupational exposure only by extrapolating from the much higher levels that have been recorded in industry. This requires two things: the measurement of a relationship between dose and response and a theory that will justify extrapolation from high doses to very low ones. We can now make an informal guess at both the former and the latter in respect of mesothelioma, but our theory in respect of lung cancer is still largely a matter of unsupported faith. We have a substantial amount of data to support the idea that the incidence of lung cancer is proportional to the cumulative dose at high and moderate levels of exposure, but unless it can be shown that asbestos is generally genotoxic or, more specifically, genotoxic to the bronchial epithelium, we have no real grounds for postulating that a linear relationship for lung cancer can be extrapolated back to the levels of dose with which we are concerned in non-occupational settings.

If we leave this difficulty aside, we are then left with a host of minor difficulties relating partly to the measurement of dose, which has been peculiarly difficult in the asbestos industry in the past, and partly to our incomplete knowledge of the biological relationships between the incidence of cancer and the duration of dose, the age at which exposure occurred, and the time since it happened. These problems have been touched on by Professor Nicholson and Professor Peto and discussed in detail by some of the committees that have produced estimates of risk and are too complex to review again here. We should note, however, that the estimates all require major assumptions, for some of which the evidence is weak and partly contradictory. All the recent ones are, however, in broad agreement that the risk attributable to exposures to concentrations of (say) 0.0005 f/ml are extremely low and of the order of a life-time risk of 1 per 100 000 or less for 10 years exposure in school or 20 years exposure in adult life. That the actual risk cannot be much higher, if indeed it can be as high, is demonstrated, as Professor McDonald pointed out, by the low and relatively steady mortality from mesothelioma not obviously attributable to occupational exposure that is still observed in North America and Europe.

Whether we should need to add to this a further risk from gastrointestinal cancer due to the presence of asbestos fibres in water supplies is still, in my opinion, a subject for research. Professor Kanarek's excellent review showed that the ecological evidence could be interpreted to mean that the exceptionally large numbers of fibres in some of the water supplied to San Francisco might increase the risk of gastric and perhaps also of oesophageal cancer by 10% and that the lack of relationship observed in other similar studies did not necessarily contradict it. I do not think, however, that we should attribute a causal significance to this finding in the absence of experimental data to show that fibres can cause gastrointestinal cancer on ingestion and, as Dr Chouroulinkov showed, this is lacking, despite the many intensive efforts that have been made to obtain it. In the absence of such evidence and the doubt about the reality of risk of gastrointestinal cancer following occupational exposure, it is, I believe, more reasonable to attribute the San Francisco findings to confounding, despite the great trouble that has been taken to exclude it, particularly as we are unable to take into account any effect of variation in the prevalence of the principal cause of gastric cancer because this principal cause is still unknown.

## Epilogue

These remarks have not done justice to all the contributions in this volume. But even if they had done so, not all the important questions would have been answered completely and for ever. Further research, which is essential, will doubtless give different answers to some of the questions which we think have been adequately answered now. Our current answers are, however, sufficiently clear and, I suggest, reliable enough for practical policies to be determined for the control of exposure to mineral fibres which will allow social benefits to be assessed in relation to social costs. The carcinogenic effects of the fibres must be weighed against the tremendous contribution that mineral fibres have made and continue to make to human welfare by their durability, insulating power, and indestructibility by fire and friction.

## References

Doll, R. (1987) Symposium on MMMF, Copenhagen, October 1986: overview and conclusions. *Ann. Occup. Hyg., 31*, 805-820

IARC (1987a) *IARC Monographs on the Evaluation of the Carcinogenic Risk of Chemicals to Humans*, Vol. 42, *Silica and Some Silicates*, Lyon, International Agency for Research on Cancer

IARC (1987b) *IARC Monographs on the Evaluation of the Carcinogenic Risk of Chemicals to Humans*, Supplement 7, *Overall Evaluations of Carcinogenicity: An Updating of IARC Monographs Volumes 1-42*, Lyon, International Agency for Research on Cancer

IARC (1988) *IARC Monographs on the Evaluation of the Carcinogenic Risk of Chemicals to Humans*, Vol. 43, *Man-made Mineral Fibres and Radon*, Lyon, International Agency for Research on Cancer

Pott, F. & Friedrichs, K.H. (1972) Tumours in rats after intraperitoneal injection of asbestos fibres. *Naturwissenschaften, 59*, 318-332

Stanton, M.F. & Wrench, C. (1972) Mechanisms of mesothelioma induction with asbestos and fibrous glass. *J. Natl Cancer Inst., 48*, 797-821

Wagner, J.C., Skidmore, J.W., Hill, R.J. & Griffiths, D.M. (1985) Erionite exposure and mesotheliomas in rats. *Br. J. Cancer, 51*, 727-730

World Health Organization (1986) *Asbestos and Other Natural Mineral Fibres*, Geneva (*Environmental Health Criteria, No. 53*)

# INDEX

Abdominal tumours, after injection of dusts 194, 195
Actinolite, tumour induction 175–177
Activation of surface-active sites 103, 107–108
Active oxygen species in inflammation 82
Adenomatous polyps and fibre ingestion 122, 131
Airborne mineral fibres
  concentrations 337–338, 361–364
  exposure 18–21
  levels in Sarihidir, Turkey 399–400
  see also Airborne particulates and under specific fibres
Airborne particulates, composition 357
  see also Airborne mineral fibres
Air
  contamination, indoor 251
    see also Buildings
  monitoring in hazard evaluation 251–253
  sampling 263
  sensitivity of contaminant measurement 241
  see also Asbestos, airborne
Algorithmic index of pollution by MMMF 293–294
Alveolar load of mineral fibres 310, 311–313
Alveolar macrophages
  release of interleukin 1, 154
  release of lactic acid dehydrogenase 192–193
  release of $O_2^-$ 85–86
Amosite
  carcinogenicity 116, 119
  in combination with azoxymethane 121
  infrared absorption 198, 199
  in workers' lungs 333, 488–489, 492
  sprayed-trowelled, removal 283–288
Amphibole asbestos bodies 412
  see also Asbestos bodies; Ferruginous bodies
Amphiboles
  as cause of mesothelioma 226
  exposure and mesothelioma 443
  lung burden 446, 490
  in mesotheliomas from Osaka 441
  surface activation 107

Antarctica, asbestos concentrations in ice 271–272
Anthracosis 356–357
Asbestiform fibres 5, 14–16
Asbestos 4
  abatement work 251, 252
  adenomatous polyp formation 122
  airborne, environmental exposure 408, 476–477
  airborne levels 357
    near building containing 371
    in buildings 254
    in Corsica 408
    in Fiji 270
    as function of distance from source and meteorological parameters 339–342
    health hazards 479–482
    in highly polluted areas 274
    in Japan 267–270
    at Ogasawara Island 270
    near a serpentine quarry 265–266, 269
  amphibole-like in drinking-water 429
  benzo[a]pyrene contamination 141
  bioassays 47–53
  bodies
    amphibole 412
    among Cape Town autopsies 237
    chrysotile in 211
    in domestically exposed women 213
    in necropsy lung tissue 237
    nature of 88
    occurrence in lung tissue 211, 212
    in Quebec mining communities 213–214
    see also Amphibole asbestos bodies; Ferruginous bodies
  brake emissions as source 209, 254–255, 269, 272–273
  in buildings
    contamination 212, 251
    under normal occupation 279–281
    removal 281–288
  carcinogenic effects 175, 278
  chronic ingestion, tumour pathology 130–131

—519—

Asbestos (*contd.*)
    coating, sprayed/trowelled 279-281
    concentrations
        in buildings with warm air heaters 279
        near factories 270
        along main roads 268-269
    consumer products, use in 347
    consumption in Japan 262
    domestic exposure 209
    dose-response relationship 177
    in drinking-water 112
        amphibole-like 429
        epidemiological studies 428
        health significance 478
        in Quebec 432
        risks of gastrointestinal cancer from 433-434, 435
    environmental exposure 463, 477-478
        braking of vehicles as source 209, 254-255, 269, 272-273
        measurement 240-245
        risk from 466, 482
    exposure in buildings 208, 254, 279-281
    fibres
        cellular response 97
        measurement by electron microscopy 240
        release from asbestos-cement 368
        size and carcinogenic hazard 434
    global circulation of fibres 275
    ingestion
        and gastrointestinal cancer 112, 120, 123, 465-466
        and tumour pathology 130-131
    Japan 231, 262
    lung burden 314, 316, 487
    lymphocyte activating factor, release on exposure 149-154
    and mesothelioma 112, 314, 376, 384-385, 486
    molecular events after treatment with 65
    neighbourhood exposure 209, 380-382
    occupational exposure
        cohort studies 458
        fibres found in pulmonary tissues 332
        fibre types and lung burden 331-335
        gas-mask work 487, 490-491
        insulation work 112, 333-334, 380, 445-447, 458
        mesothelioma
            among household contacts 380
            in workers 314, 316
        shipyard work 331-334
    oxygen free radicals produced in response to 87-88
    PAH adsorption on fibres 140-147
    para-occupational exposure 378, 486
    products, friable, in schools 471-473
    pure, microscopy 305-306
    removal from buildings 281-289, 469
    risk from environmental exposure 466, 482
    scar as precondition for tumour development 177
    slate-board factories 270
    in snow and ice samples from Tokyo and Antarctica 271-272
    and superoxide dismutase in lungs 87
    surface active sites 101
    transfection by viral DNA, ability to mediate 65
    uncoated fibres in urban dwellers' lungs 237
    weathering of asbestos-cement materials as source of air pollution 253
    whitewash in Turkey 384-385
    workers, *see* Asbestos, occupational exposure
    Zielhuis groupings of occupational exposure 220
    *see also following entries*
Asbestos-associated disease, pathogenesis 82
Asbestos-associated radiographic abnormalities 391
Asbestos-cement drinking-water pipe, exposures 429, 432
Asbestos-cement plant, air concentrations of fibres around 337-338
Asbestos-cement products
    carcinogenic potential of fibres from 190-195
    corrosion and weathering 367, 369
    fibre concentrations near buildings containing 371
    fibre identification 200
    sampling device for measurement of fibrous aerosols released by 367-368
Asbestos-containing surfacing materials, buildings with 245-251
Asbestos-exposed workers, lung cancer risk 243-244
Asbestos-induced cellular alterations, mechanisms 97-98
Asbestos-induced neoplastic diseases, risk 330

Asbestosis
  association with lung cancer 480
  association of crocidolite and amosite with 447
  degree in asbestos factory workers 445
  severity and lung fibre content 444
Asbestos-related diseases, epidemiology 447
Atmospheric fibre pollution
  by MMMF, methods of evaluating 292–293
  from geological sources 292–293
  see also Airborne mineral fibres
Atomic absorption spectrometry, determination of silicon content 134, 135
Attapulgite 5, 14
  carcinogenic potency 180–184
  effect on cell growth 182
  as replacement for asbestos 256–257
  use in consumer products 348
Azoxymethane, effect of asbestos fibres in combination with 121

Basalt fibres, induction of tumours 177
Benzo[a]pyrene, contamination of asbestos fibres 141
Bilateral pleural plaques
  characteristics of patients 408
  in Corsicans 406–409
  etiology 406
Braking of vehicles
  asbestos emission 272–273
  as source of environmental asbestos exposure 209, 254–255, 269, 272–273
  see also Friction materials
Bronchoalveolar lavage fluid, alveolar load of mineral fibres 310, 311–312
Buildings
  air concentrations of asbestos 254
  with asbestos-containing surfacing materials 245–251
  contamination by asbestos 212, 251
  exposure to asbestos in 21–22, 208
  maintenance as source of air contamination 251
  MMMF pollution 292–294
  under normal occupation, asbestos in 279–281
  removal of asbestos 281–288
  source of asbestos contamination in 251
  with warm air heaters, asbestos concentrations 279
  see also Asbestos, abatement; Asbestos, removal; Domestic exposure; Kindergartens; Schools containing friable asbestos products; Sick building syndrome

Cancer incidence and asbestos fibres in drinking-water 428
Cancer risk
  epidemiological detection 434
  estimates, lifetime 471–474
  see also Risk
Carcinogenesis
  asbestos, model of 461
  chemical, model for the genotoxic activity of phagocytosed mineral particles 101
  multi-stage, conventional models 461–462
  in rats exposed to radon and injected with chrysotile 162–164
  see also following entry and Tumours
Carcinogenicity of fibres
  fibre characteristics 501–512
  as function of fibre length 178
  intraperitoneal test for 175–179
  potency 177
  variation with length and diameter of fibres 243
  see also Carcinogenesis
Carcinoma of the lung, see Lung cancer
Case-control studies 392
  of non-occupational mesothelioma 378–382
Cell
  growth, effect of attapulgite 182
  lines, mesothelioma 168
  response to asbestos 97
  response to chrysotile 159
  transformation 64–65, 75
Ceramic fibres
  definition 299
  induction of tumours 177
  occupational exposure 300–303
  release of silicon 135
Chemical carcinogenesis, and genotoxicity of phagocytosed mineral particles 101
Childhood exposure 463
Choline
  effect of administration of chrysotile with 186, 187–188
  intraperitoneal injection 186–188
Chromosome changes, numerical, induced by fibres 59–60, 169, 170

Chrysotile
  abdominal tumours after i.p. injection 195
  adenomatous polyps in long-term ingestion 131
  airborne, 213, 245
  association with specific products and trades 334
  carcinogenesis in rats exposed to radon 162-164
  carcinogenicity in long-term ingestion studies 114-115, 119
  carcinogenic potential of fibrils 243
  choline co-administration 186-188
  complete carcinogenic potency 158
  corroded, chemical and crystallographic changes 369
  cytotoxicity 182
  dimethylhydrazine coadministration 121
  diseases associated with exposure to asbestos dust, role in causation 447
  DNA repair, stimulation 182
  in drinking-water 432-433
  and gastrointestinal tumour frequency 123-124
  haemolysis of rat erythrocytes 191-192
  industrial/mining neighbourhood air pollution 209
  infrared spectrophotometry 197-203
  ingestion, long term
    adenomatous polyps in 131
    carcinogenicity 114-115, 119
    effect 127-132
  initiation-promotion model 156-160
  LAF release by macrophages 150-153
  lung burden, related to air levels 214
  mesothelioma
    induced by 168-171, 182
    risk from 314, 317, 473, 479
    role in causation 227
  risk of neoplastic diseases 330, 465
  tumour induction after injection 175-176
  in urban air samples 207
  see also following entries
Chrysotile/crocidolite mixture, effects of long-term ingestion 127-132
Chrysotile-treated cells, tumorigenicity *in vivo* 159
Cigarette smoke, mineral particles in 323
Clastogenic effects of fibres 60-61
Cloning efficiency of untreated and treated cells 158-159

Cohort studies
  of lung cancer 389, 390
  of mesothelioma 379, 381-382
Collagen and non-collagen protein synthesis 84, 85, 90
Community exposure 9-11, 18-21
  see also Neighbourhood exposure
Complete carcinogenic potency of chrysotile and crocidolite 158
Concrete, health and safety risks associated with 501
Connecticut, asbestos fibres in drinking-water 429, 432
Consumer products containing inorganic fibrous material 347, 348
Corsica 408, 409
Crocidolite
  carcinogenic potency 158
  carcinogenicity in long-term ingestion studies 116, 120
  collagen and non-collagen protein synthesis 85
  environmental exposure, occurrence in lungs after 488, 491, 492
  initiation-promotion model, mode of action 156-160
  occupationally exposed workers, occurrence in lungs 488, 491, 492
  quantitation by IR spectrophotometry 197-203
  shipyard workers, prevalence in 333-334
  solubility 135
  sprayed, removal of 281-288
Cyprus 411-417
Cytogenetic analysis of tumour cell lines 168

Diet, role in gastrointestinal tumour increase 124
Dimethylhydrazine dihydrochloride, effect in combination with asbestos 121
Dissolution of glass and minerals, kinetics 136, 139
DMPO, as radical trapping agent 104
DNA
  damage, induction 62-63
  repair stimulated by chrysotile 182
Domestic exposure 21
  air fibre levels resulting from 209
  to asbestos 209, 253-254
  case-control studies of mesothelioma related to 378-380

Domestic exposure (*contd.*)
    cohort studies
        of lung cancer related to 389
        of mesothelioma related to 389
    lung cancer associated with 213–215
    lung fibre burden following 213–215
    to MMMF 319–322
Dose measurement 457–459
    *see also* Lung fibre burden
Dose-response relationships 459–460, 462, 466
Drinking-water
    cancer incidence and asbestos fibres in 428
    in Duluth, amphibole-like asbestos in 429
    health significance of asbestos in 478
    in Puget Sound, chrysotile fibres in 428
    in Quebec, asbestos fibres in 432
    risk of gastrointestinal cancer from asbestos fibres in 433–434, 435
    in San Francisco Bay area, chrysotile fibres in 432–433
Dry gas, adsorption of PAHs on fibrous material in 143–144
Duluth, Minnesota, asbestos in drinking-water 429
Dusts
    analysis in lungs 424–425
    environmental exposure 207–215
    haemolysis of rat erythrocytes after incubation 191–192
    incidence of abdominal tumours after i.p. injection 194, 195
    phase-contrast optical microscopy of airborne samples 305

East London asbestos factory workers 445–447
Eden, New South Wales 498–507
Emphysema, smoking history and severity 324
Environmental asbestos 463, 476–478
    analytical methods 240–241
    from braking of vehicles 209, 254–255, 269, 272–273
    non-occupational levels 267–270
    risk 466, 482
    sources 234, 447
Environmental exposure
    background, mesothelioma due to 493
    to dusts 207–215
    to fibres 512–514
    general
        asbestos in lungs as measure of 487
        assessment of effect 487
        hazard of mesothelioma 424, 425

        in industrial areas 382–384
        in non-industrial areas 384–385
    occurrence of amosite in lungs of workers after 488–489, 492
    occurrence of chrysotile in lungs after 488, 491, 492
    reflected by fibre levels in lung samples 210–215
    *see also* Environmental asbestos; Environmental pollution
Environmental pollution, in etiology of lung cancer 354
Enzyme release, *see* Lactic acid dehydrogenase
Epidemiological studies
    ability to detect low cancer risk 434
    on cancer incidence and asbestos fibres in drinking-water 428
    case-control 378, 392
    cohort 379, 389
    confounders 434–435
    indirect 435
Erionite 6, 15
    carcinogenic potency 404
    cell transformation 75, 76, 78
    *in vitro* toxicity 75
    lung cancer in Turkey associated with 390–391
    mesothelioma in Turkey associated with 385–388, 468
    mesothelioma induced by 42–43, 74, 168–171, 208
    oncogenic (carcinogenic) activity 70, 109
    size distribution and cytotoxicity 75–76
Exposure index 445
    and mean percentage of types of fibre 447
Exposure measurement, *see* Dose measurement

Ferric oxide hydrate, tumour induction 178
Ferruginous bodies 88
    definition 237
    detection 230, 231, 237
    digestion method 230
    identification of core fibres 230–231, 232, 234, 237
    in residents of Turkish villages 212
    trends in prevalence in Japan 231, 237
    *see also* Amphibole asbestos bodies; Asbestos bodies
Fibres
    counting, standard error 264
    definition 277

Fibres (*contd.*)
   dose, cumulative, pleural mesothelioma and lung cancer 402, 404
   fibre exposure, cumulative, estimation 399
   identification by IR spectrophotometry 200
   length and diameter, *see* Fibres, size
   long, electron microscopy 306
   migration 38-42
   physical chemistry 58, 66
   pollution, atmospheric
      by MMMF, methods of evaluating 292-293
      from natural sources 208
      urban 207
   shape, effects 56-58
   size, effects 34-35, 56-58
      and carcinogenic hazard of asbestos 434
      carcinogenic potency 177
      experimental evidence 463-464
   type, effects 36-37
   *see also* Inorganic fibres *and specific fibre types*
Fibrosis
   absence in intraperitoneal test 176-177
   pulmonary immunological abnormalities accompanying 149
Fibrous materials
   adsorption of PAHs 143-146
   aerosols from asbestos-cement products, sampling 367-368
   determination of solubilities 134-139
   environmental exposure 207-215
   and oxidative stress 109
Fiji, airborne asbestos 270
Friable asbestos products 471-473
Friction materials, degradation 254-255
   *see also* Braking of vehicles

Gas mask workers 487, 490-491
Gastrointestinal cancer
   chrysotile and increase in frequency 123-124
   and ingestion of asbestos fibres 123, 465-466
   possible association with asbestos fibres in drinking-water 435
   risk from asbestos fibres in drinking-water 433-434
   role of diet in increase 124
Gene mutation, induction 61-62

Genotoxicity of fibres
   *in vitro* 58-60, 63-64
   of phagocytosed mineral particle 101
Germany, Federal Republic of
   distribution of fibre sizes in 364-365
   fibre concentration in ambient air 362-364
   monitoring of inorganic fibres in ambient air 361
Glass fibres
   concentrations in ambient air 256
   induction of tumours 177
   solubility 135
Global circulation of asbestos fibres 275
Growth analysis of rat pleural mesothelial cells 181
Gypsum fibres 255

Haemolysis of rat erythrocytes 191-192
Hazard evaluation, *see* Risk
Home environment, *see* Domestic exposure
Humid gas, adsorption of PAHs on fibrous materials in 144-146
Hyalin plaques 17, 391
Hydroxyl radical 101-102

Industrial control 462
Industrial neighbourhood exposure, lung fibre burden 213-215
Inflammation
   involvement of active oxygen species 82
   in rats injected with fibres 163, 164
Infrared spectrophotometry
   fibre identification 200
   quantitation of chrysotile 197-203
Ingestion
   long-term
      adenomatous polyps 131
      carcinogenicity of chrysotile 114-115, 119
      effect 127-132
   of mineral fibres 22
Initiation-promotion model of chrysotile and crocidolite action 156-160
Inorganic fibres
   around asbestos-cement plant 337-338
   composition 343
   concentrations as function of distance from source and meteorological parameters 339-342
   in consumer products 344
   size distribution 344
   *see also* Fibres *and specific fibre types*

Insulated buildings, exposure inside 21-22
Insulating material, fibre identification 200
Insulation
    fibre concentrations after disturbance 320
    fibre concentrations after installation 321-322
    fibre concentrations in workers' lungs 333
    MMMF in materials 291
    presence of amosite in workers' lungs 334
    see also Asbestos workers; Shipyard workers
Interleukin 1, release by alveolar macrophages on exposure to asbestos 154
    see also Lymphocyte activating factor
Intraperitoneal injection of dusts 175-179, 194, 195

Japan
    airborne levels of asbestos in non-occupational environments 267-270
    asbestos imports and prevalence of ferruginous bodies 231
    consumption of asbestos 262
    import of raw asbestos 438-439

Karain, Turkey 388, 399-403
Karyotypes of tumour cell lines 168, 169
Kindergartens, MMMF and health 450-453
    see also Sick building syndrome

Lactic acid dehydrogenase release
    by macrophages 151, 153, 192-193
    by phagocytosing tumour cells 194, 195
LAF, see Lymphocyte activating factor
Latency period, effect of polyvinylpyridine-N-oxide 177
LDH, see Lactic acid dehydrogenase
Long fibres, scanning electron microscopy 306
Long-term ingestion of chrysotile 114-115, 119
Lung cancer
    and domestic exposure to asbestos 389
    and neighbourhood exposure to asbestos 390
    association with asbestosis 480
    association with crocidolite and amosite 447
    background risk 473
    cohort studies 389-390
    dose-response relationships 459-460, 466
    model for 460-461
    risk
        of asbestos-exposed workers 243-244
        background 473

    role of environmental pollution in etiology 354
Lung dust analysis 424-425
Lung fibre analysis, sources of variation 210-211
Lung fibre burden
    asbestos, and disease 314, 316, 447
    assessment at post mortem 493
    and environmental exposure 210-215
    in subjects with no occupational asbestos exposure 306, 307
    and urban and rural residence 211-212
    see also following entry
Lung parenchyma
    concentration of mineral particulates 355-356
    correlation between smoking and particulates in 359
    environmental particulates in 357-358, 359
    types of particles in 356
Lungs
    fibrous minerals as cause of oxidative stress 109
    mineral content in mesothelioma without asbestos exposure 220-227
Lung tissue
    amphibole asbestos bodies in mesothelioma case 412
    necropsy, incidence of asbestos bodies 237
    occurrence of asbestos bodies 211, 212
    of shipyard workers, prevalence of crocidolite 333-334
Lymphocyte activating factor, and asbestos 149-154

Macrophages, alveolar, see Alveolar macrophages
Man-made mineral fibres (MMMF) 7, 15-16, 37
    adsorption of polycyclic hydrocarbons on 140-147
    airborne levels in UK dwellings 319-322
    algorithmic index of pollution 293-294
    atmospheric pollution, methods of evaluating 292-293
    background outdoor concentrations 296
    in buildings, exposure 208, 292, 293-294
    carcinogenicity 56, 464-465, 468
    concentrations in relation to insulation 320-322
    dose-response relationship 177
    durability 58
    exposure limits 300

insulation materials 291, 320–322
in kindergartens 449
  assessment of pollution 450, 451
  correlation between measurements and symptoms 452
lung cancer associated with occupational exposure 391
membrane filter method for determination of concentration 292–293
mesothelioma associated with 389
use of polarizing optical microscope to identify 292
world production 291
see also Ceramic fibres; Glass fibres
Mathematical modelling, choice of parameter values 473
Measurement indices 277–279
Mechanism of action of fibres, fibre parameters for 56–58
Membrane filter method for determination of concentration of MMMF 292–293
Mesothelioma
amosite levels and 488–489, 491, 492
amphibole levels in 226
and asbestos exposure 384–385, 486
association with crocidolite and amosite 447
due to background environmental exposure 493
case-control studies 378–380
cases in Cyprus 412
cases in Karain, Turkey 388
caused by para-occupational exposure 223, 486
cell lines, preparations 168
chrysotile-induced 168–171, 317, 318
cohort studies 379
crocidolite levels and 488–489, 491, 492
diagnosis and certification 393
and domestic exposure 378–380, 486
dose-response relationships 459–460, 466
hazard from general environmental exposure 424, 425
among household contacts of asbestos workers 380
incidence, dependence on age since first exposure 289
incidence trends 421, 425
  predictive model 421–423
induced by chrysotile 168–171, 317, 318
induced by erionite 42–43, 168–171, 208
induced by zeolite 56

lung asbestos burden in workers 314, 316
lung fibre burden 317
in Metsovo, Greece 385
mineral content of lungs without asbestos exposure 220–227
mineral fibre content in Osaka 441–442
model for 461
and neighbourhood exposure to asbestos 380–382
occupationally exposed, levels of amosite and crocidolite 488–489, 491, 492
occurrence without occupational or environmental exposure 486–487
pleural, relationship with asbestos in north-western Cape Province 377–378
proportions of cases attributable to asbestos exposure 376
systematic ascertainment 420, 423–424
villages 20, 208, 386–388, 390–391
Metsovo, Greece 385, 391
Migration of fibres 38–42
Mineral fibres
airborne, exposure 18–21
alveolar load 310, 311–312, 313
carcinogenicity of long-term ingestion 113–114
ingestion 22
reducing surface activity 102
toxicity of long-term ingestion 113–114
see also specific fibre types and Inorganic fibres; Man-made mineral fibres
Mineral particles, model of genotoxic activity 101
Minerals, fibrous contamination 498
Mineral wools, see Man-made mineral fibres
MMMF, see Man-made mineral fibres
Mortality from pleural mesothelioma and lung cancer in Karain, Turkey 402
Mortality rates in Sarihidir, Turkey 399–403
Multi-stage carcinogenesis, models 461–462

Naphthalene and phenanthrene
  adsorption on various fibrous materials 143–146
  as models for polycyclic hydrocarbons 141
Neighbourhood exposure 19–20
  to asbestos 209, 380–382
  industrial, lung fibre burden following 213–215
  lung cancer related to 390
  mesothelioma related to 380–382
Neoplastic diseases, asbestos-induced 330

Non-asbestos mineral fibres
   in the general environment 255–257
   lung cancer associated with 390–391
   mesothelioma associated with 385–389
Non-malignant pleural abnormalities 391
Non-occupational exposure
   airborne levels of asbestos 267–270
   to asbestos 251, 252
   assessment 11–13
   human diseases associated with 16–17
   potential sources of asbestos 239
   see also Domestic exposure
Non-smokers' lungs, mineral particles in 325–326, 328
North-western Cape Province, relationship between asbestos and pleural mesothelioma 377–378

Occupational exposure
   to asbestos fibres and mesothelioma 112, 314
   to asbestos, Zielhuis groupings 220
   to ceramic fibres 300–303
   fibre types and lung burden 331–335
   gas-mask work 487, 490-491
   insulation work 112, 333–334, 380, 445–447, 458
   shipyard work 331–334
Ogasawara Island, asbestos in air 270
Oppenheimer effect 179
Optical microscopy
   limit of sensitivity 240
   of liquid suspension samples of pure asbestos 305
   of MMMF 292
Osaka, mineral fibre content of mesotheliomas 441–442
Osaka Mesothelioma Panel 440
Oxidative stress 109, 110
Oxygen free radicals 82, 87–89

PAH, see Polycyclic aromatic hydrocarbons
Palygorskite, see Attapulgite
Para-occupational exposure
   to asbestos 378, 486
   mesothelioma caused by 223
   see also Domestic exposure
Parenchymal abnormalities, radiological 391
Particulates
   airborne, composition 357
   release of $O_2^-$ from response of alveolar macrophages 85–86

Passivation of surface-active sites 102, 108
Phagocytosis of particles 101, 194
Phase-contrast optical microscopy 305–307
Pleural abnormalities, non-malignant 391
'Pleural arc' 208
Pleural mesothelioma in north-western Cape Province, relationship with asbestos 377–378
Pleural plaques 17
   bilateral, etiology 406
   characteristics of patients 408
   in Corsican ex-miners 406
   of non-occupational origin in Corsica 409
   prevalence in patients born in North Corsica 407
Polarizing optical microscope, to identify MMMF 292
Polycyclic aromatic hydrocarbons
   adsorption on surfaces of asbestos fibres 140–147
   naphthalene and phenanthrene as models 141
Polyvinylpyridine-$N$-oxide, effect on latency period and tumour incidence 177
Population exposure, general 20–21
Puget Sound, chrysotile fibres in drinking-water 433
Pulmonary fibrosis 149
Pulmonary tissue, see Lung tissue

Quartz, and LAF release by macrophages 150–154
Quebec, asbestos fibres in drinking-water 432
Quebec chrysotile mining region 213–215

Radiographic abnormalities, asbestos-associated 391
Radon, and subcutaneous chrysotile in rats 162–164
Rat
   erythrocyte haemolysis 191–192
   mesothelioma cell lines 169, 170
   pleural mesothelial cells 181
Refractory fibres, see Ceramic fibres
Regional lymph-nodes, translocation of fibres from injection site 160, 163, 164
Risk
   air monitoring in evaluation of 251–253
   assessment 22, 392, 393, 507, 514
   associated with quarry, residents' perception 506–507
   associated with use of concrete 501

Risk (*contd.*)
  based on degree of exposure and lung clearance 490-493
  from environmental exposure to asbestos 466
  lifetime
    of asbestos-induced neoplastic diseases 330
    for environmental exposure to asbestos 466, 482
    estimates 471-474
    of gastrointestinal cancer from asbestos fibres in drinking water 433-434
  potential, of fibres 512-513
  relative, based on amphibole lung burden 490
  to students attending schools containing friable asbestos products 471-473
  *see also* Cancer risk
Rock, fibrous contamination 489-490
Rutile, and LAF activity of macrophages 150-151

Sample preparation
  direct method 278
  'indirect transfer' method 263
Sampling of aerosols from asbestos-cement products 367-368
San Francisco Bay Area, fibres in drinking-water 432-433
Sarihidir, Turkey 399-403
Scanning electron microscopy
  of airborne dust samples 305
  of industrial and environmental samples 307
  of liquid suspension samples of pure asbestos 305
  of long fibres 306
  of pure asbestos 306
  of treated fibres 136
  use to identify fibres 241
Scar as precondition for tumour development caused by asbestos 177
Schools containing friable asbestos products, risk to students 471-474
  *see also* Kindergartens
Sepiolite 6, 14
Serum haptoglobin as marker of inflammation 163, 164-165
Sheep lung, chrysotile and tremolite in 413
Shipyard workers 331-334

Sick building syndrome 449-452
  *see also* Kindergartens
Silicates in the environment 357
Silicon, release by ceramic fibres 135
Sister chromatid exchanges, induction 62
Smokers' lungs, mineral particles in 325-326, 328
Smoking and particulates in lung parenchyma 359
Solubility of fibrous materials 134-139
Sprayed crocidolite insulation, removal 281-288
Sprayed surfacing materials, pollution by 293, 297
Sprayed-trowelled amosite insulation, removal 283-288
Stucco, chrysotile and tremolite fibres in 413
Superfine materials 299
Superoxide dismutase 83, 84, 87
Surface activation of amphiboles 107
Surface-active sites 102, 103
Surface reducing activity of particles 103
Surfacing materials, asbestos-containing, building with 245-251
Synthetic fibres, *see* Man-made mineral fibres

Taconite tailings, carcinogenicity in long-term ingestion 117, 120
Thermal or sound insulation, MMMF in materials used for 291
Thymocyte proliferation 151
Tokyo, asbestos concentration in snow 271-272
Transfection by viral DNA, ability of asbestos to mediate 65
Transformation of cultured cells *in vitro* 64-65
Translocation of fibres to regional lymph-nodes 163, 164, 166
Transmission electron microscopy
  of airborne dust samples 305
  to enumerate and size asbestos fibres 240
  to evaluate the quality of air in buildings 248-251
  to identify asbestos fibres 241
  of industrial and environmental samples 307
  of liquid suspension samples of pure asbestos 305
  measurement of fibre concentration 320
  of pure asbestos 306
Transplantation of tumour cell lines 169

Tremolite
    in lung tissue 212, 215
    relationship between air levels and lung fibre burden 214
    role in mesotheliomas produced by chrysotile ore 318
    in sheep lung 413
    in whitewash in Metsovo, Greece 385
Tumorigenicity *in vivo* of chrysotile-treated cells 159
Tumour cell lines 168, 169
Tumour incidence, effect of polyvinylpyridine-*N*-oxide 177
Tumours
    abdominal, after i.p. injection of dusts 194, 195
    gastrointestinal, role of diet in increase 124
    induction
        by basalt fibres 177
        by ceramic fibres 177
        by ferric oxide hydrate 178
        by glass fibres 177
        after injection of chrysotile 175–176
    *see also specific sites*

Turkey, mesothelioma 384–388, 399–403, 468

Uncoated fibres, enumeration 231, 234, 236
Urban atmospheric fibre pollution 207
US environmental regulatory action 242

Vinyl asbestos tile, fibre release from 253
Vitreous fibres, *see* Man-made mineral fibres

Whitewash
    tremolite-containing, in north-western Greece 385, 418
    in Turkey, use of asbestos in 384–385
Wollastonite 6, 15, 257
    induction of tumours 177
Workers, *see* Occupational exposure

Xanthine-xanthine oxidase 85

Zeolite 242
Zielhuis groupings of occupational exposure to asbestos 220–223

# PUBLICATIONS OF THE INTERNATIONAL AGENCY FOR RESEARCH ON CANCER
## SCIENTIFIC PUBLICATIONS SERIES

(Available from Oxford University Press)
through local bookshops

No. 1 LIVER CANCER
1971; 176 pages; out of print

No. 2 ONCOGENESIS AND HERPESVIRUSES
Edited by P.M. Biggs, G. de-Thé & L.N. Payne
1972; 515 pages; out of print

No. 3 N-NITROSO COMPOUNDS: ANALYSIS AND FORMATION
Edited by P. Bogovski, R. Preussmann & E. A. Walker
1972; 140 pages; out of print

No. 4 TRANSPLACENTAL CARCINOGENESIS
Edited by L. Tomatis & U. Mohr
1973; 181 pages; out of print

*No. 5 PATHOLOGY OF TUMOURS IN LABORATORY ANIMALS. VOLUME 1. TUMOURS OF THE RAT. PART 1
Editor-in-Chief V.S. Turusov
1973; 214 pages

*No. 6 PATHOLOGY OF TUMOURS IN LABORATORY ANIMALS. VOLUME 1. TUMOURS OF THE RAT. PART 2
Editor-in-Chief V.S. Turusov
1976; 319 pages
*reprinted in one volume, Price £50.00

No. 7 HOST ENVIRONMENT INTERACTIONS IN THE ETIOLOGY OF CANCER IN MAN
Edited by R. Doll & I. Vodopija
1973; 464 pages; £32.50

No. 8 BIOLOGICAL EFFECTS OF ASBESTOS
Edited by P. Bogovski, J.C. Gilson, V. Timbrell & J.C. Wagner
1973; 346 pages; out of print

No. 9 N-NITROSO COMPOUNDS IN THE ENVIRONMENT
Edited by P. Bogovski & E. A. Walker
1974; 243 pages; £16.50

No. 10 CHEMICAL CARCINOGENESIS ESSAYS
Edited by R. Montesano & L. Tomatis
1974; 230 pages; out of print

No. 11 ONCOGENESIS AND HERPESVIRUSES II
Edited by G. de-Thé, M.A. Epstein & H. zur Hausen
1975; Part 1, 511 pages; Part 2, 403 pages; £65.-

No. 12 SCREENING TESTS IN CHEMICAL CARCINOGENESIS
Edited by R. Montesano, H. Bartsch & L. Tomatis
1976; 666 pages; £12.-

No. 13 ENVIRONMENTAL POLLUTION AND CARCINOGENIC RISKS
Edited by C. Rosenfeld & W. Davis
1976; 454 pages; out of print

No. 14 ENVIRONMENTAL N-NITROSO COMPOUNDS: ANALYSIS AND FORMATION
Edited by E.A. Walker, P. Bogovski & L. Griciute
1976; 512 pages; £37.50

No. 15 CANCER INCIDENCE IN FIVE CONTINENTS. VOLUME III
Edited by J. Waterhouse, C. Muir, P. Correa & J. Powell
1976; 584 pages; out of print

No. 16 AIR POLLUTION AND CANCER IN MAN
Edited by U. Mohr, D. Schmähl & L. Tomatis
1977; 311 pages; out of print

No. 17 DIRECTORY OF ON-GOING RESEARCH IN CANCER EPIDEMIOLOGY 1977
Edited by C.S. Muir & G. Wagner
1977; 599 pages; out of print

No. 18 ENVIRONMENTAL CARCINOGENS: SELECTED METHODS OF ANALYSIS
Edited-in-Chief H. Egan
VOLUME 1. ANALYSIS OF VOLATILE NITROSAMINES IN FOOD
Edited by R. Preussmann, M. Castegnaro, E.A. Walker & A.E. Wassermann
1978; 212 pages; out of print

No. 19 ENVIRONMENTAL ASPECTS OF N-NITROSO COMPOUNDS
Edited by E.A. Walker, M. Castegnaro, L. Griciute & R.E. Lyle
1978; 566 pages; out of print

No. 20 NASOPHARYNGEAL CARCINOMA: ETIOLOGY AND CONTROL
Edited by G. de-Thé & Y. Ito
1978; 610 pages; out of print

No. 21 CANCER REGISTRATION AND ITS TECHNIQUES
Edited by R. MacLennan, C. Muir, R. Steinitz & A. Winkler
1978; 235 pages; £35.-

---

Prices, valid for October 1988, are subject to change without notice

# SCIENTIFIC PUBLICATIONS SERIES

No. 22 ENVIRONMENTAL CARCINOGENS: SELECTED METHODS OF ANALYSIS
Editor-in-Chief H. Egan
VOLUME 2. METHODS FOR THE MEASUREMENT OF VINYL CHLORIDE IN POLY(VINYL CHLORIDE), AIR, WATER AND FOODSTUFFS
Edited by D.C.M. Squirrell & W. Thain
1978; 142 pages; out of print

No. 23 PATHOLOGY OF TUMOURS IN LABORATORY ANIMALS. VOLUME II. TUMOURS OF THE MOUSE
Editor-in-Chief V.S. Turusov
1979; 669 pages; out of print

No. 24 ONCOGENESIS AND HERPESVIRUSES III
Edited by G. de-Thé, W. Henle & F. Rapp
1978; Part 1, 580 pages; Part 2, 522 pages; out of print

No. 25 CARCINOGENIC RISKS: STRATEGIES FOR INTERVENTION
Edited by W. Davis & C. Rosenfeld
1979; 283 pages; out of print

No. 26 DIRECTORY OF ON-GOING RESEARCH IN CANCER EPIDEMIOLOGY 1978
Edited by C.S. Muir & G. Wagner,
1978; 550 pages; out of print

No. 27 MOLECULAR AND CELLULAR ASPECTS OF CARCINOGEN SCREENING TESTS
Edited by R. Montesano, H. Bartsch & L. Tomatis
1980; 371 pages; £22.50

No. 28 DIRECTORY OF ON-GOING RESEARCH IN CANCER EPIDEMIOLOGY 1979
Edited by C.S. Muir & G. Wagner
1979; 672 pages; out of print

No. 29 ENVIRONMENTAL CARCINOGENS: SELECTED METHODS OF ANALYSIS
Editor-in-Chief H. Egan
VOLUME 3. ANALYSIS OF POLYCYCLIC AROMATIC HYDROCARBONS IN ENVIRONMENTAL SAMPLES
Edited by M. Castegnaro, P. Bogovski, H. Kunte & E.A. Walker
1979; 240 pages; out of print

No. 30 BIOLOGICAL EFFECTS OF MINERAL FIBRES
Editor-in-Chief J.C. Wagner
1980; Volume 1, 494 pages; Volume 2, 513 pages; £55.-

No. 31 N-NITROSO COMPOUNDS: ANALYSIS, FORMATION AND OCCURRENCE
Edited by E.A. Walker, L. Griciute, M. Castegnaro & M. Börzsönyi
1980; 841 pages; out of print

No. 32 STATISTICAL METHODS IN CANCER RESEARCH. VOLUME 1. THE ANALYSIS OF CASE-CONTROL STUDIES
By N.E. Breslow & N.E. Day
1980; 338 pages; £20.-

No. 33 HANDLING CHEMICAL CARCINOGENS IN THE LABORATORY: PROBLEMS OF SAFETY
Edited by R. Montesano, H. Bartsch, E. Boyland, G. Della Porta, L. Fishbein, R.A. Griesemer, A.B. Swan & L. Tomatis
1979; 32 pages; out of print

No. 34 PATHOLOGY OF TUMOURS IN LABORATORY ANIMALS. VOLUME III. TUMOURS OF THE HAMSTER
Editor-in-Chief V.S. Turusov
1982; 461 pages; £32.50

No. 35 DIRECTORY OF ON-GOING RESEARCH IN CANCER EPIDEMIOLOGY 1980
Edited by C.S. Muir & G. Wagner
1980; 660 pages; out of print

No. 36 CANCER MORTALITY BY OCCUPATION AND SOCIAL CLASS 1851-1971
By W.P.D. Logan
1982; 253 pages; £22.50

No. 37 LABORATORY DECONTAMINATION AND DESTRUCTION OF AFLATOXINS $B_1$, $B_2$, $G_1$, $G_2$ IN LABORATORY WASTES
Edited by M. Castegnaro, D.C. Hunt, E.B. Sansone, P.L. Schuller, M.G. Siriwardana, G.M. Telling, H.P. Van Egmond & E.A. Walker
1980; 59 pages; £6.50

No. 38 DIRECTORY OF ON-GOING RESEARCH IN CANCER EPIDEMIOLOGY 1981
Edited by C.S. Muir & G. Wagner
1981; 696 pages; out of print

No. 39 HOST FACTORS IN HUMAN CARCINOGENESIS
Edited by H. Bartsch & B. Armstrong
1982; 583 pages; £37.50

No. 40 ENVIRONMENTAL CARCINOGENS: SELECTED METHODS OF ANALYSIS
Edited-in-Chief H. Egan
VOLUME 4. SOME AROMATIC AMINES AND AZO DYES IN THE GENERAL AND INDUSTRIAL ENVIRONMENT
Edited by L. Fishbein, M. Castegnaro, I.K. O'Neill & H. Bartsch
1981; 347 pages; £22.50

No. 41 N-NITROSO COMPOUNDS: OCCURRENCE AND BIOLOGICAL EFFECTS
Edited by H. Bartsch, I.K. O'Neill, M. Castegnaro & M. Okada
1982; 755 pages; £37.50

No. 42 CANCER INCIDENCE IN FIVE CONTINENTS. VOLUME IV
Edited by J. Waterhouse, C. Muir, K. Shanmugaratnam & J. Powell
1982; 811 pages; £37.50

# SCIENTIFIC PUBLICATIONS SERIES

No. 43 LABORATORY DECONTAMINATION
AND DESTRUCTION OF CARCINOGENS IN
LABORATORY WASTES: SOME N-NITROSAMINES
Edited by M. Castegnaro, G. Eisenbrand, G. Ellen,
L. Keefer, D. Klein, E.B. Sansone, D. Spincer,
G. Telling & K. Webb
1982; 73 pages; £7.50

No. 44 ENVIRONMENTAL CARCINOGENS:.
SELECTED METHODS OF ANALYSIS
Editor-in-Chief H. Egan
VOLUME 5. SOME MYCOTOXINS
Edited by L. Stoloff, M. Castegnaro, P. Scott,
I.K. O'Neill & H. Bartsch
1983; 455 pages; £22.50

No. 45 ENVIRONMENTAL CARCINOGENS:
SELECTED METHODS OF ANALYSIS
Editor-in-Chief H. Egan
VOLUME 6. N-NITROSO COMPOUNDS
Edited by R. Preussmann, I.K. O'Neill, G. Eisenbrand,
B. Spiegelhalder & H. Bartsch
1983; 508 pages; £22.50

No. 46 DIRECTORY OF ON-GOING RESEARCH
IN CANCER EPIDEMIOLOGY 1982
Edited by C.S. Muir & G. Wagner
1982; 722 pages; out of print

No. 47 CANCER INCIDENCE IN SINGAPORE
1968-1977
Edited by K. Shanmugaratnam, H.P. Lee & N.E. Day
1982; 171 pages; out of print

No. 48 CANCER INCIDENCE IN THE USSR
Second Revised Edition
Edited by N.P. Napalkov, G.F. Tserkovny,
V.M. Merabishvili, D.M. Parkin, M. Smans & C.S. Muir,
1983; 75 pages; £12.-

No. 49 LABORATORY DECONTAMINATION AND
DESTRUCTION OF CARCINOGENS IN
LABORATORY WASTES: SOME POLYCYCLIC
AROMATIC HYDROCARBONS
Edited by M. Castegnaro, G. Grimmer, O. Hutzinger,
W. Karcher, H. Kunte, M. Lafontaine, E.B. Sansone,
G. Telling & S.P. Tucker
1983; 81 pages; £9.-

No. 50 DIRECTORY OF ON-GOING RESEARCH
IN CANCER EPIDEMIOLOGY 1983
Edited by C.S. Muir & G. Wagner
1983; 740 pages; out of print

No. 51 MODULATORS OF EXPERIMENTAL
CARCINOGENESIS
Edited by V. Turusov & R. Montesano
1983; 307 pages; £22.50

No. 52 SECOND CANCER IN RELATION TO
RADIATION TREATMENT FOR CERVICAL
CANCER
Edited by N.E. Day & J.D. Boice, Jr
1984; 207 pages; £20.-

No. 53 NICKEL IN THE HUMAN ENVIRONMENT
Editor-in-Chief F.W. Sunderman, Jr
1984: 530 pages; £32.50

No. 54 LABORATORY DECONTAMINATION
AND DESTRUCTION OF CARCINOGENS IN
LABORATORY WASTES: SOME HYDRAZINES
Edited by M. Castegnaro, G. Ellen, M. Lafontaine,
H.C. van der Plas, E.B. Sansone & S.P. Tucker
1983; 87 pages; £9.-

No. 55 LABORATORY DECONTAMINATION
AND DESTRUCTION OF CARCINOGENS IN
LABORATORY WASTES: SOME N-NITROSAMIDES
Edited by M. Castegnaro, M. Benard,
L.W. van Broekhoven, D. Fine, R. Massey,
E.B. Sansone, P.L.R. Smith, B. Spiegelhalder,
A. Stacchini, G. Telling & J.J. Vallon
1984; 65 pages; £7.50

No. 56 MODELS, MECHANISMS AND ETIOLOGY
OF TUMOUR PROMOTION
Edited by M. Börszönyi, N.E. Day, K. Lapis
& H. Yamasaki
1984; 532 pages; £32.50

No. 57 N-NITROSO COMPOUNDS:
OCCURRENCE, BIOLOGICAL EFFECTS
AND RELEVANCE TO HUMAN CANCER
Edited by I.K. O'Neill, R.C. von Borstel, C.T. Miller,
J. Long & H. Bartsch
1984; 1011 pages; £80.-

No. 58 AGE-RELATED FACTORS IN
CARCINOGENESIS
Edited by A. Likhachev, V. Anisimov & R. Montesano
1985; 288 pages; £20.-

No. 59 MONITORING HUMAN EXPOSURE TO
CARCINOGENIC AND MUTAGENIC AGENTS
Edited by A. Berlin, M. Draper, K. Hemminki
& H. Vainio
1984; 457 pages; £27.50

No. 60 BURKITT'S LYMPHOMA: A HUMAN
CANCER MODEL
Edited by G. Lenoir, G. O'Conor & C.L.M. Olweny
1985; 484 pages; £22.50

No. 61 LABORATORY DECONTAMINATION
AND DESTRUCTION OF CARCINOGENS IN
LABORATORY WASTES: SOME HALOETHERS
Edited by M. Castegnaro, M. Alvarez, M. Iovu,
E.B. Sansone, G.M. Telling & D.T. Williams
1984; 53 pages; £7.50

No. 62 DIRECTORY OF ON-GOING RESEARCH
IN CANCER EPIDEMIOLOGY 1984
Edited by C.S. Muir & G. Wagner
1984; 728 pages; £26.-

No. 63 VIRUS-ASSOCIATED CANCERS IN AFRICA
Edited by A.O. Williams, G.T. O'Conor, G.B. de-Thé
& C.A. Johnson
1984; 774 pages; £22.-

# SCIENTIFIC PUBLICATIONS SERIES

No. 64 LABORATORY DECONTAMINATION AND DESTRUCTION OF CARCINOGENS IN LABORATORY WASTES: SOME AROMATIC AMINES AND 4-NITROBIPHENYL
Edited by M. Castegnaro, J. Barek, J. Dennis, G. Ellen, M. Klibanov, M. Lafontaine, R. Mitchum, P. Van Roosmalen, E.B. Sansone, L.A. Sternson & M. Vahl
1985; 85 pages; £6.95

No. 65 INTERPRETATION OF NEGATIVE EPIDEMIOLOGICAL EVIDENCE FOR CARCINOGENICITY
Edited by N.J. Wald & R. Doll
1985; 232 pages; £20.-

No. 66 THE ROLE OF THE REGISTRY IN CANCER CONTROL
Edited by D.M. Parkin, G. Wagner & C. Muir
1985; 155 pages; £10.-

No. 67 TRANSFORMATION ASSAY OF ESTABLISHED CELL LINES: MECHANISMS AND APPLICATION
Edited by T. Kakunaga & H. Yamasaki
1985; 225 pages; £20.-

No. 68 ENVIRONMENTAL CARCINOGENS: SELECTED METHODS OF ANALYSIS VOLUME 7. SOME VOLATILE HALOGENATED HYDROCARBONS
Edited by L. Fishbein & I.K. O'Neill
1985; 479 pages; £20.-

No. 69 DIRECTORY OF ON-GOING RESEARCH IN CANCER EPIDEMIOLOGY 1985
Edited by C.S. Muir & G. Wagner
1985; 756 pages; £22.

No. 70 THE ROLE OF CYCLIC NUCLEIC ACID ADDUCTS IN CARCINOGENESIS AND MUTAGENESIS
Edited by B. Singer & H. Bartsch
1986; 467 pages; £40.-

No. 71 ENVIRONMENTAL CARCINOGENS: SELECTED METHODS OF ANALYSIS VOLUME 8. SOME METALS: As, Be, Cd, Cr, Ni, Pb, Se, Zn
Edited by I.K. O'Neill, P. Schuller & L. Fishbein
1986; 485 pages; £20.

No. 72 ATLAS OF CANCER IN SCOTLAND 1975-1980: INCIDENCE AND EPIDEMIOLOGICAL PERSPECTIVE
Edited by I. Kemp, P. Boyle, M. Smans & C. Muir
1985; 282 pages; £35.-

No. 73 LABORATORY DECONTAMINATION AND DESTRUCTION OF CARCINOGENS IN LABORATORY WASTES: SOME ANTINEOPLASTIC AGENTS
Edited by M. Castegnaro, J. Adams, M. Armour, J. Barek, J. Benvenuto, C. Confalonieri, U. Goff, S. Ludeman, D. Reed, E.B. Sansone & G. Telling
1985; 163 pages; £10.-

No. 74 TOBACCO: A MAJOR INTERNATIONAL HEALTH HAZARD
Edited by D. Zaridze & R. Peto
1986; 324 pages; £20.-

No. 75 CANCER OCCURRENCE IN DEVELOPING COUNTRIES
Edited by D.M. Parkin
1986; 339 pages; £20.-

No. 76 SCREENING FOR CANCER OF THE UTERINE CERVIX
Edited by M. Hakama, A.B. Miller & N.E. Day
1986; 315 pages; £25.-

No. 77 HEXACHLOROBENZENE: PROCEEDINGS OF AN INTERNATIONAL SYMPOSIUM
Edited by C.R. Morris & J.R.P. Cabral
1986; 668 pages; £50.-

No. 78 CARCINOGENICITY OF ALKYLATING CYTOSTATIC DRUGS
Edited by D. Schmähl & J. M. Kaldor
1986; 338 pages; £25.-

No. 79 STATISTICAL METHODS IN CANCER RESEARCH. VOLUME III. THE DESIGN AND ANALYSIS OF LONG-TERM ANIMAL EXPERIMENTS
By J.J. Gart, D. Krewski, P.N. Lee, R.E. Tarone & J. Wahrendorf
1986; 219 pages; £20.-

No. 80 DIRECTORY OF ON-GOING RESEARCH IN CANCER EPIDEMIOLOGY 1986
Edited by C.S. Muir & G. Wagner
1986; 805 pages; £22.-

No. 81 ENVIRONMENTAL CARCINOGENS: METHODS OF ANALYSIS AND EXPOSURE MEASUREMENT. VOLUME 9. PASSIVE SMOKING
Edited by I.K. O'Neill, K.D. Brunnemann, B. Dodet & D. Hoffmann
1987; 379 pages; £30.-

No. 82 STATISTICAL METHODS IN CANCER RESEARCH. VOLUME II. THE DESIGN AND ANALYSIS OF COHORT STUDIES
By N.E. Breslow & N.E. Day
1987; 404 pages; £30.-

No. 83 LONG-TERM AND SHORT-TERM ASSAYS FOR CARCINOGENS: A CRITICAL APPRAISAL
Edited by R. Montesano, H. Bartsch, H. Vainio, J. Wilbourn & H. Yamasaki
1986; 575 pages; £32.50

No. 84 THE RELEVANCE OF N-NITROSO COMPOUNDS TO HUMAN CANCER: EXPOSURES AND MECHANISMS
Edited by H. Bartsch, I.K. O'Neill & R. Schulte-Hermann
1987; 671 pages; £50.-

# SCIENTIFIC PUBLICATIONS SERIES

No. 85 ENVIRONMENTAL CARCINOGENS: METHODS OF ANALYSIS AND EXPOSURE MEASUREMENT. VOLUME 10. BENZENE AND ALKYLATED BENZENES
Edited by L. Fishbein & I.K. O'Neill
1988; 318 pages; £35.-

No. 86 DIRECTORY OF ON-GOING RESEARCH IN CANCER EPIDEMIOLOGY 1987
Edited by D.M. Parkin & J. Wahrendorf
1987; 685 pages; £22.-

No. 87 INTERNATIONAL INCIDENCE OF CHILDHOOD CANCER
Edited by D.M. Parkin, C.A. Stiller, G.J. Draper, C.A. Bieber, B. Terracini & J.L. Young
1988; 402 pages; £35.-

No. 88 CANCER INCIDENCE IN FIVE CONTINENTS. VOLUME V
Edited by C. Muir, J. Waterhouse, T. Mack, J. Powell & S. Whelan
1988; 1004 pages; £50.-

No. 89 METHODS FOR DETECTING DNA DAMAGING AGENTS IN HUMANS: APPLICATIONS IN CANCER EPIDEMIOLOGY AND PREVENTION
Edited by H. Bartsch, K. Hemminki & I.K. O'Neill
1988; 518 pages; £45.-

No. 90 NON-OCCUPATIONAL EXPOSURE TO MINERAL FIBRES
Edited by J. Bignon, J. Peto & R. Saracci
1988; 530 pages; £45.-

No. 91 TRENDS IN CANCER INCIDENCE IN SINGAPORE 1968-1982
Edited by H.P. Lee, N.E. Day & K. Shanmugaratnam
1988; 160 pages; £25.-

No. 92 CELL DIFFERENTIATION, GENES AND CANCER
Edited by T. Kakunaga, T. Sugimura, L. Tomatis and H. Yamasaki
1988; 204 pages; £25.-

No. 93 DIRECTORY OF ON-GOING RESEARCH IN CANCER EPIDEMIOLOGY 1988
Edited by M. Coleman & J. Wahrendorf
1988; 662 pages; £26.-

# IARC MONOGRAPHS ON THE EVALUATION OF THE CARCINOGENIC RISK OF CHEMICALS TO HUMANS
*(English editions only)*

(Available from booksellers through the network of WHO Sales Agents*)

**Volume 1**
*Some inorganic substances, chlorinated hydrocarbons, aromatic amines, N-nitroso compounds, and natural products*
1972; 184 pages; out of print

**Volume 2**
*Some inorganic and organometallic compounds*
1973; 181 pages; out of print

**Volume 3**
*Certain polycyclic aromatic hydrocarbons and heterocyclic compounds*
1973; 271 pages; out of print

**Volume 4**
*Some aromatic amines, hydrazine and related substances, N-nitroso compounds and miscellaneous alkylating agents*
1974; 286 pages;
Sw. fr. 18.-

**Volume 5**
*Some organochlorine pesticides*
1974; 241 pages; out of print

**Volume 6**
*Sex hormones*
1974; 243 pages;
out of print

**Volume 7**
*Some anti-thyroid and related substances, nitrofurans and industrial chemicals*
1974; 326 pages; out of print

**Volume 8**
*Some aromatic azo compounds*
1975; 357 pages; Sw.fr. 36.-

**Volume 9**
*Some aziridines, N-, S- and O-mustards and selenium*
1975; 268 pages; Sw. fr. 27.-

**Volume 10**
*Some naturally occurring substances*
1976; 353 pages; out of print

**Volume 11**
*Cadmium, nickel, some epoxides, miscellaneous industrial chemicals and general considerations on volatile anaesthetics*
1976; 306 pages; out of print

**Volume 12**
*Some carbamates, thiocarbamates and carbazides*
1976; 282 pages; Sw. fr. 34.-

**Volume 13**
*Some miscellaneous pharmaceutical substances*
1977; 255 pages; Sw. fr. 30.-

**Volume 14**
*Asbestos*
1977; 106 pages; out of print

**Volume 15**
*Some fumigants, the herbicides 2,4-D and 2,4,5-T, chlorinated dibenzodioxins and miscellaneous industrial chemicals*
1977; 354 pages; Sw. fr. 50.-

**Volume 16**
*Some aromatic amines and related nitro compounds — hair dyes, colouring agents and miscellaneous industrial chemicals*
1978; 400 pages; Sw. fr. 50.-

**Volume 17**
*Some N-nitroso compounds*
1978; 365 pages; Sw. fr. 50.

**Volume 18**
*Polychlorinated biphenyls and polybrominated biphenyls*
1978; 140 pages; Sw. fr. 20.-

**Volume 19**
*Some monomers, plastics and synthetic elastomers, and acrolein*
1979; 513 pages; Sw. fr. 60.-

**Volume 20**
*Some halogenated hydrocarbons*
1979; 609 pages; Sw. fr. 60.-

**Volume 21**
*Sex hormones (II)*
1979; 583 pages; Sw. fr. 60.-

**Volume 22**
*Some non-nutritive sweetening agents*
1980; 208 pages; Sw. fr. 25.-

**Volume 23**
*Some metals and metallic compounds*
1980; 438 pages; Sw. fr. 50.-

**Volume 24**
*Some pharmaceutical drugs*
1980; 337 pages; Sw. fr. 40.-

**Volume 25**
*Wood, leather and some associated industries*
1981; 412 pages; Sw. fr. 60.-

**Volume 26**
*Some antineoplastic and immunosuppressive agents*
1981; 411 pages; Sw. fr. 62.-

*A list of these Agents may be obtained by writing to the World Health Organization, Distribution and Sales Service, 1211 Geneva 27, Switzerland

# IARC MONOGRAPHS SERIES

Volume 27
*Some aromatic amines, anthraquinones and nitroso compounds, and inorganic fluorides used in drinking-water and dental preparations*
1982; 341 pages; Sw. fr. 40.-

Volume 28
*The rubber industry*
1982; 486 pages; Sw. fr. 70.-

Volume 29
*Some industrial chemicals and dyestuffs*
1982; 416 pages; Sw. fr. 60.-

Volume 30
*Miscellaneous pesticides*
1983; 424 pages; Sw. fr. 60.-

Volume 31
*Some food additives, feed additives and naturally occurring substances*
1983; 14 pages; Sw. fr. 60.-

Volume 32
*Polynuclear aromatic compounds, Part 1, Chemical, environmental and experimental data*
1984; 477 pages; Sw. fr. 60.-

Volume 33
*Polynuclear aromatic compounds, Part 2, Carbon blacks, mineral oils and some nitroarenes*
1984; 245 pages; Sw. fr. 50.-

Volume 34
*Polynuclear aromatic compounds, Part 3, Industrial exposures in aluminium production, coal gasification, coke production, and iron and steel founding*
1984; 219 pages; Sw. fr. 48.-

Volume 35
*Polynuclear aromatic compounds, Part 4, Bitumens, coal-tars and derived products, shale-oils and soots*
1985; 271 pages; Sw. fr.70.-

Volume 36
*Allyl compounds, aldehydes, epoxides and peroxides*
1985; 369 pages; Sw. fr. 70.-

Volume 37
*Tobacco habits other than smoking; betel-quid and areca-nut chewing; and some related nitrosamines*
1985; 291 pages; Sw. fr. 70.-

Volume 38
*Tobacco smoking*
1986; 421 pages; Sw. fr. 75.-

Volume 39
*Some chemicals used in plastics and elastomers*
1986; 403 pages; Sw. fr. 60.-

Volume 40
*Some naturally occurring and synthetic food components, furocoumarins and ultraviolet radiation*
1986; 444 pages; Sw. fr. 65.-

Volume 41
*Some halogenated hydrocarbons and pesticide exposures*
1986; 434 pages; Sw. fr. 65.-

Volume 42
*Silica and some silicates*
1987; 289 pages; Sw. fr. 65.-

*Volume 43
*Man-made mineral fibres and radon*
1988; 300 pages; Sw. fr. 65.-

*Volume 44
*Alcohol drinking*
1988; 416 pages; Sw. fr. 65.-

Supplement No. 1
*Chemicals and industrial processes associated with cancer in humans (IARC Monographs, Volumes 1 to 20)*
1979; 71 pages; out of print

Supplement No. 2
*Long-term and short-term screening assays for carcinogens: a critical appraisal*
1980; 426 pages; Sw. fr. 40.-

Supplement No. 3
*Cross index of synonyms and trade names in Volumes 1 to 26*
1982; 199 pages; Sw. fr. 60.-

Supplement No. 4
*Chemicals, industrial processes and industries associated with cancer in humans (IARC Monographs, Volumes 1 to 29)*
1982; 292 pages; Sw. fr. 60.-

Supplement No. 5
*Cross index of synonyms and trade names in Volumes 1 to 36*
1985; 259 pages; Sw. fr. 60.-

*Supplement No. 6
*Genetic and related effects: An updating of selected IARC Monographs from Volumes 1-42*
1987; 730 pages; Sw. fr. 80.-

*Supplement No. 7
*Overall evaluations of carcinogenicity: An updating of IARC Monographs Volumes 1-42*
1987; 440 pages; Sw. fr. 65.-

*From Volume 43 onwards, the series title has been changed to IARC MONOGRAPHS ON THE EVALUATION OF CARCINOGENIC RISKS TO HUMANS

# INFORMATION BULLETINS ON THE SURVEY OF CHEMICALS BEING TESTED FOR CARCINOGENICITY*

No. 8 (1979)
Edited by M.-J. Ghess, H. Bartsch
& L. Tomatis
604 pages; Sw. fr. 40.-

No. 9 (1981)
Edited by M.-J. Ghess, J.D. Wilbourn,
H. Bartsch & L. Tomatis
294 pages; Sw. fr. 41.-

No. 10 (1982)
Edited by M.-J. Ghess, J.D. Wilbourn
& H. Bartsch
362 pages; Sw. fr. 42.-

No. 11 (1984)
Edited by M.-J. Ghess, J.D. Wilbourn,
H. Vainio & H. Bartsch
362 pages; Sw. fr. 50.-

No. 12 (1986)
Edited by M.-J. Ghess, J.D. Wilbourn,
A. Tossavainen & H. Vainio
385 pages; Sw. fr. 50.-

No. 13 (1988)
Edited by M.-J. Ghess, J.D. Wilbourn
& A. Aitio
404 pages; Sw. fr. 43.-

## NON-SERIAL PUBLICATIONS

### (Available from IARC)

ALCOOL ET CANCER
By A. Tuyns (in French only)
1978; 42 pages; Fr. fr. 35.-

CANCER MORBIDITY AND CAUSES OF
DEATH AMONG DANISH BREWERY
WORKERS
By O.M. Jensen
1980; 143 pages; Fr. fr. 75.-

DIRECTORY OF COMPUTER SYSTEMS
USED IN CANCER REGISTRIES
By H.R. Menck & D.M. Parkin
1986; 236 pages; Fr. fr. 50.-

*Available from IARC; or the World Health Organization Distribution and Sales Services, 1211 Geneva 27 Switzerland or WHO Sales Agents.